Benford's Law: Theory and Applications

Edited by Steven J. Miller

PRINCETON UNIVERSITY PRESS

PRINCETON AND OXFORD

In the United Kingdom: Princeton University Press, 6 Oxford Street, Woodstock, Oxfordshire OX20 1TW

press.princeton.edu

ISBN 978-0-691-14761-1

British Library Cataloging-in-Publication Data is available

This book has been composed in Times Roman in LaTeX

The publisher would like to acknowledge the authors of this volume for providing the camera-ready copy from which this book was printed.

Printed on acid-free paper. ∞

Printed in the United States of America

10 9 8 7 6 5 4 3 2 1

Dedicated to my colleagues and students for many fun years exploring Benford's Law together, to my parents Arlene and William for their support and encouragement over the years, and to the number 1 for being such a good, frequent companion. — SJM, Williamstown, MA, 2015.

Contents

Foreword

Perhaps I should immediately explain to the reader that this foreword is neither a voice from the grave, nor a remarkable display of anticipation. The "Frank A. Benford" who's the author of this foreword is a grandson of the "Frank A. Benford" for whom Benford's Law is named. As this relationship by itself hardly qualifies me to write this foreword, let me hasten to add that I'm a professional applied mathematician with a Ph.D. from Harvard.

That my grandfather's "Law of Anomalous Numbers" became known as "Benford's Law" instead of "Newcomb's Law" is, of course, a historical accident. I'm not complaining, obviously, but descendants of Simon Newcomb have a legitimate beef. I recently learned that my possibly distant cousin Gregory Benford, the well known physicist and science fiction author, had a colleague William Newcomb who was Simon Newcomb's grandson. Although Gregory and William worked closely together, the Benford/Newcomb connection never seems to have come up! Well, they're physicists, not mathematicians, so maybe I shouldn't be too surprised.

Figure 1   Frank Benford and his family, 1946.

I'd like to be able to claim that I remember grandfather, but that would not be true; he died when I was only three. I know that I met him, however, because a photograph (see Figure 1) from 1946 shows him and grandmother, their four sons, four daughters-in-law, and eight grandchildren at a family reunion. My mother was impressed by grandfather's ability to work calculus problems amid the hubbub of a family gathering.

My grandfather worked as a physicist in the Research Laboratory of the General Electric Company in Schenectady, NY, and most of his work for G.E. concerned optics. I'm sure that grandfather, being a conscientious employee, did all the research and writing of "The Law of Anomalous Numbers" on his own time. In a three-page autobiographical sketch he wrote for Leonard Clark of Union College in 1939, only one short paragraph concerns his Law of Anomalous Numbers.

Steven Miller, the editor of this volume, suggested that I include "stories about your grandfather, anything you know about the reception of his work, how he felt about it, how he would feel to see his name attached to something arising in so many different fields." As I don't have any first-hand information about grandfather, I passed this request along to my father, who wrote in his reply,

> My father was extremely modest and had little to say about his publications. He certainly never boasted. He was, indeed, interested in the phenomenon of first digits, but not excessively so. He would truly be surprised to learn of the interest that seems currently alive.

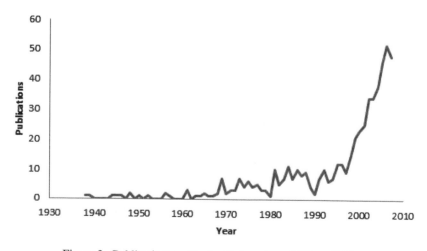

Number of Benford Law Publications by Year

Figure 2   Publications on Benford's Law from 1938 to 2007.

I think my grandfather was proud of his paper, and fond of it, but he wasn't the sort to brag, and I suppose he was resigned to the idea that his Law of Anomalous

Numbers would soon disappear into obscurity. He would certainly be gratified (and possibly astonished) that the first-digit phenomenon he'd rediscovered seems to be a topic of perennial interest. As Ken Ross puts it, writing papers about Benford's Law appears to be "a growth industry." I offer as evidence some data I've compiled from the Benford Online Bibliography [BerH2]. Figure 2 shows the number of known "Benford relevant" publications per year for the first 70 years: 1938 until 2007; the last few years are omitted as there is a delay between when some papers are published and when they are added to the online bibliography. In Table 1 we give the frequencies of the first digits of the publication data from Figure 2 (omitting years where no relevant papers were published). The pattern of first digits should look familiar![1]

First digit	Frequency	Percentage	Benford's Law
1	20	35.7	30.1
2	9	16.1	17.6
3	8	14.3	12.5
4	5	8.9	9.7
5	3	5.4	7.9
6	2	3.6	6.7
7	6	10.7	5.8
8	1	1.8	5.1
9	2	3.6	4.6

Table 1   Frequencies in publication data of papers on Benford's Law: 1938 to 2007.

Roughly speaking, publications dealing with Benford's Law may be sorted into three classes: theoretical (extensions of Benford's Law, and investigations into the circumstances where Benford's Law does, or does not, apply), applications of Benford's Law, and popularizations (i.e., expository pieces aimed at the "intelligent layman"). I expect that almost all papers published between 1940 and 1965 are theoretical in nature. Starting with Ralph Raimi's 1969 Scientific American article [Rai], popular accounts of Benford's Law have appeared at a steady rate. This is attributable, of course, to the counterintuitive nature of the phenomenon. While theoretical papers on the first digit phenomenon have continued to appear, the publication of Mark Nigrini's 1992 dissertation [Nig1] marks the beginning of a wave of publications concerned with applications of Benford's Law. This is reflected, appropriately, in the contents of this book. There are 5 chapters that are theoretical in nature, and 13 chapters concerned with applications in accounting, vote fraud, economics, psychology, the natural sciences, and image processing.

It's been 75 years since my grandfather published "The Law of Anomalous Numbers," and it seems like a propitious time to publish a summary of the current state

---

[1]The chi-square statistic (comparing to the Benford frequencies) is 5.8 with 8 degrees of freedom (the 5% threshold value is 15.5, and thus the data is consistent with Benford behavior).

of affairs, i.e., the book you're reading. In the 75 years since grandfather wrote his article, the first-digit phenomenon has gone from an obscure curiosity to a fairly well-known and useful "law." Who knows what another 75 years will bring? There may be departments of Pure and Applied Benfordology at most major universities!

Frank A. Benford
`BenfordAppliedMath.com`
Salem, Oregon
June, 2014

Preface

One of the greatest beauties in mathematics is how the same equations can describe phenomena in widely different fields. Benford's Law of digit bias is an outstanding example of this. Briefly, it asserts that for many natural data sets we are more likely to see numbers with small leading digits than large ones. More precisely, our system is Benford if the probability of a first digit of d is $\log_{10} \frac{d+1}{d}$; we often consider the related but stronger statement that the probability of a significand being at most s is $\log_{10} s$, or the natural generalizations to other number bases. Base 10, the probabilities range from having a leading digit of 1 almost 30% of the time, to only about a 4.6% chance of starting with a 9.

Benford's Law arises in a variety of disciplines, including accounting, computer science, dynamical systems, economics, engineering, medicine, number theory, probability, psychology and statistics, to name just a few, and provides a wonderful opportunity for a common meeting ground for people with diverse interests and backgrounds. My first encounter with it was in Serre's *A Course in Arithmetic* [Ser]. On page 76 he remarks that Bombieri showed him a proof that the analytic density of the set of primes with leading digit 1 is $\log_{10} 2$, which is the Benford probability; a short argument using Poisson Summation yields the proof. I next saw it in Knuth's *The Art of Computer Programming, Volume 2: Seminumerical Algorithms* ([Knu], page 255), where he discusses applications of Benford's Law to analyzing floating point operations, especially the fact that Benford behavior implies the relative error from rounding is typically higher than one would expect. Once aware (or perhaps I should say doubly aware) of its existence, I saw it more and more often.

Our purposes here are to show students and researchers useful techniques from a variety of subjects, highlight the connections between the different areas and encourage research and cross-departmental collaboration on these problems. To do this, we develop much of the general theory in the first few chapters (concentrating on the methods which are applicable to a variety of problems), and then conclude with numerous chapters on applications written by world-experts in that field. Though there are common themes and methods throughout the applications, these chapters are self-contained, needing only the introductory chapters and some standard material. **For those wishing to use this as a textbook, numerous exercises and supplemental material are collected in the final chapter, and additionally are posted online (where more problems can easily be added, and links to relevant material for that chapter are collected); see** `http://web.williams.edu/Mathematics/sjmiller/public_html/benford/`.

One advantage of posting problems online is that this need not be a static list, and thus please feel free to email suggestions for additional exercises.

Below we briefly outline the major themes of the book.

- **Part I: General Theory I: Basis of Benford's Law:** We begin our study of Benford's Law with a brief introduction by Miller in Chapter 1. We concentrate on the history and some possible explanations, and briefly discuss a few of the many applications and central questions in the field.

  While for many readers this level of depth suffices, the subject can (and should!) be built on firm foundations. We do this in Chapter 2, where Berger and Hill rigorously derive many results through the use of appropriate σ-algebras. There are many approaches to proving a system satisfies Benford's Law. One of the most important is the Fundamental Equivalence (also called the uniform distribution characterization), which says a system $\{x_n\}$ satisfies Benford's Law base B if and only if its logarithm modulo 1 (i.e., $y_n = \log_B x_n \bmod 1$) is uniformly distributed. In other words, in the limit, the probability the logarithm modulo 1 lies in a subinterval $[a, b]$ of $[0, 1]$ is just $b - a$. The authors describe this and additional characterizations of Benford's Law (including the scale-invariance characterization and the base-invariance characterization), and prove many deterministic and random processes satisfy Benford's Law, as well as discussing flaws of other proposed explanations (such as the spread distribution approach).

  For the uniform distribution characterization to be useful, however, we need ways to show these logarithms are uniformly distributed. Often techniques from Fourier analysis are well suited for such an analysis. The Fundamental Equivalence reduces the Benfordness of $\{x_n\}$ to the distribution of the fractional parts of its logarithms $\{y_n\}$. Fourier analysis is built on the functions $e_m(t) := \exp(2\pi i m t)$ (where $i = \sqrt{-1}$); note that the painful modulo condition in y_n vanishes when it is the argument of e_m, as $e_m(y_n) = e_m(y_n \bmod 1)$. Chapter 3 by Miller is devoted to developing Fourier analytic techniques to prove Benford behavior. We demonstrate the power of this machinery by applying it to a variety of problems, including products and chains of random variables, L-functions, special densities and the infamous $3x + 1$ problem. For example, using techniques from Fourier analysis (especially Poisson Summation), one can show that the standard exponential random variable is very close to satisfying Benford's Law. The exponential is a special case of the three-parameter Weibull distribution. A similar analysis shows that, so long as the shape exponent of the Weibull is not too large, it too is close to being Benford. There are numerous applications of these results. The closeness of the standard exponential to Benford implies that order statistics are almost Benford as well. The Weibull distribution arises in many survival models, and thus the analysis here provides another explanation of the prevalence of Benford behavior in many diverse systems.

- **Part II: General Theory II: Distributions and Rates of Convergence:** Combinations of data sets or random variables are often closer to satisfying Benford's Law than the individual data sets or distributions. This suggests a natural problem: looking for distributions that are exactly or at least close to being Benford. One of the most important examples of a distribution that exhibits Benford behavior is that of a geometric random variable. Numerous phenomena obey a geometric growth law; in particular, the solution to almost any linear difference equations is a linear combination of geometric series. We then investigate other important distributions and see how close they are to Benford. Although Benford's Law applies to a wide variety of data sets, none of the popular parametric distributions, such as the exponential and normal distributions, conforms exactly. Chapter 4 highlights the failure of several well-known probability distributions, then delves into the geometry associated with probability distributions that obey Benford's Law exactly. The starting point of these constructions is the fact that if U is a uniform random variable on $[a, a + n]$ for some integer n, then $T = 10^U$ is Benford base 10.

As the exponential and Weibull distributions are not exactly Benford, it is important to obtain estimates on the size of the deviations. There are many ways to obtain such bounds. In Chapter 3 these bounds were obtained from Poisson Summation and the Fourier transform; in Chapter 5 Dümbgen and Leuenberger derive bounds from the total variation of the density (and its derivatives). These results are applied to numerous distributions, such as exponential, normal and Weibull random variables.

This part concludes with Chapter 6 by Schürger. Earlier in the book we showed geometric Brownian motions are Benford. While processes such as the stock market were initially modeled by Brownian motions, such models have several defects, and current work must incorporate jumps and heavy tails. This leads to the study of Lévy Processes. These processes are described in detail, and their convergence to Benford behavior is shown. The techniques required are similar to those for geometric Brownian motion. On the other hand, the class of Lévy processes is much more general than just geometric Brownian motion, with applications in stochastic processes and finance; in particular, these and related processes model financial data, which has long been known to closely follow Benford's Law.

The final parts of this book deal with just some of the many applications of Benford's Law. Due to space constraints it is impossible to discuss all of the places Benford's Law appears. We have therefore chosen to focus on just a few situations, going for depth over breadth. We encourage the reader to peruse the many resources, such as the searchable online bibliography at [BerH2] or the large compilation [Hu], for a tour through additional areas to explore.

- **Part III: Applications I: Accounting and Vote Fraud:** Though initially an amusing observation about the distribution of digits in various data sets, since then Benford's Law has found numerous applications in many diverse fields. We briefly survey some of these. Probably the most famous application is to detecting tax fraud, though of course it is fruitfully used elsewhere too. We start in Chapter 7 with some of the basics of accounting, where Cleary and Thibodeau describe how Benford's Law can be integrated into business statistics and accounting courses. In particular, in the American Statistical Association's 2005 report *Guidelines for Assessment and Instruction in Statistics Education*, the following four goals (among others) are listed for what students should know after a first statistics course: (1) that variability is natural, predictable and quantifiable; (2) that random sampling allows results of surveys and experiments to be extended to the population from which the sample was taken; (3) how to interpret statistical results in context; (4) how to critique news stories and journal articles that include statistical information, including identifying what's missing in the presentation and the flaws in the studies or methods used to generate the information. The rest of the chapter shows how incorporating Benford's Law realizes these objectives.

  Chapter 8 by Nigrini describes one of the most important applications of Benford's Law: detecting fraud. Many diverse systems approximately obey the law, and thus deviations often indicate fraud. The chapter begins by examining some data sets that follow the law (tax returns, the 2000 census, stream flow data and accounts payable data), and concludes by showing how Benford's Law successfully detected fraud in accounts payable amounts, payroll data and corporate numbers (such as Enron).

  We continue with another important example where Benford's Law has successfully detected fraud. Chapters 9 by Mebane and 10 by Roukema discuss how Benford's Law can detect vote fraud; the first chapter develops tests based on the second digit and explores its use in practice, while the second concentrates on a recent Iranian election whose official vote counts were claimed to be invalid. .

- **Part IV: Applications II: Economics:** While there is no dearth of interesting topics to explore, we have chosen to devote this part of the book to economics because of the huge impact of recent events. A spectacular example of this is given by European Union (EU) policy, and the situation in Greece. We begin in Chapter 11 by Rauch, Göttsche, Brähler and Engel with a description of EU practices and data from several countries. As the stakes are high, there is enormous pressure to misreport statistics to avoid being hit with EU deficit procedures. We continue in Chapter 12 by Tödter with additional analysis, especially of published economics research papers. A surprisingly large proportion of first digits of regression coefficients and standard errors violate Benford's Law, in contrast to second digits. Routine applications of Benford tests would increase the efficiency of replication exercises and raise the risk of scientific misconduct. Another issue discussed is fitting data to a Generalized Benford Law, a topic Lee, Cho and Judge address in Chapter

17; both of these chapters deal with the issues facing the public arising from researchers falsifying data. We conclude this part with an analysis of data from the U.S. financial sector. The main finding is that Benford's Law fits the data from before the housing crisis well, but not the data afterwards.

- **Part V: Applications III: Sciences:** In previous chapters we discussed which distributions fit (and which don't fit) Benford's Law, as well as tests to detect fraud. In this part we take a different approach, and explore the psychology behind the people generating numbers. Chapters 14 by Burns and Krygier and 15 by Chou, Kong, Teo and Zheng explore patterns and tendencies in number generation, and the resulting implications, followed by Hoyle's chapter on the prevalence of Benford's Law in the natural sciences, including a summary of its occurrences and a discussion of the consequences. We end in Chapter 17 by Lee, Cho and Judge with a nice mix of theory and application. The authors consider a generalization of Benford's Law, developing the theory and analyzing known cases of fraud. They study the related Stigler distribution, and describe how it may be found from information-theoretic methods. This leads to alternative digit distributions based on maximum entropy principles. The chapter ends by using these new distributions in an analysis of some medical data which was known to be falsified, where the falsified data is detected. An important application of the material of this part is in developing tests to detect whether researchers are submitting fraudulent data. Similar to the chapters from economics, as the costs to society from incorrectly adopting conclusions of faulty research can be high, these tests provide a valuable tool to check the veracity of claims.

- **Part VI: Applications IV: Images:** Our final part deals with whether or not images follow Benford's Law. Chiverton and Wells, in Chapter 18, explore the relationship between intensities in medical images and Benford behavior. They describe a simple classifier based on Bayes theory which uses the Benford Partial Volume (PV) distribution as a prior; the results show experimentally that the Benford PV distribution is a reasonable modeling tool for the classification of imaging data affected by the PV artifact. The fraud-based applications of Benford's Law have grown from financial data sets to others as well. The last chapter, Chapter 19 by Pérez-González, Quach, Abdallah, Heileman and Miller, explores whether or not Benford's Law can detect modifications in images. Specifically, while images in the pixel domain are not close to Benford, the result after applying the Discrete Cosine Transform is. These results can be used to look for hidden messages in pictures, as well as to test whether or not the image has been compressed.

We are extremely grateful to Princeton University Press, especially to our editor Vickie Kearn and to Betsy Blumenthal and Jill Harris, for all their help and aid, to our copyeditor Alison Durham who did a terrific job, especially in standardizing the exposition across chapters, to Meghan Kanabay for assistance with many of the illustrations, and Amanda Weiss for help with the jacket design. Many people proofread the book, looking not just for grammatical issues but also making sure it was a coherent whole with widely accessible expositions; it is a pleasure to thank them, especially John Bihn and Jaclyn Porfilio.

The editor was partially supported by NSF Grants DMS0600848, DMS0970067 and DMS1265673; some of his students assisting with the project were supported by NSF Grants DMS0850577 and DMS1347804, the Clare Boothe Luce Program, and Williams College. Some of this book is based on a conference organized by Chaouki T. Abdallah, Gregory L. Heileman, Steven J. Miller and Fernando Pérez-González and assisted by Ted Hill: *Conference on the Theory and Applications of Benford's Law* (16–18 December 2007, Santa Fe, NM). This conference was supported in part by Brown University, IEEE, NSF Grant DMS-0753043, the New Mexico Consortium's Institute for Advanced Study, Universidade de Vigo and the University of New Mexico, and it is a pleasure to thank them and the participants.

Steven J. Miller
Williams College
Williamstown, MA
October 2013

sjm1@williams.edu, Steven.Miller.MC.96@aya.yale.edu

Notation

□ : indicates the end of a proof.

≡: $x \equiv y \mod n$ means there exists an integer a such that $x = y + an$.

∃ : there exists.

∀ : for all.

$|\cdot|$: $|S|$ (or $\#S$) is the number of elements in the set S.

$\lceil \cdot \rceil$: $\lceil x \rceil$ is the smallest integer greater than or equal to x, read "the ceiling of x."

$\lfloor \cdot \rfloor$ or $[\cdot]$: $\lfloor x \rfloor$ (also written $[x]$) is the greatest integer less than or equal to x, read "the floor of x."

$\{\cdot\}$ or $\langle \cdot \rangle$: $\{x\}$ is the fractional part of x; note $x = [x] + \{x\}$.

$\ll, \gg$: see big-Oh notation.

$\vee$: $a \vee b$ is the maximum of a and b.

$\wedge$: $a \wedge b$ is the minimum of a and b.

$\mathbb{1}_A$ (or I_A) : the indicator function of set A; thus $\mathbb{1}_A(x)$ is 1 if $x \in A$ and 0 otherwise.

δ_a : Dirac probability measure concentrated at $a \in \Omega$.

λ : Lebesgue measure on $(\mathbb{R}, \mathcal{B})$ or parts thereof.

$\lambda_{a,b}$: normalized Lebesgue measure (uniform distribution) on $([a, b), \mathcal{B}[a, b))$.

$\sigma(f)$: the σ-algebra generated by the function $f : \Omega \to \mathbb{R}$.

$\sigma(A)$: the spectrum (set of eigenvalues) of a $d \times d$-matrix A.

A^c : the complement of A in some ambient space Ω clear from the context; i.e., $A^c = \{\omega \in \Omega : \omega \notin A\}$.

$A \backslash B$: the set of elements of A not in B, i.e., $A \backslash B = A \cap B^c$.

$A \Delta B$: the symmetric difference of A and B, i.e., $A \Delta B = A \backslash B \cup B \backslash A$.

a.e. : (Lebesgue) almost every (or almost everywhere).

a.s. : almost surely, i.e., with probability one.

$\mathbb{B}$: Benford distribution on $(\mathbb{R}^+, \mathcal{S})$.

$\mathcal{B}$: Borel σ-algebra on $\mathbb{R}$ or parts thereof.

Big-Oh notation : $A(x) = O(B(x))$, read "$A(x)$ is of order (or big-Oh) $B(x)$," means there exists a $C > 0$ and an x_0 such that for all $x \geq x_0$, $|A(x)| \leq CB(x)$. This is also written $A(x) \ll B(x)$ or $B(x) \gg A(x)$.

$\mathbb{C}$: the set of complex numbers: $\{z : z = x + iy, \, x, y \in \mathbb{R}\}$.

C^ℓ : the set of all ℓ times continuously differentiable functions, $\ell \in \mathbb{N}_0$.

C^∞ : the set of all smooth (i.e., infinitely differentiable) functions; $C^\infty = \bigcap_{\ell \geq 0} C^\ell$.

$D_1, D_2, D_3, \ldots$: the first, second, third, $\ldots$ significant decimal digit.

$D_m^{(b)}$: the mth significant digit base b.

$\mathbb{E}[X]$ (or $\mathbb{E}X$) : the expectation of X.

$e(x)$: $e(x) = e^{2\pi i x}$.

$f_* \mathbb{P}$: a probability measure on $\mathbb{R}$ induced by $\mathbb{P}$ and the measurable function $f : \Omega \to \mathbb{R}$, via $f_* \mathbb{P}(\cdot) := \mathbb{P}(f^{-1}(\cdot))$.

F_n: $\{F_n\}$ is the sequence of Fibonacci numbers, $\{F_n\} = \{0, 1, 1, 2, 3, 5, 8, \ldots\}$ ($F_{n+2} = F_{n+1} + F_n$ with $F_0 = 0$ and $F_1 = 1$).

F_P, F_X : the distribution functions of P and X.

$i : i = \sqrt{-1}$.

i.i.d. : independent, identically distributed (sequence or family of random variables); often one writes i.i.d.r.v.

$\Im z$: see $\Re z$.

infimum : the infimum of a set, denoted $\inf_n x_n$, is the largest number c (if one exists) such that $x_n \geq c$ for all n, and for any $\epsilon > 0$ there is some n_0 such that $x_{n_0} < c + \epsilon$. If the sequence has finitely many terms, the infimum is the same as the minimum value.

j : in some chapters $j = \sqrt{-1}$ (this convention is frequently used in engineering).

Leb : Lebesgue measure.

Little-Oh notation : $A(x) = o(B(x))$, read "$A(x)$ is little-Oh of $B(x)$," means $\lim_{x \to \infty} A(x)/B(x) = 0$.

$L^1(\mathbb{R})$: all $f : \mathbb{R} \longrightarrow \mathbb{C}$ which are measurable and Lebesgue integrable.

log : usually the natural logarithm, though in some chapters it is the logarithm base 10.

ln : the natural logarithm.

$\mathbb{N}$: the set of natural numbers: $\{0, 1, 2, 3, \ldots\}$.

$\mathbb{N}_0$: the set of positive natural number: $\{1, 2, 3, \ldots\}$.

N_f : the Newton map associated with a differentiable function f.

$o(\cdot)$, $O(\cdot)$: see "little-Oh" and "big-Oh" notation, respectively.

$O_T(x_0)$: the orbit of x_0 under the map T, possibly nonautonomous.

$\{p_n\}$: the set of prime numbers: 2, 3, 5, 7, 11, 13, $\ldots$.

P : probability measure on $(\mathbb{R}, \mathcal{B})$, possibly random.

P_X : the distribution of the random variable X.

Prob (or Pr) : a probability function on a probability space.

$\mathbb{Q}$: the set of rational numbers: $\{x : x = \frac{p}{q}, p, q \in \mathbb{Z}, q \neq 0\}$.

$\mathbb{R}$: the set of real numbers.

$\mathbb{R}^+$: the set of positive real numbers.

$\Re z$, $\Im z$: the real and imaginary parts of $z \in \mathbb{C}$; if $z = x + iy$ then $\Re z = x$ and $\Im z = y$.

S : the significand function: if $x > 0$ then $x = S(x) \cdot 10^{k(x)}$, where $S(x) \in [1, 10)$ and $k(x) \in \mathbb{Z}$; more generally one can study the significand function S_B in base B.

$\mathcal{S}$: the significand σ-algebra.

supremum : given a sequence $\{x_n\}_{n=1}^{\infty}$, the supremum of the set, denoted $\sup_n x_n$, is the smallest number c (if one exists) such that $x_n \leq c$ for all n, and for any $\epsilon > 0$ there is some n_0 such that $x_{n_0} > c - \epsilon$. If the sequence has finitely many terms, the supremum is the same as the maximum value.

u.d. mod 1 : uniformly distributed modulo 1.

$\text{Var}(X)$ (or $\text{var}(X)$) : the variance of the random variable X, assuming the expected value of X is finite; $\text{Var}(X) = \mathbb{E}[(X - \mathbb{E}[X])^2]$.

$\mathbb{W}$: the set of whole numbers: $\{1, 2, 3, 4, \dots\}$.

$X_n \overset{\mathcal{D}}{\to} X$: (X_n) converges in distribution to X.

$X_n \overset{\text{a.s.}}{\to} X$: (X_n) converges to X almost surely.

$\bar{z}$, $|z|$: the conjugate and absolute value of $z \in \mathbb{C}$.

$\mathbb{Z}$: the set of integers: $\{\dots, -2, -1, 0, 1, 2, \dots\}$.

$\mathbb{Z}^+$: the set of non-negative integers, $\{0, 1, 2, \dots\}$.

PART I
General Theory I: Basis of Benford's Law

Chapter One

A Quick Introduction to Benford's Law

Steven J. Miller[1]

The history of Benford's Law is a fascinating and unexpected story of the interplay between theory and applications. From its beginnings in understanding the distribution of digits in tables of logarithms, the subject has grown enormously. Currently hundreds of papers are being written by accountants, computer scientists, engineers, mathematicians, statisticians and many others. In this chapter we start by stating Benford's Law of digit bias and describing its history. We discuss its origins and give numerous examples of data sets that follow this law, as well as some that do not. From these examples we extract several explanations as to the prevalence of Benford's Law, which are described in greater detail later in the book. We end by quickly summarizing many of the diverse situations in which Benford's Law holds, and why an observation that began in looking at the wear and tear in tables of logarithms has become a major tool in subjects as diverse as detecting tax fraud and building efficient computers. We then continue in the next chapters with rigorous derivations, and then launch into a survey of some of the many applications. In particular, in the next chapter we put Benford's Law on a solid foundation. There we explore several different categorizations of Benford's Law, and rigorously prove that certain systems satisfy these conditions.

1.1 OVERVIEW

We live in an age when we are constantly bombarded with massive amounts of data. Satellites orbiting the Earth daily transmit more information than is in the entire Library of Congress; researchers must quickly sort through these data sets to find the relevant pieces. It is thus not surprising that people are interested in patterns in data. One of the more interesting, and initially surprising, is Benford's Law on the distribution of the first or the leading digits.

In this chapter we concentrate on a mostly non-technical introduction to the subject, saving the details for later. Before we can describe the law, we must first set notation. At some point in secondary school, we are introduced to **scientific notation**: any positive number x may be written as $S(x) \cdot 10^k$, where $S(x) \in [1, 10)$ is the **significand** and k is an integer (called the **exponent**). The integer part of the

[1] Department of Mathematics and Statistics, Williams College, Williamstown, MA 01267. The author was partially supported by NSF grants DMS0970067 and DMS1265673.

significand is called the **leading digit** or the **first digit**. Some people prefer to call $S(x)$ the mantissa and not the significand; unfortunately this can lead to confusion, as the **mantissa** is the fractional part of the logarithm, and this quantity too will be important in our investigations. As always, examples help clarify the notation. The number 1701.24601 would be written as $1.70124601 \cdot 10^3$ in scientific notation. The significand is 1.70124601, the exponent is 3 and the leading digit is 1. If we take the logarithm base 10, we find $\log_{10} 1701.24601 \approx 3.2307671196444460726$, so the mantissa is approximately .2307671196444460726.

There are many advantages to studying the first digits of a data set. One reason is that it helps us compare apples and apples and not apples and oranges. By this we mean the following: two different data sets could have very different scales; one could be masses of subatomic particles while another could be closing stock prices. While the units are different and the magnitudes differ greatly, every number has a unique leading digit, and thus we can compare the distribution of the first digits of the two data sets.

The most natural guess would be to assert that for a generic data set, all numbers are equally likely to be the leading digit. We would then posit that we should observe about 11% of the time a leading digit of 1, 2, ..., 9 (note that we would guess each number occurs one-ninth of the time and not one-tenth of the time, as 0 is the leading digit for only one number, namely 0). The content of Benford's Law is that this is frequently not so; specifically, in many situations we expect the leading digit to be d with probability approximately $\log_{10}\left(\frac{d+1}{d}\right)$, which means the probability of a first digit of 1 is about 30% while a first digit of 9 happens about 4.6% of the time.

1.2 NEWCOMB

Though it is called Benford's Law, he was not the first to observe this digit bias. Our story begins with the astronomer–mathematician Simon Newcomb, who observed this behavior more than 50 years before Benford. Newcomb was born in Nova Scotia in 1835 and died in Washington, DC in 1909. In 1881 he published a short article in the American Journal of Mathematics, *Note on the Frequency of Use of the Different Digits in Natural Numbers* (see [New]). The article begins,

> That the ten digits do not occur with equal frequency must be evident to any one making much use of logarithmic tables, and noticing how much faster the first pages wear out than the last ones. The first significant figure is oftener 1 than any other digit, and the frequency diminishes up to 9. The question naturally arises whether the reverse would be true of logarithms. That is, in a table of anti-logarithms, would the last part be more used than the first, or would every part be used equally? The law of frequency in the one case may be deduced from that in the other. The question we have to consider is, what is the probability that if a natural number be taken at random its first significant digit will be n, its second n', etc.

As natural numbers occur in nature, they are to be considered as the ratios of quantities. Therefore, instead of selecting a number at random, we must select two numbers, and inquire what is the probability that the first significant digit of their ratio is the digit n. To solve the problem we may form an indefinite number of such ratios, taken independently; and then must make the same inquiry respecting their quotients, and continue the process so as to find the limit towards which the probability approaches.

In this short article two very important properties of the distribution of digits are noted. The first is that all digits are not equally likely. The article ends with a quantification of how oftener the first digit is a 1 than a 9, with Newcomb stating,

The law of probability of the occurrence of numbers is such that all mantissæ of their logarithms are equally probable.

Specifically, Newcomb gives a table (see Table 1.1) for the probabilities of first and second digits.

d	Probability first digit d	Probability second digit d
0		0.1197
1	0.3010	0.1139
2	0.1761	0.1088
3	0.1249	0.1043
4	0.0969	0.1003
5	0.0792	0.0967
6	0.0669	0.0934
7	0.0580	0.0904
8	0.0512	0.0876
9	0.0458	0.0850

Table 1.1 Newcomb's conjecture for the probabilities of observing a first digit of d or a second digit of d; all probabilities are reported to four decimal digits.

The second key observation of his paper is noting the importance of scale. The numerical value of a physical quantity clearly depends on the scale used, and thus Newcomb suggests that the correct items to study are ratios of measurements.

1.3 BENFORD

The next step forward in studying the distribution of the leading digits of numbers was Frank Benford's *The Law of Anomalous Numbers*, published in the Proceedings of the American Philosophical Society in 1938 (see [Ben]). In addition to advancing explanations as to why digits have this distribution, he also presents some justification as to why this is a problem worthy of study.

It has been observed that the pages of a much used table of common logarithms show evidences of a selective use of the natural numbers. The pages containing the logarithms of the low numbers 1 and 2 are apt to be more stained and frayed by use than those of the higher numbers 8 and 9. Of course, no one could be expected to be greatly interested in the condition of a table of logarithms, but the matter may be considered more worthy of study when we recall that the table is used in the building up of our scientific, engineering, and general factual literature. There may be, in the relative cleanliness of the pages of a logarithm table, data on how we think and how we react when dealing with things that can be described by means of numbers.

Benford studied the distribution of leading digits of 20 sets of data, including rivers, areas, populations, physical constants, mathematical sequences (such as $\sqrt{n}$, $n!$, n^2, ...), sports, an issue of Reader's Digest and the first 342 street addresses given in the (then) current American Men of Science. We reproduce his observations in Table 1.2.

Title	1	2	3	4	5	6	7	8	9	Count
Rivers, Area	31.0	16.4	10.7	11.3	7.2	8.6	5.5	4.2	5.1	335
Population	33.9	20.4	14.2	8.1	7.2	6.2	4.1	3.7	2.2	3259
Constants	41.3	14.4	4.8	8.6	10.6	5.8	1.0	2.9	10.6	104
Newspapers	30.0	18.0	12.0	10.0	8.0	6.0	6.0	5.0	5.0	100
Spec. Heat	24.0	18.4	16.2	14.6	10.6	4.1	3.2	4.8	4.1	1389
Pressure	29.6	18.3	12.8	9.8	8.3	6.4	5.7	4.4	4.7	703
H.P. Lost	30.0	18.4	11.9	10.8	8.1	7.0	5.1	5.1	3.6	690
Mol. Wgt.	26.7	25.2	15.4	10.8	6.7	5.1	4.1	2.8	3.2	1800
Drainage	27.1	23.9	13.8	12.6	8.2	5.0	5.0	2.5	1.9	159
Atomic Wgt.	47.2	18.7	5.5	4.4	6.6	4.4	3.3	4.4	5.5	91
n^{-1}, $\sqrt{n}$	25.7	20.3	9.7	6.8	6.6	6.8	7.2	8.0	8.9	5000
Design	26.8	14.8	14.3	7.5	8.3	8.4	7.0	7.3	5.6	560
Digest	33.4	18.5	12.4	7.5	7.1	6.5	5.5	4.9	4.2	308
Cost Data	32.4	18.8	10.1	10.1	9.8	5.5	4.7	5.5	3.1	741
X-Ray Volts	27.9	17.5	14.4	9.0	8.1	7.4	5.1	5.8	4.8	707
Am. League	32.7	17.6	12.6	9.8	7.4	6.4	4.9	5.6	3.0	1458
Black Body	31.0	17.3	14.1	8.7	6.6	7.0	5.2	4.7	5.4	1165
Addresses	28.9	19.2	12.6	8.8	8.5	6.4	5.6	5.0	5.0	342
$n, n^2, \ldots, n!$	25.3	16.0	12.0	10.0	8.5	8.8	6.8	7.1	5.5	900
Death Rate	27.0	18.6	15.7	9.4	6.7	6.5	7.2	4.8	4.1	418
Average	30.6	18.5	12.4	9.4	8.0	6.4	5.1	4.9	4.7	1011
Benford's Law	30.1	17.6	12.5	9.7	7.9	6.7	5.8	5.1	4.6	

Table 1.2 Distribution of leading digits from the data sets of Benford's paper [Ben]; the amalgamation of all observations is denoted by "Average." Note that the agreement with Benford's Law is better for some examples than others, and the amalgamation of all examples is fairly close to Benford's Law.

Benford's paper contains many of the key observations in the subject. One of the most important is that while individual data sets may fail to satisfy Benford's Law, amalgamating many different sets of data leads to a new sequence whose behavior is typically closer to Benford's Law. This is seen both in the row corresponding to $n, n^2, \ldots$ (where we can prove that each of these is non-Benford) as well as in the average over all data sets.

Benford's article suffered a much better fate than Newcomb's paper, possibly in part because it immediately preceded a physics article by Bethe, Rose and Smith on the multiple scattering of electrons. Whereas it was decades before there was another article building on Newcomb's work, the next article after Benford's paper was six years later (by S. A. Goutsmit and W. H. Furry, *Significant Figures of Numbers in Statistical Tables*, in Nature), and after that the papers started occurring more and more frequently. See Hurlimann's extensive bibliography [Hu] for a list of papers, books and reports on Benford's Law from 1881 to 2006, as well as the online bibliography maintained by Arno Berger and Ted Hill [BerH2].

1.4 STATEMENT OF BENFORD'S LAW

We are now ready to give precise statements of Benford's Law.

Definition 1.4.1 (Benford's Law for the Leading Digit). *A set of numbers satisfies Benford's Law for the Leading Digit if the probability of observing a first digit of d is* $\log_{10}\left(\frac{d+1}{d}\right)$.

While clean and easy to state, the above definition has several problems when we apply it to real data sets. The most glaring is that the numbers $\log_{10}\left(\frac{d+1}{d}\right)$ are irrational. If we have a data set with N observations, then the number of times the first digit is d must be an integer, and hence the observed frequencies are always rational numbers.

One solution to this issue is to consider only infinite sets. Unfortunately this is not possible in many cases of interest, as most real-world data sets are finite (i.e., there are only finitely many counties or finitely many trading days). Thus, while Definition 1.4.1 is fine for mathematical investigations of sequences and functions, it is not practical for many sets of interest. We therefore adjust the definition to

Definition 1.4.2 (Benford's Law for the Leading Digit (Working Definition)). *We say a data set satisfies Benford's Law for the Leading Digit if the probability of observing a first digit of d is approximately* $\log_{10}\left(\frac{d+1}{d}\right)$.

Note that the above definition is vague, as we need to clarify what is meant by "approximately." It is a non-trivial task to find good statistical tests for large data sets. The famous and popular chi-square tests, for example, frequently cannot be used with extensive data sets as this test becomes very sensitive to small deviations when there are many observations. For now, we shall use the above definition and interpret "approximately" to mean a good visual fit. This approach works quite well for many applications. For example, in Chapter 8 we shall see that many corporate and other financial data sets follow Benford's Law, and thus if the distribution is visually far from Benford, it is quite likely that the data's integrity has been compromised.

Finally, instead of studying just the leading digit we could study the entire significand. Thus in place of asking for the probability of a first digit of 1 or 2 or 3, we now ask for the probability of observing a significand between 1 and 2, or between π and e. This generalization is frequently called the **Strong Benford's Law**.

Definition 1.4.3 (Strong Benford's Law for the Leading Digits (Working Definition)). *We say a data set satisfies the Strong Benford's Law if the probability of observing a significand in* $[1, s)$ *is* $\log_{10} s$.

Note that Strong Benford behavior implies Benford behavior; the probability of a first digit of d is just the probability the significand is in $[d, d+1)$. Writing $[d, d+1)$ as $[1, d+1)\backslash[1, d)$, we see this probability is just $\log_{10}(d+1) - \log_{10} d = \log_{10} \frac{d+1}{d}$.

1.5 EXAMPLES AND EXPLANATIONS

In this section we briefly give some explanations for why so many different and diverse data sets satisfy Benford's Law, saving for later chapters more detailed explanation. It's worthwhile to take a few minutes to reflect on how Benford's Law was discovered, and to see whether or not similar behavior might be lurking in other systems. The story is that Newcomb was led to the law by observing that the pages in logarithm tables corresponding to numbers beginning with 1 were significantly more worn than the pages corresponding to numbers with higher first digit. A reasonable explanation for the additional wear and tear is that numbers with a low first digit are more common than those with a higher first digit. It is thus quite fortunate for the field that there were no calculators back then, as otherwise the law could easily have been missed. Though few (if any) of us still use logarithm tables, it is possible to see a similar phenomenon in the real world today. Our analysis of this leads to one of the most important theorems in probability and statistics, the Central Limit Theorem, which plays a role in understanding the ubiquity of Benford's Law.

Instead of looking at logarithm tables, we can look at the steps in an old building, or how worn the grass is on college campuses. Assuming the steps haven't been replaced and that there is a reasonable amount of traffic in and out of the building, then lots of people will walk up and down these stairs. Each person causes a small amount of wear and tear on the steps; though each person's contribution is small, if there are enough people over a long enough time period then the cumulative effect will be visually apparent. Typically the steps are significantly more worn towards the center and less so as one moves towards the edges. A little thought suggests the obvious answer: people typically walk up the middle of a flight of stairs unless someone else is coming down. Similar to carbon dating, one could attempt to determine the age of a building by the indentation of the steps. Looking at these patterns, we would probably see something akin to the normal distribution, and if we were fortunate we might "discover" the Central Limit Theorem. There are many other examples from everyday life. We can also observe this in looking at lawns. Everyone knows the shortest distance between two points is a line, and people frequently leave the sidewalks and paths and cut across the grass, wearing

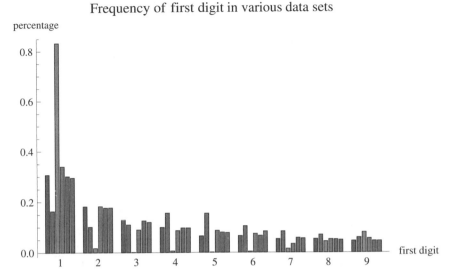

Figure 1.1 Frequencies of leading digits for (a) U.S. county populations (from 2000 census); (b) U.S. county land areas in square miles (from 2000 census); (c) daily volume of NYSE trades from 2000 through 2003; (d) fundamental constants (from NIST); (e) first 3219 Fibonacci numbers; (f) first 3219 factorials. Note the census data includes Puerto Rico and the District of Columbia.

it down to dirt in some places and leaving it untouched in others. Another example is to look at keyboards, and compare the well-worn "E" to the almost pristine "Q." Or the wear and tear on doors. The list is virtually endless.

In Figure 1.1 we look at the leading digits of the several "natural" data sets. Four arise from the real world, coming from the 2000 census in the United States (population and area in square miles of U.S. counties), daily volumes of transactions on the New York Stock Exchange (NYSE) from 2000 through 2003 and the physical constants posted on the homepage of the National Institute for Standards and Technology (NIST); the remaining two data sets are popular mathematical sequences: the first 3219 Fibonacci numbers and factorials (we chose this number so that we would have as many entries as we do counties).

If these are "generic" data sets, then we see that no one law describes the behavior of each set. Some of the sets are quite close to following Benford's Law, others are far off; none are close to having each digit equally likely to be the leading digit. Except for the second and third sets, the rest of the data behaves similarly; this is easier to see if we remove these two examples, which we do in Figure 1.2.

Before launching into explanations of why so many data sets are Benford (or at least close to it), it's worth briefly remarking why many are not. There are several reasons and ways a data set can fail to be Benford; we quickly introduce some of these reasons now, and expand on them more when we advance explanations for Benford's Law below. For example, imagine we are recording hourly temperatures

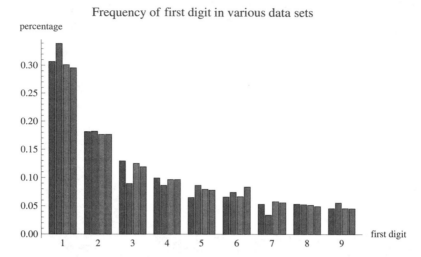

Figure 1.2 Frequencies of leading digits for (a) U.S. county populations (from 2000 census);
(b) fundamental constants (from NIST); (c) first 3219 Fibonacci numbers; (d)
first 3219 factorials. Note the census data includes Puerto Rico and the District
of Columbia.

in May at London Heathrow Airport. In Fahrenheit the temperatures range from
lows of around 40 degrees to highs of around 80. As all digits are not accessible,
it's impossible to be Benford, though perhaps *given this restriction, the relative
probabilities of the digits are Benford.*

For another issue, we have many phenomena that are given by specific, concentrated distributions that will not be Benford. The Central Limit Theorem is often a
good approximation for the behavior of numerous processes, ranging from heights
and weights of people to batting averages to scores on exams. In these situations
we clearly do not expect Benford behavior, though we will see below that processes whose *logarithms* are normally distributed (with large standard deviations)
are close to Benford.

Thus, in looking for data sets that are close to Benford, it is natural to concentrate
on situations where the values are not given by a distribution concentrated in a small
interval. We now explore some possibilities below.

1.5.1 The Spread Explanation

We drew the examples in Figure 1.1 from very different fields; why do so many
of them behave similarly, and why do others violently differ? While the first question still confounds researchers, we can easily explain why two data sets had such
different behavior, and this reason has been advanced by many as a source of Benford's Law (though there are issues with it, which we'll comment on shortly). Let's
look at the first two sets of data: the population in U.S. counties in 2000 and daily
volume of the NYSE from 2000 through 2003. You can see from the histogram in

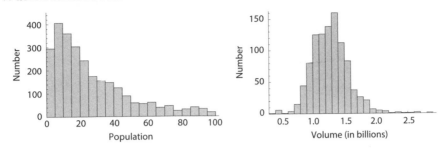

Figure 1.3 (Left) The population (in thousands) of U.S. counties under 250,000 (which is about 84% of all counties). (Right) The daily volume of the NYSE from 2000 through 2003. Note the population spans two orders of magnitude while the stock volumes are mostly within a factor of 2 of each other.

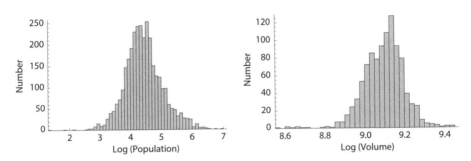

Figure 1.4 (Left) The population of U.S. counties. (Right) The daily volume of the NYSE from 2000 through 2003.

Figure 1.3 the stock market transactions are clustered around one value and span only one order of magnitude. Thus it is not surprising that there is little variation in these first digits. For the county populations, however, the data is far more spread out. These effects are clearer if we look at a histogram of the log-plot of the data, which we do in Figure 1.4. A detailed analysis of the other data sets shows similar behavior; the four data sets that behave similarly are spread out on a logarithmic plot over several orders of magnitude, while the two sets that exhibit different behavior are more clustered on a log-plot.

Our discussion above leads to our first explanation for Benford's Law, the **spread hypothesis**. The spread hypothesis states that if a data set is distributed over several orders of magnitude, then the leading digits will approximately follow Benford's Law. Of course, a little thought shows that we need to assume far more than the data just being spread out over several orders of magnitude. For example, if our set of observations were

$$\{1, 10, 100, 1000, \ldots, 10^{2015}\}$$

then clearly it is non-Benford, even though it does cover over 2000 orders of magnitude! As remarked above, our purpose in this introduction is to just briefly introduce the various ideas and approaches, saving the details for later. There are many issues with the **spread hypothesis**; see Chapter 2 and [BerH3] for an excellent analysis of these problems.

1.5.2 The Geometric Explanation

Our next attempt to explain the prevalence of Benford's Law goes back to Benford's paper [Ben], whose second part is titled *Geometric Basis of the Law*. The idea is that if we have a process with a constant growth rate, then more time will be spent at lower digits than higher digits. For definiteness, imagine we have a stock that increases at 4% per year. The amount of time it takes to move from \$1 to \$2 is the same as it would take to move from \$10,000 to \$20,000 or from \$100,000,000 to \$200,000,000. If n_d is the number of years it takes to move from d dollars to $d+1$ dollars then $d \cdot (1.04)^{n_d} = (d+1)$, or

$$n_d = \frac{\log\left(\frac{d+1}{d}\right)}{\log 1.04}. \tag{1.1}$$

In Table 1.3 we consider the (happy) situation of a stock that rises 4% each and every year. Notice that it takes over 17 years to move from being worth \$1 to being worth \$2, but less than 3 years to move from being worth \$9 to \$10.

First digit	Years	Percentage of time	Benford's Law
1	17.6730	0.30103	0.30103
2	10.3380	0.17609	0.17609
3	7.3350	0.12494	0.12494
4	5.6894	0.09691	0.09691
5	4.6486	0.07918	0.07918
6	3.9303	0.06695	0.06695
7	3.4046	0.05799	0.05799
8	3.0031	0.05115	0.05115
9	2.6863	0.04576	0.04576

Table 1.3 How long the first digit of a stock has leading digit d, given that the stock rises 4% each year. It takes the stock approximately 58.7084 years to increase from \$1 to \$10.

A little algebra shows that this implies Benford behavior. If n is the amount of time it takes to move from \$1 to \$10, then $1 \cdot (1.04)^n = 10$ or $n = \frac{\log 10}{\log 1.04}$. Thus by (1.1), we see the percentage of the time spent with a first digit of d is

$$\frac{\log\left(\frac{d+1}{d}\right)}{\log 1.04} \bigg/ \frac{\log 10}{\log 1.04} = \frac{\log\left(\frac{d+1}{d}\right)}{\log 10} = \log_{10}\left(\frac{d+1}{d}\right), \tag{1.2}$$

which is just Benford's Law! There is nothing special about 4%; the same analysis works in general *provided* that at each moment we grow by the same, fixed rate. The analysis is more interesting if at each instance the growth percentage is a random variable, say drawn from a Gaussian. For more on such processes see Chapter 6.

This is not an isolated example. Many natural and mathematical phenomena are governed by geometric growth. Examples range from radioactive decay and bacteria populations to the Fibonacci numbers. One reason for this is that solutions to many difference equations are given by linear combinations of geometric series; as difference equations are just discrete analogues of differential equations, it is thus not surprising that they model many situations. For example, the Fibonacci numbers satisfy the second order linear recurrence relation

$$F_{n+2} = F_{n+1} + F_n. \tag{1.3}$$

Once the first two Fibonacci numbers are known, the recurrence (1.3) determines the rest. If we start with $F_0 = 0$ and $F_1 = 1$, we find $F_2 = 1$, $F_3 = 2$, $F_4 = 3$, $F_5 = 5$ and so on. Moreover, there is an explicit formula for the nth term, namely

$$F_n = \frac{1}{\sqrt{5}} \left(\frac{1 + \sqrt{5}}{2} \right)^n - \frac{1}{\sqrt{5}} \left(\frac{1 - \sqrt{5}}{2} \right)^n; \tag{1.4}$$

known as Binet's formula; generalizations of it hold for solutions to linear recurrence relations. As $\left| \frac{1+\sqrt{5}}{2} \right| > 1$ and $\left| \frac{1-\sqrt{5}}{2} \right| < 1$, for large n this implies $F_n \approx \frac{1}{\sqrt{5}} \left(\frac{1+\sqrt{5}}{2} \right)^n$. Note that $F_{n+1} \approx \frac{1+\sqrt{5}}{2} F_n$, or $F_{n+1} \approx 1.61803 F_n$. This means that the Fibonacci numbers are well approximated by what would be a highly desirable stock rising about 61.803% each year, and hence by our previous analysis it is reasonable to expect the Fibonacci numbers will be Benford as well.

While the discreteness of the Fibonacci numbers makes the analysis a bit more complicated than the continuous growth rate problem, a generalization of these methods proves that the Fibonacci numbers, as well as the solution to many difference equations, are Benford. Again, our purpose here is to merely provide some evidence as to why so many different, diverse systems satisfy Benford's Law. It is not the case that every recurrence relation leads to Benford behavior. To see this, consider $a_{n+2} = 2a_{n+1} - a_n$ with either $a_0 = a_1 = 1$ (which implies $a_n = 1$ for all n) or $a_0 = 0$ and $a_1 = 1$ (which implies $a_n = n$ for all n). While there are examples of recurrence relations that are non-Benford, a "generic" one will satisfy Benford's Law, and thus studying these systems provides another path to Benford.

1.5.3 The Scale-Invariance Explanation

For our next explanation, we return to a comment from Newcomb's [New] paper:

> As natural numbers occur in nature, they are to be considered as the ratios of quantities. Therefore, instead of selecting a number at random, we must select two numbers, and inquire what is the probability that the first significant digit of their ratio is the digit n.

The import of this comment is that the behavior should be independent of the units used. For example, if we look at the value of stocks in our portfolio then the magnitudes will change if we measure their worth in dollars or euros or yen or bars of gold pressed latinum, though the physical quantities are unchanged. Similarly we can use the metric system or the (British) imperial system in measuring physical constants. As the universe doesn't care what units we use for our experiments, it is natural to expect that the distribution of leading digits should be unchanged if we change our units.

For definiteness, let's consider the areas of the countries in the world. There are almost 200 countries; if we measure area in square kilometers then about 28.49% have a first digit of 1 and 18.99% have a first digit of 2, while if we measure in square miles it is 34.08% have a first digit of 1 and 16.20% have a first digit of 2, which should be compared to the Benford probabilities of approximately 30.10% and 17.61%; one observes a similar closeness with the other digits.

The assumption that there *is* a distribution of the first digit and that this distribution is independent of scale implies the first digits follow Benford's Law. The analysis of this involves introducing a σ-algebra and studying scale-invariant probability measures on this space. Without going into these details now, we can at least show that Benford's Law is consistent with scale invariance.

Let's assume our data set satisfies the Strong Benford Law (see Definition 1.4.3). Then the probability the significand is in $[a, b] \subset [1, 10)$ is $\log_{10}(b/a)$. Assume now we rescale every number in our set by multiplying by a fixed constant C. For definiteness we take $C = \sqrt{3}$ and compute the probability that numbers in the scaled data set have leading digit 1. Note that multiplying $[1, 10)$ by $\sqrt{3}$ gives us the interval $[\sqrt{3}, 10\sqrt{3}) \approx [1.73, 17.32)$. The parts of this new interval with a leading digit of 1 are $[\sqrt{3}, 2)$ and $[10, 10\sqrt{3})$, which come from $[1, 2/\sqrt{3})$ and $[10/\sqrt{3}, 10)$. As we are assuming the strong form of Benford's Law, the probabilities of these two intervals are $\log_{10} \frac{2/\sqrt{3}}{1}$ and $\log_{10} \frac{10}{10\sqrt{3}}$. Summing yields the probability of the first digit of the scaled set being 1 is

$$\log_{10}\left(\frac{2/\sqrt{3}}{1}\right) + \log_{10}\left(\frac{10}{10\sqrt{3}}\right) = \log_{10} 2,$$

which is the Benford probability! A similar analysis works for the other leading digits and other choices of C.

We close this section by noting that scale invariance fits naturally with the other explanations introduced to date. If our initial data set were spread out over several orders of magnitude, so too would the scaled data. Similarly, if we return to our hypothetical stock increasing by 4% per year, the effect of changing the units of our currency can be viewed as changing our principal; however, what governs how long our stock spends with a leading digit of d is not the principal but rather the rate of growth, and that is unchanged.

1.5.4 The Central Limit Explanation

We need to introduce some machinery for our last heuristic explanation. If $y \geq 0$ is a real number, by $y \bmod 1$ we mean the fractional part of y. Other notations for this are $\{y\}$ or $y - \lfloor y \rfloor$. If $y < 0$ then $y \bmod 1$ is $1 - (-y \bmod 1)$. In other words, $y \bmod 1$ is the unique number in $[0, 1)$ such that $y - (y \bmod 1)$ is an integer. Thus $3.14 \bmod 1$ is .14, while $-3.14 \bmod 1$ is .86. We say y **modulo** 1 for $y \bmod 1$.

Recall that any positive number x may be written in **scientific notation** as $x = S(x) \cdot 10^k$, where $S(x) \in [1, 10)$ and k is an integer. The real number $S(x)$, called the **significand**, encodes all the information about the digits of x; the effect of k is to specify the decimal point's location. Thus, if we are interested in either the first digit or the significand, the value of k is immaterial. This suggests that rather than studying our data as given, it might be worthwhile to transform the data as follows:

$$x \mapsto \log_{10} x \bmod 1. \tag{1.5}$$

A little algebra shows that two positive numbers have the same leading digits if and only if their signficands have the same first digit. Thus if we have a set of values $\{x_1, x_2, x_3, \ldots\}$ then the subset with leading digit d is $\{x_i : S(x_i) \in [d, d+1)\}$, which is equivalent to $\{x_i : \log_{10} S(x_i) \in [\log_{10} d, \log_{10}(d+1))\}$.

This innocent-looking reformulation turns out to be not only one of the most fruitful ways of exploring Benford's Law, but also highlights what is going on. We first explain the new perspective gained by transforming the data. According to Benford's Law, the probability of observing a first digit of d is $\log_{10} \frac{d+1}{d}$. This is $\log_{10}(d+1) - \log_{10} d$, which is the length of the interval $[\log_{10} d, \log_{10}(d+1))$! In other words, consider a data set satisfying Benford's Law, and transform the set as in (1.5). The new set lives in $[0, 1)$ and is uniformly distributed there. Specifically, the probability that we have a value in the interval $[\log_{10} d, \log_{10}(d + 1))$ is the length of that interval.

While it may not seem natural to take the logarithm base 10 of each number, and then look at the result modulo 1, under such a process the resulting values are uniformly distributed if the initial set obeys Benford's Law. Another way of looking at this is that there is a natural transformation which takes a set satisfying Benford's Law and returns a new set of numbers that is uniformly distributed.

We briefly comment on why this is a natural process. We replace x with $\log_{10} x$ mod1. If we write $x = S(x) \cdot 10^k$, then $\log_{10} x \bmod 1$ is just $\log_{10} S(x)$. Thus taking the logarithm modulo 1 is a way to get our hands on the significand (actually, its logarithm), which is what we want to understand. While the logarithm function is a nice function, removing the integer part *in general* is messy and leads to complications; however, there is a very important situation where it is painless to remove the integer part. Recall the exponential function

$$e(x) := e^{2\pi i x} = \cos(2\pi x) + i \sin(2\pi x), \tag{1.6}$$

where $i = \sqrt{-1}$. As $e(x + 1) = e(x)$, we see

$$e(x \bmod 1) = e(x). \tag{1.7}$$

The utility of the above becomes apparent when we apply Fourier analysis. In Fourier analysis one uses sines, cosines or exponential functions to understand more complicated functions. From our analysis above, we may either include the modulo 1 or not in the argument of the exponential function. While we will elaborate on this at great length later, the key takeaway is that the transformed data is ideally suited for Fourier analysis.

We can now sketch how this is related to Benford's Law. There are many data sets in the world whose values are the product of numerous measurements. For example, the monetary value of a gold brick is a product of the brick's length, width, height, density and value of gold per pound. Imagine we have some quantity X which is a product of n values, so

$$X = X_1 \cdot X_2 \cdot \ldots \cdot X_n.$$

We assume the X_i's are nice random variables. From our discussion above, to show that X obeys Benford's Law it suffices to know that the distribution of the logarithm of X modulo 1 is uniformly distributed. Thus we are led to study

$$\log_{10} X = \log_{10}(X_1 \cdot X_2 \cdot \ldots \cdot X_n) = \log_{10} X_1 + \cdots + \log_{10} X_n.$$

By the Central Limit Theorem, if n is large then the above sum is approximately normally distributed, and the variance will grow with n; however, what we are really interested in is not this sum but rather this sum modulo 1:

$$\log_{10} X \bmod 1 = (\log_{10} X_1 + \cdots + \log_{10} X_n) \bmod 1.$$

A nice computation shows that as the variance σ tends to infinity, if we look at the probability density of a normal with variance σ modulo 1 then that is approximately uniformly distributed on $[0, 1]$. Explicitly, let Y be normally distributed with some mean μ and very large variance σ. If we look at the probability density of the new random variable $Y \bmod 1$, then this is approximately uniformly distributed on $[0, 1)$. This means that the probability that $Y \in [\log_{10} d, \log_{10}(d + 1))$ is just $\log_{10}(d + 1) - \log_{10} d$, or $\log_{10} \frac{d+1}{d}$; however, note that these are just the Benford probabilities!

While we have chosen to give the argument for multiplying random variables, similar results hold for other combinations (such as addition, exponentiation, etc.). The Central Limit Theorem is lurking in the background, and if we adjust our viewpoint we can see its effect.

1.6 QUESTIONS

Our goal in this book is to explain the prevalence of Benford's Law, and discuss its implications and applications. The question of leading digits is but one of many that we could ask. There are many generalizations; below we state the two most common.

1. *Instead of studying the distribution of the first digit, we may study the distribution of the first two, three, or more generally the significand, of our number. The Strong Benford's Law is that the probability of observing a significand of at most s is $\log_{10} s$.*

2. *Instead of working in base 10, we may work in base B, in which case the Benford probabilities become $\log_B \left(\frac{d+1}{d}\right)$ for the distribution of the first digit, and $\log_B s$ for a significand of at most s.*

Incorporating these two generalizations, we are led to our final definition of Benford's Law.

Definition 1.6.1 (Strong Benford's Law Base B). *A data set satisfies the Strong Benford's Law Base B if the probability of observing a significand of at most s in base B is $\log_B s$. We shall often refer to the distribution of just the first digit as Benford's Law, as well as the distribution of the entire significand.*

We end the introduction by briefly summarizing the goals of this book and what follows. We address two central questions:

1. *Which data sets (mathematical expressions, physical data, financial transactions) follow this law, and why?*

2. *What are the practical implications of this law?*

There are several different arguments for the first question, depending on the structure of the data. Our studies will show that the answer is deeply connected to results in subjects ranging from probability to Fourier analysis to dynamical systems to number theory. We shall develop enough of these topics for our investigations, recalling standard results in each when needed.

The second question leads to many surprising characters entering the scene. The reason Benford's Law is not just a curiosity of pure mathematics is due to the wealth of applications, in particular to data integrity and fraud tests. There have (sadly) been numerous examples of researchers and corporations tinkering with data; if undetected, the consequences could be severe, ranging from companies not paying their fair share of taxes, to unsafe medical treatments being approved, to unscrupulous researchers being funded at the expense of their honest peers, to electoral fraud and the effective disenfranchisement of voters. With a large enough data set, the laws of probability and statistics state that certain patterns should emerge. Some of these consequences are well known, and thus are easily incorporated by people modifying data. For example, while everyone knows that if you simulate flipping a fair coin 1,000,000 times then there should be about 500,000 heads, fewer know how likely it is to have 100 consecutive heads in the sequence of tosses. The situation is similar with Benford's Law. Almost anyone unfamiliar with Benford's Law would, if asked to simulate data, create a set where either the first digits are equally likely to be anything from 1 to 9, or else clustered around 5. As many real-world data sets follow Benford's Law, this leads to a quick and easy test for fraud. Such tests are now routinely used by the IRS to detect tax fraud, while generalizations may be used in the future to detect whether or not an image has been modified.

What better way to end the introduction than with notes from a talk that Frank Benford gave on the law that now bears his name! While this was one of the earliest talks in the subject, it was by no means the last. As the online bibliography [BerH2] shows, Benford's Law has become a very active research area with numerous applications across disciplines, many of which are described in the following chapters. Enjoy!

Nearly everyone that gives any casual attention to the numbers one, two, three up to nine assumes that we use all nine digits equally and without preference. Actually we have some peculiar habits in our use of the familiar digits, and this evening I would like to point out that we have strong preferences, and I hope that I can indicate why we prefer some digits more than others, and also show that we have logical reasons for our preferences.

This matter first came to my attention when I was in high school, but until four years ago I paid no further attention to it, and it remained in my recollection as a faintly aggravating conundrum. About four years ago I had occasion to order some of the gummed letters and numbers that librarians paste on the back of books, and then my ancient conundrum bobbed up with such force that I could hardly have avoided discovering the relations I would now like to describe.

The preliminary exploration into the matter consisted of observing how frequently the digits one, two, three, etc., were used at the beginning of numbers that consisted of three or more digits. The sources of these numbers were newspapers, magazines, handbooks, and technical encyclopedias. For an example of the method, let us assume that we are using a telephone directory to investigate the numbers used in the street addresses of the subscribers. We observe that a subscriber lives at 5643 Main Street and we tabulate the 5 and ignore the remaining digits. If a group of, say, 500 such numbers are taken from a random collection, such as a telephone book contains, we will invariably find the following

-2-

relations:

The digit one is used 30 per cent of the time as the first digit of numbers and the digit two is used 18 per cent of the time. The percentage falls as we go to higher digits, and the last digit nine is used less than 5 per cent of the time. This is far from an even distribution, and let us yield to a natural curiosity and see if we can discover why the digit one is nearly seven times as popular as nine as a digit with which to begin a number. Since every large group of numbers follows the same rule, we can begin with a generalization about numbers as we use them.

Our natural numbers, one, two, three, etc., are not distributed in the same manner as are nature phenomena and events. In going from one to two, we double the quantity while if we go from nine to ten, we add only eleven per cent to the size of the nine. If, in going from one to two, we take successive increments of eleven per cent of the value of the last number, we find that seven increments or steps are necessary, and this is exactly the occurrence ratio of numbers beginning with the digit one over numbers beginning with the digit nine. Looked at in this way, our so-called natural numbers are poorly adapted for measuring things that can be subdivided into smaller units. Even in our methods of thought, we use a number scale that is a wide departure from the one, two, three, that we are taught is the natural way of counting.

-2-

To illustrate how we actually think when considering sizes, let us assume that you own a row boat that is ten feet long that you think is too small and you therefore want a slightly larger boat. If I offered you an eleven-foot boat, you would possibly be satisfied with this increase. But if you owned a fifty-foot yacht and wished a larger one, I feel quite certain that a fifty-one-foot yacht would not seem significantly longer, but a fifty-five-foot yacht might do. To carry it one step further, an increase of five feet to a five hundred-foot ship would be insignificant, but a fifty-foot increase would be definitely a distinct size of ship. In all three cases the next larger size was a fixed percentage of the previous size, and this is our customary way of thinking of and determining sizes. A logical system of numbers might then run 10, 11 --- 50, 55--- 500, 550 ---- 1000, 1100 ----

This type of number scale would make arithmetic even less popular in certain quarters than it is now, but people with high school educations are familiar with such a scale. They call it a geometric series, and another name is logarithmic scale.

Our information about the existence of this type of a working scale is based on a tabulation of our twenty-thousand numbers. It was found that used numbers from every source investigated conformed much more closely to a logarithmic scale than to the natural scale. The final re-

-4-

sult of this tabulation was that the numbers were found
to follow the distribution law

$$F_a = \log \frac{a + 1}{a}$$

with a high degree of fidelity.

One of the interesting facts brought out by this investigation was that purely random and unrelated numbers, such as
are found in the news items of a newspaper, give better
agreement with the formal law than do numbers that are themselves the result of some mathematical law. In the latter
class, we can include the tabulations of squares, cubes and
square roots of natural numbers given in handbooks. As a
result of completely random numbers being more orderly in
arrangement than closely related numbers, it has been suggested that this theory be called the Law of Lawless Numbers,
and this is essentially the meaning of its formal title,
The Law of Anomalous Numbers.

Chapter Two

A Short Introduction to the Mathematical Theory of Benford's Law

Arno Berger and T. P. Hill[1]

This chapter is an abbreviated version of [BerH4], which can be consulted for additional details. Many of the results presented here, notably those in Sections 2.5 and 2.6, can be strengthened considerably; the interested reader may want to consult [BerH5] in this regard.

2.1 INTRODUCTION

Benford's Law, or **BL** for short, is the observation that in many collections of numbers, be they mathematical tables, real-life data, or combinations thereof, the leading significant digits are not uniformly distributed, as might be expected, but are heavily skewed toward the smaller digits. The reader may find many formulations and applications of BL in the online database [BerH2].

More specifically, BL says that the significant digits in many data sets follow a very particular logarithmic distribution. In its most common formulation, namely the special case of first significant *decimal* (i.e., base-10) digits, BL is also known as the **First-Digit Phenomenon** and reads

$$\text{Prob}\,(D_1 = d_1) \; = \; \log_{10}\left(1 + d_1^{-1}\right) \quad \text{for all } d_1 = 1, 2, \ldots, 9 ; \tag{2.1}$$

here D_1 denotes the first significant decimal digit [Ben, New]. For example, (2.1) asserts that

$$\text{Prob}\,(D_1 = 1) \; = \; \log_{10} 2 \; = \; 0.3010\ldots,$$

$$\text{Prob}\,(D_1 = 9) \; = \; \log_{10} \frac{10}{9} \; = \; 0.04575\ldots. \tag{2.2}$$

In a form more complete than (2.1), BL is a statement about the joint distribution of *all* decimal digits: For every positive integer m,

$$\text{Prob}\,\big((D_1, D_2, \ldots, D_m) \; = \; (d_1, d_2, \ldots, d_m)\big)$$

$$= \; \log_{10}\left(1 + \left(\sum_{j=1}^{m} 10^{m-j} d_j\right)^{-1}\right) \tag{2.3}$$

[1]Department of Mathematical & Statistical Sciences, University of Alberta, and School of Mathematics, Georgia Institute of Technology.

holds for all m-tuples $(d_1, d_2, \ldots, d_m)$, where d_1 is an integer in $\{1, 2, \ldots, 9\}$ and for $j \geq 2$, d_j is an integer in $\{0, 1, \ldots, 9\}$; here D_2, D_3, D_4, etc. represent the second, third, fourth, etc. significant decimal digit. Thus, for example, (2.3) implies that

$$\text{Prob}\left((D_1, D_2, D_3) = (3, 1, 4)\right) = \log_{10} \frac{315}{314} = 0.001380 \ldots .$$

Note. Throughout this overview of the basic theory of BL, attention will more or less exclusively be restricted to significant *decimal* (i.e., base-10) digits. *From now on in this chapter, therefore, $\log x$ will always denote the logarithm base 10 of x, while $\ln x$ is the natural logarithm of x. For convenience, the convention $\log 0 := 0$ will be adopted.*

2.2 SIGNIFICANT DIGITS AND THE SIGNIFICAND

Since Benford's Law is a statement about the statistical distribution of significant (decimal) digits, a natural starting point for any study of BL is the formal definition of *significant digits* and the *significand (function)*.

2.2.1 Significant Digits

Definition 2.2.1 (First significant decimal digit). *For every non-zero real number x, the **first significant decimal digit** of x, denoted by $D_1(x)$, is the unique integer $j \in \{1, 2, \ldots, 9\}$ satisfying $10^k j \leq |x| < 10^k(j+1)$ for some (necessarily unique) $k \in \mathbb{Z}$.*

*Similarly, for every $m \geq 2$, $m \in \mathbb{N}$, the m**th significant decimal digit** of x, denoted by $D_m(x)$, is defined inductively as the unique integer $j \in \{0, 1, \ldots, 9\}$ such that*

$$10^k \left(\sum_{i=1}^{m-1} D_i(x) 10^{m-i} + j \right) \leq |x| < 10^k \left(\sum_{i=1}^{m-1} D_i(x) 10^{m-i} + j + 1 \right)$$

for some (necessarily unique) $k \in \mathbb{Z}$; for convenience, $D_m(0) := 0$ for all $m \in \mathbb{N}$.

Note that, by definition, the first significant digit $D_1(x)$ of $x \neq 0$ is never zero, whereas the second, third, etc. significant digits may be any integers in $\{0, 1, \ldots, 9\}$.

Example 2.2.2. *Since $\sqrt{2} \approx 1.414$ and $1/\pi \approx 0.3183$,*

$$D_1(\sqrt{2}) = D_1(-\sqrt{2}) = D_1(10\sqrt{2}) = 1, \quad D_2(\sqrt{2}) = 4, \quad D_3(\sqrt{2}) = 1;$$

$$D_1(\pi^{-1}) = D_1(10\pi^{-1}) = 3, \quad D_2(\pi^{-1}) = 1, \quad D_3(\pi^{-1}) = 8.$$

2.2.2 The Significand

The **significand** of a real number is its coefficient when it is expressed in floating point ("scientific notation") form, more precisely

Definition 2.2.3. *The* (decimal) significand function $S : \mathbb{R} \to [1, 10)$ *is defined as follows: If* $x \neq 0$ *then* $S(x) = t$, *where* t *is the unique number in* $[1, 10)$ *with* $|x| = 10^k t$ *for some (necessarily unique)* $k \in \mathbb{Z}$; *if* $x = 0$ *then, for convenience,* $S(0) := 0$.

Example 2.2.4.

$$S(\sqrt{2}) = S(10\sqrt{2}) = \sqrt{2} = 1.414\ldots,$$

$$S(\pi^{-1}) = S(10\pi^{-1}) = 10\pi^{-1} = 3.183\ldots.$$

The significand uniquely determines the significant digits, and vice versa. This relationship is recorded in the next proposition which immediately follows from Definitions 2.2.1 and 2.2.3. Here and throughout the **floor function**, $\lfloor x \rfloor$, denotes the largest integer not larger than x.

Proposition 2.2.5. *For every real number* x,

1. $S(x) = \sum_{m \in \mathbb{N}} 10^{1-m} D_m(x)$;

2. $D_m(x) = \lfloor 10^{m-1} S(x) \rfloor - 10 \lfloor 10^{m-2} S(x) \rfloor$ *for every* $m \in \mathbb{N}$.

Since the significant digits determine the significand, and are in turn determined by it, the informal version (2.3) of BL in the introduction has an immediate and very concise counterpart in terms of the significand function, namely

$$\text{Prob}\,(S \leq t) = \log t \quad \text{for all } 1 \leq t < 10. \tag{2.4}$$

2.2.3 The Significand σ-Algebra

The informal statements (2.1), (2.3), and (2.4) of BL involve *probabilities*. The key step in formulating BL precisely is identifying the appropriate probability space, and hence in particular the correct σ-algebra. As it turns out, in the significant digit framework there is only one natural candidate which is both intuitive and easy to describe.

Definition 2.2.6. *The* **significand σ-algebra** $\mathcal{S}$ *is the* σ-algebra on $\mathbb{R}^+$ *generated by the significand function* S, *i.e.,* $\mathcal{S} = \mathbb{R}^+ \cap \sigma(S)$.

The importance of the σ-algebra $\mathcal{S}$ comes from the fact that for every event $A \in \mathcal{S}$ and every $x > 0$, knowing $S(x)$ is enough to decide whether $x \in A$ or $x \notin A$. Worded slightly more formally, this observation reads as follows, where $\sigma(f)$ denotes the σ-algebra generated by f, i.e., the smallest σ-algebra containing all sets of the form $\{x : a \leq f(x) \leq b\}$, and $\mathcal{B}(I)$ denotes the real **Borel σ-algebra** restricted to an interval I. If $I = \mathbb{R}$ or $I = \mathbb{R}^+ = \{t \in \mathbb{R} : t > 0\}$ then, for convenience, instead of $\mathcal{B}(I)$ simply write $\mathcal{B}$ and $\mathcal{B}^+$, respectively. Also, here and throughout, for every set $C \subset \mathbb{R}$ and $t \in \mathbb{R}$, let $tC := \{tc : c \in C\}$.

Lemma 2.2.7. *For every function* $f : \mathbb{R}^+ \to \mathbb{R}$ *the following statements are equivalent:*

1. *f can be described completely in terms of S, that is, $f(x) = \varphi(S(x))$ holds for all $x \in \mathbb{R}^+$, with some function $\varphi : [1, 10) \to \mathbb{R}$ satisfying $\sigma(\varphi) \subset \mathcal{B}[1, 10)$.*

2. $\sigma(f) \subset \mathcal{S}$.

Proof. Routine. □

Theorem 2.2.8 ([Hi4]). *For every $A \in \mathcal{S}$,*

$$A = \bigcup_{k \in \mathbb{Z}} 10^k S(A) \tag{2.5}$$

where $S(A) = \{S(x) : x \in A\} \subset [1, 10)$. Moreover,

$$\mathcal{S} = \mathbb{R}^+ \cap \sigma(D_1, D_2, D_3, \ldots) = \left\{ \bigcup_{k \in \mathbb{Z}} 10^k B : B \in \mathcal{B}[1, 10) \right\}. \tag{2.6}$$

Proof. By definition,

$$\mathcal{S} = \mathbb{R}^+ \cap \sigma(S) = \mathbb{R}^+ \cap \{S^{-1}(B) : B \in \mathcal{B}\} = \mathbb{R}^+ \cap \{S^{-1}(B) : B \in \mathcal{B}[1, 10)\}.$$

Thus, given any $A \in \mathcal{S}$, there exists a set $B \in \mathcal{B}[1, 10)$ with $A = \mathbb{R}^+ \cap S^{-1}(B) = \bigcup_{k \in \mathbb{Z}} 10^k B$. Since $S(A) = B$, it follows that (2.5) holds for all $A \in \mathcal{S}$.

To prove (2.6), first observe that by Proposition 2.2.5(1) the significand function S is completely determined by the significant digits $D_1, D_2, D_3, \ldots$, so $\sigma(S) \subset \sigma(D_1, D_2, D_3, \ldots)$ and hence $\mathcal{S} \subset \mathbb{R}^+ \cap \sigma(D_1, D_2, D_3, \ldots)$. Conversely, according to Proposition 2.2.5(2), every D_m is determined by S, thus $\sigma(D_m) \subset \sigma(S)$ for all $m \in \mathbb{N}$, showing that $\sigma(D_1, D_2, D_3, \ldots) \subset \sigma(S)$ as well. To verify the remaining equality in (2.6), note that for every $A \in \mathcal{S}$, $S(A) \in \mathcal{B}[1, 10)$ and hence $A = \bigcup_{k \in \mathbb{Z}} 10^k B$ for $B = S(A)$, by (2.5). Conversely, every set of the form $\bigcup_{k \in \mathbb{Z}} 10^k B = \mathbb{R}^+ \cap S^{-1}(B)$ with $B \in \mathcal{B}[1, 10)$ obviously belongs to $\mathcal{S}$. □

Note that for every $A \in \mathcal{S}$ there is a *unique* $B \in \mathcal{B}[1, 10)$, the Borel subsets of $[1, 10)$, such that $A = \bigcup_{k \in \mathbb{Z}} 10^k B$, and (2.5) shows that in fact $B = S(A)$.

Example 2.2.9. *The set A_4 of positive numbers with*

$$A_4 = \{10^k : k \in \mathbb{Z}\} = \{\ldots, 0.01, 0.1, 1, 10, 100, \ldots\}$$

belongs to $\mathcal{S}$. This can be seen either by observing that A_4 is the set of positive reals with significand exactly equal to 1, i.e., $A_4 = \mathbb{R}^+ \cap S^{-1}(\{1\})$, or by noting that $A_4 = \{x > 0 : D_1(x) = 1, D_m(x) = 0 \text{ for all } m \geq 2\}$, or by using (2.6) and the fact that $A_4 = \bigcup_{k \in \mathbb{Z}} 10^k\{1\}$ and $\{1\} \in \mathcal{B}[1, 10)$.

Example 2.2.10. *The singleton set $\{1\}$ and the interval $[1, 2]$ do not belong to $\mathcal{S}$, since the number 1 cannot be distinguished from the number 10, for instance, using only significant digits. Nor can the interval $[1, 2]$ be distinguished from $[10, 20]$. Formally, neither of these sets is of the form $\bigcup_{k \in \mathbb{Z}} 10^k B$ for any $B \in \mathcal{B}[1, 10)$.*

The next lemma establishes some basic closure properties of the significand σ-algebra that will be essential later in studying characteristic aspects of BL such as scale and base invariance. To concisely formulate these properties, for every $C \subset \mathbb{R}^+$ and $n \in \mathbb{N}$, let $C^{1/n} := \{t > 0 : t^n \in C\}$.

Lemma 2.2.11. *The following properties hold for the significand σ-algebra $\mathcal{S}$:*

1. *$\mathcal{S}$ is self-similar with respect to multiplication by integer powers of 10, i.e.,*

$$10^k A \;=\; A \quad \text{for every } A \in \mathcal{S} \text{ and } k \in \mathbb{Z}.$$

2. *$\mathcal{S}$ is closed under multiplication by a scalar, i.e.,*

$$\alpha A \in \mathcal{S} \quad \text{for every } A \in \mathcal{S} \text{ and } \alpha > 0.$$

3. *$\mathcal{S}$ is closed under integral roots, i.e.,*

$$A^{1/n} \in \mathcal{S} \quad \text{for every } A \in \mathcal{S} \text{ and } n \in \mathbb{N}.$$

Proof. (1) This is obvious from (2.5) since $S(10^k A) = S(A)$ for every k.

(2) Given $A \in \mathcal{S}$, by (2.6) there exists $B \in \mathcal{B}[1, 10)$ such that $A = \bigcup_{k \in \mathbb{Z}} 10^k B$. In view of (1), assume without loss of generality that $1 < \alpha < 10$. Then

$$
\begin{aligned}
\alpha A &= \bigcup_{k \in \mathbb{Z}} 10^k \alpha B \\
&= \bigcup_{k \in \mathbb{Z}} 10^k \left((\alpha B \cap [\alpha, 10)) \cup \left(\tfrac{\alpha}{10} B \cap [1, \alpha) \right) \right) \\
&= \bigcup_{k \in \mathbb{Z}} 10^k C,
\end{aligned}
$$

with $C = (\alpha B \cap [\alpha, 10)) \cup \left(\tfrac{\alpha}{10} B \cap [1, \alpha) \right) \in \mathcal{B}[1, 10)$, showing that $\alpha A \in \mathcal{S}$.

(3) Since intervals of the form $[1, t]$ generate $\mathcal{B}[1, 10)$, i.e., since $\mathcal{B}[1, 10) = \sigma(\{[1, t] : 1 < t < 10\})$, it is enough to verify the claim for the special case $A = \bigcup_{k \in \mathbb{Z}} 10^k [1, 10^s]$ for every $0 < s < 1$. In this case

$$
\begin{aligned}
A^{1/n} &= \bigcup_{k \in \mathbb{Z}} 10^{k/n} [1, 10^{s/n}] \\
&= \bigcup_{k \in \mathbb{Z}} 10^k \bigcup_{j=0}^{n-1} [10^{j/n}, 10^{(j+s)/n}] \\
&= \bigcup_{k \in \mathbb{Z}} 10^k C,
\end{aligned}
$$

with $C = \bigcup_{j=0}^{n-1} [10^{j/n}, 10^{(j+s)/n}] \in \mathcal{B}[1, 10)$. Hence $A^{1/n} \in \mathcal{S}$. $\square$

Since, by Theorem 2.2.8, the significand σ-algebra $\mathcal{S}$ is the same as the significant digit σ-algebra $\sigma(D_1, D_2, D_3, \cdots)$, the closure properties established in Lemma 2.2.11 carry over to sets determined by significant digits. The next example illustrates closure under multiplication by a scalar and integral roots, and that $\mathcal{S}$ is *not* closed under taking integer powers.

Example 2.2.12. *Let A_5 be the set of positive real numbers with first significant digit 1, i.e.,*

$$A_5 \;=\; \{x > 0 : D_1(x) = 1\} \;=\; \{x > 0 : 1 \le S(x) < 2\} \;=\; \bigcup_{k \in \mathbb{Z}} 10^k [1, 2).$$

Then

$$2A_5 = \{x > 0 : D_1(x) \in \{2,3\}\} = \{x > 0 : 2 \le S(x) < 3\}$$
$$= \bigcup_{k \in \mathbb{Z}} 10^k[2,4) \in \mathcal{S},$$

(2.7)

and also

$$A_5^{1/2} = \{x > 0 : S(x) \in [1, \sqrt{2}) \cup [\sqrt{10}, \sqrt{20})\}$$
$$= \bigcup_{k \in \mathbb{Z}} 10^k \left([1, \sqrt{2}) \cup [\sqrt{10}, 2\sqrt{5})\right) \in \mathcal{S},$$

whereas on the other hand clearly

$$A_5^2 = \bigcup_{k \in \mathbb{Z}} 10^{2k}[1,4) \notin \mathcal{S},$$

since e.g. $[1,4) \subset A_5^2$ *but* $[10, 40) \not\subset A_5^2$.

The next lemma provides a very convenient framework for studying probabilities on the significand σ-algebra by translating them into probability measures on the classical space of Borel subsets of $[0,1)$, that is, on $\left([0,1), \mathcal{B}[0,1)\right)$.

Lemma 2.2.13. *The function* $\ell : \mathbb{R}^+ \to [0,1)$ *defined by* $\ell(x) = \log S(x)$ *establishes a one-to-one and onto correspondence (measure isomorphism) between probability measures on* $(\mathbb{R}^+, \mathcal{S})$ *and on* $\left([0,1), \mathcal{B}[0,1)\right)$.

Proof. Routine.   □

2.3 THE BENFORD PROPERTY

In order to translate the informal versions (2.1), (2.3), and (2.4) of BL into more precise statements about various types of mathematical objects, it is necessary to specify exactly what the Benford property means for any one of these objects. For the purpose of the present section, the objects of interest fall into three categories: sequences of real numbers; real-valued functions defined on $[0, +\infty)$; and probability distributions associated with random variables. Accordingly, denote by $\#M$ the cardinality of a finite set M, and let λ symbolize Lebesgue measure on $(\mathbb{R}, \mathcal{B})$ (or parts thereof).

2.3.1 Benford Sequences

Definition 2.3.1. *A sequence* (x_n) *of real numbers is a **Benford sequence**, or **Benford** for short, if*

$$\lim_{N \to \infty} \frac{\#\{1 \le n \le N : S(x_n) \le t\}}{N} = \log t \quad \text{for all } t \in [1, 10),$$

or equivalently, if for all $m \in \mathbb{N}$, all $d_1 \in \{1, 2, \cdots, 9\}$ and all $d_j \in \{0, 1, \cdots, 9\}$, $j \geq 2$,

$$\lim_{N \to \infty} \frac{\#\{1 \leq n \leq N : D_j(x_n) = d_j \text{ for } j = 1, 2, \ldots\}}{N}$$

$$= \log\left(1 + \left(\sum_{j=1}^{m} 10^{m-j} d_j\right)^{-1}\right).$$

Two specific sequences of positive integers will be used repeatedly to illustrate key concepts concerning BL: the Fibonacci numbers and the prime numbers. Both sequences play prominent roles in many areas of mathematics. As will be seen in Example 2.4.12, the sequence $(F_n) = (1, 1, 2, 3, 5, 8, 13, \ldots)$ of Fibonacci numbers, where every entry is simply the sum of its two predecessors, and $F_1 = F_2 = 1$, is Benford. In Example 2.4.11(v), it will be shown that the sequence $(p_n) = (2, 3, 5, 7, 11, 13, 17, \ldots)$ of prime numbers is *not* Benford.

2.3.2 Benford Functions

BL also appears frequently in real-valued functions such as those arising as solutions of initial value problems for differential equations (see Section 2.5.3 below). Thus, the starting point is to define what it means for a function to follow BL.

Definition 2.3.2. *A (Borel measurable) function $f : [0, +\infty) \to \mathbb{R}$ is **Benford** if*

$$\lim_{T \to +\infty} \frac{\lambda(\{\tau \in [0, T) : S(f(\tau)) \leq t\})}{T} = \log t \quad \text{for all } t \in [1, 10),$$

or equivalently, if for all $m \in \mathbb{N}$, all $d_1 \in \{1, 2, \ldots, 9\}$ and all $d_j \in \{0, 1, \ldots, 9\}$, $j \geq 2$,

$$\lim_{T \to +\infty} \frac{\lambda(\{\tau \in [0, T) : D_j(f(\tau)) = d_j \text{ for } j = 1, 2, \ldots, m\})}{T}$$

$$= \log\left(1 + \left(\sum_{j=1}^{m} 10^{m-j} d_j\right)^{-1}\right).$$

As will be seen below, the function $f(t) = e^{\alpha t}$ is Benford whenever $\alpha \neq 0$, but $f(t) = t$ and $f(t) = \sin^2 t$, for instance, are not.

2.3.3 Benford Distributions and Random Variables

This section lays the foundations for analyzing the Benford property for probability distributions and random variables.

Definition 2.3.3. *A Borel probability measure P on $\mathbb{R}$ is **Benford** if*

$$P(\{x \in \mathbb{R} : S(x) \leq t\}) = \log t \quad \text{for all } t \in [1, 10).$$

*A random variable X on a probability space $(\Omega, \mathcal{A}, \mathbb{P})$ is **Benford** if its distribution P_X on $\mathbb{R}$ is Benford, i.e., if*

$$\mathbb{P}\big(S(X) \le t\big) = P_X\big(\{x \in \mathbb{R} : S(x) \le t\}\big) = \log t \quad \text{for all } t \in [1, 10),$$

or equivalently, if for all $m \in \mathbb{N}$, all $d_1 \in \{1, 2, \ldots, 9\}$ and all $d_j \in \{0, 1, \ldots, 9\}$, $j \ge 2$,

$$\mathbb{P}\big(D_j(X) = d_j \text{ for } j = 1, 2, \ldots, m\big) = \log\left(1 + \left(\sum_{j=1}^{m} 10^{m-j} d_j\right)^{-1}\right).$$

Example 2.3.4. *If X is a Benford random variable on a probability space $(\Omega, \mathcal{A}, \mathbb{P})$, then*

$$\mathbb{P}(D_1(X) = 1) = \mathbb{P}(1 \le S(X) < 2) = \log 2 = 0.3010\ldots,$$

$$\mathbb{P}(D_1(X) = 9) = \log \frac{10}{9} = 0.04575\ldots,$$

$$\mathbb{P}\big((D_1(X), D_2(X), D_3(X)) = (3, 1, 4)\big) = \log \frac{315}{314} = 0.001380\ldots.$$

As the following example shows, there are many Benford probability measures on the positive real numbers, and thus many positive random variables that are Benford.

Example 2.3.5. *For every integer k, the probability measure P_k with density $f_k(x) = 1/(x \ln 10)$ on $[10^k, 10^{k+1})$ is Benford, and so is $\frac{1}{2}(P_k + P_{k+1})$. In fact, every convex combination of the $(P_k)_{k \in \mathbb{Z}}$, i.e., every probability measure $\sum_{k \in \mathbb{Z}} q_k P_k$ with $0 \le q_k \le 1$ for all k and $\sum_{k \in \mathbb{Z}} q_k = 1$, is Benford.*

As will be seen in Example 2.6.4 below, if U is a random variable uniformly distributed on $[0, 1)$, then the random variable $X = 10^U$ is Benford, but the random variable $X^{\log 2} = 2^U$ is not.

Definition 2.3.6 (Benford distribution). *The Benford distribution $\mathbb{B}$ is the unique probability measure on $(\mathbb{R}^+, \mathcal{S})$ with*

$$\mathbb{B}(S \le t) = \mathbb{B}\left(\bigcup_{k \in \mathbb{Z}} 10^k [1, t]\right) = \log t \quad \text{for all } t \in [1, 10),$$

or equivalently, for all $m \in \mathbb{N}$, all $d_1 \in \{1, 2, \ldots, 9\}$ and all $d_j \in \{0, 1, \ldots, 9\}$, $j \ge 2$,

$$\mathbb{B}\big(D_j = d_j \text{ for } j = 1, 2, \ldots, m\big) = \log\left(1 + \left(\sum_{j=1}^{m} 10^{m-j} d_j\right)^{-1}\right).$$

The combination of Definitions 2.3.3 and 2.3.6 gives

Proposition 2.3.7. *A Borel probability measure P on $\mathbb{R}^+$ is Benford if and only if*

$$P(A) = \mathbb{B}(A) \quad \text{for all } A \in \mathcal{S}.$$

(For probability measures on all of $\mathbb{R}$, an analogous result holds via Definition 2.3.3; cf. [BerH4].)

Example 2.3.8. (i) *If X is distributed according to $U(0,1)$, the uniform distribution on $[0,1)$, then for every $1 \le t < 10$,*

$$\mathbb{P}\big(S(X) \le t\big) \;=\; \sum_{n \in \mathbb{N}} 10^{-n}(t-1) \;=\; \frac{t-1}{9} \not\equiv \log t \,,$$

showing that $S(X)$ is uniform on $[1,10)$, and hence is not Benford.

(ii) *If X is distributed according to $\exp(1)$, the exponential distribution with mean 1, whose distribution function is given by $F_{\exp(1)}(t) = \mathbb{P}(\exp(1) \le t) = \max(0, 1 - e^{-t})$, then*

$$\mathbb{P}(D_1(X) = 1) \;=\; \mathbb{P}\left(X \in \bigcup_{k \in \mathbb{Z}} 10^k[1,2) \right) \;=\; \sum_{k \in \mathbb{Z}} \left(e^{-10^k} - e^{-2 \cdot 10^k} \right)$$

$$> \left(e^{-1/10} - e^{-2/10} \right) + \left(e^{-1} - e^{-2} \right) + \left(e^{-10} - e^{-20} \right)$$

$$= 0.3186 \ldots > \log 2 \,,$$

and hence $\exp(1)$ is not Benford either. (See [EngLeu, MiNi2] for a detailed analysis of the exponential distribution's relation to BL.)

2.4 CHARACTERIZATIONS OF BENFORD'S LAW

The purpose of this section is to establish and illustrate four useful characterizations of the Benford property in the context of sequences, functions, distributions, and random variables, respectively. These characterizations will be instrumental in demonstrating that certain data sets are, or are not, Benford, and helpful for predicting which empirical data are likely to follow BL closely.

2.4.1 The Uniform Distribution Characterization

Here and throughout, denote by $\langle t \rangle$ the **fractional part** of any real number t, that is, $\langle t \rangle = t - \lfloor t \rfloor$. For example, $\langle \pi \rangle = \langle 3.1415 \ldots \rangle = 0.1415 \ldots = \pi - 3$. Recall that $\lambda_{a,b}$, for any $a < b$, denotes (normalized) Lebesgue measure on $\big([a,b), \mathcal{B}[a,b)\big)$.

Definition 2.4.1. *A sequence (x_n) of real numbers is **uniformly distributed modulo one**, abbreviated henceforth as **u.d.** $\mathrm{mod}\,1$, if*

$$\lim_{N \to \infty} \frac{\#\{1 \le n \le N : \langle x_n \rangle \le s\}}{N} \;=\; s \quad \text{for all } s \in [0,1) \,;$$

*a (Borel measurable) function $f : [0, +\infty) \to \mathbb{R}$ is **u.d.** $\mathrm{mod}\,1$ if*

$$\lim_{T \to +\infty} \frac{\lambda\{\tau \in [0,T) : \langle f(\tau) \rangle \le s\}}{T} \;=\; s \quad \text{for all } s \in [0,1) \,;$$

*a random variable X on a probability space $(\Omega, \mathcal{A}, \mathbb{P})$ is **u.d.** $\mathrm{mod}\,1$ if*

$$\mathbb{P}(\langle X \rangle \le s) \;=\; s \quad \text{for all } s \in [0,1) \,;$$

*and a probability measure P on $(\mathbb{R}, \mathcal{B})$ is **u.d.** mod 1 if*

$$P(\{x : \langle x \rangle \leq s\}) = P\left(\bigcup_{k \in \mathbb{Z}} [k, k+s]\right) = s \quad \text{for all } s \in [0, 1).$$

The next simple theorem (cf. [Dia, MiT-B]) is one of the main tools in the theory of BL because it allows application of the powerful theory of uniform distribution mod 1, as developed e.g. in [KuiNi]. (Recall the convention $\log 0 := 0$.)

Theorem 2.4.2 (Uniform distribution characterization). *A sequence of real numbers (a Borel measurable function, a random variable, a Borel probability measure) is Benford if and only if the decimal logarithm of its absolute value is uniformly distributed modulo 1.*

Proof. Let X be a random variable and, without loss of generality, assume that $\mathbb{P}(X = 0) = 0$. Then, for all $s \in [0, 1)$,

$$\mathbb{P}(\langle \log |X| \rangle \leq s) = \mathbb{P}\left(\log |X| \in \bigcup_{k \in \mathbb{Z}} [k, k+s]\right) = \mathbb{P}\left(|X| \in \bigcup_{k \in \mathbb{Z}} [10^k, 10^{k+s}]\right)$$

$$= \mathbb{P}(S(X) \leq 10^s).$$

Hence, by Definitions 2.3.3 and 2.4.1, X is Benford if and only if $\mathbb{P}(S(X) \leq 10^s) = \log 10^s = s$ for all $s \in [0, 1)$, i.e., if and only if $\log |X|$ is u.d. mod 1. The proofs for sequences, functions, and probability distributions are completely analogous. $\qquad\qquad\square$

Next, several tools from the basic theory of uniform distribution mod 1 will be recorded that will be useful, via Theorem 2.4.2, in establishing the Benford property for many sequences, functions, and random variables; for proofs, see [BerH4].

Lemma 2.4.3.

1. *The sequence (x_n) is u.d. mod 1 if and only if the sequence $(kx_n + b)$ is u.d. mod 1 for every non-zero integer k and every $b \in \mathbb{R}$. Also, (x_n) is u.d. mod 1 if and only if (y_n) is u.d. mod 1 whenever $\lim_{n \to \infty} |y_n - x_n| = 0$.*

2. *The function f is u.d. mod 1 if and only if $t \mapsto kf(t) + b$ is u.d. mod 1 for every non-zero integer k and every $b \in \mathbb{R}$.*

3. *The random variable X is u.d. mod 1 if and only if $kX + b$ is u.d. mod 1 for every non-zero integer k and every $b \in \mathbb{R}$.*

Example 2.4.4. **(i)** *The sequence $(n\pi) = (\pi, 2\pi, 3\pi, \ldots)$ is u.d. mod 1, by Weyl's Equidistribution Theorem; see Proposition 2.4.8(1) below. Similarly, the sequence $(x_n) = (n\sqrt{2})$ is u.d. mod 1, whereas $(x_n\sqrt{2}) = (2n) = (2, 4, 6, \ldots)$ clearly is not, as $\langle 2n \rangle = 0$ for all n. Thus the requirement in Lemma 2.4.3(i) that k be an integer cannot be removed.*

(ii) *The sequence $(\log n)$ is not u.d. mod 1. A straightforward calculation shows that, for every $s \in [0, 1)$, the sequence $\left(N^{-1}\#\{1 \leq n \leq N : \langle \log n \rangle \leq s\}\right)_{N \in \mathbb{N}}$ has*

$$\frac{1}{9}(10^s - 1) \quad \text{and} \quad \frac{10}{9}(1 - 10^{-s})$$

as its limit inferior and limit superior, respectively.

Example 2.4.5. **(i)** *The function* $f(t) = at + b$ *with real* a, b *is u.d.* mod 1 *if and only if* $a \neq 0$. *As a consequence, although the function* $f(t) = \alpha t$ *is not Benford for any* α, *the function* $f(t) = e^{\alpha t}$ *is Benford whenever* $\alpha \neq 0$, *via Theorem 2.4.2, since* $\log f(t) = \alpha t / \ln 10$ *is u.d.* mod 1.

(ii) *The function* $f(t) = \log |at + b|$ *is not u.d.* mod 1 *for any* $a, b \in \mathbb{R}$. *Similarly,* $f(t) = -\log(1 + t^2)$ *is not u.d.* mod 1, *and hence* $f(t) = (1 + t^2)^{-1}$ *is not Benford.*

(iii) *The function* $f(t) = e^t$ *is u.d.* mod 1. *As a consequence, the* **superexponential function** $f(t) = e^{e^{\alpha t}}$ *is also Benford if* $\alpha \neq 0$.

Example 2.4.6. **(i)** *If the random variable* X *is uniformly distributed on* $[0, 2)$ *then it is clearly u.d.* mod 1. *However, if* X *is uniform on, say* $[0, \pi)$, *then* X *is not u.d.* mod 1.

(ii) *No exponential random variable is u.d.* mod 1 (cf. [BerH3, BerH4, LeScEv, MiNi2]).

(iii) *If* X *is a normal random variable then* X *is not u.d.* mod 1, *and neither is* $|X|$ *or* $\max(0, X)$. *While this is easily checked by a direct calculation, it is illuminating to obtain more quantitative information. To this end, assume that* X *is a normal variable with mean* 0 *and variance* σ^2. *By means of Fourier series [Pin], it can be shown that*

$$\Delta(\sigma) := \max_{0 \le s < 1} \left| F_{\langle X \rangle}(s) - s \right| \le \frac{1}{\pi} \sum_{n=1}^{\infty} n^{-1} e^{-2\sigma^2 \pi^2 n^2} ,$$

where $F_{\langle X \rangle}(s) = \mathbb{P}(\langle X \rangle \le s)$. *In particular* $\Delta(\sigma) = (e^{-2\sigma^2 \pi^2})/\pi + \mathcal{O}(e^{-8\sigma^2 \pi^2})$ *as* σ *tends to infinity, showing that* $\Delta(\sigma)$, *the deviation of* $\langle X \rangle$ *from uniformity, goes to zero very rapidly as* $\sigma \to +\infty$. *Already for* $\sigma = 1$ *one finds that* $\Delta(1) < 8.516 \cdot 10^{-10}$. *Thus even though a standard normal random variable* X *is not u.d.* mod 1, *the distribution of* $\langle X \rangle$ *is extremely close to uniform. Consequently, a log-normal random variable with large variance is practically indistinguishable from a Benford random variable.*

Corollary 2.4.7.

1. *A sequence* (x_n) *is Benford if and only if, for all* $\alpha \in \mathbb{R}$ *and* $k \in \mathbb{Z}$ *with* $\alpha k \neq 0$, *the sequence* (αx_n^k) *is also Benford.*

2. *A function* $f : [0, +\infty) \to \mathbb{R}$ *is Benford if and only if* $1/f$ *is Benford.*

3. *A random variable* X *is Benford if and only if* $1/X$ *is Benford.*

The next two statements, recorded here for ease of reference, list several key tools concerning uniform distribution mod 1, which via Theorem 2.4.2 will be used to determine Benford properties of sequences, functions, and random variables. Conclusion (1) in Proposition 2.4.8 is Weyl's classical uniform distribution result [KuiNi, Thm.3.3], conclusion (2) is an immediate consequence of Weyl's criterion [KuiNi, Thm.2.1], conclusion (3) is [Ber2, Lem.2.8], and conclusion (4) is [BerBH, Lem.2.4.(i)].

Proposition 2.4.8. *Let (x_n) be a sequence of real numbers.*

1. *If $\lim_{n\to\infty}(x_{n+1} - x_n) = \theta$ for some irrational θ, then (x_n) is u.d. mod 1.*

2. *If (x_n) is periodic, i.e., $x_{n+p} = x_n$ for some $p \in \mathbb{N}$ and all n, then $(n\theta + x_n)$ is u.d. mod 1 if and only if θ is irrational.*

3. *The sequence (x_n) is u.d. mod 1 if and only if $(x_n + \alpha \log n)$ is u.d. mod 1 for all $\alpha \in \mathbb{R}$.*

4. *If (x_n) is u.d. mod 1 and non-decreasing, then $(x_n / \log n)$ is unbounded.*

Another very useful result is **Koksma's metric theorem** [KuiNi, Thm.4.3]. For its formulation, recall that a property of real numbers is said to hold for **almost every** (*a.e.*) $x \in [a, b)$ if there exists a set $N \in \mathcal{B}[a, b)$ with $\lambda_{a,b}(N) = 0$ such that the property holds for every $x \notin N$. The probabilistic interpretation of a given property of real numbers holding for a.e. x is that this property holds **almost surely** (*a.s.*), which means that with probability one for every random variable that has a density (i.e., is absolutely continuous).

Proposition 2.4.9. *Let f_n be continuously differentiable on $[a, b]$ for all $n \in \mathbb{N}$. If $f'_m - f'_n$ is monotone and $|f'_m(x) - f'_n(x)| \geq \alpha > 0$ for all $m \neq n$, where α does not depend on x, m, and n, then $(f_n(x))$ is u.d. mod 1 for almost every $x \in [a, b]$.*

Theorem 2.4.10 ([BerHKR]). *If a, b, α, β are real numbers with $a \neq 0$ and $|\alpha| > |\beta|$ then $(\alpha^n a + \beta^n b)$ is Benford if and only if $\log |\alpha|$ is irrational.*

Proof. Since $a \neq 0$ and $|\alpha| > |\beta|$, $\lim_{n\to\infty} \dfrac{\beta^n b}{\alpha^n a} = 0$, and therefore

$$\log |\alpha^n a + \beta^n b| - \log |\alpha^n a| = \log \left| 1 + \frac{\beta^n b}{\alpha^n a} \right| \to 0 ,$$

showing that $(\log |\alpha^n a + \beta^n b|)$ is u.d. mod 1 if and only if $(\log |\alpha^n a|) = (\log |a| + n \log |\alpha|)$ is. According to Proposition 2.4.8(1), this is the case whenever $\log |\alpha|$ is irrational. On the other hand, if $\log |\alpha|$ is rational then $\langle \log |a| + n \log |\alpha| \rangle$ attains only finitely many values and hence $(\log |a| + n \log |\alpha|)$ is not u.d. mod 1. An application of Theorem 2.4.2 therefore completes the proof. $\square$

Example 2.4.11. *(i) By Theorem 2.4.10 the sequence (2^n) is Benford since $\log 2$ is irrational, but (10^n) is not Benford since $\log 10 = 1 \in \mathbb{Q}$. Similarly, (0.2^n), (3^n), (0.3^n), $(0.01 \cdot 0.2^n + 0.2 \cdot 0.01^n)$ are Benford, whereas (0.1^n), $\left(\sqrt{10}^n\right)$, $(0.1 \cdot 0.02^n + 0.02 \cdot 0.1^n)$ are not.*

(ii) The sequence $(0.2^n + (-0.2)^n)$ is not Benford, since all odd terms are zero, but $(0.2^n + (-0.2)^n + 0.03^n)$ is Benford—although this does not follow directly from Theorem 2.4.10.

(iii) By Proposition 2.4.9, the sequence $(x, 2x, 3x, \ldots) = (nx)$ is u.d. mod 1 for almost every real x, but clearly not for every x, as for example $x = 1$ shows. Consequently, by Theorem 2.4.2, (10^{nx}) is Benford for almost all real x, but not e.g. for $x = 1$ or, more generally, whenever x is rational.

(iv) *By Proposition 2.4.8(4) or Example 2.4.4(ii), the sequence* $(\log n)$ *is not u.d. mod 1, so the sequence* (n) *of positive integers is not Benford, and neither is* (αn) *for any* $\alpha \in \mathbb{R}$.

(v) *Consider the sequence* (p_n) *of prime numbers. By the Prime Number Theorem,* $p_n = \mathcal{O}(n \log n)$ *as* $n \to \infty$. *Hence it follows from Proposition 2.4.8(4) that* (p_n) *is not Benford.*

Example 2.4.12. *Consider the sequence* $(F_n) = (1, 1, 2, 3, 5, 8, 13, \ldots)$ *of Fibonacci numbers, defined inductively as* $F_{n+2} = F_{n+1} + F_n$ *for all* $n \in \mathbb{N}$, *with* $F_1 = F_2 = 1$. *It is well known (and easy to check) that*

$$F_n = \frac{1}{\sqrt{5}}\left(\left(\frac{1+\sqrt{5}}{2}\right)^n - \left(\frac{1-\sqrt{5}}{2}\right)^n\right) = \frac{\varphi^n - (-\varphi^{-1})^n}{\sqrt{5}} \quad \text{for all } n \in \mathbb{N},$$

where $\varphi = \frac{1}{2}(1 + \sqrt{5}) \approx 1.618$. *Since* $\varphi > 1$ *and* $\log \varphi$ *is irrational,* (F_n) *is Benford, by Theorem 2.4.10. Sequences such as* (F_n) *which are generated by linear recurrence relations will be studied in detail in Section 2.5.2.*

Theorem 2.4.13. *Let* X, Y *be random variables. Then*

1. *if* X *is u.d.* mod 1 *and* Y *is independent of* X, *then* $X + Y$ *is u.d.* mod 1;

2. *if* $\langle X \rangle$ *and* $\langle X + \alpha \rangle$ *have the same distribution for some irrational* α *then* X *is u.d.* mod 1;

3. *if* (X_n) *is an i.i.d. sequence of random variables and* X_1 *is not purely atomic (i.e.,* $\mathbb{P}(X_1 \in C) < 1$ *for every countable set* $C \subset \mathbb{R}$), *then*

$$\lim_{n \to \infty} \mathbb{P}\left(\left\langle \sum_{j=1}^{n} X_j \right\rangle \leq s\right) = s \quad \text{for every } 0 \leq s < 1, \tag{2.8}$$

that is, $\left\langle \sum_{j=1}^{n} X_j \right\rangle \to U(0, 1)$ *in distribution as* $n \to \infty$.

Proof. Elementary Fourier analysis; see [BerH4, Thm.4.13]. □

None of the familiar classical probability distributions or random variables, such as normal, uniform, exponential, beta, binomial, or gamma distributions are Benford. Specifically, *no* uniform distribution is even close to BL, no matter how large its range or where it is centered. This statement can be quantified explicitly as follows.

Proposition 2.4.14 ([BerH3]). *For every uniformly distributed random variable* X,

$$\max_{0 \leq s < 1} \left|F_{\langle \log X \rangle}(s) - s\right| \geq \frac{-9 + \ln 10 + 9 \ln 9 - 9 \ln \ln 10}{18 \ln 10} = 0.1334 \ldots,$$

and this bound is sharp.

Similarly, all exponential and normal random variables are uniformly bounded away from BL, as is explained in detail in [BerH3]. However, some distributions, such as the exponential distribution with mean 1, and the standard normal distribution, do come fairly close to being Benford.

The next result says that every random variable X with a density is asymptotically uniformly distributed on lattices of intervals as the size of the intervals goes to zero. Equivalently, $\langle nX \rangle$ is asymptotically uniform, as $n \to \infty$. This result has been the basis for several recent fallacious arguments claiming that if a random variable X has a density with very large "**spread**" then $\log X$ must also have a density with large spread and thus, by the theorem, must be close to u.d. mod 1, implying in turn that X must be close to Benford. The error in those arguments is that, regardless of which notion of "spread" is used, the variable X may have large spread and at the same time the variable $\log X$ may have small spread; for details, the reader is referred to [BerH3].

Theorem 2.4.15. *If X has a density then*

$$\lim_{n \to \infty} \mathbb{P}(\langle nX \rangle \leq s) = s \quad \text{for all } 0 \leq s < 1, \tag{2.9}$$

that is, $\langle nX \rangle \to U(0, 1)$ in distribution as $n \to \infty$.

Proof. Since $\langle nX \rangle = \langle n \langle X \rangle \rangle$, it can be assumed that X only takes values in $[0, 1)$. Let f be the density of X, i.e., $f : [0, 1] \to \mathbb{R}$ is a non-negative measurable function with $\mathbb{P}(X \leq s) = \int_0^s f(\sigma) \, d\sigma$ for all $s \in [0, 1)$. From

$$\mathbb{P}(\langle nX \rangle \leq s) = \mathbb{P}\left(X \in \bigcup_{l=0}^{n-1} \left[\frac{l}{n}, \frac{l+s}{n} \right] \right) = \sum_{l=0}^{n-1} \int_{l/n}^{(l+s)/n} f(\sigma) \, d\sigma$$

$$= \int_0^s \frac{1}{n} \sum_{l=0}^{n-1} f\left(\frac{l+\sigma}{n} \right) d\sigma,$$

it follows that the density of $\langle nX \rangle$ is given by

$$f_{\langle nX \rangle}(s) = \frac{1}{n} \sum_{l=0}^{n-1} f\left(\frac{l+s}{n} \right), \quad 0 \leq s < 1.$$

Note that if f is continuous, or merely Riemann integrable, then, as $n \to \infty$,

$$f_{\langle nX \rangle}(s) \to \int_0^1 f(\sigma) \, d\sigma = 1 \quad \text{for all } s \in [0, 1).$$

In general, for any $\varepsilon > 0$ there exists a continuous density g_ε with $\int_0^1 |f(\sigma) - g_\varepsilon(\sigma)| \, d\sigma$

$< \varepsilon$ and hence

$$\int_0^1 |f_{\langle nX\rangle}(\sigma) - 1| \, d\sigma \leq \int_0^1 \left| \frac{1}{n} \sum_{l=0}^{n-1} f\left(\frac{l+\sigma}{n}\right) - \frac{1}{n} \sum_{l=0}^{n-1} g_\varepsilon\left(\frac{l+\sigma}{n}\right) \right| \, d\sigma$$

$$+ \int_0^1 \left| \frac{1}{n} \sum_{l=0}^{n-1} g_\varepsilon\left(\frac{l+\sigma}{n}\right) - 1 \right| \, d\sigma$$

$$\leq \int_0^1 |f(\sigma) - g_\varepsilon(\sigma)| \, d\sigma$$

$$+ \int_0^1 \left| \frac{1}{n} \sum_{l=0}^{n-1} g_\varepsilon\left(\frac{l+\sigma}{n}\right) - \int_0^1 g(\tau) \, d\tau \right| \, d\sigma \, ,$$

which in turn shows that

$$\limsup_{n\to\infty} \int_0^1 |f_{\langle nX\rangle}(\sigma) - 1| \, d\sigma \leq \varepsilon \, ,$$

and since $\varepsilon > 0$ was arbitrary, $\int_0^1 |f_{\langle nX\rangle}(\sigma) - 1| \, d\sigma \to 0$ as $n \to \infty$. From this, the claim follows immediately because, for every $0 \leq s < 1$,

$$\left| \mathbb{P}(\langle nX\rangle \leq s) - s \right| = \left| \int_0^s (f_{\langle nX\rangle}(\sigma) - 1) \, d\sigma \right| \leq \int_0^1 |f_{\langle nX\rangle}(\sigma) - 1| \, d\sigma \to 0 \, .$$

$\square$

2.4.2 The Scale-Invariance Characterization

One popular hypothesis often related to BL is that of **scale invariance**. Informally put, scale invariance captures the intuitively attractive notion that any universal law should be independent of units. For instance, if a sufficiently large aggregation of data is converted from meters to feet, US dollars to euros, etc., then while the individual numbers change, the statements about the overall distribution of significant digits should not be affected by this change.

While a positive random variable X cannot be scale invariant, it may nevertheless have *scale-invariant significant digits*. For this, however, X has to be Benford. In fact, Theorem 2.4.18 below shows that being Benford is (not only necessary but) also sufficient for X to have scale-invariant significant digits. The result will first be stated in terms of probability distributions. For every function $f : \Omega \to \mathbb{R}$ with $\mathcal{A} \supset \sigma(f)$ and every probability measure $\mathbb{P}$ on $(\Omega, \mathcal{A})$, let $f_*\mathbb{P}$ denote the probability measure on $(\mathbb{R}, \mathcal{B})$ defined according to

$$f_*\mathbb{P}(B) = \mathbb{P}(f^{-1}(B)) \quad \text{for all } B \in \mathcal{B} \, . \tag{2.10}$$

Definition 2.4.16. *Let $\mathcal{A} \supset \mathcal{S}$ be a σ-algebra on $\mathbb{R}^+$. A probability measure P on $(\mathbb{R}^+, \mathcal{A})$ has **scale-invariant significant digits** if*

$$P(\alpha A) = P(A) \quad \text{for all } \alpha > 0 \text{ and } A \in \mathcal{S} \, ,$$

or equivalently if for all $m \in \mathbb{N}$, all $d_1 \in \{1, 2, \ldots, 9\}$ and all $d_j \in \{0, 1, \ldots, 9\}$, $j \geq 2$,

$$
\begin{aligned}
P(\{x : D_j(\alpha x) &= d_j \text{ for } j = 1, 2, \ldots, m\}) \\
&= P(\{x : D_j(x) = d_j \text{ for } j = 1, 2, \ldots, m\}) \quad\quad (2.11)
\end{aligned}
$$

holds for every $\alpha > 0$.

Example 2.4.17. **(i)** *The Benford probability measure $\mathbb{B}$ on $(\mathbb{R}^+, \mathcal{S})$ has scale-invariant significant digits. This follows from Theorem 2.4.18 below.*

(ii) *The **Dirac probability measure** δ_1 concentrated at the constant 1 does not have scale-invariant significant digits, since $\delta_2 = 2_*\delta_1$ yet $\delta_1(D_1 = 1) = 1 \neq 0 = \delta_2(D_1 = 1)$.*

(iii) *The uniform distribution on $[0, 1)$ does not have scale-invariant digits, since if X is distributed according to $\lambda_{0,1}$ then, for example*

$$
\mathbb{P}(D_1(X) = 1) = \frac{1}{9} < \frac{11}{27} = \mathbb{P}\left(D_1\left(\frac{3}{2}X\right) = 1\right).
$$

As mentioned earlier, the Benford distribution is the only probability measure (on the significand σ-algebra) having scale-invariant significant digits.

Theorem 2.4.18 (Scale-invariance characterization [Hi3]). *A probability measure P on $(\mathbb{R}^+, \mathcal{A})$ with $\mathcal{A} \supset \mathcal{S}$ has scale-invariant significant digits if and only if $P(A) = \mathbb{B}(A)$ for every $A \in \mathcal{S}$, i.e., if and only if P is Benford.*

Proof. Fix any probability measure P on $(\mathbb{R}^+, \mathcal{A})$, denote by P_0 its restriction to $(\mathbb{R}^+, \mathcal{S})$, and let $Q := \ell_* P_0$ with ℓ given by Lemma 2.2.13. According to Lemma 2.2.13, Q is a probability measure on $([0, 1), \mathcal{B}[0, 1))$. Moreover, under the correspondence established by ℓ,

$$
P_0(\alpha A) = P_0(A) \quad \text{for all } \alpha > 0, A \in \mathcal{S} \quad\quad (2.12)
$$

is equivalent to

$$
Q(\langle t + B \rangle) = Q(B) \quad \text{for all } t \in \mathbb{R}, B \in \mathcal{B}[0, 1), \quad\quad (2.13)
$$

where $\langle t + B \rangle = \{\langle t + x \rangle : x \in B\}$. Pick a random variable X such that the distribution of X is given by Q. With this, (2.13) simply means that, for every $t \in \mathbb{R}$, the distributions of $\langle X \rangle$ and $\langle t + X \rangle$ coincide. By Theorem 2.4.13(1) and (2) this is the case if and only if X is u.d. mod 1, i.e., $Q = \lambda_{0,1}$. (For the "if" part, note that a *constant* random variable is independent from every random variable.) Hence (2.12) is equivalent to $P_0 = (\ell^{-1})_* \lambda_{0,1} = \mathbb{B}$. $\square$

The next example is an elegant and entertaining application of the ideas underlying Theorem 2.4.18 to the mathematical theory of games. The game may be easily understood by a schoolchild, yet it has proven a challenge for game theorists not familiar with BL.

Example 2.4.19 ([Morr]). *Consider a two-person game where Player A and Player B each independently choose a (real) number greater than or equal to 1, and Player*

A wins if the product of their two numbers starts with a 1, 2, or 3; otherwise, Player B wins. Using the tools presented in this section, it may easily be seen that there is a strategy for Player A to choose her numbers so that she wins with probability at least $\log 4 \cong 60.2\%$*, no matter what strategy Player B uses. Conversely, there is a strategy for Player B so that Player A will win no more than* $\log 4$ *of the time, no matter what strategy Player A uses.*

The idea is simple, using the scale-invariance property of BL discussed above. If Player A chooses her number X *randomly according to BL, then since BL is scale invariant, it follows from Theorem 2.4.13(1) and Example 2.4.17(i) that* $X \cdot y$ *is still Benford no matter what number* y *Player B chooses, so Player A will win with the probability that a Benford random variable has first significant digit less than 4, i.e., with probability exactly* $\log 4$*. Conversely, if Player B chooses his number* Y *according to BL then, using scale invariance again,* $x \cdot Y$ *is Benford, so Player A will again win with the probability exactly* $\log 4$*.*

Theorem 2.4.18 showed that for a probability measure P on $(\mathbb{R}^+, \mathcal{B}^+)$ to have scale-invariant significant digits it is necessary (and sufficient) that P be Benford. In fact, as noted in [Sm], this conclusion already follows from a much weaker assumption: It is enough to require that the probability of a single significant digit remain unchanged under scaling.

Theorem 2.4.20. *For every random variable* X *with* $\mathbb{P}(X = 0) = 0$ *the following statements are equivalent:*

1. *X is Benford.*

2. *There exists a number* $d \in \{1, 2, \ldots, 9\}$ *such that*
$$\mathbb{P}(D_1(\alpha X) = d) = \mathbb{P}(D_1(X) = d) \quad \text{for all } \alpha > 0.$$

In particular, (2) implies that $\mathbb{P}(D_1(X) = d) = \log(1 + d^{-1})$*.*

Example 2.4.21 ("Ones-scaling test" [Sm])**.** *In view of the last theorem, to informally test whether a sample of data comes from a Benford distribution, simply compare the proportion of the sample that has first significant digit 1 with the proportion after the data has been rescaled, i.e., multiplied by* $\alpha, \alpha^2, \alpha^3, \ldots$*, where* $\log \alpha$ *is irrational, e.g.* $\alpha = 2$*.*

2.4.3 The Base-Invariance Characterization

The idea behind **base invariance of significant digits** is simply this: A base-10 significand event A corresponds to the base-100 event $A^{1/2}$, since the new base $b = 100$ is the square of the original base $b = 10$. As a concrete example, denote by A the set of positive reals with first significant digit 1, i.e.,
$$A = \{x > 0 : D_1(x) = 1\} = \{x > 0 : S(x) \in [1, 2)\}.$$
It is easy to see that $A^{1/2}$ is the set
$$A^{1/2} = \{x > 0 : S(x) \in [1, \sqrt{2}) \cup [\sqrt{10}, \sqrt{20})\}.$$

Consider now the base-100 significand function S_{100}, i.e., for any $x \neq 0$, $S_{100}(x)$ is the unique number in $[1, 100)$ such that $|x| = 100^k S_{100}(x)$ for some, necessarily unique, $k \in \mathbb{Z}$. (To emphasize that the usual significand function S is taken relative to base 10, it will be denoted S_{10} throughout this section.) Clearly,

$$A = \{x > 0 : S_{100}(x) \in [1, 2) \cup [10, 20)\}.$$

Hence, letting $a = \log 2$,

$$\{x > 0 : S_b(x) \in [1, b^{a/2}) \cup [b^{1/2}, b^{(1+a)/2})\} = \begin{cases} A^{1/2} & \text{if } b = 10, \\ A & \text{if } b = 100. \end{cases}$$

Thus, if a distribution P on the significand σ-algebra $\mathcal{S}$ has base-invariant significant digits, then $P(A)$ and $P(A^{1/2})$ should be the same, and similarly for other integral roots (corresponding to other integral powers of the original base $b = 10$). Thus $P(A) = P(A^{1/n})$ should hold for all n. (Recall from Lemma 2.2.11(3) that $A^{1/n} \in \mathcal{S}$ for all $A \in \mathcal{S}$ and $n \in \mathbb{N}$, so those probabilities are well defined.) This motivates the following definition.

Definition 2.4.22. *Let $\mathcal{A} \supset \mathcal{S}$ be a σ-algebra on $\mathbb{R}^+$. A probability measure P on $(\mathbb{R}^+, \mathcal{A})$ has* base-invariant significant digits *if $P(A) = P(A^{1/n})$ holds for all $A \in \mathcal{S}$ and $n \in \mathbb{N}$.*

Example 2.4.23. **(i)** *Recall that δ_a denotes the* **Dirac measure** *concentrated at the point a, that is, $\delta_a(A) = 1$ if $a \in A$, and $\delta_a(A) = 0$ if $a \notin A$. The probability measure δ_1 clearly has base-invariant significant digits since $1 \in A$ if and only if $1 \in A^{1/n}$. Similarly, δ_{10^k} has base-invariant significant digits for every $k \in \mathbb{Z}$. On the other hand, δ_2 does not have base-invariant significant digits since, with $A = \{x > 0 : S_{10}(x) \in [1, 3)\}$, $\delta_2(A) = 1$ yet $\delta_2(A^{1/2}) = 0$.*

(ii) *It is easy to see that the Benford distribution $\mathbb{B}$ has base-invariant significant digits. Indeed, for any $0 \leq s < 1$, let*

$$A = \{x > 0 : S_{10}(x) \in [1, 10^s)\} = \bigcup_{k \in \mathbb{Z}} 10^k [1, 10^s) \in \mathcal{S}.$$

Then, as seen in the proof of Lemma 2.2.11(3),

$$A^{1/n} = \bigcup_{k \in \mathbb{Z}} 10^k \bigcup_{j=0}^{n-1} [10^{j/n}, 10^{(j+s)/n})$$

and therefore

$$\mathbb{B}(A^{1/n}) = \sum_{j=0}^{n-1} \left(\log 10^{(j+s)/n} - \log 10^{j/n} \right) = \sum_{j=0}^{n-1} \left(\frac{j+s}{n} - \frac{j}{n} \right) = s = \mathbb{B}(A).$$

(iii) *The uniform distribution $\lambda_{0,1}$ on $[0, 1)$ does not have base-invariant significant digits. For instance, again taking $A = \{x > 0 : D_1(x) = 1\}$ leads to*

$$\lambda_{0,1}(A^{1/2}) = \sum_{n \in \mathbb{N}} 10^{-n} (\sqrt{2} - 1 + \sqrt{20} - \sqrt{10}) = \frac{1}{9} + \frac{(\sqrt{5} - 1)(2 - \sqrt{2})}{9}$$

$$> \frac{1}{9} = \lambda_{0,1}(A).$$

The next theorem is the main result for base-invariant significant digits.

Theorem 2.4.24 (Base-invariance characterization [Hi3]**).** *A probability measure P on $(\mathbb{R}^+, \mathcal{A})$ with $\mathcal{A} \supset \mathcal{S}$ has base-invariant significant digits if and only if, for some $q \in [0, 1]$,*

$$P(A) = q\delta_1(A) + (1 - q)\mathbb{B}(A) \quad \text{for every } A \in \mathcal{S}. \tag{2.14}$$

Corollary 2.4.25. *A continuous probability measure P on $\mathbb{R}^+$ has base-invariant significant digits if and only if $P(A) = \mathbb{B}(A)$ for all $A \in \mathcal{S}$, i.e., if and only if P is Benford.*

Recall that $\lambda_{0,1}$ denotes Lebesgue measure on $\big([0, 1), \mathcal{B}[0, 1)\big)$. For every $n \in \mathbb{N}$, denote the map $x \mapsto \langle nx \rangle$ of $[0, 1)$ into itself by T_n. Generally, if $T : [0, 1) \to \mathbb{R}$ is measurable, and $T\big([0, 1)\big) \subset [0, 1)$, a probability measure P on $\big([0, 1), \mathcal{B}[0, 1)\big)$ is said to be T-*invariant*, or T is P-*preserving*, if $T_*P = P$. Which probability measures are T_n-invariant for all $n \in \mathbb{N}$? A complete answer to this question is provided by

Lemma 2.4.26. *A probability measure P on $\big([0, 1), \mathcal{B}[0, 1)\big)$ is T_n-invariant for all $n \in \mathbb{N}$ if and only if $P = q\delta_0 + (1 - q)\lambda_{0,1}$ for some $q \in [0, 1]$.*

Proof. Recall the definition of the *Fourier coefficients* of P,

$$\widehat{P}(k) = \int_0^1 e^{2\pi iks} dP(s), \quad k \in \mathbb{Z},$$

and observe that

$$\widehat{T_n P}(k) = \widehat{P}(nk) \quad \text{for all } k \in \mathbb{Z}, n \in \mathbb{N}.$$

Assume first that $P = q\delta_0 + (1 - q)\lambda_{0,1}$ for some $q \in [0, 1]$. From $\widehat{\delta_0}(k) \equiv 1$ and $\widehat{\lambda_{0,1}}(k) = 0$ for all $k \neq 0$, it follows that

$$\widehat{P}(k) = \begin{cases} 1 & \text{if } k = 0, \\ q & \text{if } k \neq 0. \end{cases}$$

For every $n \in \mathbb{N}$ and $k \in \mathbb{Z}\backslash\{0\}$, therefore, $\widehat{T_n P}(k) = q$, and clearly $\widehat{T_n P}(0) = 1$. Thus $\widehat{T_n P} = \widehat{P}$ and since the Fourier coefficients determine P uniquely, $T_{n*}P = P$ for all $n \in \mathbb{N}$.

Conversely, assume that P is T_n-invariant for all $n \in \mathbb{N}$. In this case, $\widehat{P}(n) = \widehat{T_n P}(1) = \widehat{P}(1)$, and similarly $\widehat{P}(-n) = \widehat{T_n P}(-1) = \widehat{P}(-1)$. Since generally $\widehat{P}(-k) = \overline{\widehat{P}(k)}$, there exists $q \in \mathbb{C}$ such that

$$\widehat{P}(k) = \begin{cases} q & \text{if } k > 0, \\ 1 & \text{if } k = 0, \\ \overline{q} & \text{if } k < 0. \end{cases}$$

Also, observe that for every $t \in \mathbb{R}$,

$$\lim_{n \to \infty} \frac{1}{n} \sum_{j=1}^{n} e^{2\pi itj} = \begin{cases} 1 & \text{if } t \in \mathbb{Z}, \\ 0 & \text{if } t \notin \mathbb{Z}. \end{cases}$$

Using this and the Dominated Convergence Theorem, it follows from

$$P(\{0\}) = \int_0^1 \lim_{n\to\infty} \frac{1}{n} \sum_{j=1}^n e^{2\pi \imath s j} \mathrm{d}P(s) = \lim_{n\to\infty} \frac{1}{n} \sum_{j=1}^n \widehat{P}(j) = q\,,$$

that q is real, and in fact $q \in [0,1]$. Hence the Fourier coefficients of P are exactly the same as those of $q\delta_0 + (1-q)\lambda_{0,1}$. By uniqueness, therefore, $P = q\delta_0 + (1-q)\lambda_{0,1}$. $\qquad\square$

Proof. As in the proof of Theorem 2.4.18, fix a probability measure P on $(\mathbb{R}^+, \mathcal{A})$, denote by P_0 its restriction to $(\mathbb{R}^+, \mathcal{S})$, and let $Q = \ell_* P_0$. Observe that P_0 has base-invariant significant digits if and only if Q is T_n-invariant for all $n \in \mathbb{N}$. Indeed, with $0 \le s < 1$ and $A = \{x > 0 : S_{10}(x) < 10^s\}$,

$$T_{n*}Q([0,s)) = Q\left(\bigcup_{j=0}^{n-1}\left[\frac{j}{n}, \frac{j+s}{n}\right)\right)$$

$$= P_0\left(\bigcup_{k\in\mathbb{Z}} 10^k \bigcup_{j=0}^{n-1} [10^{j/n}, 10^{(j+s)/n})\right) = P_0(A^{1/n}) \quad (2.15)$$

and hence $T_{n*}Q = Q$ for all n precisely if P_0 has base-invariant significant digits. In this case, by Lemma 2.4.26, $Q = q\delta_0 + (1-q)\lambda_{0,1}$ for some $q \in [0,1]$, which in turn implies that $P_0(A) = q\delta_1(A) + (1-q)\mathbb{B}(A)$ for every $A \in \mathcal{S}$. $\qquad\square$

Corollary 2.4.27. *If a probability measure on $\mathbb{R}^+$ has scale-invariant significant digits then it also has base-invariant significant digits.*

2.4.4 The Sum-Invariance Characterization

As first observed by M. Nigrini [Nig1], if a table of real data approximately follows BL, then the sum of the significands of all entries in the table with first significant digit 1 is very close to the sum of the significands of all entries with first significant digit 2, and to the sum of the significands of entries with the other possible first significant digits as well. This clearly implies that the table must contain more entries starting with 1 than with 2, more entries starting with 2 than with 3, and so forth. This motivates the following definition.

Definition 2.4.28. *A sequence (x_n) of real numbers has **sum-invariant significant digits** if, for every $m \in \mathbb{N}$, the limit*

$$\lim_{N\to\infty} \frac{\sum_{n=1}^N S_{d_1,\ldots,d_m}(x_n)}{N}$$

exists and is independent of $d_1, \ldots, d_m$.

The definitions of sum invariance of significant digits for functions, distributions, and random variables are similar, and it is in the context of distributions and random variables that the sum-invariance characterization of BL will be stated.

Definition 2.4.29. *A random variable X has* sum-invariant significant digits *if, for every $m \in \mathbb{N}$, the value of $\mathbb{E}S_{d_1,\ldots,d_m}(X)$ is independent of $d_1, \ldots, d_m$.*

Example 2.4.30. **(i)** *If X is uniformly distributed on $[0, 1)$, then X does not have sum-invariant significant digits. This follows from Theorem 2.4.31 below.*

(ii) *Similarly, if $\mathbb{P}(X = 1) = 1$ then X does not have sum-invariant significant digits, as*

$$\mathbb{E}S_d(X) = \begin{cases} 1 & \text{if } d = 1, \\ 0 & \text{if } d \geq 2. \end{cases}$$

(iii) *Assume X is Benford. For every $m \in \mathbb{N}$, $d_1 \in \{1, 2, \ldots, 9\}$ and $d_j \in \{0, 1, \ldots, 9\}$, $j \geq 2$,*

$$\mathbb{E}S_{d_1,\ldots,d_m}(X) = \int_{d_1+10^{-1}d_2+\cdots+10^{1-m}d_m}^{d_1+10^{-1}d_2+\cdots+10^{1-m}(d_m+1)} t \cdot \frac{1}{t \ln 10} \, \mathrm{d}t = \frac{10^{1-m}}{\ln 10}.$$

Thus X has sum-invariant significant digits.

According to Example 2.4.30(iii) every Benford random variable has sum-invariant significant digits. As hinted at earlier, the converse is also true, i.e., sum-invariant significant digits characterize BL.

Theorem 2.4.31 (Sum-invariance characterization [Al]). *A random variable X with $\mathbb{P}(X = 0) = 0$ has sum-invariant significant digits if and only if it is Benford.*

Proof. See [Al] or [BerH4, Thm.4.37]. □

2.5 BENFORD'S LAW FOR DETERMINISTIC PROCESSES

The goal of this section is to present the basic theory of BL in the context of deterministic processes, such as iterates of maps, powers of matrices, and solutions of differential equations. Except for somewhat artificial examples, processes with linear growth are not Benford, and among the others, there is a clear distinction between those with exponential growth or decay, and those with superexponential growth or decay. In the exponential case, processes typically are Benford for all starting points in a region, but are not Benford with respect to other bases. In contrast, superexponential processes typically are Benford for all bases, but have small sets (of measure zero) of exceptional points whose orbits or trajectories are not Benford.

2.5.1 One-Dimensional Discrete-Time Processes

Let $T : C \to C$ be a (measurable) map that maps $C \subset \mathbb{R}$ into itself, and for every $n \in \mathbb{N}$ denote by T^n the n-fold iterate of T, i.e., $T^1 := T$ and $T^{n+1} := T^n \circ T$; also let T^0 be the identity map id_C on C, that is, $T^0(x) = x$ for all $x \in C$. The **orbit** of $x_0 \in C$ is the sequence

$$O_T(x_0) := \left(T^{n-1}(x_0)\right)_{n \in \mathbb{N}} = \left(x_0, T(x_0), T^2(x_0), \ldots\right).$$

Example 2.5.1. **(i)** *If $T(x) = 2x$ then $O_T(x_0) = (x_0, 2x_0, 2^2 x_0, \ldots) = (2^{n-1} x_0)$ for all x_0. Hence $\lim_{n \to \infty} |x_n| = +\infty$ whenever $x_0 \neq 0$.*

(ii) *If $T(x) = x^2$ then $O_T(x_0) = (x_0, x_0^2, x_0^{2^2}, \ldots) = \left(x_0^{2^{n-1}} \right)$ for all x_0. Here x_n approaches 0 or $+\infty$ depending on whether $|x_0| < 1$ or $|x_0| > 1$. Moreover, $O_T(\pm 1) = (\pm 1, 1, 1, \ldots)$.*

(iii) *If $T(x) = 1 + x^2$ then $O_T(x_0) = (x_0, 1 + x_0^2, 2 + 2x_0^2 + x_0^4, \ldots)$. Since $x_n \geq n$ for all x_0 and $n \in \mathbb{N}$, $\lim_{n \to \infty} x_n = +\infty$ for every x_0.*

Recall from Example 2.4.11(i) that (2^n) is Benford, and in fact $(2^n x_0)$ is Benford for every $x_0 \neq 0$. In other words, Example 2.5.1(i) says that with $T(x) = 2x$, the orbit $O_T(x_0)$ is Benford whenever $x_0 \neq 0$. The goal of the present subsection is to extend this observation to a much wider class of maps T. The main result (Theorem 2.5.5) rests upon three lemmas.

Lemma 2.5.2. *Let $T(x) = ax$ with $a \in \mathbb{R}$. Then $O_T(x_0)$ is Benford for every $x_0 \neq 0$ or for no x_0 at all, depending on whether $\log |a|$ is irrational or rational, respectively.*

Proof. By Theorem 2.4.10, $O_T(x_0) = (a^{n-1} x_0)$ is Benford for every $x_0 \neq 0$ or none, depending on whether $\log |a|$ is irrational or not. $\square$

Clearly, the simple proof of Lemma 2.5.2 works only for maps that are *exactly* linear. The same argument would for instance not work for $T(x) = 2x + e^{-x}$ even though $T(x) \approx 2x$ for large x. To establish the Benford behavior of maps like this, a simple version of **shadowing** will be used.

Lemma 2.5.3 (Shadowing Lemma). *Let $T : \mathbb{R} \to \mathbb{R}$ be a map, and β a real number with $|\beta| > 1$. If $\sup_{x \in \mathbb{R}} |T(x) - \beta x| < +\infty$ then there exists, for every $x \in \mathbb{R}$, one and only one point $\overline{x}$ such that the sequence $(T^n(x) - \beta^n \overline{x})$ is bounded.*

Proof. See [BerBH]. $\square$

The next lemma enables application of Lemma 2.5.3 to establish the Benford property for orbits of a wide class of maps.

Lemma 2.5.4.

1. *Assume that (a_n) and (b_n) are sequences of real numbers with $|a_n| \to +\infty$ and $\sup_{n \in \mathbb{N}} |a_n - b_n| < +\infty$. Then (b_n) is Benford if and only if (a_n) is Benford.*

2. *Suppose that the measurable functions $f, g : [0, +\infty) \to \mathbb{R}$ are such that $|f(t)| \to +\infty$ as $t \to +\infty$, and $\sup_{t \geq 0} |f(t) - g(t)| < +\infty$. Then f is Benford if and only if g is Benford.*

Proof. To prove (1), let $c := \sup_{n \in \mathbb{N}} |a_n - b_n| + 1$. By discarding finitely many terms if necessary, it can be assumed that $|a_n|, |b_n| \geq 2c$ for all n. From

$$- \log\left(1 + \frac{c}{|a_n| - c}\right) \leq \log \frac{|b_n|}{|b_n| + c} \leq \log \frac{|b_n|}{|a_n|}$$

$$\leq \log \frac{|a_n| + c}{|a_n|} \leq \log\left(1 + \frac{c}{|a_n| - c}\right),$$

it follows that

$$\left| \log|b_n| - \log|a_n| \right| = \left| \log \frac{|b_n|}{|a_n|} \right| \leq \log\left(1 + \frac{c}{|a_n| - c}\right) \to 0 \quad \text{as } n \to \infty.$$

Lemma 2.4.3(1) now shows that $(\log|b_n|)$ is u.d. mod 1 if and only $(\log|a_n|)$ is. The proof of (2) is completely analogous.   □

Lemmas 2.5.3 and 2.5.4 can now easily be combined to produce the desired general result. The theorem is formulated for orbits converging to zero. As explained in the subsequent Example 2.5.6, a reciprocal version holds for orbits converging to $\pm\infty$.

Theorem 2.5.5 ([BerBH]). *Let $T : \mathbb{R} \to \mathbb{R}$ be a C^2-map with $T(0) = 0$. Assume that $0 < |T'(0)| < 1$. Then $O_T(x_0)$ is Benford for all $x_0 \neq 0$ sufficiently close to 0 if and only if $\log|T'(0)|$ is irrational. If $\log|T'(0)|$ is rational then $O_T(x_0)$ is not Benford for any x_0 sufficiently close to 0.*

Proof. Let $\alpha := T'(0)$ and observe that there exists a continuous function $f : \mathbb{R} \to \mathbb{R}$ such that $T(x) = \alpha x \big(1 - x f(x)\big)$. In particular, $T(x) \neq 0$ for all $x \neq 0$ sufficiently close to 0. Define

$$\widetilde{T}(x) := T(x^{-1})^{-1} = \frac{x^2}{\alpha\big(x - f(x^{-1})\big)},$$

and note that

$$\widetilde{T}(x) - \alpha^{-1}x = \frac{x}{\alpha} \cdot \frac{f(x^{-1})}{x - f(x^{-1})} = \frac{f(x^{-1})}{\alpha} + \frac{f(x^{-1})^2}{\alpha\big(x - f(x^{-1})\big)}.$$

From this it is clear that $\sup_{|x|>\xi} |\widetilde{T}(x) - \alpha^{-1}x|$ is finite, provided that ξ is sufficiently large. Hence Lemma 2.5.3 shows that for every x with $|x|$ sufficiently large, $\big(|\widetilde{T}^n(x) - \alpha^{-n}\overline{x}|\big)$ is bounded with an appropriate $\overline{x} \neq 0$. Lemma 2.5.4 implies that $O_{\widetilde{T}}(x_0)$ is Benford if and only if $(\alpha^{1-n}\overline{x_0})$ is, which in turn is the case precisely if $\log|\alpha|$ is irrational. The result then follows from noting that, for all $x_0 \neq 0$ with $|x_0|$ sufficiently small, $O_T(x_0) = \big(\widetilde{T}^{n-1}(x_0^{-1})^{-1}\big)_{n\in\mathbb{N}}$, and Corollary 2.4.7(1) which shows that (x_n^{-1}) is Benford whenever (x_n) is.   □

Example 2.5.6. (i) *For $T(x) = \frac{1}{2}x + \frac{1}{4}x^2$, the orbit $O_T(x_0)$ is Benford for every $x_0 \neq 0$ sufficiently close to 0. A simple graphical analysis shows that $\lim_{n\to\infty} T^n(x) = 0$ if and only if $-4 < x < 2$. Thus for every $x_0 \in (-4,2)\backslash\{0\}$, $O_T(x_0)$ is Benford. Clearly, $O_T(-4) = (-4,2,2,\ldots)$ and $O_T(2) = (2,2,2,\ldots)$ are not Benford.*

(ii) *To see that Theorem 2.5.5 applies to the map* $T(x) = 2x + e^{-x}$, *let*

$$\widetilde{T}(x) := T(x^{-2})^{-1/2} = \frac{x}{\sqrt{2 + x^2 e^{-1/x^2}}}, \quad x \neq 0.$$

With $\widetilde{T}(0) := 0$, *the map* $\widetilde{T} : \mathbb{R} \to \mathbb{R}$ *is smooth, and* $\widetilde{T}'(0) = \frac{1}{\sqrt{2}}$. *Moreover,* $\lim_{n \to \infty} \widetilde{T}^n(x) = 0$ *for every* $x \in \mathbb{R}$. *By Theorem 2.5.5,* $O_{\widetilde{T}}(x_0)$ *is Benford for every* $x_0 \neq 0$, *and hence* $O_T(x_0)$ *is Benford for every* $x_0 \neq 0$ *as well, because* $T^n(x) = \widetilde{T}^n(|x|^{-1/2})^{-2}$ *for all* n.

Processes with Superexponential Growth or Decay

The following is an analog of Lemma 2.5.2 in the doubly exponential setting. Recall that a statement holds for *almost every* x if there is a set of Lebesgue measure zero that contains all x for which the statement does *not* hold.

Lemma 2.5.7. *Let* $T(x) = \alpha x^\beta$ *for some* $\alpha > 0$ *and* $\beta > 1$. *Then* $O_T(x_0)$ *is Benford for almost every* $x_0 > 0$, *but there also exist uncountably many exceptional points, i.e.,* $x_0 > 0$ *for which* $O_T(x_0)$ *is not Benford.*

Proof. Note first that letting $\widetilde{T}(x) = cT(c^{-1}x)$ for any $c > 0$ implies $O_T(x) = c^{-1}O_{\widetilde{T}}(cx)$, and with $c = \alpha^{(\beta-1)^{-1}}$ one finds $\widetilde{T}(x) = x^\beta$. Without loss of generality, it can therefore be assumed that $\alpha = 1$, i.e., $T(x) = x^\beta$. Define $R : \mathbb{R} \to \mathbb{R}$ as $R(y) = \log T(10^y) = \beta y$. Since $x \mapsto \log x$ establishes a bijective correspondence between both the points and the nullsets in $\mathbb{R}^+$ and $\mathbb{R}$, respectively, all that has to be shown is that $O_R(y)$ is u.d. mod 1 for a.e. $y \in \mathbb{R}$, but also that $O_R(y)$ fails to be u.d. mod 1 for at least uncountably many y. To see the former, let $f_n(y) = R^n(y) = \beta^n y$. Clearly, $f_n'(y) - f_m'(y) = \beta^{n-m}(\beta^m - 1)$ is monotone, and $|f_n' - f_m'| \geq \beta - 1 > 0$ whenever $m \neq n$. By Proposition 2.4.9, therefore, $O_R(y)$ is u.d. mod 1 for a.e. $y \in \mathbb{R}$.

The statement concerning exceptional points will be proved here only under the additional assumption that β is an *integer*; see [Ber4] for the remaining cases. Given an integer $\beta \geq 2$, let (η_n) be any sequence of 0s and 1s such that $\eta_n \eta_{n+1} = 0$ for all $n \in \mathbb{N}$, that is, (η_n) does not contain two consecutive 1s. With this, consider

$$y_0 := \sum_{j=1}^{\infty} \eta_j \beta^{-j}$$

and observe that, for every $n \in \mathbb{N}$,

$$0 \leq \langle \beta^n y_0 \rangle = \sum_{j=n+1}^{\infty} \eta_j \beta^{n-j} \leq \frac{1}{\beta} + \frac{1}{\beta^2(\beta-1)} < 1,$$

from which it is clear that $(\beta^n y_0)$ is not u.d. mod 1. The proof is completed by noting that there are uncountably many different sequences (η_n), and each sequence defines a different point y_0. □

The following is an analog of Theorem 2.5.5 for the case when T is dominated by power-like terms.

Theorem 2.5.8 ([BerBH]). *Let T be a smooth map with $T(0) = 0$, and assume that $T'(0) = 0$ but $T^{(p)}(0) \neq 0$ for some $p \in \mathbb{N}\backslash\{1\}$. Then $O_T(x_0)$ is Benford for almost every x_0 sufficiently close to 0, but there are also uncountably many exceptional points.*

Proof. Without loss of generality, assume that $p = \min\{j \in \mathbb{N} : T^{(j)}(0) \neq 0\}$. The map T can be written in the form $T(x) = \alpha x^p (1 + f(x))$ where f is a C^∞-function with $f(0) = 0$, and $\alpha \neq 0$. As in the proof of Lemma 2.5.7, it may be assumed that $\alpha = 1$. Let $R(y) = -\log T(10^{-y}) = py - \log(1 + f(10^{-y}))$, so that $O_T(x_0)$ is Benford if and only if $O_R(-\log x_0)$ is u.d. mod 1. As the proof of Lemma 2.5.7 has shown, $(p^n y)$ is u.d. mod 1 for a.e. $y \in \mathbb{R}$. Moreover, Lemma 2.5.3 applies to R, and it can be checked by term-by-term differentiation that the shadowing map

$$h : y \mapsto \overline{y} = y - \sum_{j=1}^{\infty} p^{-j} \log\left(1 + f\left(10^{-R^j(y)}\right)\right)$$

is a C^∞-diffeomorphism on $[y_0, +\infty)$ for y_0 sufficiently large. For a.e. sufficiently large y, therefore, $O_R(y)$ is u.d. mod 1. As explained earlier, this means that $O_T(x_0)$ is Benford for a.e. x_0 sufficiently close to 0. The existence of exceptional points follows similarly as in the proof of Lemma 2.5.7. $\square$

Example 2.5.9. **(i)** *Consider the map $T(x) = \frac{1}{2}(x^2 + x^4)$ and note that $\lim_{n\to\infty} T^n(x) = 0$ if and only if $|x| < 1$. Theorem 2.5.8 shows that $O_T(x_0)$ is Benford for a.e. $x_0 \in (-1, 1)$. If $|x| > 1$ then $\lim_{n\to\infty} T^n(x) = +\infty$, and Theorem 2.5.8 applies to the reciprocal version $\widetilde{T}$ of T, namely*

$$\widetilde{T}(x) := T(x^{-1})^{-1} = \frac{2x^4}{1 + x^2}$$

near $x = 0$. Overall, therefore, $O_T(x_0)$ is Benford for a.e. $x_0 \in \mathbb{R}$.

(ii) *Let $T(x) = 1 + x^2$. Again Theorem 2.5.8 applied to*

$$\widetilde{T}(x) = T(x^{-1})^{-1} = \frac{x^2}{1 + x^2},$$

shows that $O_T(x_0)$ is Benford for a.e. $x_0 \in \mathbb{R}$.

An Application: Newton's Method and Related Algorithms

In scientific calculations using digital computers and floating point arithmetic, round-off errors are inevitable, thus, for the problem of finding numerically the root of a function by means of **Newton's Method**, it is important to study the distribution of significant digits (or significands) of the approximations generated by the method.

Throughout this subsection, let $f : I \to \mathbb{R}$ be a differentiable function defined on some open interval $I \subset \mathbb{R}$, and denote by N_f the map associated with f by Newton's Method, that is,

$$N_f(x) := x - \frac{f(x)}{f'(x)} \quad \text{for all } x \in I \text{ with } f'(x) \neq 0.$$

For N_f to be defined wherever f is, set $N_f(x) := x$ if $f'(x) = 0$.

If $f : I \to \mathbb{R}$ is real-analytic and $x^* \in I$ is a root of f, i.e., if $f(x^*) = 0$, then $f(x) = (x - x^*)^m g(x)$ for some $m \in \mathbb{N}$ and some real-analytic $g : I \to \mathbb{R}$ with $g(x^*) \neq 0$. The number m is the **multiplicity** of the root x^*; if $m = 1$ then x^* is referred to as a **simple** root.

Theorem 2.5.10 ([BerH1]). *Let $f : I \to \mathbb{R}$ be real-analytic with $f(x^*) = 0$, and assume that f is not linear. Then*

1. *if x^* is a simple root, then $(x_n - x^*)$ and $(x_{n+1} - x_n)$ are both Benford for (Lebesgue) almost every, but not every x_0 in a neighborhood of x^*;*

2. *if x^* is a root of multiplicity at least two, then $(x_n - x^*)$ and $(x_{n+1} - x_n)$ are Benford for all $x_0 \neq x^*$ sufficiently close to x^*.*

Here (x_n) denotes the sequence of iterates of N_f starting at x_0, that is, $(x_n) = O_{N_f}(x_0)$.

The full proof of Theorem 2.5.10 can be found in [BerH1]. It uses the following lemma which may be of independent interest for studying BL in other numerical approximation procedures. Part (1) is an analog of Lemma 2.5.4, and (2) and (3) follow directly from Theorems 2.5.8 and 2.5.5, respectively.

Lemma 2.5.11. *Let $T : I \to I$ be C^∞ with $T(y^*) = y^*$ for some $y^* \in I$.*

1. *If $T'(y^*) \neq 1$, then for all y_0 such that $\lim_{n \to \infty} T^n(y_0) = y^*$, the sequence $(T^n(y_0) - y^*)$ is Benford precisely when $(T^{n+1}(y_0) - T^n(y_0))$ is Benford.*

2. *If $T'(y^*) = 0$ but $T^{(p)}(y^*) \neq 0$ for some $p \in \mathbb{N}\backslash\{1\}$, then $(T^n(y_0) - y^*)$ is Benford for (Lebesgue) almost every, but not every y_0 in a neighborhood of y^*.*

3. *If $0 < |T'(y^*)| < 1$, then $(T^n(y_0) - y^*)$ is Benford for all $y_0 \neq y^*$ sufficiently close to y^* precisely when $\log |T'(y^*)|$ is irrational.*

Example 2.5.12. **(i)** *Let $f(x) = x/(1 - x)$ for $x < 1$. Then f has a simple root at $x^* = 0$, and $N_f(x) = x^2$. By Theorem 2.5.10(1), the sequences (x_n) and $(x_{n+1} - x_n)$ are both Benford sequences for (Lebesgue) almost every x_0 in a neighborhood of 0.*

(ii) *Let $f(x) = x^2$. Then f has a double root at $x^* = 0$ and $N_f(x) = x/2$, so by Theorem 2.5.10(2), the sequence of iterates (x_n) of N_f as well as $(x_{n+1} - x_n)$ are both Benford for all starting points $x_0 \neq 0$. (They are not, however, 2-Benford.)*

Utilizing Lemma 2.5.11, an analog of Theorem 2.5.10 can be established for other root-finding algorithms as well (see [BerH1]).

Time-Dependent Systems

So far, the sequences considered in this section have been generated by the iteration of a single map T. Beyond this setting there has been, in the recent past, an increased interest in systems that are **non-autonomous**, i.e., explicitly time dependent in one way or the other.

Throughout, let (T_n) be a sequence of maps that map $\mathbb{R}$ or parts thereof into itself, and for every $n \in \mathbb{N}$ denote by T^n the n-fold composition $T^n := T_n \circ \cdots \circ T_1$; also let T^0 be the identity map on $\mathbb{R}$. Given x_0, it makes sense to consider the sequence $O_T(x_0) := \left(T^{n-1}(x_0)\right)_{n \in \mathbb{N}} = \left(x_0, T_1(x_0), T_2(T_1(x_0)), \ldots\right)$.

The following is a non-autonomous variant of Theorem 2.5.5. A proof (of a substantially more general version) can be found in [BerBH]. It relies heavily on a non-autonomous version of the Shadowing Lemma, Lemma 2.5.3.

Theorem 2.5.13 ([BerBH]). *Let $T_j : \mathbb{R} \to \mathbb{R}$ be C^2-maps with $T_j(0) = 0$ and $T_j'(0) \neq 0$ for all $j \in \mathbb{N}$, and set $\alpha_j := T_j'(0)$. Assume that $\sup_j \max_{|x| \leq 1} |T_j''(x)|$ and $\sum_{n=1}^{\infty} \prod_{j=1}^{n} |\alpha_j|$ are both finite. If $\lim_{j \to \infty} \log |\alpha_j|$ exists and is irrational, then $O_T(x_0)$ is Benford for all $x_0 \neq 0$ sufficiently close to 0.*

Example 2.5.14. **(i)** *Let $R_j(x) = (2 + j^{-1})x$ for $j = 1, 2, \ldots$. It is easy to see that all assumptions of Theorem 2.5.13 are met for*

$$T_j(x) = R_j(x^{-1})^{-1} = \frac{j}{2j+1}x,$$

with $\lim_{j \to \infty} \log |\alpha_j| = -\log 2$. Hence $O_R(x_0)$ is Benford for all $x_0 \neq 0$.

(ii) *Let $T_j(x) = F_{j+1}/F_j x$ for all $j \in \mathbb{N}$, where F_j denotes the jth Fibonacci number. Since $\lim_{j \to \infty} \log(F_{j+1}/F_j) = \log \frac{1+\sqrt{5}}{2}$ is irrational, and by taking reciprocals as in (i), Theorem 2.5.13 shows that $O_T(x_0)$ is Benford for all $x_0 \neq 0$. In particular, $O_T(F_1) = (F_n)$ is Benford, as was already seen in Example 2.4.12. Note that the same argument would not work to show that $(n!)$ is Benford.*

In situations where most of the maps T_j are power-like or even more strongly expanding, the following generalization of Lemma 2.5.7 may be useful. (In its fully developed form, the result also extends Theorem 2.5.8; see [BerBH, Thm.5.5] and [Ber3, Thm.3.7].) Again the reader is referred to [Ber4] for a proof.

Theorem 2.5.15 ([Ber4]). *Assume the maps $T_j : \mathbb{R}^+ \to \mathbb{R}^+$ satisfy, for some $\xi > 0$ and all $j \in \mathbb{N}$, the following conditions:*

1. $x \mapsto \ln T_j(e^x)$ is convex on $[\xi, +\infty)$;

2. $xT_j'(x)/T_j(x) \geq \beta_j > 0$ for all $x \geq \xi$.

If $\liminf_{j \to \infty} \beta_j > 1$ then $O_T(x_0)$ is Benford for almost every sufficiently large x_0, but there are also uncountably many exceptional points.

Example 2.5.16. **(i)** *To see that Theorem 2.5.15 does indeed generalize Lemma 2.5.7, let $T_j(x) = \alpha x^\beta$ for all $j \in \mathbb{N}$. Then $x \mapsto \ln T_j(e^x) = \beta x + \ln \alpha$ clearly is convex, and $xT_j'(x)/T_j(x) = \beta > 1$ for all $x > 0$.*

(ii) *Theorem 2.5.15 also shows that $O_T(x_0)$ with $T(x) = e^x$ is Benford for almost every, but not every $x_0 \in \mathbb{R}$, as $x \mapsto \ln T(e^x) = e^x$ is convex, and*

$xT'(x)/T(x) = x$ *as well as* $T^3(x) > e$ *holds for all* $x \in \mathbb{R}$. *Similarly, the theorem applies to* $T(x) = 1 + x^2$.

(iii) *For a truly non-autonomous example consider*

$$T_j(x) = \begin{cases} x^2 & \text{if } j \text{ is even}, \\ 2^x & \text{if } j \text{ is odd}, \end{cases} \quad \text{or} \quad T_j(x) = (j+1)^x.$$

In both cases, $O_T(x_0)$ *is Benford for almost every, but not every* $x_0 \in \mathbb{R}$.

(iv) *Finally, it is important to note that Theorem 2.5.15 may fail if one of its hypotheses is violated even for a single* j. *For example,*

$$T_j(x) = \begin{cases} 10 & \text{if } j = 1, \\ x^2 & \text{if } j \geq 2, \end{cases}$$

satisfies (1) and (2) for all $j > 1$, *but does not satisfy assumption (2) for* $j = 1$. *Clearly,* $O_T(x_0)$ *is not Benford for any* $x_0 \in \mathbb{R}$, *since* $D_1\big(T^n(x_0)\big) \equiv 1$ *for all* $n \in \mathbb{N}$.

2.5.2 Multidimensional Discrete-Time Processes

The purpose of this subsection is to extend the basic results of the previous section to multidimensional systems, notably to linear, as well as some non-linear recurrence relations. Recall from Example 2.4.12 that the Fibonacci sequence (F_n) is Benford. Hence the linear recurrence relation $x_{n+1} = x_n + x_{n-1}$ generates a Benford sequence when started from $x_0 = x_1 = 1$. As will be seen shortly, many, but not all linear recurrence relations generate Benford sequences.

Example 2.5.17. (i) *Let the sequence* (x_n) *be defined recursively as*

$$x_{n+1} = x_n - x_{n-1}, \quad n = 1, 2, \ldots, \tag{2.16}$$

with given $x_0, x_1 \in \mathbb{R}$. *By using the matrix* $\begin{bmatrix} 0 & 1 \\ -1 & 1 \end{bmatrix}$ *associated with (2.16), it is straightforward to derive an explicit representation for* (x_n),

$$x_n = x_0 \cos\left(\tfrac{1}{3}\pi n\right) + \frac{2x_1 - x_0}{\sqrt{3}} \sin\left(\tfrac{1}{3}\pi n\right), \quad n = 0, 1, \ldots.$$

From this it is clear that $x_{n+6} = x_n$ *for all* n, *i.e.,* (x_n) *is 6-periodic. For no choice of* x_0, x_1, *therefore, is* (x_n) *Benford.*

(ii) *Consider the linear 3-step recursion*

$$x_{n+1} = 2x_n + 10x_{n-1} - 20x_{n-2}, \quad n = 2, 3, \ldots. \tag{2.17}$$

Clearly, $\lim_{n\to\infty} |x_n| = +\infty$ *unless* $x_0 = x_1 = x_2 = 0$, *so unlike in (i) the sequence* (x_n) *is not bounded or oscillatory. However, if* $|c_2| \neq |c_3|$ *then*

$$\log |x_n| = \frac{n}{2} + \log\left|c_1 10^{-n(\frac{1}{2} - \log 2)} + c_2 + (-1)^n c_3\right| \approx \frac{n}{2} + \log |c_2 + (-1)^n c_3|,$$

showing that $\big(S(x_n)\big)$ *is asymptotically 2-periodic and hence* (x_n) *is not Benford. Similarly, if* $|c_2| = |c_3| \neq 0$ *then* $\big(S(x_n)\big)$ *is convergent along even (if* $c_2 = c_3$) *or odd (if* $c_2 = -c_3$) *indices* n, *and again* (x_n) *is not Benford. Only if* $c_2 = c_3 = 0$ *yet* $c_1 \neq 0$, *or equivalently if* $\tfrac{1}{4}x_2 = \tfrac{1}{2}x_1 = x_0 \neq 0$, *is* (x_n) *Benford.*

The above recurrence relations (2.16) and (2.17) are linear and have constant coefficients. Hence they can be rewritten and analyzed using matrix–vector notation. For instance, in Example 2.5.17(i)

$$\begin{bmatrix} x_n \\ x_{n+1} \end{bmatrix} = \begin{bmatrix} 0 & 1 \\ -1 & 1 \end{bmatrix} \begin{bmatrix} x_{n-1} \\ x_n \end{bmatrix},$$

so that, with $A = \begin{bmatrix} 0 & 1 \\ -1 & 1 \end{bmatrix} \in \mathbb{R}^{2\times 2}$, the sequence (x_n) is simply given by

$$x_n = \begin{bmatrix} 1 & 0 \end{bmatrix} A^n \begin{bmatrix} x_0 \\ x_1 \end{bmatrix}, \quad n = 0, 1, \dots .$$

It is natural, therefore, to study the Benford property of more general sequences $(x^\top A^n y)$ for any $A \in \mathbb{R}^{d\times d}$ and $x, y \in \mathbb{R}^d$. Linear recurrence relations like the ones in Example 2.5.17 are then merely special cases.

Recall complex numbers $z_1, z_2, \dots, z_m$ are **rationally independent** if $\sum_{j=1}^m q_j z_j = 0$ with rational $q_1, q_2, \dots, q_m$ implies that $q_j = 0$ for all $j = 1, 2, \dots, m$. Let $Z \subset \mathbb{C}$ be any set such that all elements of Z have the same modulus ζ, i.e., Z is contained in the periphery of a circle with radius ζ centered at the origin of the complex plain. Call the set Z **resonant** if either $\#(Z \cap \mathbb{R}) = 2$ or the numbers $1, \log \zeta$, and the elements of $\frac{1}{2\pi} \arg Z$ are rationally dependent, where $\frac{1}{2\pi} \arg Z = \{\frac{1}{2\pi} \arg z : z \in Z\} \setminus \{-\frac{1}{2}, 0\}$.

Definition 2.5.18. *A matrix $A \in \mathbb{R}^{d\times d}$ is **Benford regular (base** 10) if $\sigma(A)^+$ (the subset of the spectrum of A with non-negative imaginary components) contains no resonant set.*

Note that in the simplest case, i.e., for $d = 1$, the matrix $A = [a]$ is Benford regular if and only if $\log |a|$ is irrational. Hence Benford regularity may be considered a generalization of this irrationality property. Also note that A is regular (invertible) whenever it is Benford regular.

Example 2.5.19. *None of the matrices associated with the recurrence relations in Example 2.5.17 are Benford regular. Indeed, in (i), $A = \begin{bmatrix} 0 & 1 \\ -1 & 1 \end{bmatrix}$, hence $\sigma(A)^+ = \{e^{i\pi/3}\}$, and clearly $\log |e^{i\pi/3}| = 0$ is rational. Similarly, in (ii), $A = \begin{bmatrix} 0 & 1 & 0 \\ 0 & 0 & 1 \\ -10 & 10 & 2 \end{bmatrix}$, and $\sigma(A)^+ = \{-\sqrt{10}, 2, \sqrt{10}\}$ contains the resonant set $\{-\sqrt{10}, \sqrt{10}\}$.*

Example 2.5.20. *Let $A = \begin{bmatrix} 1 & -1 \\ 1 & 1 \end{bmatrix} \in \mathbb{R}^{2\times 2}$, with characteristic polynomial $p_A(\lambda) = \lambda^2 - 2\lambda + 2$, and hence $\sigma(A)^+ = \{\sqrt{2}e^{i\pi/4}\}$. As $1, \log \sqrt{2}$, and $\frac{1}{2\pi} \cdot \frac{\pi}{4} = \frac{1}{8}$ are rationally dependent, the matrix A is not Benford regular.*

Example 2.5.21. *Consider $A = \begin{bmatrix} 0 & 1 \\ 1 & 1 \end{bmatrix} \in \mathbb{R}^{2\times 2}$. The characteristic polynomial of A is $p_A(\lambda) = \lambda^2 - \lambda - 1$, and so, with $\varphi = \frac{1}{2}(1 + \sqrt{5})$, the eigenvalues of A are*

φ and $-\varphi^{-1}$. Since p_A is irreducible and has two roots of different absolute value, it follows that $\log \varphi$ is irrational (in fact, even transcendental). Thus A is Benford regular.

With the one-dimensional result (Lemma 2.5.2), as well as Example 2.5.17 and Definition 2.5.18 in mind, it seems realistic to hope that iterating (i.e., taking powers of) any matrix $A \in \mathbb{R}^{d \times d}$ produces many Benford sequences, provided that A is Benford regular. This is indeed the case. To concisely formulate the pertinent result, call a sequence (z_n) of complex numbers **terminating** if $z_n = 0$ for all sufficiently large n.

Theorem 2.5.22 ([Ber2]). *Assume that $A \in \mathbb{R}^{d \times d}$ is Benford regular. Then, for every $x, y \in \mathbb{R}^d$, the sequence $(x^\top A^n y)$ is either Benford or terminating. Also, $(\|A^n x\|)$ is Benford for every $x \neq 0$.*

Proof. Apply Theorem 2.4.2 and the following proposition, a variant of [Ber2, Lem.2.9].   □

Proposition 2.5.23. *Assume that the real numbers $1, \rho_0, \rho_1, \ldots, \rho_m$ are rationally independent. Let (z_n) be a convergent sequence in $\mathbb{C}$, and at least one of the numbers $c_1, c_2, \ldots, c_m \in \mathbb{C}$ non-zero. Then (x_n) given by*

$$x_n = n\rho_0 + \log \left| \Re \left(c_1 e^{2\pi i n \rho_1} + \cdots + c_m e^{2\pi i n \rho_m} + z_n \right) \right|$$

is u.d. mod 1.

Example 2.5.24. *According to Example 2.5.21, the matrix $\begin{bmatrix} 0 & 1 \\ 1 & 1 \end{bmatrix}$ is Benford regular. By Theorem 2.5.22, every solution of the difference equation $x_{n+1} = x_n + x_{n-1}$ is Benford, except for the trivial solution $x_n \equiv 0$ resulting from $x_0 = x_1 = 0$. In particular, therefore, the sequences of Fibonacci and Lucas numbers, $(F_n) = (1, 1, 2, 3, 5, \ldots)$ and $(L_n) = (-1, 2, 1, 3, 4, \ldots)$, generated respectively from the initial values $\begin{bmatrix} x_0 & x_1 \end{bmatrix} = \begin{bmatrix} 1 & 1 \end{bmatrix}$ and $\begin{bmatrix} x_0 & x_1 \end{bmatrix} = \begin{bmatrix} -1 & 2 \end{bmatrix}$, are Benford. For the former sequence, this has already been seen in Example 2.4.12. Note that (F_n^2), for instance, is Benford as well by Corollary 2.4.7(1).*

Example 2.5.25. *Recall from Example 2.5.20 that $A = \begin{bmatrix} 1 & -1 \\ 1 & 1 \end{bmatrix}$ is not Benford regular. Hence Theorem 2.5.22 does not apply, and $(x^\top A^n y)$ may, for some $x, y \in \mathbb{R}^2$, be neither Benford nor terminating. Indeed, pick for example $x = y = \begin{bmatrix} 1 & 0 \end{bmatrix}^\top$ and note that for $n = 0, 1, \ldots,$*

$$x^\top A^n y = \begin{bmatrix} 1 & 0 \end{bmatrix} 2^{n/2} \begin{bmatrix} \cos(\tfrac{1}{4}\pi n) & -\sin(\tfrac{1}{4}\pi n) \\ \sin(\tfrac{1}{4}\pi n) & \cos(\tfrac{1}{4}\pi n) \end{bmatrix} \begin{bmatrix} 1 \\ 0 \end{bmatrix} = 2^{n/2} \cos\left(\tfrac{1}{4}\pi n\right)$$

is clearly not Benford as $x^\top A^n y = 0$ whenever $n = 2 + 4l$ for some $l \in \mathbb{N}_0$.

The present section closes with an example of a non-linear system. The sole purpose is to hint at possible extensions of the results presented earlier; for more details the interested reader is referred to [Ber2].

Example 2.5.26. *Consider the non-linear map* $T : \mathbb{R}^2 \to \mathbb{R}^2$ *given by*

$$T : \begin{bmatrix} x_1 \\ x_2 \end{bmatrix} \mapsto \begin{bmatrix} 2 & 0 \\ 0 & 2 \end{bmatrix} \begin{bmatrix} x_1 \\ x_2 \end{bmatrix} + \begin{bmatrix} f(x_1) \\ f(x_2) \end{bmatrix},$$

with the bounded continuous function

$$f(t) = \frac{3}{2}|t+2| - 3|t+1| + 3|t-1| - \frac{3}{2}|t-2| = \begin{cases} 0 & \text{if } |t| \geq 2, \\ 3t+6 & \text{if } -2 < t < -1, \\ -3t & \text{if } -1 \leq t < 1, \\ 3t-6 & \text{if } 1 \leq t < 2. \end{cases}$$

Sufficiently far away from the x_1- and x_2-axes, i.e., for $\min\{|x_1|, |x_2|\}$ sufficiently large, the dynamics of T is governed by the matrix $\begin{bmatrix} 2 & 0 \\ 0 & 2 \end{bmatrix}$, and since the latter is Benford regular, one may reasonably expect that $(x^\top T^n(y))$ should be Benford. This is indeed the case.

2.5.3 Differential Equations

By presenting a few results on, and examples of, differential equations, i.e., deterministic continuous-time processes, this section aims at convincing the reader that the emergence of BL is not at all restricted to discrete-time dynamics. Rather, solutions of ordinary or partial differential equations often turn out to be Benford as well. Recall that a (Borel measurable) function $f : [0, +\infty) \to \mathbb{R}$ is Benford if and only if $\log|f|$ is u.d. mod 1.

Consider the **initial value problem (IVP)**

$$\dot{x} = F(x), \quad x(0) = x_0, \tag{2.18}$$

where $F : \mathbb{R} \to \mathbb{R}$ is continuously differentiable with $F(0) = 0$, and $x_0 \in \mathbb{R}$. In the simplest case, $F(x) \equiv \alpha x$ with some $\alpha \in \mathbb{R}$. In this case, the unique solution of (2.18) is $x(t) = x_0 e^{\alpha t}$. Unless $\alpha x_0 = 0$, therefore, every solution of (2.18) is Benford. As in the discrete-time setting, this feature persists for arbitrary C^2-functions F with $F'(0) < 0$. The direct analog of Theorem 2.5.5 is

Theorem 2.5.27 ([BerBH]). *Let $F : \mathbb{R} \to \mathbb{R}$ be C^2 with $F(0) = 0$. Assume that $F'(0) < 0$. Then, for every $x_0 \neq 0$ sufficiently close to 0, the unique solution of (2.18) is Benford.*

Proof. Pick $\delta > 0$ so small that $xF(x) < 0$ for all $0 < |x| \leq \delta$. As F is C^2, the IVP (2.18) has a unique local solution whenever $|x_0| \leq \delta$; see [Walt]. Since the interval $[-\delta, \delta]$ is forward invariant, this solution exists for all $t \geq 0$. Fix any x_0 with $0 < |x_0| \leq \delta$ and denote the unique solution of (2.18) as $x = x(t)$. Clearly, $\lim_{t \to +\infty} x(t) = 0$. With $y : [0, +\infty) \to \mathbb{R}$ defined as $y = x^{-1}$ therefore $y(0) = x_0^{-1} =: y_0$ and $\lim_{t \to +\infty} |y(t)| = +\infty$. Let $\alpha := -F'(0) > 0$ and note that there exists a continuous function $g : \mathbb{R} \to \mathbb{R}$ such that $F(x) = -\alpha x + x^2 g(x)$. From

$$\dot{y} = -\frac{\dot{x}}{x^2} = \alpha y - g(y^{-1}),$$

it follows via the variation of constants formula that, for all $t \geq 0$,

$$y(t) = e^{\alpha t} y_0 - \int_0^t e^{\alpha(t-\tau)} g\big(y(\tau)^{-1}\big) \, \mathrm{d}\tau \, .$$

As $\alpha > 0$ and g is continuous, the number

$$\overline{y_0} := y_0 - \int_0^{+\infty} e^{-\alpha\tau} g\big(y(\tau)^{-1}\big) \, \mathrm{d}\tau$$

is well defined. Moreover, for all $t > 0$,

$$\big| y(t) - e^{\alpha t} \overline{y_0} \big| = \left| \int_t^{+\infty} e^{\alpha(t-\tau)} g\big(y(\tau)^{-1}\big) \, \mathrm{d}\tau \right|$$

$$\leq \int_0^{+\infty} e^{-\alpha\tau} \big| g\big(y(t+\tau)^{-1}\big) \big| \, \mathrm{d}\tau \leq \frac{\|g\|_\infty}{\alpha} \, ,$$

where $\|g\|_\infty = \max_{|x| \leq \delta} |g(x)|$, and Lemma 2.5.4(2) shows that y is Benford if and only if $t \mapsto e^{\alpha t} \overline{y_0}$ is. An application of Corollary 2.4.7(2) therefore completes the proof. $\qquad\square$

Example 2.5.28. (i) *The function $F(x) = -x + x^4 e^{-x^2}$ satisfies the assumptions of Theorem 2.5.27. Thus except for the trivial solution $x = 0$, every solution of $\dot{x} = -x + x^4 e^{-x^2}$ is Benford.*

(ii) *The function $F(x) = -x^3 + x^4 e^{-x^2}$ is also smooth with $xF(x) < 0$ for all $x \neq 0$. Hence for every $x_0 \in \mathbb{R}$, the IVP (2.18) has a unique solution with $\lim_{t \to +\infty} x(t) = 0$. However, $F'(0) = 0$, and it is not hard to see that this causes x to approach 0 rather slowly. In fact, $\lim_{t \to +\infty} 2tx(t)^2 = 1$ whenever $x_0 \neq 0$, and this prevents x from being Benford.*

Similar results follow for the linear d-dimensional ordinary differential equations $\dot{x} = Ax$, where A is a real $d \times d$-matrix; see [Ber2].

Finally, it should be mentioned that at present little seems to be known about the Benford property for solutions of *partial* differential equations or more general functional equations such as e.g. delay or integro-differential equations. Quite likely, it will be very hard to decide in any generality whether many, or even most, solutions of such systems exhibit the Benford property in one form or another.

Example 2.5.29. *A fundamental example of a partial differential equation is the so-called one-dimensional **heat (or diffusion) equation***

$$\frac{\partial u}{\partial t} = \frac{\partial^2 u}{\partial x^2} \, , \tag{2.19}$$

a linear second-order equation for $u = u(t, x)$. Physically, (2.19) describes e.g. the diffusion over time of heat in a homogeneous one-dimensional medium. Without further conditions, (2.19) has many solutions of which for instance

$$u(t, x) = cx^2 + 2ct \, ,$$

with any constant $c \neq 0$, is neither Benford in t ("time") nor in x ("space"), whereas

$$u(t, x) = e^{-c^2 t} \sin(cx)$$

is Benford (or identically zero) in t but not in x,

$$u(t, x) = \frac{1}{\sqrt{t}} e^{-x^2/(4t)} \quad (t > 0)$$

is Benford in x but not in t, and

$$u(t, x) = e^{c^2 t + cx}$$

is Benford in both t and x.

2.6 BENFORD'S LAW FOR RANDOM PROCESSES

The purpose of this section is to show how BL arises naturally in a variety of stochastic settings, including products of independent random variables, mixtures of random samples from different distributions, and iterations of random maps. Perhaps not surprisingly, BL arises in many other important fields of stochastics as well, such as geometric Brownian motion, random matrices, Lévy processes, and Bayesian models. The present section may also serve as a preparation for the specialized literature on these advanced topics [EngLeu, JaKKKM, LeScEv, MiNi1, MiNi2, Schür2].

2.6.1 Independent Random Variables

Recall that a sequence (X_n) of random variables **converges in distribution** to a random variable X, symbolically $X_n \overset{\mathcal{D}}{\to} X$, if $\lim_{n\to\infty} \mathbb{P}(X_n \leq t) = \mathbb{P}(X \leq t)$ holds for every $t \in \mathbb{R}$ for which $\mathbb{P}(X = t) = 0$. By a slight abuse of terminology, say that (X_n) **converges in distribution to BL** if $S(X_n) \overset{\mathcal{D}}{\to} S(X)$, where X is a Benford random variable, or equivalently if

$$\lim_{n\to\infty} \mathbb{P}(S(X_n) \leq t) = \log t \quad \text{for all } t \in [1, 10).$$

An especially simple way of generating a sequence of random variables is this: Fix a random variable X, and set $X_n := X^n$ for every $n \in \mathbb{N}$. While the sequence (X_n) thus generated is clearly not i.i.d. unless $X = 0$ a.s. or $X = 1$ a.s., Theorems 2.4.10 and 2.4.15 imply

Theorem 2.6.1. *Assume that the random variable X has a density. Then*

1. X^n converges in distribution to BL;

2. with probability one, (X^n) is Benford.

Proof. To prove (1), note that the random variable $\log |X|$ has a density as well. Hence, by Theorem 2.4.15,

$$\mathbb{P}(S(X_n) \leq t) = \mathbb{P}(\langle \log |X^n| \rangle \leq \log t) = \mathbb{P}(\langle n \log |X| \rangle \leq \log t) \to \log t$$

as $n \to \infty$ holds for all $t \in [1, 10)$, i.e., (X_n) converges in distribution to BL.

To see (2), simply note that $\log |X|$ is irrational with probability one. By Theorem 2.4.10, therefore, $\mathbb{P}((X^n) \text{ is Benford}) = 1$. $\qquad \square$

Example 2.6.2. (i) *Let X be uniformly distributed on $[0, 1)$. For every $n \in \mathbb{N}$,*

$$F_{S(X^n)}(t) = \frac{t^{1/n} - 1}{10^{1/n} - 1}, \quad 1 \le t < 10,$$

and a short calculation, together with the elementary estimate $\dfrac{e^t - 1 - t}{e^t - 1} < \dfrac{t}{2}$ for all $t > 0$ shows that

$$\left| F_{S(X^n)}(t) - \log t \right| \le \frac{10^{1/n} - 1 - \frac{\ln 10}{n}}{10^{1/n} - 1} < \frac{\ln 10}{2n} \to 0 \quad \text{as } n \to \infty,$$

and hence (X^n) converges in distribution to BL. Since $\mathbb{P}(\log X$ is rational$) = 0$, the sequence (X^n) is Benford with probability one.

(ii) *Assume that $X = 2$ a.s. Thus $P_X = \delta_2$, and X does not have a density. For every n, $S(X^n) = 10^{\langle n \log 2 \rangle}$ with probability one, so (X^n) does not converge in distribution to BL. On the other hand, (X^n) is Benford a.s.*

The sequence of random variables considered in Theorem 2.6.1 is very special in that X^n is the product of n quantities that are identical, and hence dependent *in extremis*. Note that X^n is Benford for all n if and only if X is Benford. This invariance property of BL persists if, unlike the case in Theorem 2.6.1, products of *independent* factors are considered.

Theorem 2.6.3. *Let X, Y be two independent random variables with $\mathbb{P}(XY = 0) = 0$. Then*

1. *if X is Benford then so is XY;*

2. *if $S(X)$ and $S(XY)$ have the same distribution, then either $\log S(Y)$ is rational with probability one, or X is Benford.*

Proof. As in the proof of Lemma 2.4.26, the argument becomes short and transparent through the usage of Fourier coefficients. Note first that $\log S(XY) = \langle \log S(X) + \log S(Y) \rangle$ and, since the random variables $X_0 := \log S(X)$ and $Y_0 := \log S(Y)$ are independent,

$$\widehat{P_{\log S(XY)}} = \widehat{P_{\langle X_0 + Y_0 \rangle}} = \widehat{P_{X_0}} \cdot \widehat{P_{Y_0}}. \tag{2.20}$$

To prove (1), simply recall that X being Benford is equivalent to $P_{X_0} = \lambda_{0,1}$, and hence $\widehat{P_{X_0}}(k) = 0$ for every integer $k \ne 0$. Consequently, $\widehat{P_{\log S(XY)}}(k) = 0$ as well, i.e., XY is Benford.

To see (2), assume that $S(X)$ and $S(XY)$ have the same distribution. In this case, (2.20) implies that

$$\widehat{P_{X_0}}(k) \left(1 - \widehat{P_{Y_0}}(k) \right) = 0 \quad \text{for all } k \in \mathbb{Z}.$$

If $\widehat{P_{Y_0}}(k) \ne 1$ for all non-zero k, then $\widehat{P_{X_0}} = \lambda_{0,1}$, i.e., X is Benford. Alternatively, if $\widehat{P_{Y_0}}(k_0) = 1$ for some $k_0 \ne 0$ then $P_{Y_0}(\frac{1}{|k_0|}\mathbb{Z}) = 1$, hence $|k_0|Y_0 = |k_0| \log S(Y)$ is an integer with probability one. $\qquad\square$

Example 2.6.4. *Let V, W be independent random variables distributed according to $U(0,1)$. Then $X := 10^V$ and $Y := W$ are independent and, by Theorem 2.6.3(1), XY is Benford even though Y is not. If, on the other hand, $X := 10^V$ and $Y := 10^{1-V}$ then X and Y are both Benford, yet XY is not. Hence the independence of X and Y is crucial in Theorem 2.6.3(1). It is essential in assertion (2) as well, as can be seen by letting X equal either $10^{\sqrt{2}-1}$ or $10^{2-\sqrt{2}}$ with probability $\frac{1}{2}$ each, and choosing $Y := X^{-2}$. Then $S(X)$ and $S(XY) = S(X^{-1})$ have the same distribution, but neither X is Benford nor $\log S(Y)$ is rational with probability one.*

Theorem 2.6.5. *Let (X_n) be an i.i.d. sequence of random variables that are not purely atomic, i.e., $\mathbb{P}(X_1 \in C) < 1$ for every countable set $C \subset \mathbb{R}$. Then*

1. *$\left(\prod_{j=1}^n X_j\right)$ converges in distribution to BL;*

2. *with probability one, $\left(\prod_{j=1}^n X_j\right)$ is Benford.*

Proof. Let $Y_n = \log|X_n|$. Then (Y_n) is an i.i.d. sequence of random variables that are not purely atomic. By Theorem 2.4.13(3), the sequence of $\left\langle \sum_{j=1}^n Y_j \right\rangle = \left\langle \log\left|\prod_{j=1}^n X_j\right| \right\rangle$ converges in distribution to $U(0,1)$. Thus $\left(\prod_{j=1}^n X_j\right)$ converges in distribution to BL.

To prove (2), let Y_0 be u.d. mod 1 and independent of $(Y_n)_{n\in\mathbb{N}}$, and define
$$S_j := \langle Y_0 + Y_1 + \cdots + Y_j \rangle, \quad j \in \mathbb{N}_0.$$
Recall from Theorem 2.4.13(1) that S_j is u.d. mod 1 for every $j \geq 0$. Also note that, by definition, the random variables $Y_{j+1}, Y_{j+2}, \ldots$ are independent of S_j. The following argument is most transparent when formulated in ergodic theory terminology. To this end, endow $\mathbb{T}_\infty := [0,1)^{\mathbb{N}_0} = \{(x_j)_{j\in\mathbb{N}_0} : x_j \in [0,1) \text{ for all } j\}$ with the σ-algebra
$$\mathcal{B}_\infty := \sigma\big(\{B_0 \times B_1 \times \cdots \times B_j \times [0,1) \times [0,1) \times \cdots : j \in \mathbb{N}_0,$$
$$B_0, B_1, \ldots, B_j \in \mathcal{B}[0,1)\}\big) \tag{2.21}$$
$$= \bigotimes_{j\in\mathbb{N}_0} \mathcal{B}[0,1).$$
A probability measure P_∞ is uniquely defined on $(\mathbb{T}_\infty, \mathcal{B}_\infty)$ by setting
$$P_\infty(B_0 \times B_1 \times \cdots \times B_j \times [0,1) \times [0,1) \times \cdots)$$
$$= \mathbb{P}(S_0 \in B_0, S_1 \in B_1, \ldots, S_j \in B_j)$$
for all $j \in \mathbb{N}_0$ and $B_0, B_1, \ldots, B_j \in \mathcal{B}[0,1)$.

The map $\sigma_\infty : \mathbb{T}_\infty \to \mathbb{T}_\infty$ with $\sigma_\infty((x_j)) = (x_{j+1})$, often referred to as the *(one-sided) left shift* on $\mathbb{T}_\infty$, is clearly measurable, i.e., $\sigma_\infty^{-1}(A) \in \mathcal{B}_\infty$ for every $A \in \mathcal{B}_\infty$. As a consequence, $(\sigma_\infty)_* P_\infty$ is a well-defined probability measure on $(\mathbb{T}_\infty, \mathcal{B}_\infty)$. In fact, since S_1 is u.d. mod 1 and (Y_n) is an i.i.d. sequence,
$$(\sigma_\infty)_* P_\infty(B_0 \times B_1 \times \cdots \times B_j \times [0,1) \times [0,1) \times \cdots)$$
$$= P_\infty([0,1) \times B_0 \times B_1 \times \cdots \times B_j \times [0,1) \times [0,1) \times \cdots)$$
$$= \mathbb{P}(S_1 \in B_0, S_2 \in B_1, \ldots, S_{j+1} \in B_j)$$
$$= \mathbb{P}(S_0 \in B_0, S_1 \in B_1, \ldots, S_j \in B_j)$$
$$= P_\infty(B_0 \times B_1 \times \cdots \times B_j \times [0,1) \times [0,1) \times \cdots),$$

showing that $(\sigma_\infty)_* P_\infty = P_\infty$, i.e., σ_∞ is P_∞-preserving. (In probabilistic terms, this is equivalent to saying that the random process $(S_j)_{j \in \mathbb{N}_0}$ is **stationary**; see [Shi, Def.V.1.1].) It will now be shown that σ_∞ is even **ergodic** with respect to P_∞. Recall that this simply means that every invariant set $A \in \mathcal{B}_\infty$ has measure zero or one, or, more formally, that $P_\infty(\sigma_\infty^{-1}(A)\Delta A) = 0$ implies $P_\infty(A) \in \{0, 1\}$; here the symbol Δ denotes the symmetric difference of two sets, i.e., $A\Delta B = A\backslash B \cup B\backslash A$. Assume, therefore, that $P_\infty(\sigma_\infty^{-1}(A)\Delta A) = 0$ for some $A \in \mathcal{B}_\infty$. Given $\varepsilon > 0$, there exists a number $N \in \mathbb{N}$ and sets $B_0, B_1, \ldots, B_N \in \mathcal{B}[0, 1)$ such that

$$ P_\infty \big(A \Delta (B_0 \times B_1 \times \cdots \times B_N \times [0, 1) \times [0, 1) \times \cdots) \big) < \varepsilon. $$

For notational convenience, let $A_\varepsilon := B_0 \times B_1 \times \cdots \times B_N \times [0, 1) \times [0, 1) \times \cdots \in \mathcal{B}_\infty$, and note that $P_\infty\big(\sigma_\infty^{-j}(A)\Delta\sigma_\infty^{-j}(A_\varepsilon)\big) < \varepsilon$ for all $j \in \mathbb{N}_0$. Recall now from Theorem 2.4.13(3) that, given $S_0, S_1, \ldots, S_N$, the random variables S_n converge in distribution to $U(0, 1)$. Thus, for all sufficiently large M,

$$ \big|P_\infty\big(A_\varepsilon^c \cap \sigma_\infty^{-M}(A_\varepsilon)\big) - P_\infty(A_\varepsilon^c)P_\infty\big(\sigma_\infty^{-M}(A_\varepsilon)\big)\big| $$
$$ = \big|P_\infty\big(A_\varepsilon^c \cap \sigma_\infty^{-M}(A_\varepsilon)\big) - P_\infty(A_\varepsilon^c)P_\infty(A_\varepsilon)\big| < \varepsilon, $$

and similarly $\big|P_\infty\big(A_\varepsilon \cap \sigma_\infty^{-M}(A_\varepsilon^c)\big) - P_\infty(A_\varepsilon)P_\infty(A_\varepsilon^c)\big| < \varepsilon$. (Note that (2.22) may not hold if X_1, and hence also Y_1, is purely atomic.) Overall, therefore,

$$ 2P_\infty(A_\varepsilon)\big(1 - P_\infty(A_\varepsilon)\big) \leq 2\varepsilon + P_\infty\big(A_\varepsilon\Delta\sigma_\infty^{-M}(A_\varepsilon)\big) $$
$$ \leq 2\varepsilon + P_\infty(A_\varepsilon\Delta A) + P_\infty\big(A\Delta\sigma_\infty^{-M}(A)\big) $$
$$ + P_\infty\big(\sigma_\infty^{-M}(A)\Delta\sigma_\infty^{-M}(A_\varepsilon)\big) \tag{2.22} $$
$$ < 4\varepsilon, $$

and consequently $P_\infty(A)\big(1 - P_\infty(A)\big) < 4\varepsilon + \varepsilon^2$. Since $\varepsilon > 0$ was arbitrary, $P_\infty(A) \in \{0, 1\}$, which in turn shows that σ_∞ is ergodic. (Again, this is equivalent to saying, in probabilistic parlance, that the random process $(S_j)_{j \in \mathbb{N}_0}$ is *ergodic*; see [Shi, Def.V.3.2].) By the **Birkhoff Ergodic Theorem** (e.g. [Ber1]), for every (measurable) function $f : [0, 1) \to \mathbb{C}$ with $\int_0^1 |f(x)| \, dx < +\infty$,

$$ \frac{1}{n}\sum_{j=0}^n f(x_j) \to \int_0^1 f(x) \, dx \quad \text{as } n \to \infty $$

holds for all $(x_j)_{j \in \mathbb{N}_0} \in \mathbb{T}_\infty$, with the possible exception of a set of P_∞-measure zero. In probabilistic terms, this means that

$$ \lim_{n \to \infty} \frac{1}{n}\sum_{j=0}^n f(S_j) = \int_0^1 f(x) \, dx \quad \text{a.s.} \tag{2.23} $$

Assume from now on that f is actually *continuous* with $\lim_{x \uparrow 1} f(x) = f(0)$, e.g. $f(x) = e^{2\pi i x}$. For any such f, as well as any $t \in [0, 1)$ and $m \in \mathbb{N}$, let

$$ \Omega_{f, t, m} := $$
$$ \left\{ \omega \in \Omega : \limsup_{n \to \infty} \left| \frac{1}{n}\sum_{j=1}^n f\big(\langle t + Y_1(\omega) + \cdots + Y_j(\omega)\rangle\big) - \int_0^1 f(x) \, dx \right| < \frac{1}{m} \right\}. $$

According to (2.23), $1 = \int_0^1 \mathbb{P}(\Omega_{f,t,m})\,dt$, and hence $\mathbb{P}(\Omega_{f,t,m}) = 1$ for a.e. $t \in [0,1)$. Since f is uniformly continuous, for every $m \geq 2$ there exists $t_m > 0$ such that $\mathbb{P}(\Omega_{f,t_m,m}) = 1$ and $\Omega_{f,t_m,m} \subset \Omega_{f,0,\lfloor m/2 \rfloor}$. From

$$1 = \mathbb{P}\left(\bigcap_{m \geq 2} \Omega_{f,t_m,m} \right) \leq \mathbb{P}\left(\bigcap_{m \geq 2} \Omega_{f,0,\lfloor m/2 \rfloor} \right) \leq 1,$$

it is clear that

$$\lim_{n \to \infty} \frac{1}{n} \sum_{j=1}^n f\big(\langle Y_1 + \cdots + Y_j \rangle \big) = \int_0^1 f(x)\,dx \quad \text{a.s.} \tag{2.24}$$

As the intersection of countably many sets of full measure has itself full measure, choosing $f(x) = e^{2\pi \imath k x}$, $k \in \mathbb{Z}$ in (2.24) shows that, with probability one,

$$\lim_{n \to \infty} \frac{1}{n} \sum_{j=1}^n e^{2\pi \imath k (Y_1 + \cdots + Y_j)} = \int_0^1 e^{2\pi \imath k x}\,dx = 0 \quad \text{for all } k \in \mathbb{Z}, k \neq 0. \tag{2.25}$$

By Weyl's criterion [KuiNi, Thm.2.1], (2.25) is equivalent to

$$\mathbb{P}\left(\left(\sum_{j=1}^n Y_j \right) \text{ is u.d. mod } 1 \right) = 1.$$

In other words, $(\prod_{j=1}^n X_j)$ is Benford with probability one. $\qquad\square$

Example 2.6.6. (i) *Let (X_n) be an i.i.d. sequence with X_1 distributed according to $U(0,a)$, the uniform distribution on $[0,a]$ with $a > 0$. The kth Fourier coefficient of $P_{\langle \log X_1 \rangle}$ is*

$$\widehat{P_{\langle \log X_1 \rangle}}(k) = e^{2\pi \imath k \log a} \frac{\ln 10}{\ln 10 + 2\pi \imath k}, \quad k \in \mathbb{Z},$$

so that, for every $k \neq 0$,

$$\left| \widehat{P_{\langle \log X_1 \rangle}}(k) \right| = \frac{\ln 10}{\sqrt{(\ln 10)^2 + 4\pi^2 k^2}} < 1.$$

*As seen in the proof of Theorem 2.4.13(3), this implies that $(\prod_{j=1}^n X_j)$ converges in distribution to BL, a fact apparently first recorded in [AdhSa]. Note also that $\mathbb{E} \log X_1 = \log \frac{a}{e}$. Thus with probability one, $(\prod_{j=1}^n X_j)$ converges to 0 or $+\infty$, depending on whether $a < e$ or $a > e$. In fact, by the **Strong Law of Large Numbers** [ChT],*

$$\sqrt[n]{\prod_{j=1}^n X_j} \overset{a.s.}{\to} \frac{a}{e}$$

holds for every $a > 0$. If $a = e$ then

$$\mathbb{P}\left(\liminf_{n \to \infty} \prod_{j=1}^n X_j = 0 \text{ and } \limsup_{n \to \infty} \prod_{j=1}^n X_j = +\infty \right) = 1,$$

showing that in this case the product $\prod_{j=1}^{n} X_j$ does not converge but rather attains, with probability one, arbitrarily small as well as arbitrarily large positive values. By Theorem 2.6.5(2), the sequence $\left(\prod_{j=1}^{n} X_j\right)$ is a.s. Benford, regardless of the value of a.

(ii) *Consider an i.i.d. sequence (X_n) with X_1 distributed according to a lognormal distribution such that $\log X_1$ is standard normal. Denote by f_n the density of $\langle \log \prod_{j=1}^{n} X_j \rangle$. Since $\log \prod_{j=1}^{n} X_j = \sum_{j=1}^{n} \log X_j$ is normal with mean zero and variance n,*

$$f_n(s) = \frac{1}{\sqrt{2\pi n}} \sum_{k \in \mathbb{Z}} e^{-(k+s)^2/(2n)}, \quad 0 \le s < 1,$$

from which it is straightforward to deduce that

$$\lim_{n \to \infty} f_n(s) = 1, \quad \text{uniformly in } 0 \le s < 1.$$

Consequently, for all $t \in [1, 10)$,

$$\mathbb{P}\left(S\left(\prod_{j=1}^{n} X_j \right) \le t \right) = \mathbb{P}\left(\left\langle \log \prod_{j=1}^{n} X_j \right\rangle \le \log t \right)$$

$$= \int_0^{\log t} f_n(s)\, \mathrm{d}s \to \int_0^{\log t} 1\, \mathrm{d}s = \log t,$$

i.e., $\left(\prod_{j=1}^{n} X_j \right)$ converges in distribution to BL. By Theorem 2.6.5(2) also

$$\mathbb{P}\left(\left(\prod_{j=1}^{n} X_j \right) \text{ is Benford} \right) = 1,$$

even though $\mathbb{E} \log \prod_{j=1}^{n} X_j = \sum_{j=1}^{n} \mathbb{E} \log X_j = 0$, and hence, as in the previous example, the sequence $\left(\prod_{j=1}^{n} X_j \right)$ a.s. oscillates forever between 0 and $+\infty$.

Having seen Theorem 2.6.5, the reader may wonder whether there is an analogous result for *sums* of i.i.d. random variables. After all, the focus in classical probability theory is on sums much more than on products. Unfortunately, the statistical behavior of the significands is much more complex for sums than for products. The main basic reason is that the significand of the sum of two or more numbers depends not only on the significand of each number (as in the case of products), but also on their *exponents*. For example, observe that

$$S\left(3 \cdot 10^3 + 2 \cdot 10^2\right) = 3.2 \ne 5 = S\left(3 \cdot 10^2 + 2 \cdot 10^2\right),$$

while clearly

$$S\left(3 \cdot 10^3 \times 2 \cdot 10^2\right) = 6 = S\left(3 \cdot 10^2 \times 2 \cdot 10^2\right).$$

Practically, this difficulty is reflected in the fact that for positive real numbers u, v, the value of $\log(u + v)$, relevant for conformance with BL via Theorem 2.4.2, is not easily expressed in terms of $\log u$ and $\log v$, whereas $\log(uv) = \log u + \log v$.

In view of these difficulties, it is perhaps not surprising that the analog of Theorem 2.6.5 for sums arrives at a radically different conclusion.

Theorem 2.6.7. *Let* (X_n) *be an i.i.d. sequence of random variables with finite variance, that is,* $\mathbb{E}X_1^2 < +\infty$. *Then*

1. *not even a subsequence of* $\left(\sum_{j=1}^n X_j\right)$ *converges in distribution to BL;*

2. *with probability one,* $\left(\sum_{j=1}^n X_j\right)$ *is not Benford.*

Proof. See [BerH4, Thm.6.8]. □

Example 2.6.8. *Let* (X_n) *be an i.i.d. sequence with* $\mathbb{P}(X_1 = 0) = \mathbb{P}(X_1 = 1) = \frac{1}{2}$. *Then* $\mathbb{E}X_1 = \mathbb{E}X_1^2 = \frac{1}{2}$, *and by Theorem 2.6.7(1) neither* $\left(\sum_{j=1}^n X_j\right)$ *nor any of its subsequences converges in distribution to BL. Note that* $\sum_{j=1}^n X_j$ *is binomial with parameters* n *and* $\frac{1}{2}$, *i.e., for all* $n \in \mathbb{N}$,

$$\mathbb{P}\left(\sum_{j=1}^n X_j = l\right) = 2^{-n} \binom{n}{l}, \quad l = 0, 1, \ldots, n.$$

The law of the iterated logarithm [ChT] asserts that

$$\sum_{j=1}^n X_j = \frac{n}{2} + Y_n \sqrt{n \ln \ln n} \quad \text{for all } n \geq 3, \tag{2.26}$$

where the sequence (Y_n) *of random variables is bounded; in fact* $|Y_n| \leq 1$ *a.s. for all* n. *From (2.26) it is clear that, with probability one, the sequence* $\left(\sum_{j=1}^n X_j\right)$ *is not Benford.*

2.6.2 Mixtures of Distributions

The main goal of this section is to provide a statistical derivation of BL, in the form of a Central-Limit-like theorem that says that if random samples are taken from different distributions, and the results combined, then—provided the sampling is "unbiased" as to scale or base—the resulting combined samples will converge to the Benford distribution.

Denote by $\mathcal{M}$ the set of all probability measures on $(\mathbb{R}, \mathcal{B})$. Recall that a (real Borel) **random probability measure**, abbreviated henceforth as **r.p.m.**, is a function $P : \Omega \to \mathcal{M}$, defined on some underlying probability space $(\Omega, \mathcal{A}, \mathbb{P})$, such that for every $B \in \mathcal{B}$ the function $\omega \mapsto P(\omega)(B)$ is a random variable. Thus, for every $\omega \in \Omega$, $P(\omega)$ is a probability measure on $(\mathbb{R}, \mathcal{B})$, and, given any real numbers a, b and any Borel set B,

$$\{\omega : a \leq P(\omega)(B) \leq b\} \in \mathcal{A};$$

see e.g. [Ka] for an authoritative account on random probability measures.

Example 2.6.9. **(i)** *Let* P *be an r.p.m. that is,* $U(0, 1)$ *with probability* $\frac{1}{2}$, *and otherwise is* $\exp(1)$, *i.e., exponential with mean 1, hence* $\mathbb{P}(X > t) = \min(1, e^{-t})$ *for all* $t \in \mathbb{R}$, *see Example 2.3.8(i,ii). Thus, for every* $\omega \in \Omega$, *the probability measure* $P(\omega)$ *is either* $U(0, 1)$ *or* $\exp(1)$, *and* $\mathbb{P}(P(\omega) = U(0, 1)) = \mathbb{P}(P(\omega) = \exp(1)) = \frac{1}{2}$. *For a practical realization of* P *simply flip a fair coin—if it comes* up heads, $\mathbb{P}(\omega)$ is a $U(0, 1)$-distribution, and if it comes up tails, then $P(\omega)$ is an exp(1)-distribution.

(ii) *Let* X *be distributed according to* exp(1), *and let* P *be an r.p.m. where, for each* $\omega \in \Omega$, $P(\omega)$ *is the normal distribution with mean* $X(\omega)$ *and variance* 1. *In contrast to the example in (i), here* P *is continuous, i.e.,* $\mathbb{P}(P = Q) = 0$ *for each probability measure* $Q \in \mathcal{M}$.

The following example of an r.p.m. is a variant of a classical construction due to L. Dubins and D. Freedman which, as will be seen below, is an r.p.m. leading to BL.

Example 2.6.10. *Let* P *be the r.p.m. with support on* $[1, 10)$, *i.e.,* $P\big([1, 10)\big) = 1$ *with probability one, defined by its (random) cumulative distribution function* F_P, *i.e.,*

$$F_P(t) := F_{P(\omega)}(t) = P(\omega)\big([1, t]\big), \quad 1 \le t < 10,$$

as follows: Set $F_P(1) = 0$ *and* $F_P(10) = 1$. *Next pick* $F_P(10^{1/2})$ *according to the uniform distribution on* $[0, 1)$. *Then pick* $F_P(10^{1/4})$ *and* $F_P(10^{3,4})$ *independently, uniformly on* $\big[0, F_P(10^{1/2})\big)$ *and* $\big[F_P(10^{1/2}), 1\big)$, *respectively, and continue in this manner. This construction is known to generate an r.p.m. a.s. [DuFr, Lem.9.28], and as can easily be seen, is dense in the set of all probability measures on* $\big([1, 10), \mathcal{B}[1, 10)\big)$, *i.e., it generates probability measures that are arbitrarily close to any Borel probability measure on* $[1, 10)$.

The next definition formalizes the notion of combining data from different distributions. Essentially, it mimics what Benford did in combining baseball statistics with square-root tables and numbers taken from newspapers etc. This definition is key to everything that follows. It rests upon using an r.p.m. to generate a random sequence of probability distributions, and then successively selecting random samples from each of those distributions.

Definition 2.6.11. *Let* m *be a positive integer and* P *an r.p.m. A sequence of* P-*random* m-*samples is a sequence* (X_n) *of random variables on* $(\Omega, \mathcal{A}, \mathbb{P})$ *such that, for all* $j \in \mathbb{N}$ *and some i.i.d. sequence* (P_n) *of r.p.m.s with* $P_1 = P$, *the following two properties hold:*

Given that $P_j = Q$, *the random variables* $X_{(j-1)m+1}, X_{(j-1)m+2}, \ldots, X_{jm}$
are i.i.d. with distribution Q. $\hspace{4cm}$ (2.27)

The random variables $X_{(j-1)m+1}, X_{(j-1)m+2}, \ldots, X_{jm}$ *are independent of*
$P_i, X_{(i-1)m+1}, X_{(i-1)m+2}, \ldots, X_{im}$ *for every* $i \ne j$. $\hspace{2cm}$ (2.28)

Thus for any sequence (X_n) of P-random m-samples, for each $\omega \in \Omega$ in the underlying probability space, the first m random variables are a random sample (i.e., i.i.d.) from $P_1(\omega)$, a random probability distribution chosen according to the r.p.m. P; the second m-tuple of random variables is a random sample from $P_2(\omega)$ and so on. Note the two levels of randomness here: First a probability is selected at random, and then a random sample is drawn from this distribution, and this two-tiered process is continued.

Example 2.6.12. *Let P be the r.p.m. in Example 2.6.9(i), and let $m = 3$. Then a sequence of P-random 3-samples is a sequence (X_n) of random variables such that with probability $\frac{1}{2}$, X_1, X_2, X_3 are i.i.d. and distributed according to $U(0,1)$, and otherwise they are i.i.d. but distributed according to $\exp(1)$; the random variables X_4, X_5, X_6 are again equally likely to be i.i.d. $U(0,1)$ or $\exp(1)$, and they are independent of X_1, X_2, X_3, etc. Clearly the (X_n) are all identically distributed as they are all generated by exactly the same process. Note, however, that for instance X_1 and X_2 are dependent: Given that $X_1 > 1$, for example, the random variable X_2 is $\exp(1)$-distributed with probability one, whereas the unconditional probability that X_2 is $\exp(1)$-distributed is only $\frac{1}{2}$.*

Although sequences of P-random m-samples have a fairly simple structure, they do not fit into any of the familiar categories of sequences of random variables. For example, they are not in general independent, exchangeable, Markov, martingale, or stationary sequences. (See [Hi4]).

Recall that, given an r.p.m. P and any Borel set B, the quantity $P(B)$ is a random variable with values between 0 and 1. The following property of the expectation of $P(B)$, as a function of B, is easy to check.

Proposition 2.6.13. *Let P be an r.p.m. Then $\mathbb{E}P$, defined as*

$$(\mathbb{E}P)(B) := \mathbb{E}P(B) = \int_\Omega P(\omega)(B)\, d\mathbb{P}(\omega) \quad \text{for all } B \in \mathcal{B},$$

is a probability measure on $(\mathbb{R}, \mathcal{B})$.

Example 2.6.14. *Let P be the r.p.m. of Example 2.6.9(i). Then $\mathbb{E}P$ is the Borel probability measure with density*

$$f_{\mathbb{E}P}(t) = \left\{ \begin{array}{ll} 0 & \text{if } t < 0, \\ \frac{1}{2} + \frac{1}{2}e^{-t} & \text{if } 0 \leq t < 1, \\ \frac{1}{2}e^{-t} & \text{if } t \geq 1, \end{array} \right\} = \frac{1}{2}\mathbf{1}_{[0,1)}(t) + \frac{1}{2}e^{-t}\mathbf{1}_{[0,+\infty)}, \quad t \in \mathbb{R}.$$

The next lemma shows that the limiting proportion of times that a sequence of P-random m-samples falls in a (Borel) set B is, with probability one, the average $\mathbb{P}$-value of the set B, i.e., the limiting proportion equals $\mathbb{E}P(B)$. Note that this is not simply a direct corollary of the classical Strong Law of Large Numbers as the random variables in the sequence are not in general independent.

Lemma 2.6.15. *Let P be an r.p.m., and let (X_n) be a sequence of P-random m-samples for some $m \in \mathbb{N}$. Then, for every $B \in \mathcal{B}$,*

$$\frac{\#\{1 \leq n \leq N : X_n \in B\}}{N} \overset{a.s.}{\to} \mathbb{E}P(B) \quad \text{as } N \to \infty.$$

Proof. Fix $B \in \mathcal{B}$ and $j \in \mathbb{N}$, and let $Y_j = \#\{1 \leq i \leq m : X_{(j-1)m+i} \in B\}$. It is clear that

$$\lim_{N \to \infty} \frac{\#\{1 \leq n \leq N : X_n \in B\}}{N} = \frac{1}{m} \lim_{n \to \infty} \frac{1}{n} \sum_{j=1}^{n} Y_j, \qquad (2.29)$$

whenever the limit on the right exists. By (2.27), given P_j, the random variable Y_j is binomially distributed with parameters m and $\mathbb{E}(P_j(B))$, hence a.s.

$$\mathbb{E}Y_j = \mathbb{E}(\mathbb{E}(Y_j|P_j)) = \mathbb{E}(mP_j(B)) = m\mathbb{E}P(B) \qquad (2.30)$$

since P_j has the same distribution as P. By (2.28), the Y_j are independent. They are also uniformly bounded, as $0 \leq Y_j \leq m$ for all j, and hence $\sum_{j=1}^{\infty} \mathbb{E}Y_j^2/j^2 < +\infty$. Moreover, by (2.30) all Y_j have the same mean value $m\mathbb{E}P(B)$. Thus by [ChT, Cor.5.1]

$$\frac{1}{n} \sum_{j=1}^{n} Y_j \overset{a.s.}{\to} m\mathbb{E}P(B) \quad \text{as } n \to \infty, \qquad (2.31)$$

and the conclusion follows by (2.29) and (2.31). $\square$

The stage is now set to give a statistical limit law (Theorem 2.6.18 below) that is, a Central-Limit-like theorem for significant digits mentioned above. Roughly speaking, this law says that if probability distributions are selected at random, and random samples are then taken from each of these distributions in such a way that the overall process is scale or base neutral, then the significant digit frequencies of the combined sample will converge to the logarithmic distribution. This theorem may help explain and predict the appearance of BL in significant digits in mixtures of tabulated data such as the combined data from Benford's individual data sets, and also his individual data set of numbers gleaned from newspapers.

Definition 2.6.16. *An r.p.m. P has* scale-unbiased (decimal) significant digits *if, for every significand event A, i.e., for every $A \in \mathcal{S}$, the expected value of $P(A)$ is the same as the expected value $P(\alpha A)$ for every $\alpha > 0$, that is, if*

$$\mathbb{E}(P(\alpha A)) - \mathbb{E}(P(A)) \quad \text{for all } \alpha > 0, A \in \mathcal{S}.$$

Equivalently, the Borel probability measure $\mathbb{E}P$ has scale-invariant significant digits.

Similarly, P has base-unbiased significant (decimal) digits *if, for every $A \in \mathcal{S}$ the expected value of $P(A)$ is the same as the expected value of $P(A^{1/n})$ for every $n \in \mathbb{N}$, that is, if*

$$\mathbb{E}(P(A^{1/n})) = \mathbb{E}(P(A)) \quad \text{for all } n \in \mathbb{N}, A \in \mathcal{S},$$

i.e., if $\mathbb{E}P$ has base-invariant significant digits.

An immediate consequence of Theorems 2.4.18 and 2.4.24 is

Proposition 2.6.17. *Let P be an r.p.m. with $\mathbb{E}P(\{0\}) = 0$. Then the following statements are equivalent:*

   1. P has scale-unbiased significant digits.

2. $P(\{\pm 10^k : k \in \mathbb{Z}\}) = 0$, or equivalently $S_* P(\{1\}) = 0$, holds with probability one, and P has base-unbiased significant digits.

3. $\mathbb{E}P(A) = \mathbb{B}(A)$ for all $A \in \mathcal{S}$, i.e., $\mathbb{E}P$ is Benford.

As will be seen in the next theorem, scale- or base-unbiasedness of an r.p.m. implies that a sequence of P-random samples are Benford a.s. A crucial point in the definition of an r.p.m. P with scale- or base-unbiased significant digits is that it does not require individual realizations of P to have scale- or base-invariant significant digits. In fact, it is often the case (see Benford's original data in [Ben] and Example 2.6.20 below) that a.s. *none* of the random probabilities has either of these properties, and it is only on average that the sampling process does not favor one scale or base over another. Recall from the notation introduced above that $S_* P(\{1\}) = 0$ is the event $\{\omega \in \Omega : P(\omega)(S = 1) = 0\}$.

Theorem 2.6.18 ([Hi4]). *Let P be an r.p.m. Assume that P either has scale-unbiased significant digits, or else has base-unbiased significant digits and $S_* P(\{1\})$ $= 0$ with probability one. Then, for every $m \in \mathbb{N}$, every sequence (X_n) of P-random m-samples is Benford with probability one, that is, for all $t \in [1, 10)$,*

$$\frac{\#\{1 \le n \le N : S(X_n) < t\}}{N} \overset{a.s.}{\to} \log t \quad \text{as } N \to \infty.$$

Proof. Assume first that P has scale-unbiased significant digits, i.e., the probability measure $\mathbb{E}P$ has scale-invariant significant digits. According to Theorem 2.4.18, $\mathbb{E}P$ is Benford. Consequently, Lemma 2.6.15 implies that for every sequence (X_n) of P-random m-samples and every $t \in [1, 10)$,

$$\frac{\#\{1 \le n \le N : S(X_n) < t\}}{N}$$
$$= \frac{\#\left\{1 \le n \le N : X_n \in \bigcup_{k \in \mathbb{Z}} 10^k ((-t, -1] \cup [1, t))\right\}}{N}$$
$$\overset{a.s.}{\to} \mathbb{E}P\left(\bigcup_{k \in \mathbb{Z}} 10^k ((-t, -1] \cup [1, t))\right) = \log t$$

as $N \to \infty$. Assume in turn that $S_* P(\{1\}) = 0$ with probability one, and that P has base-unbiased significant digits. Then

$$S_* \mathbb{E}P(\{1\}) = \mathbb{E}P(S^{-1}(\{1\})) = \int_\Omega S_* P(\omega)(\{1\}) \, d\mathbb{P}(\omega) = 0.$$

Hence $q = 0$ holds in (2.14) with P replaced by $\mathbb{E}P$, proving that $\mathbb{E}P$ is Benford, and the remaining argument is the same as before. □

Corollary 2.6.19. *If an r.p.m. P has scale-unbiased significant digits, then for every $m \in \mathbb{N}$, every sequence (X_n) of P-random m-samples, and every $d \in \{1, 2, \ldots, 9\}$,*

$$\frac{\#\{1 \le n \le N : D_1(X_n) = d\}}{N} \overset{a.s.}{\to} \log(1 + d^{-1}) \quad \text{as } N \to \infty.$$

Justification of the hypothesis of scale- or base-unbiasedness of significant digits in practice is akin to justification of the hypothesis of independence (and identical distribution) when applying the Strong Law of Large Numbers or the Central Limit Theorem to real-life processes: Neither hypothesis can be formally proved, yet in many real-life sampling procedures, they appear to be reasonable assumptions.

Many of the standard constructions of r.p.m. automatically have scale- and base-unbiased significant digits, and thus satisfy BL in the sense of Theorem 2.6.18.

Example 2.6.20. *Recall the classical Dubins–Freedman construction of an r.p.m. P described in Example 2.6.10. It follows from [DuFr, Lem.9.28] that $\mathbb{E}P$ is Benford. Hence P has scale- and base-unbiased significant digits. Note, however, that with probability one P will* not *have scale- or base-invariant significant digits. It is* only *on average that these properties hold but, as demonstrated by Theorem 2.6.18, this is enough to guarantee that random sampling from P will generate Benford sequences a.s.*

2.6.3 Random Maps

The purpose of this brief concluding section is to illustrate one basic theorem that combines the deterministic aspects of BL studied in Section 2.5 with the stochastic considerations of the present section. Specifically, it is shown how applying randomly selected maps successively may generate Benford sequences with probability one. Random maps constitute a wide and intensely studied field, and for stronger results than the one discussed here the interested reader is referred e.g. to [Ber3].

For a simple example, first consider the map $T : \mathbb{R} \to \mathbb{R}$ with $T(x) = \sqrt{|x|}$. Since $T^n(x) = |x|^{2^{-n}} \to 1$ as $n \to \infty$ whenever $x \neq 0$, the orbit $O_T(x_0)$ is not Benford for any x_0. More generally, consider the randomized map

$$T(x) = \begin{cases} \sqrt{|x|} & \text{with probability } p\,, \\ x^3 & \text{with probability } 1 - p\,, \end{cases} \tag{2.32}$$

and assume that, at each step, the iteration of T is independent of the entire past process. If $p = 1$, this is simply the map studied before, and hence for every $x_0 \in \mathbb{R}$, the orbit $O_T(x_0)$ is not Benford. On the other hand, if $p = 0$ then Theorem 2.5.8 implies that, for almost every $x_0 \in \mathbb{R}$, $O_T(x_0)$ is Benford. It is plausible to expect that the latter situation persists for small $p > 0$. As the following theorem shows, this is indeed that case even when the non-Benford map $\sqrt{|x|}$ occurs more than half of the time: If

$$p < \frac{\log 3}{\log 2 + \log 3} = 0.6131\ldots\,, \tag{2.33}$$

then, for a.e. $x_0 \in \mathbb{R}$, the (random) orbit $O_T(x_0)$ is Benford with probability one. To concisely formulate the result from which this follows, recall that for any (deterministic or random) sequence (T_n) of maps mapping $\mathbb{R}$ or parts thereof into itself, the orbit $O_T(x_0)$ of $x_0 \in \mathbb{R}$ simply denotes the sequence $\left(T_{n-1} \circ \cdots \circ T_1(x_0)\right)_{n \in \mathbb{N}}$.

Theorem 2.6.21 ([Ber3]). *Let (β_n) be an i.i.d. sequence of positive random variables, and assume that $\log \beta_1$ has finite variance, i.e., $\mathbb{E}(\log \beta_1)^2 < +\infty$. For the sequence (T_n) of random maps given by $T_n : x \mapsto x^{\beta_n}$ and a.e. $x_0 \in \mathbb{R}$, the orbit $O_T(x_0)$ is Benford with probability one or zero, depending on whether $\mathbb{E} \log \beta_1 > 0$ or $\mathbb{E} \log \beta_1 \leq 0$.*

Proof. See [Ber3]. □

Statements in the spirit of Theorem 2.6.21 are true also for more general random maps, not just monomials [Ber3].

Chapter Three

Fourier Analysis and Benford's Law

Steven J. Miller[1]

This chapter continues the development of the theory of Benford's Law. We use Fourier analysis (in particular, Poisson Summation) to prove many systems either satisfy or almost satisfy the Fundamental Equivalence, and hence either obey Benford's Law, or are well approximated by it. Examples range from geometric Brownian motions to random matrix theory to products and chains of random variables to special distributions (this latter topic is explored further in Part II). We develop the notion of a Benford-good system, which describes many of our systems. Unfortunately one of the conditions concerns the cancelation in sums of translated errors related to the cumulative distribution function, and proving the required cancelation often requires techniques specific to the system of interest; for example, the $3x + 1$ problem is related to the irrationality exponent of $\log_{10} 2$.

3.1 INTRODUCTION

Chapter 2 introduced a rigorous definition of Benford's Law, and discussed approaches to proving Benford behavior. The purpose of this chapter is to expand on a theme briefly touched there, describing applications of Fourier analysis in general, and Poisson Summation in particular, to Benford problems.

Fourier analysis is concerned with expanding periodic functions as a sum of complex exponentials. For a "nice" function $f : [0, 1] \to \mathbb{C}$ (f twice continuously differentiable suffices), we have the **Fourier series** expansion

$$f(x) = \sum_{n=-\infty}^{\infty} \widehat{f}(n)e^{2\pi inx}, \tag{3.1}$$

where

$$\widehat{f}(n) = \int_0^1 f(x)e^{-2\pi inx}dx \tag{3.2}$$

(and of course $i = \sqrt{-1}$, $e^{i\theta} = \cos\theta + i\sin\theta$). The $\widehat{f}(n)$'s are the **Fourier coefficients** of f, and inherit decay properties from the smoothness of f (if f is

---

[1]Department of Mathematics and Statistics, Williams College, Williamstown, MA 01267. The author was partially supported by NSF grants DMS0970067 and DMS1265673.

k times continuously differentiable, then $\widehat{f}(n) \ll 1/n^k$). For more on Fourier analysis, see [MiT-B, StSh1].

The reason Fourier analysis is so beautifully suited to Benford investigations is due to the **Uniform Distribution Characterization** (Theorem 2.4.2). There we see that a random variable X (or a sequence of data) is Benford base B if and only if $\log_B X$ is equidistributed modulo 1 (see [BrDu, Wash] for applications of these ideas to proving the Benfordness of Fibonacci numbers). This leads us to study $\log_B X \bmod 1$, the fractional part of $\log_B X$. While typically the modulo 1 function interacts poorly with compositions of functions, this is not the case with the exponential functions of Fourier analysis. There we have $e^{2\pi i n x} = e^{2\pi i n (x \bmod 1)}$. The upshot of this is that we may drop the modulo 1 function if it occurs as the argument of an exponential function, and this is why Fourier analysis is ideally suited to attack these problems. In particular, many convergence theorems, as well as Poisson Summation, are now available to aid us in analyzing Benford behavior.

A major theme of this chapter is the advantage of different perspectives in simplifying the computations. Sometimes it is more convenient to work with the density of X than that of $\log_B X$; in this case, instead of the Fourier transform one uses the **Mellin transform**. These two transforms are the same after a logarithmic change of variables, which of course exactly mirrors the passage from X to $\log_B X$. Finally, instead of a Fourier series we might have the **Fourier transform**, given by

$$\widehat{f}(y) = \int_{-\infty}^{\infty} f(x) e^{-2\pi i x y} dx. \tag{3.3}$$

After converting a given Benford problem to one in Fourier analysis, frequently Poisson Summation is useful in showing Benford behavior. The following is not the most general statement of the theorem, but suffices for most applications. We first set some notation. We say a function $f(x)$ decays like x^{-a} if there are constants x_0 and C such that for all $|x| > x_0$, $|f(x)| \leq C/|x|^a$.

Theorem 3.1.1 (**Poisson Summation**). *Assume f is twice continuously differentiable and that f, f' and f'' decay like $|x|^{-(1+\eta)}$ for some $\eta > 0$. Then*

$$\sum_{n=-\infty}^{\infty} f(n) = \sum_{n=-\infty}^{\infty} \hat{f}(n). \tag{3.4}$$

The power of Poisson Summation is that it converts one infinite sum with *slow* decay to another sum with *rapid* decay; because of this, Poisson Summation is an extremely useful technique for a variety of problems.

This chapter is based on the papers [CuLeMi, JaKKKM, KonMi, MiNi1, MiNi2], and is meant to give a flavor of how Fourier analysis can be successfully applied to a diverse set of systems. The main idea is the following. Frequently we have some process whose logarithm is a Gaussian, and the variance of the Gaussian is growing as some parameter grows. For example, perhaps we have a product of independent random variables, and the parameter is the number of random variables. Using Poisson Summation, one can show that as the variance tends to infinity, the normal distribution modulo 1 converges to the uniform distribution on $[0, 1]$. This is plausible, as the density changes very little on scales of size 1 when the variance is enormous, and thus locally it looks constant. As the logarithm modulo 1 is converging to the uniform distribution, by the Uniform Distribution Characterization the original process converges to Benford's Law.

We first prove that a **geometric Brownian motion** (roughly, this means the logarithm is normally distributed with growing variance; see the next section for a complete definition) is Benford. While some important systems do exhibit such behavior, frequently only the main term is given by a Brownian motion. This requires us to deal with the complications that arise from the error term. Unfortunately, while we can isolate the key steps and general features a system must have to be solvable by these methods, the actual analysis of these error terms is often specific to the problem, and thus different techniques are needed depending on the structure of the problem. We briefly describe methods to overcome such issues, and refer the reader to the literature for the details. We next discuss some interesting examples from random matrix theory, L-functions and the $3x + 1$ problem. We then attack products and chains of random variables (which are often equivalent after a change of variables), finding conditions which yield convergence to Benford behavior as the number of variables grows (with explicit error estimates), as well as when these products or chains are never Benford. We end with an analysis of Weibull random variables. This is a three-parameter distribution which is very useful in survival analysis (the exponential distribution is a special case) and order statistics, and we quantify how close it is to Benford as a function of its parameters.

3.2 BENFORD-GOOD PROCESSES

In this section we first formally define a geometric Brownian motion, and show that it converges to Benford behavior. We then generalize to what we call a Benford-Good Process, where typically the main term of the behavior is a geometric Brownian motion. We show how to prove many Benford-Good Processes, as the name suggests, do indeed converge to Benford behavior. We then discuss some systems that exhibit this behavior, primarily from number theory and random matrix theory.

3.2.1 Geometric Brownian Motion

A **Brownian motion** (or a **Wiener process**) is a continuous process with independent, normally distributed increments. If W is a Brownian motion, then $W_t - W_s$ is a random variable having the Gaussian distribution with mean zero and variance $t - s$, and is independent of the random variable $W_s - W_u$ provided $u < s < t$. A standard realization of Brownian motion is as the scaled limit of a random walk.

Let $X_1, X_2, X_3, \ldots$ be independent Bernoulli trials (taking the values $+1$ and -1 with equal probability) and let $S_n = \sum_{i=1}^n X_i$ denote the partial sum. Then the normalized process $W_t^{(n)} = S_{nt}/\sqrt{n}$ (extended to a continuous process by linear interpolation) converges, as $n \to \infty$, to the Wiener process. See [Bi2] or Chapter 2.4 of [KaSh] for further details. The continuous, scaled limit is significantly easier to investigate, as there are neither error terms nor discreteness issues, and thus we study this important example first.

A **geometric Brownian motion** is a process X such that the process $Y = \log X$ is a Brownian motion. Often the stock market is modeled by geometric Brownian motion (though the work of Mandelbrot and others suggests this model is incorrect and a fractal model would do better; see for example [ManHu]), and there is an extensive literature on stock prices and Benford's Law. As the Uniform Distribution Characterization (Theorem 2.4.2) implies that a data set is Benford base B if and only if its logarithms base B are equidistributed modulo 1, to show a geometric Brownian motion X is Benford it suffices to show that $Y = \log X$ is equidistributed modulo 1.

We are thus reduced to studying the random variables Y_N whose densities are

$$f_{Y_N}(y) \ = \ \frac{1}{\sqrt{2\pi\sigma_N^2}} \exp\left(-(y - \mu_N)^2/2\sigma_N^2\right), \tag{3.5}$$

where μ_N is the mean at time N and σ_N^2 is the variance at time N. To simplify the resulting algebra, we assume that we are working base e, that the mean is zero (there is no drift), and that the variance at time N is $N/2\pi$. We must show that, as $N \to \infty$, $Y_N = \log X_N$ becomes equidistributed modulo 1. This is equivalent to showing, for any $[a, b] \subset [0, 1]$, that

$$\lim_{N\to\infty} \text{Prob}(Y \bmod 1 \in [a, b]) \ = \ b - a, \tag{3.6}$$

or equivalently that

$$\lim_{N\to\infty} \sum_{n=-\infty}^{\infty} \int_a^b \frac{1}{N} \exp\left(-(y + n)^2/N^2\right) \ = \ b - a. \tag{3.7}$$

The first step is to determine which n contribute significantly to the probability above. We first show that we may restrict our sum to n at most a little more than the standard deviation by showing the contribution from large $|n|$ is negligible.

Lemma 3.2.1. *We have*

$$\frac{2}{\sqrt{2\pi\sigma^2}} \int_{\sigma^{1+\delta}}^{\infty} e^{-x^2/2\sigma^2}\, dx \ \ll \ e^{-\sigma^{2\delta}/2}. \tag{3.8}$$

Proof. Change the variable of integration to $w = \frac{x}{\sigma\sqrt{2}}$. Denoting the above integral by I, we find

$$I \ = \ \frac{2}{\sqrt{2\pi\sigma^2}} \int_{\sigma^\delta/\sqrt{2}}^{\infty} e^{-w^2} \cdot \sigma\sqrt{2}\, dw \ = \ \frac{2}{\sqrt{\pi}} \int_{\sigma^\delta/\sqrt{2}}^{\infty} e^{-w^2}\, dw. \tag{3.9}$$

The integrand is monotonically decreasing. For $w \in \left[\frac{\sigma^\delta}{\sqrt{2}}, \frac{\sigma^\delta}{\sqrt{2}} + 1\right]$, the integrand is bounded by substituting in the left endpoint, and the region of integration is of length 1. Thus,

$$I \ < \ 1 \cdot \frac{2}{\sqrt{\pi}} e^{-\sigma^{2\delta}/2} + \frac{2}{\sqrt{\pi}} \int_{\frac{\sigma^\delta}{\sqrt{2}}+1}^{\infty} e^{-w^2}\, dw \ < \ 4e^{-\sigma^{2\delta}/2}. \tag{3.10}$$

$\square$

Using the above lemma, we now prove a geometric Brownian motion converges to Benford behavior (see Corollary 5.4.7 for an alternate proof).

Theorem 3.2.2. *As $N \to \infty$, $f_{Y_N}(y) = \frac{e^{-\pi y^2/N}}{\sqrt{N}}$ becomes equidistributed modulo 1, implying that $X_N = \exp(Y_N)$ converges to Benford behavior base e.*

Proof. For any $[a, b] \subset [0, 1]$, we must show that as $N \to \infty$ we have the probability of Y_N mod 1 in $[a, b]$ tends to $b - a$; i.e., we must show

$$\int_{\substack{y=-\infty \\ y \bmod 1 \in [a,b]}}^{\infty} p_N(y) dy = \frac{1}{\sqrt{N}} \sum_{n \in \mathbb{Z}} \int_{y=a}^{b} e^{-\pi(y+n)^2/N} dy \qquad (3.11)$$

converges to $b - a$ as $N \to \infty$. For $y \in [a, b]$, the Taylor series expansion of the Gaussian density gives

$$e^{-\pi(y+n)^2/N} = e^{-\pi n^2/N} + O\left(\frac{\max(1, |n|)}{N} e^{-n^2/N}\right). \qquad (3.12)$$

It's sufficient to restrict the summation in (3.11) to $|n| \le N^{5/4}$. The proof is immediate from Lemma 3.2.1: we increase the integration by expanding to $y \in [0, 1]$, and then trivially estimate. Thus, up to negligible terms, all the contribution is from $|n| \le N^{5/4}$.

We now use the Poisson Summation (Theorem 3.1.1), which in this case yields

$$\frac{1}{\sqrt{N}} \sum_{n=-\infty}^{\infty} e^{-\pi n^2/N} = \sum_{n=-\infty}^{\infty} e^{-\pi n^2 N} \qquad (3.13)$$

(we chose our normalization to make the Fourier transform of the Gaussian factor nice). The exponential terms on the left of (3.13) are all of size 1 for $n \le \sqrt{N}$, and are not small until $n \gg \sqrt{N}$ (for instance, once $n > \sqrt{N} \log N$, the exponential terms are small for large N); however, almost all of the contribution on the right comes from $n = 0$. The power of Poisson Summation is it often allows us to approximate well long sums with short sums. We therefore have

$$\frac{1}{\sqrt{N}} \sum_{|n| \le N^{5/4}} \int_{y=a}^{b} e^{-\pi(y+n)^2/N} dy$$

$$= \frac{1}{\sqrt{N}} \sum_{|n| \le N^{5/4}} \int_{y=a}^{b} \left[e^{-\pi n^2/N} + O\left(\frac{\max(1, |n|)}{N} e^{-n^2/N}\right) \right] dy$$

$$= \frac{b-a}{\sqrt{N}} \sum_{|n| \le N^{5/4}} e^{-\pi n^2/N} + O\left(\frac{1}{N} \sum_{n=0}^{N^{5/4}} \frac{n+1}{\sqrt{N}} e^{-\pi(n/\sqrt{N})^2}\right)$$

$$= \frac{b-a}{\sqrt{N}} \sum_{|n| \le N^{5/4}} e^{-\pi n^2/N} + O\left(\frac{1}{N} \int_{w=0}^{N^{3/4}} (w+1) e^{-\pi w^2} \sqrt{N} dw\right)$$

$$= \frac{b-a}{\sqrt{N}} \sum_{|n| \le N^{5/4}} e^{-\pi n^2/N} + O\left(N^{-1/2}\right). \qquad (3.14)$$

By Lemma 3.2.1 we can extend all sums to $n \in \mathbb{Z}$ in (3.14) with negligible error. We now apply Poisson Summation and find that up to lower-order terms,

$$\frac{1}{\sqrt{N}} \sum_{n \in \mathbb{Z}} \int_{y=a}^{b} e^{-\pi(y+n)^2/N} dy \approx (b-a) \cdot \sum_{n \in \mathbb{Z}} e^{-\pi n^2 N}. \qquad (3.15)$$

For $n = 0$ the right-hand side of (3.15) is $b - a$. For all other n, we trivially estimate the sum:

$$\sum_{n \neq 0} e^{-\pi n^2 N} \leq 2 \sum_{n \geq 1} e^{-\pi n N} \leq \frac{2e^{-\pi N}}{1 - e^{-\pi N}}, \tag{3.16}$$

which is less than $4e^{-\pi N}$ for N sufficiently large. $\square$

3.2.2 Benford-Good Processes

We generalize our analysis from geometric Brownian motions to handle error terms (i.e., processes that are not purely Brownian motions). We closely follow the arguments in Kontorovich–Miller [KonMi]. We develop a general framework that is applicable for a variety of problems, and quickly summarize three interesting applications: the distribution of values of L-functions, characteristic polynomials of random matrices, and iterates of the $3x + 1$ map. These three problems live in three very different fields, and illustrate the power of these methods.

We first set the notation. At time T the values of our system are described by the random variable X_T with log-process $Y_{T,B} = \log_B X_T$ with density $f_{Y_{T,B}}$; we are interested in the behavior as $T \to \infty$. As always, by the Uniform Distribution Characterization (Theorem 2.4.2) proving X_T converges to Benford behavior base B is equivalent to showing that as $T \to \infty$, $Y_{T,B}$ becomes equidistributed modulo 1.

Throughout we let $f(x)$ be a fixed probability density with cumulative distribution function $F(x) = \int_{-\infty}^{x} f(t)\, dt$. The specific form of f is often unimportant (and, in fact, is frequently the Gaussian density). In the examples below the probability densities of $Y_{T,B}$ are approximately a spread version of f such as $f_T(x) = \frac{1}{T} f\left(\frac{x}{T}\right)$. The difficulty is in dealing with the error terms, not the specific form of f. Because of the error term, the log-process $Y_{T,B}$ has a cumulative distribution function given by

$$\begin{aligned}
F_T(y) &= \mathbb{P}\{Y_{T,B} \leq y\} \\
&= \int_{-\infty}^{y} \frac{1}{T} f\left(\frac{t}{T}\right) dt + E_T(y) \\
&= F\left(\frac{y}{T}\right) + E_T(y),
\end{aligned} \tag{3.17}$$

where E_T is an error term. For many systems the error term is negligible and $f_T(x)$ spreads to make $Y_{T,B}$ equidistributed modulo 1 as $T \to \infty$, implying that X is Benford. Analyzing the error term is the most difficult part of the proofs, and frequently requires techniques specific to the problem of interest. The following conditions are fairly weak and frequently met.

Definition 3.2.3 (Benford-good). *We say $Y_{T,B}$ with cumulative distribution functions F_T is **Benford-good** if F_T satisfies (3.17), the probability density f satisfies sufficient conditions for Poisson Summation and there is a monotone increasing function $h(T)$ with $\lim_{T \to \infty} h(T) = \infty$ such that f and E_T satisfy the following conditions:*

Condition 1. *Small tails:*

$$F_T(\infty) - F_T(Th(T)) = o(1), \quad F_T(-Th(T)) - F_T(-\infty) = o(1). \quad (3.18)$$

Condition 2. *Rapid decay of the characteristic function:*

$$S(T) = \sum_{\substack{k \in \mathbb{Z} \\ k \neq 0}} \left| \frac{\widehat{f}(Tk)}{k} \right| = o(1). \quad (3.19)$$

Condition 3. *Small truncated translated error: for all* $0 \leq a < b \leq 1$,

$$\mathcal{E}_T(a, b) = \sum_{|k| \leq Th(T)} [E_T(b + k) - E_T(a + k)] = o(1). \quad (3.20)$$

The Poisson Summation Formula holds if f is a Gaussian density or it and its first two derivatives decay sufficiently fast. We briefly comment on the conditions.

Condition 1 asserts that essentially all of the mass lies in $[-Th(T), Th(T)]$. In applications T is the standard deviation, and this follows from Central-Limit-type convergence.

Condition 2 is quite weak and is satisfied in all cases of interest. For example, if f is differentiable and f' is integrable (as is the case if f is the Gaussian density), then $|\widehat{f}(y)| \leq \frac{1}{|y|} \int |f'(x)| dx = O\left(\frac{1}{|y|}\right)$, which suffices to show $S(T) = o(1)$.

Condition 3 is the most difficult to prove for a system. Unfortunately there are some systems, such as Bernoulli trials, where the best attainable estimate is $E_T(x) = O\left(\frac{1}{T}\right)$. Errors this large lead to $\mathcal{E}_T(a, b) = O(1)$ (see [Fel] for details). In our investigation of the $3x + 1$ problem we'll see that dealing with this issue for that system requires us to use a quantified version of the Kronecker–Weyl equidistribution theorem, which depends on how well a relevant irrational number can be approximated by rationals.

The main result is

Theorem 3.2.4 (Kontorovich–Miller [KonMi]). *Let* X_T *have corresponding log-process* $Y_{T,B}$, *and assume* $Y_{T,B}$ *is Benford-good. Then* $Y_{T,B} \to Y_B$, *where* Y_B *is equidistributed modulo 1, and thus* X *is Benford base* B.

Sketch of the proof: The argument is similar to that used in proving geometric Brownian motions tend to Benford behavior. There are two differences: for general f we need to use Condition 2 to show that the sum of the Fourier transform at n with $0 < |n| \leq Th(T)$ tends to zero, and Condition 3 to handle the sums of the errors in replacing $F_T(y)$ with $F(y/T)$. See [KonMi] for details. $\quad \square$

3.2.3 Applications

We now give a few examples of Benford-good systems. The challenge in each analysis is proving Condition 3 holds.

3.2.3.1 Random Matrix Theory

Random matrix theory (RMT) began in the early 1900s in the study of the statistics of population characteristics [Wis]. The field developed rapidly in the 1950s when it was found to describe the spacing distributions of adjacent resonances (of the same spin and parity) observed in the interaction of low energy neutrons with nuclei [Wig1], and it flourished in the 1970s following a chance encounter between Hugh Montgomery and Freeman Dyson [Mon] (when they saw it also predicted answers to many of the most difficult problems in number theory). Since then numerous problems have been shown to be well modeled by random matrix theory, covering everything from mathematical physics to bus routes in Cuernevaca, Mexico [BaBoDS]. As random matrix theory describes so many phenomena, it is worth investigating whether or not it, and the phenomena it models, satisfy Benford's Law.

We briefly describe the basics of random matrix theory; the discussion below is paraphrased from [FiMil], where complete details are given. Let H be the Hamiltonian of some system, with energy eigenstates ψ_N with corresponding energy levels E_N: $H\psi_N = E_N\psi_N$. In practice it is impossible to compute the E_N's and ψ_N's, as H is an infinite-dimensional matrix with extremely complicated rules governing its entries. Wigner's great insight was that this enormous complexity is similar to what happens in statistical mechanics, and actually helps us. The interactions are so complex we might as well regard each entry as some randomly chosen number. Thus instead of considering the true Hamiltonian H for the system, we consider $N \times N$ real symmetric matrices with entries independently chosen from nice probability distributions. We compute whatever statistics we are interested in for these matrices, average over all matrices, and then take the $N \to \infty$ scaling limit. The main result is that the behavior of the eigenvalues of an arbitrary matrix is often well approximated by the behavior obtained by averaging over all matrices, and this is a good model for many systems.

While the eigenvalues of these ensembles have been found to describe many systems, the values of the characteristic polynomials are also of use and interest. For example, if U is a unitary (which implies its eigenvalues are of the form $e^{i\theta_n}$ for θ_n real) then

$$Z(U,\theta) := \det(I - Ue^{-i\theta}) = \prod_{n=1}^{N}\left(1 - e^{i(\theta_n - \theta)}\right) \qquad (3.21)$$

is the **characteristic polynomial** of U. The values of characteristic polynomials have been shown to model well the values of L-functions (described later in this section).

The following result connects Benford's Law and random matrix theory (see [KonMi] for the proof); as random matrix theory describes numerous natural phenomena, this is another indication that Benford behavior should be prevalent.

Theorem 3.2.5 (Kontorovich–Miller [KonMi]). *As $N \to \infty$, the distribution of digits of the absolute values of the characteristic polynomials of $N \times N$ unitary matrices (with respect to Haar measure) converges to the Benford probabilities.*

3.2.3.2 L-Functions

We now turn to number theory. One of the central objects of study is the **Riemann zeta function** $\zeta(s)$:

$$\zeta(s) = \sum_{n=1}^{\infty} \frac{1}{n^s} = \prod_{p \text{ prime}} \left(1 - \frac{1}{p^s}\right)^{-1}; \qquad (3.22)$$

both the sum and the product converge for $\Re(s) > 1$, and the equivalence of the two follows from the **unique factorization** property of the integers. This means every integer n can be written uniquely as a product of prime powers, where the primes are the integers $\{2, 3, 5, 7, 11, \dots\}$ which are divisible only by themselves and 1. The primes are the building blocks of the integers, and not surprisingly number theorists want to know as much about them as possible. While there is much we don't know about the primes, the integers are quite well understood, and we can use the Riemann zeta function to pass from knowledge of the integers to knowledge of the primes through the product–sum equivalence. For example, as $s \to 1$ from above we have $\zeta(s)$ approaches the harmonic sum $\sum_n 1/n$, which diverges. Thus there must be infinitely many primes, as otherwise the product would have a finite limit as $s \to 1$ from above.

This is but one of many results about primes which can be deduced by studying $\zeta(s)$. It turns out $\zeta(s)$ can be meromorphically continued to the entire complex plane, yielding a new function $\xi(s) = \pi^{-s/2}\Gamma(s/2)\zeta(s)$ which satisfies the functional equation $\xi(s) = \xi(1 - s)$. As the functional equation relates values at s to values at $1 - s$, it suffices to study $\zeta(s)$ for $\Re(s) \geq 1/2$. One can show that $\zeta(s) \neq 0$ for $\Re(s) > 1$. The fact that $\zeta(s) \neq 0$ for $\Re(s) = 1$ is equivalent to the **Prime Number Theorem**, which states the number of primes at most x is $x/\log x + o(x/\log x)$. Finer questions about the primes turn out to involve the distribution of zeros of $\zeta(s)$. The celebrated **Riemann Hypothesis** asserts these zeros either have real part equal to $1/2$ or are a negative even integer.

The Riemann zeta function can be generalized. One typically studies L-functions

$$L(s, f) = \sum_{n=1}^{\infty} \frac{a_n(f)}{n^s} = \prod_{p \text{ prime}} L_p\left(p^{-s}, f\right)^{-1}, \qquad (3.23)$$

where $L_p\left(p^{-s}, f\right)$ is polynomial. For a general choice of coefficients $a_n(f)$ we will of course not have a product expansion; however, this product expansion (called the **Euler product**) holds for many arithmetically interesting choices of $a_n(f)$. Good choices include Dirichlet L-functions (where the $a_n(f)$ are related to characters from groups such as $(\mathbb{Z}/q\mathbb{Z})^*$) and elliptic curve L-functions (where $a_p(f)$ is related to the number of solutions to $y^2 \equiv x^3 + Ax + B \bmod p$). As the Riemann zeta function yields information about primes, these L-functions provide information about other systems: the Dirichlet L-functions are useful in studying primes in arithmetic progression, and the elliptic curve L-functions provide information about the group of rational solutions to an elliptic curve (these groups are frequently used in cryptographic applications). For more details on L-functions, see [IwKo, MiT-B, Ser].

We assume our L-functions below are "good" (see [KonMi] for the precise, technical definition of good in this context, which is known to include the Riemann zeta function). Using Hejhal's [Hej] estimates on the error term in the log-normal law for the distribution of values of L-functions near the critical line, one finds

Theorem 3.2.6 (Kontorovich–Miller [KonMi]). *Let $L(s, f)$ be a good L-function; for example we may take $\zeta(s)$. Fix a $\delta \in (0, 1)$. For each T, let $\sigma_T = \frac{1}{2} + \frac{1}{\log^\delta T}$. Then*

$$\lim_{T \to \infty} \frac{\mu\{t \in [T, 2T] : 1 \leq M_B(|L(\sigma_T + it, f)|) \leq \tau\}}{T} = \log_B \tau \qquad (3.24)$$

(with μ Lebesgue measure). Thus the values of the L-function satisfy Benford's Law in the limit (with the limit taken as described above) for any base B.

As values of L-functions are believed to be modeled by the characteristic polynomials in random matrix theory, seeing Benford behavior in both areas is a good consistency check. The above result is just one of many that relate L-functions and Benford's Law. For others, see [AnRoSt, JaThYe] on coefficients of modular forms and partition functions, or Barrale, Hendel and Sluys [BaHeSl] on sequences related to the Fibonacci numbers.

3.2.3.3 The $3x + 1$ Problem

While there are a multitude of interesting problems that we could study, the $3x + 1$ problem is a particularly good choice as the analysis of its convergence to Benford behavior highlights many of difficulties that can arise in applying the tools from Fourier analysis (especially in showing Condition 3 holds). This leads to a lattice supported distribution. We can surmount the challenges this causes by using a quantified version of the Kronecker–Weyl theorem (see [HaWr, MiT-B]). The standard version says that if α is irrational, then the set $\{n\alpha \bmod 1 : n \leq N\}$ becomes equidistributed as $N \to \infty$; for this problem, we need some control over how rapidly the equidistribution sets in. We obtain the needed control by studying how well α may be approximated by rationals.

The $3x + 1$ **problem** is one of the most captivating problems in mathematics (see [Lag1, Lag2, Lag3] for more details, especially the latter). There are several variants; we define the $3x + 1$ **map** by

$$a_{n+1} = \frac{3a_n + 1}{2^k}, \qquad (3.25)$$

where k is the largest power of 2 that divides $3a_n + 1$. For example,

$$11 \to 17 \to 13 \to 5 \to 1 \to 1 \to 1 \cdots. \qquad (3.26)$$

If we start with 21 and iterate we get to 1 in 1 step, if we start with 23 we iterate to 1 in 4 steps, if we start with 25 we iterate to 1 in 7 steps, and if we start with 29 we iterate to 1 in 5 steps. The astute reader might notice we skipped 27; for 27 we

Digit	Number	Observed	Benford	Number	Observed	Benford
1	24251	0.301	0.301	72924	0.302	0.301
2	14156	0.176	0.176	42357	0.176	0.176
3	10227	0.127	0.125	30201	0.125	0.125
4	7931	0.099	0.097	23507	0.097	0.097
5	6359	0.079	0.079	18928	0.078	0.079
6	5372	0.067	0.067	16296	0.068	0.067
7	4476	0.056	0.058	13702	0.057	0.058
8	4092	0.051	0.051	12356	0.051	0.051
9	3650	0.045	0.046	11073	0.046	0.046

Table 3.1   The 80,514 (left columns) iterations of a random 10,000 digit number (the predicted number of iterations to reach 1 is 80,319) under the $3x + 1$ map, removing as many powers of 2 as possible in each step. The chi-square value is 13.5 (the critical value at the 5% level is 15.5). The three right columns are the result of removing at most one power of 2 in a given step. There are 241,344 iterations with a chi-square of 11.4.

have

$27 \rightarrow 41 \rightarrow 31 \rightarrow 47 \rightarrow 71 \rightarrow 107 \rightarrow 161 \rightarrow 121 \rightarrow 91 \rightarrow$

$137 \rightarrow 103 \rightarrow 155 \rightarrow 233 \rightarrow 175 \rightarrow 263 \rightarrow 395 \rightarrow 593 \rightarrow$

$445 \rightarrow 167 \rightarrow 251 \rightarrow 377 \rightarrow 283 \rightarrow 425 \rightarrow 319 \rightarrow 479 \rightarrow$

$719 \rightarrow 1079 \rightarrow 1619 \rightarrow 2429 \rightarrow 911 \rightarrow 1367 \rightarrow 2051 \rightarrow 3077 \rightarrow$

$577 \rightarrow 433 \rightarrow 325 \rightarrow 61 \rightarrow 23 \rightarrow 35 \rightarrow 53 \rightarrow 5 \rightarrow 1$

(42 steps); it takes longer, but it does iterate to 1. The famous $3x + 1$ **conjecture** is that, no matter what integer we start with, it will iterate to 1 eventually. Kakutani described this problem as a Soviet conspiracy to slow down American mathematics, while Erdös stated that mathematics is not yet ready for problems such as this!

As this is a book on Benford's Law, the natural question for us to ask is whether or not the distribution of leading digits satisfies Benford's Law. Even though 27 yields only 41 iterates, the agreement with Benford's Law is already quite striking. We list the observed versus predicted percentages for each digit: 28.6% (30.1%), 19.0% (17.6%), 16.7% (12.4%), 14.3% (9.7%), 9.5% (7.9%), 2.4% (6.7%), 4.8% (5.8%), 0% (5.1%), 4.8% (4.6%). The fit improves as our number grows; in Table 3.1 we explore the iterates of a random 10,000 digit number under the $3x + 1$ map, while in Table 3.2 we do the same but for a generalization: the $5x + 1$ map.

We state two results about Benford behavior for the $3x+1$ problem (the definition of finite irrationality type is given later in the section). The first requires some notation. Let x_0 denote the starting seed, and let x_m be its mth iterate. By the Structure Theorem of Sinai [Sin] and Kontorovich–Sinai [KonSi], we expect the sequence of x_i's to behave like a geometric Brownian motion with drift $\log(3/4) < 0$, which implies that it should iterate to 1 and that $x_m \approx (3/4)^m x_0$; as we know geometric Brownian motions are Benford, we expect Benford behavior here as well. In the first theorem we see Benford behavior by looking at the ratio of the

Digit	Number	Observed	Benford	Number	Observed	Benford
1	8154	0.302	0.301	72652	0.301	0.301
2	4770	0.177	0.176	42499	0.176	0.176
3	3405	0.126	0.125	30153	0.125	0.125
4	2634	0.098	0.097	23388	0.097	0.097
5	2105	0.078	0.079	19110	0.079	0.079
6	1787	0.066	0.067	16159	0.067	0.067
7	1568	0.058	0.058	13995	0.058	0.058
8	1357	0.050	0.051	12345	0.051	0.051
9	1224	0.045	0.046	11043	0.046	0.046

Table 3.2   The first 27,004 (left columns) iterations of a random 10,000 digit number under the $5x + 1$ map, removing as many powers of 2 as possible in each step. The chi-square value is 1.8 (the critical value at the 5% level is 15.5). The three right columns are the result of removing at most one power of 2 in a given step. The results are for the first 241,344 iterations, with a chi-square of $3 \cdot 10^{-4}$.

observed mth iterate to the expected value of the mth iterate, while in the second theorem we look at the distribution of digits in the iterates of one given seed.

Theorem 3.2.7 (Kontorovich–Miller [KonMi]). *Let B be any real number such that $\log_B 2$ is irrational of type $\kappa < \infty$ (for example, one may take any integer B which is not a perfect power of 2). Then for any $[a, b] \subset [0, 1]$,*

$$\lim_{m \to \infty} \mathbb{P}\left(\log_B \left[\frac{x_m}{\left(\frac{3}{4}\right)^m x_0}\right] \mod 1 \in [a, b]\right) = b - a. \tag{3.27}$$

As $(3/4)^m x_0$ is the expected value of x_m, this implies the distribution of the ratio of the actual versus predicted value after m iterates obeys Benford's Law (base B). If $B = 2^n$ for some integer n, in the limit $\log_B \left[\frac{x_m}{(3/4)^m x_0}\right] \mod 1$ takes on the values $0, \frac{1}{n}, \frac{2}{n}, \ldots, \frac{n-1}{n}$ with equal probability, leading to a non-Benford digit bias depending only on n.

These Benford results can be generalized to other, related iterative processes (see [KonMi] for details).

Theorem 3.2.8 (Lagarias–Soundararajan [LagSo]). *Let $B \geq 2$ be a fixed integer base. For each $N \geq 1$ and each $X \geq 2^N$, for all but at most $c(B)N^{-1/36}X$ initial seeds (where $c(B)$ is a positive constant depending only on B) the distribution of the first N iterates of the $3x + 1$ map are within $2N^{-1/36}$ of the Benford probabilities.*

We conclude with a few words about the proof of Theorem 3.2.7 (see [KonMi] for the details). The key step is using the finite irrationality type of $\log_B 2$ in the **Erdös–Turan Theorem** to get a quantified form of equidistribution of $n \log_B 2 \mod 1$. We describe these ingredients in some detail as this is the heart of the proof, and these techniques should be useful for other related investigations. For many applications, it is essential to understand the rate of convergence. For example, in Chapter 8 we see the IRS uses Benford's Law to detect tax fraud; arguments like this let us know how many data points are needed before the limiting behavior sets in.

We need a few preliminary facts and notation before sketching the proof. A number α is of **irrationality type** κ if κ is the supremum of all γ with

$$\varliminf_{q \to \infty} q^{\gamma+1} \min_p \left| \alpha - \frac{p}{q} \right| = 0. \tag{3.28}$$

Measuring the quality of the rational approximation in terms of the denominator is very natural. For instance, it is trivial to approximate $\pi \approx 3.1415926536$ to 6 decimal places by $31415926/10000000$; however, we get a much better approximation from $355/113 \approx 3.141592920$ (for an error of about $2.7 \cdot 10^{-7}$). It should be clear that the latter approximation is preferred; it's of comparable quality but has a significantly smaller denominator and is thus much "simpler."

The most frequently encountered irrationals are the algebraic numbers, which are solutions to polynomials with rational coefficients. By Roth's theorem, every algebraic irrational is of irrationality type 1; however, there do exist irrationals that can be arbitrarily well approximated by rationals. The standard example is the Liouville numbers, such as $\sum_{n=1}^{\infty} 10^{-n!}$. These numbers have infinite irrationality type, and from a computational perspective are very hard to distinguish from rational numbers. See Chapters 5 and 6 of [MiT-B] for more details.

For our purposes, we need to know that $\log_B 2$ is of finite type if B is an integer not of the form 2^n. This is proved in [KonMi] by using Baker's theory of linear forms in logarithms [Bak]. Crudely applying Baker's results gives an irrationality exponent for $\log_{10} 2$ of at most $\kappa = 2.3942 \times 10^{602}$ (though a more careful analysis would lead to a lower number!).

We now show the connection between the irrationality type of α and equidistribution of $n\alpha \mod 1$; see Theorem 3.3 on page 124 of [KuiNi] for complete details. Define the **discrepancy of a sequence** $\{x_n\}$ by

$$D_N = \frac{1}{N} \sup_{[a,b] \subset [0,1]} |N(b-a) - \#\{n \le N : x_n \mod 1 \in [a,b]\}|; \tag{3.29}$$

if our sequence is equidistributed then D_N should be $o(1)$ as we expect the number of valid n to be on the order of $(b-a)N$. Finally, let $||x||$ denote the distance from x to the nearest integer.

Theorem 3.2.9 (Erdős–Turan Inequality). *There exists a C such that for all m,*

$$D_N \le C \left(\frac{1}{m} + \sum_{h=1}^{m} \frac{1}{h} \left| \frac{1}{N} \sum_{n=1}^{N} e^{2\pi i h x_n} \right| \right). \tag{3.30}$$

See [KuiNi], page 112 for a proof.

In our Benford investigations, we frequently have sequences of the form $x_n = n\alpha$. We sketch the bound from the Erdős–Turan inequality in this case. Using the geometric series formula to evaluate the exponential sum, we find it is less than $\frac{1}{|\sin(\pi h\alpha)|} \leq \frac{1}{2||h\alpha||}$. We must control $\sum_{h=1}^{m} \frac{1}{h||h\alpha||}$, and this is where the irrationality type enters. One can show that if α has irrationality type κ then $\sum_{h=1}^{m} \frac{1}{h||h\alpha||} = O\left(m^{\kappa-1+\epsilon}\right)$. We then balance the two terms in the inequality by taking $m = \lfloor N^{1/\kappa} \rfloor$, which gives $D_N = O(N^{-1/\kappa+\epsilon'})$ where ϵ' can be made arbitrarily small. We are thus left with a power savings in the discrepancy; the power savings might be small (as it is with our bounds for $\log_{10} 2$), but there is a power savings in N.

3.3 PRODUCTS OF INDEPENDENT RANDOM VARIABLES

The Uniform Distribution Characterization (Theorem 2.4.2) states that X is Benford base B if and only if its logarithm base B is uniformly distributed modulo 1. Earlier in this chapter we saw how this observation, combined with results from Fourier analysis, lead to the limiting behavior of many processes being Benford. We now see another example of the power of this perspective. If $\mathcal{X}_N = X_1 \cdots X_N$ is the product of N independent random variables, then taking logarithms gives

$$\log_B \mathcal{X}_N = Y_1 + \cdots + Y_N, \tag{3.31}$$

where $Y_m = \log_B X_m$. Thus the Benford behavior of a product of independent random variables is reduced to understanding a sum, which is far more familiar.

In Chapter 2, using Fourier analytic methods, Berger and Hill proved the product of independent, *identically* distributed random variables converges to Benford's Law provided the random variables are not purely atomic. Our goal in this section is to generalize to the case of non-identical random variables. While the "typical" outcome is Benford behavior, we'll construct a sequence of random variables with different densities such that their product converges to non-Benford behavior (but if we took all our random variables from any of these densities we would obtain Benford's Law).

If the $\log X_m$ are sufficiently nice, we can appeal to the Central Limit Theorem. The resulting Gaussian has increasing variance. Using Poisson Summation, we find it becomes uniformly distributed modulo 1 as $N \to \infty$; see §3.2. Our goal in this section is to prove Benford behavior under weaker conditions than are required for the Central Limit Theorem to be applicable. For instance, if we want to use the Central Limit Theorem then we need the variances of the $\log_B P_k$ to be finite; however, for Benford's Law we need to understand the resulting sum modulo 1 only, and thus this condition (though often met in practice) is needlessly restrictive (we give an example illustrating this in §3.3.4).

So far in this chapter we've chosen to give proofs based on Poisson Summation. We take a different approach in this section. Using elementary results from Fourier analysis, we give a necessary and sufficient condition for the sum of M independent continuous random variables modulo 1 to converge to the uniform distribution in $L^1([0, 1])$ (similar results hold for sums of discrete random variables). A consequence is that if $X_1, \ldots, X_M$ are independent continuous random variables with densities $f_1, \ldots, f_M$, for any base B as $M \to \infty$ for many choices of the densities the distribution of the digits of $X_1 \cdots X_M$ converges to Benford's Law base

B. The rate of convergence can be quantified in terms of the Fourier coefficients of the densities, which we do in Corollary 3.4.2 in the next section). These results provide another explanation for the prevalence of Benford behavior in many diverse systems, as many observations arise as the product of independent random variables.

This section is a modification of [JaKKKM, MiNi1], which the reader may consult for additional details and examples.

3.3.1 Introduction

Many authors [Sak, SpTh, AdhSa, Adh, Ha, Tur] have observed that the product (and more generally, any nice arithmetic operation) of two random variables is often closer to satisfying Benford's Law than the input random variables; further, as the number of terms increases, the resulting expression seems to approach Benford's Law. While many of the previous works are concerned with exact formulas for the distribution of $X_1 \cdots X_M$, to understand the distribution of the digits by the Uniform Distribution Characterization (Theorem 2.4.2) it suffices to understand $\log_B |X_1 \cdots X_M| \bmod 1$. This is because a sequence is Benford base B if and only if its base B logarithms are equidistributed modulo 1.

This is an ideal problem for Fourier analysis, as

$$\log_B |X_1 \cdots X_M| = \log_B |X_1| + \cdots + \log_B |X_M|. \tag{3.32}$$

Letting $Y_m = \log_B |X_m|$, we see we have a sum of independent random variables, which can often be analyzed by appealing to the Central Limit Theorem. There are, however, some differences between our situation and standard applications of the Central Limit Theorem. Typically one studies $\frac{\sum_m Y_m - \mathbb{E}[\sum_m Y_m]}{\mathrm{StDev}(\sum_m Y_m)}$, where we have subtracted off the mean and divided by the standard deviation to obtain a quantity which will be finite as $M \to \infty$. There is no need to do either in our case, as we are considering sums modulo 1.

The main result is a variant of the Central Limit Theorem, which in this context states that for "nice" random variables, as $M \to \infty$ the sum of M independent random variables modulo 1 tends to the uniform distribution; as remarked by simple exponentiation this is equivalent to Benford's Law for the product. To emphasize the similarity to the standard Central Limit Theorem and the fact that the sums are modulo 1, we refer to such results as Modulo 1 Central Limit Theorems. Many authors [Bh, Boy, Hol, JanRu, Lév2, Loy, Rob, Sc1, Sc2, Sc3] have analyzed this problem in various settings and generalizations, obtaining sufficient conditions on the random variables (often identically distributed) as well as estimates on the rate of convergence.

We let $\widehat{g_m}(n)$ denote the nth Fourier coefficient of a probability density g_m on $[0, 1]$:

$$\widehat{g_m}(n) = \int_0^1 g_m(x) e^{-2\pi i n x} dx. \tag{3.33}$$

Theorem 3.3.1 (Miller–Nigrini [MiNi1]: The Modulo 1 Central Limit Theorem for Independent Continuous Random Variables). *Let $\{Y_m\}$ be independent continuous* *random variables on* $[0, 1)$*, not necessarily identically distributed, with densities* $\{g_m\}$*. A necessary and sufficient condition for the sum* $Y_1 + \cdots + Y_M$ *modulo* 1 *to converge to the uniform distribution as* $M \to \infty$ *in* $L^1([0, 1])$ *is that for each* $n \neq 0$ *we have* $\lim_{M \to \infty} \widehat{g_1}(n) \cdots \widehat{g_M}(n) = 0$*. In particular,* $X_1 \cdots X_M$ *converges to Benford's Law base* B*, where* $Y_m = \log_B X_m$*.*

As other authors have noticed, the importance of results such as Theorem 3.3.1 is that they give an explanation of why so many data sets follow Benford's Law (or at least a close approximation to it). Specifically, if we can consider the observed values of a system to be the product of many independent processes with reasonable densities, then the distribution of the digits of the resulting product will be close to Benford's Law.

We briefly compare this approach with other proofs of results such as Theorem 3.3.1 (where the random variables are often taken as identically distributed). If the random variables are identically distributed with density g, our condition reduces to $|\widehat{g}(n)| < 1$ for $n \neq 0$. For a probability distribution, $|\widehat{g}(n)| = 1$ for $n \neq 0$ if and only if there exists $\alpha \in \mathbb{R}$ such that all the mass is contained in the set $\{\alpha, \alpha + \frac{1}{n}, \ldots, \alpha + \frac{n-1}{n}\}$. As we are assuming our random variables are continuous and not discrete, the corresponding densities are in $L^1([0, 1])$ and this condition is not met. In other words, the sum of identically distributed random variables modulo 1 converges to the uniform distribution if and only if the support of the distribution is not contained in a coset of a finite subgroup of the circle group $[0, 1)$. Interestingly, Lévy [Lév2] proved this just one year after Benford's paper [Ben], though his paper does not study digits. Lévy's result has been generalized to other compact groups, with estimates on the rate of convergence [Bh]. Stromberg [Str] proved that[2] *the* n*-fold convolution of a regular probability measure on a compact Hausdorff group* G *converges to the normalized Haar measure in the weak-star topology if and only if the support of the distribution is not contained in a coset of a proper normal closed subgroup of* G*.*

The arguments in the proof of Theorem 3.3.1 may be generalized to independent discrete random variables, at the cost of replacing L^1-convergence with weak convergence; see [MiNi1] for statements and proofs.

3.3.2 Fourier Analysis Preliminaries

We quickly recall some standard facts from Fourier analysis (see [MiT-B, StSh1] for example). The convolution of two functions in $L^1([0, 1])$ is

$$(f * g)(x) = \int_0^1 f(y)g(x - y)dy = \int_0^1 f(x - y)g(y)dy. \qquad (3.34)$$

Convolution is commutative and associative, and the nth Fourier coefficient of a convolution is the product of the two nth Fourier coefficients. Note if X and Y are two independent random variables, the convolution of their densities is the density of $X + Y$.

---

[2]The following formulation is taken almost verbatim from the first paragraph of [Bh].

Definition 3.3.2 (Fejér kernel, Fejér series). *Let* $f \in L^1([0,1])$. *The Nth Fejér kernel is*

$$F_N(x) = \sum_{n=-N}^{N} \left(1 - \frac{|n|}{N}\right) e^{2\pi inx}, \tag{3.35}$$

and the Nth Fejér series of f is

$$T_N f(x) = (f * F_N)(x) = \sum_{n=-N}^{N} \left(1 - \frac{|n|}{N}\right) \widehat{f}(n) e^{2\pi inx}. \tag{3.36}$$

The Fejér kernels are an approximation to the identity (they are non-negative, integrate to 1, *and for any* $\delta \in (0, 1/2)$ *we have* $\lim_{N\to\infty} \int_{\delta}^{1-\delta} F_N(x) dx = 0$).

Instead of looking at Fejér kernels and series, one could study Dirichlet kernels and series, which lead to approximating $f(x)$ by $\sum_{n=-N}^{N} \widehat{f}(n) e^{2\pi inx}$. While this unweighted Fourier series is the more natural object to study, it has significantly worse convergence properties. For example, Kolmogorov [Kol] constructed an $L^1([0,1])$ function f whose Fourier series diverges at each point! Though the Fourier series is a fascinating object worthy of study, for us it's merely a tool to investigate Benford behavior. As such, we use the weighted Fourier series (the Fejér series). A major result in the subject is Fejér's Theorem, which we recall.

Fejér's Theorem: If f is continuous and periodic on $[0, 1]$ then as $N \to \infty$, $T_N f$ converges uniformly to f.

For our purposes, Fejér's Theorem is just barely insufficient as we wish to consider random variables whose densities are $L^1([0, 1])$ and not just those whose density functions are continuous. Fortunately it is easy to generalize Fejér's Theorem to handle this case.

Theorem 3.3.3 (Lebesgue's Theorem). *Let* $f \in L^1([0,1])$. *As* $N \to \infty$, $T_N f$ *converges to* f *in* $L^1([0,1])$, *i.e.,* $\lim_{N\to\infty} \int_0^1 |f(x) - T_N f(x)| dx = 0$.

Lebesgue's Theorem follows from Fejér's Theorem by a standard three-epsilon argument. Given f, we find a continuous g that is close to f. We then apply Fejér's Theorem to g, and then show the Fejér series of g is close to the Fejér series of f.

The following fact is useful in proving the Modulo 1 Central Limit Theorem.

Lemma 3.3.4. *Let* $f, g \in L^1([0,1])$. *Then* $T_N(f * g) = (T_N f) * g$.

Proof. One first shows that convolution is associative, so $h_1 * (h_2 * h_3) = (h_1 * h_2) * h_3$. As $T_N h = F_N * h$, we have

$$T_N(f * g) = F_N * (f * g) = (F_N * f) * g = (T_N f) * g. \tag{3.37}$$

$\square$

3.3.3 Proof of the Modulo 1 Central Limit Theorem

We can now prove Theorem 3.3.1.

Proof. We first show our condition is sufficient. The density of the sum modulo 1 is $h_M = g_1 * \cdots * g_M$. It suffices to show that, for any $\epsilon > 0$,

$$\lim_{M \to \infty} \int_0^1 |h_M(x) - 1| dx < \epsilon. \tag{3.38}$$

Using Lebesgue's Theorem (Theorem 3.3.3), choose N sufficiently large so that

$$\int_0^1 |h_1(x) - T_N h_1(x)| dx < \frac{\epsilon}{2}. \tag{3.39}$$

While N was chosen so that (3.39) holds with h_1, in fact this N works for *all* h_M (with the same ϵ). This follows by induction. The base case is immediate (this is just our choice of N). Assume now that (3.39) holds with h_1 replaced by h_M; we must show it holds with h_1 replaced by $h_{M+1} = h_M * g_{M+1}$. By Lemma 3.3.4 we have

$$T_N h_{M+1} = T_N(h_M * g_{M+1}) = (T_N h_M) * g_{M+1}. \tag{3.40}$$

This implies

$$\int_0^1 |h_{M+1}(x) - T_N h_{M+1}(x)| dx$$

$$= \int_0^1 |(h_M * g_{M+1})(x) - (T_N h_M) * g_{M+1}(x)| dx$$

$$= \int_0^1 \left| \int_0^1 (h_M(y) - T_N h_M(y)) \cdot g_{M+1}(x - y) \right| dy \, dx$$

$$\le \int_0^1 \int_0^1 |h_M(y) - T_N h_M(y)| \cdot g_{M+1}(x - y) dx \, dy$$

$$= \int_0^1 |h_M(y) - T_N h_M(y)| dy \cdot 1 < \frac{\epsilon}{2}; \tag{3.41}$$

the interchange of integration above is justified by the absolute value being integrable in the product measure, and the x-integral is 1 as g_{M+1} is a probability density.

To show h_M converges to the uniform distribution in $L^1([0, 1])$, we must show $\lim_{M \to \infty} \int_0^1 |h_M(x) - 1| dx = 0$. Let N and ϵ be as above. By the triangle inequality we have

$$\int_0^1 |h_M(x) - 1| dx \le \int_0^1 |h_M(x) - T_N h_M(x)| dx + \int_0^1 |T_N h_M(x) - 1| dx. \tag{3.42}$$

From our choices of N and ϵ, $\int_0^1 |h_M(x) - T_N h_M(x)| dx < \epsilon/2$; thus we need show only that $\int_0^1 |T_N h_M(x) - 1| dx < \epsilon/2$ to complete the proof. As $\widehat{h_M}(0) = 1$,

$$\int_0^1 |T_N h_M(x) - 1| dx = \int_0^1 \left| \sum_{\substack{n=-N \\ n \neq 0}}^{N} \left(1 - \frac{|n|}{N}\right) \widehat{h_M}(n) e^{2\pi i n x} \right| dx$$

$$\leq \sum_{\substack{n=-N \\ n \neq 0}}^{N} \left(1 - \frac{|n|}{N}\right) |\widehat{h_M}(n)|. \qquad (3.43)$$

However, $\widehat{h_M}(n) = \widehat{g_1}(n) \cdots \widehat{g_M}(n)$, and by assumption tends to zero as $M \to \infty$ (as each $\widehat{g_m}(n)$ is at most 1 in absolute value, for each n the absolute value of the product is non-increasing in M). For fixed N and ϵ, we may choose M sufficiently large so that $|\widehat{h_M}(n)| < \epsilon/4N$ whenever $n \neq 0$ and $|n| \leq N$. Thus

$$\int_0^1 |T_N h_M(x) - 1| dx < 2N \cdot \frac{\epsilon}{4N} = \frac{\epsilon}{2}, \qquad (3.44)$$

which implies

$$\int_0^1 |h_M(x) - 1| dx < \epsilon \qquad (3.45)$$

for M sufficiently large. As ϵ is arbitrary, this completes the proof of the sufficiency; we now prove this condition is necessary.

Assume for some $n_0 \neq 0$ that $\lim_{M \to \infty} |\widehat{h_M}(n_0)| \neq 0$ (where as always $h_M = g_1 * \cdots * g_M$). As the g_m are probability densities, $|\widehat{g_m}(n)| \leq 1$; thus the sequence $\{|\widehat{h_M}(n)|\}_{M=1}^{\infty}$ is non-increasing for each n, and hence by assumption converges to some number $c_n \in (0, 1]$.

Let $E_M(x) = h_M(x) - 1$; note $\widehat{E_M}(n) = \widehat{h_M}(n)$ for $n \neq 0$. To show h_M does not converge to the uniform distribution on $[0, 1]$, it suffices to show that E_M does not converge almost everywhere to the zero function on $[0, 1]$. Let n_0 be as above. We have

$$\left| \widehat{h_M}(n_0) \right| = \left| \widehat{E_M}(n_0) \right| = \left| \int_0^1 E_M(x) e^{2\pi i n_0 x} dx \right| \geq c_{n_0} > 0. \qquad (3.46)$$

Therefore at least one of the following integrals is at least $c_{n_0}/2$:

$$\int_{\substack{x \in [0,1] \\ \mathrm{Re}(E_M(x)) \geq 0}} \mathrm{Re}\,(E_M(x))\,dx, \qquad \int_{\substack{x \in [0,1] \\ \mathrm{Re}(E_M(x)) \leq 0}} \mathrm{Re}\,(-E_M(x))\,dx,$$

$$\int_{\substack{x \in [0,1] \\ \mathrm{Im}(E_M(x)) \geq 0}} \mathrm{Im}\,(E_M(x))\,dx, \qquad \int_{\substack{x \in [0,1] \\ \mathrm{Im}(E_M(x)) < 0}} \mathrm{Im}\,(-E_M(x))\,dx, \qquad (3.47)$$

and h_M cannot converge to the zero function in $L^1([0, 1])$; further, we obtain an estimate on the L^1-distance between the uniform distribution and h_M. $\qquad \square$

The behavior is non-Benford if the conditions of Theorem 3.3.1 are violated. It is enough to show that we can find a sequence of densities $g_{B,m}$ such that $\lim_{M \to \infty} \prod_{m=1}^{M} \widehat{g_{B,m}}(1) \neq 0$. We are reduced to searching for an infinite product that

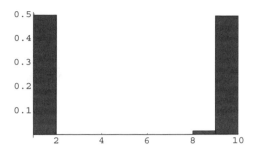

Figure 3.1   Distribution of digits (base 10) of 1000 products $X_1 \cdots X_{1000}$, where $g_{10,m} = \phi_{11^m}$.

is non-zero; we also need each term to be at most 1, as the Fourier coefficients of a probability density are dominated by 1. One solution is $\prod_{m=1}^{\infty} \frac{m^2+2m}{(m+1)^2}$, which equals $\frac{1}{2}$. Thus as long as $\widehat{g}_{B,m}(1) \geq \frac{m^2+2m}{(m+1)^2}$, the conclusion of Theorem 3.3.1 will not hold for the products of the associated random variables; analogous reasoning yields a sum of independent random variables modulo 1 which does not converge to the uniform distribution.

Example 3.3.5 (Non-Benford behavior of products). *Consider*

$$\phi_m = \begin{cases} m & \text{if } |x - \frac{1}{8}| \leq \frac{1}{2m}, \\ 0 & \text{otherwise}; \end{cases} \tag{3.48}$$

ϕ_m is non-negative and integrates to 1. As $m \to \infty$ we have $|\widehat{\phi}_m(1)| \to 1$ because the density becomes concentrated at $\frac{1}{8}$ (direct calculation gives $\widehat{\phi}_m(1) = e^{2\pi i/8} + O(m^{-2})$). Let $X_1, \ldots, X_M$ be independent random variables where the associated densities $g_{B,m}$ of $\log_B M(|X_m|)$ are ϕ_{11^m}. The behavior is non-Benford (see Figure 3.1). Note, however, that if each X_m had the common distribution ϕ_i for any fixed i, then in the limit the product will satisfy Benford's Law.

3.3.4 Comparison with Alternate Techniques

We discuss an alternate approach to proving Theorem 3.3.1. One could try to apply the standard Central Limit Theorem to the sum $\log_B |X_1| + \cdots + \log_B |X_M|$, noting that as the variance of a Gaussian increases to infinity, the Gaussian becomes uniformly distributed modulo 1. A significant drawback of a proof by the Central Limit Theorem is the requirement (at a minimum) that the variance of each $\log_B |X_m|$ be finite. This is a very weak condition, and in fact many random variables X with infinite variance (such as Pareto or modified Cauchy distributions) do have $\log_B |X|$ having finite variance; however, there are distributions where $\log_B |X|$ has infinite variance. One example is to let X be the random variable with density

$$f_\alpha(x) = \begin{cases} \alpha/(x \log^{\alpha+1} x)^{-1} & \text{if } x \geq e, \\ 0 & \text{otherwise}. \end{cases} \tag{3.49}$$

This is a probability distribution for $\alpha > 0$, and is a modification of a Pareto distribution (which arises in power law behavior). See [Mi1] for some applications and properties of this distribution. For $\alpha \in (0, 2]$ the logarithm modulo 1 has infinite variance, and thus the Central Limit Theorem is not applicable, though the resulting behavior does converge to Benford's Law (see [MiNi1] for the proof). The reason the Central Limit Theorem fails for densities such as that in (3.49) is that it tries to provide *too much* information. The Central Limit Theorem tries to give us the limiting distribution of $\log_B |X_1 \cdots X_M| = \log_B |X_1| + \cdots + \log_B |X_M|$; however, as we are interested in only the distribution of the digits of $X_1 \cdots X_M$, this is more information than we need.

3.4 CHAINS OF RANDOM VARIABLES

In this section we examine **chains of random variables**, and give conditions implying Benford behavior. Through a change of variables, these chains are often equivalent to products of related random variables. In the previous section we obtained convergence of products to Benford's Law, but without good estimates on the rate of convergence. We remedy this below, quantifying the convergence in terms of the Mellin and Fourier transforms of densities related to the random variables, depending on whether or not we are studying the original random variables or their logarithms. Both viewpoints are useful. Each leads to a different transform entering the error expressions, and one of the goals here is to highlight both approaches.

3.4.1 Introduction

We describe how Benford's Law arises in chains of probability distributions and hierarchical Bayesian models. This allows us to construct tests (based on Benford's Law) of certain models. We may interpret these results as saying that in many Markov chain Monte Carlo problems, the distribution of first digits of the stationary distribution satisfies Benford's Law, and the chain has rapid mixing (i.e., few iterations are required to have excellent agreement with Benford's Law).

Since the early work of Newcomb [New] and Benford [Ben], there have been numerous theoretical advances as to why various data sets and operations yield Benford behavior. One reason for the immense amount of interest generated by this law is the observation that, in many cases, combining two data sets yields a new set which is closer to Benford's Law (see for example [Ha]). A common example is street addresses. If one studies the distribution of leading digits on a long street, the result is clearly non-Benford; depending on the length of the street, the probability of a first digit of 1 can oscillate between $1/9$ and $5/9$. However, if we consider many streets and amalgamate the data (as Benford [Ben] did), the result is quite close to Benford's Law. We may interpret the above as first choosing a street length from some distribution, so the street addresses say are integers in $[1, X]$ for some random variable X. Then for each choice of X we study the distribution of the leading digits on that street, and then calculate the expected frequencies as X

varies.

In [Koss1, Koss2], Kossovsky suggested such an interpretation and proposed that generalizations of the above procedure will rapidly lead to convergence to Benford behavior. Explicitly, he studied the distribution of leading digits of chained probability distributions, and conjectured that as the length of the chain increases then the behavior tends to Benford's Law. Let $\mathcal{D}_i(\theta)$ denote a one-parameter distribution with parameter θ and density function $f_{\mathcal{D}_i(\theta)}$; thus by $X \sim \mathcal{D}_i(\theta)$ we mean

$$\mathrm{Prob}(X \in [a,b]) = \int_a^b f_{\mathcal{D}_i(\theta)}(x)dx. \tag{3.50}$$

We can create a chain of random variables as follows. Let $p : \mathbb{N} \to \mathbb{N}$. Let $X_1 = \mathcal{D}_{p(1)}(1)$ and define X_m inductively by $X_m \sim \mathcal{D}_{p(m)}(X_{m-1})$. Computer simulations and other considerations led Kossovsky to conjecture that if our underlying distributions are "nice," then as $n \to \infty$ the distribution of the leading digits of X_n converges to Benford's Law.[3] Note that our example of street addresses is just a special case with a chain length of two and uniform distributions. Another way of stating our results is that for certain Markov chain Monte Carlo processes, Benford's Law is absorbing for the distribution of first digits (and in fact the system is rapidly mixing as well).

While it is possible to attack chained random variables directly, these problems can often be recast in terms of products of independent random variables. *Assume the density* $f_{\mathcal{D}_{p(m)}(\theta)}(x) = \theta^{-1}f_{p(m)}(x/\theta)$ *for some* $f_{p(m)}$ *(with antiderivative F).* Notation as above, the density of X_n is exactly that of the density of $W_1 \cdots W_n$, where the W_m are independent random variables with $W_m \sim \mathcal{D}_{p(m)}(1)$. For example, the density of the random variable $W_1 \cdot W_2$ is given by

$$\int_0^\infty f_{\mathcal{D}_{p(2)}(1)}\left(\frac{x}{t}\right) f_{\mathcal{D}_{p(1)}(1)}(t)\frac{dt}{t} \tag{3.51}$$

(the generalization to more products is straightforward). To see this, we calculate the probability that $W_1 \cdot W_2 \in [0,x]$ and then differentiate with respect to x. Thus

$$\mathrm{Prob}(W_1 \cdot W_2 \in [0,x]) = \int_{t=0}^\infty \mathrm{Prob}\left(W_2 \in \left[0,\frac{x}{t}\right]\right) f_{\mathcal{D}_{p(1)}(1)}(t)dt$$

$$= \int_{t=0}^\infty F_{\mathcal{D}_{\nu(2)}(1)}\left(\frac{x}{t}\right) f_{\mathcal{D}_{p(1)}(1)}(t)dt. \tag{3.52}$$

Differentiating gives the density of $W_1 \cdot W_2$, which equals

$$\int_{t=0}^\infty f_{\mathcal{D}_{p(2)}(1)}\left(\frac{x}{t}\right) f_{\mathcal{D}_{p(1)}(1)}(t)\frac{dt}{t} = \int_{t=0}^\infty f_{\mathcal{D}_{p(2)}(t)}(x) f_{\mathcal{D}_{p(1)}(1)}(t)dt, \tag{3.53}$$

which from the definition of conditional probability is the density for the chained variable X_2. An advantage of the approach below is the good, explicit error estimates. Further, it allows us to view this problem in a more familiar language.

---

[3]The conjecture may fail if we chain arbitrary parameters of arbitrary distributions. A good test case is to consider chaining the shape parameter γ of a Weibull distribution: $f(x) = \gamma x^{\gamma-1}\exp(-x^\gamma)$ for $x \geq 0$. The difficulty with numerics here is that very quickly we end up with a shape parameter very small (say less than 10^{-20}), and thus the numerics become suspect.

The analysis is significantly easier if one of the random variables is Benford; it is straightforward to show that if W_1 is Benford then $W_1 \cdots W_n$ is Benford (essentially this is due to the fact that if we add the uniform distribution modulo 1 to any distribution, the result is also the uniform distribution modulo 1); see [Wój] for a converse involving knowing the distributions of X and XY for Y independent of X.

3.4.2 Mellin Transform Preliminaries

We introduce some notation before giving precise statements of convergence of chains and products to Benford behavior. Let $f(x)$ be a continuous real-valued function on $[0, \infty)$. We define its **Mellin transform**, $(\mathcal{M}f)(s)$, by

$$(\mathcal{M}f)(s) \;=\; \int_0^\infty f(x) x^s \frac{dx}{x}. \tag{3.54}$$

Note $(\mathcal{M}f)(s) = \mathbb{E}[x^{s-1}]$, and thus results about expected values translate to results on Mellin transforms; for example, $(\mathcal{M}f)(1) = 1$ for any distribution supported on $[0, \infty)$. Additionally, if we let $x = e^{2\pi u}$ and $s = \sigma - i\xi$, then we find

$$(\mathcal{M}f)(\sigma - i\xi) \;=\; 2\pi \int_{-\infty}^\infty \left(f(e^{2\pi u}) e^{2\pi \sigma u} \right) e^{-2\pi i u \xi} du, \tag{3.55}$$

which is the Fourier transform of $g(u) = 2\pi f(e^{2\pi u}) e^{2\pi \sigma u}$. The Mellin and Fourier transforms are thus related; in fact, it is this logarithmic change of variables which explains why both enter into Benford's Law problems. For proofs of the Mellin transform properties one can therefore just mimic the proofs of the corresponding statements for the Fourier transform; a good reference is [StSh1].

In particular, properties of the Fourier transform have Mellin analogues. If we set $g(s) = (\mathcal{M}f)(s)$ then $f(x) = (\mathcal{M}^{-1}g)(x)$. In this setting, it's now convenient to define the convolution of two functions f_1 and f_2 by

$$(f_1 \star f_2)(x) \;=\; \int_0^\infty f_2\left(\frac{x}{t}\right) f_1(t) \frac{dt}{t} \;=\; \int_0^\infty f_1\left(\frac{x}{t}\right) f_2(t) \frac{dt}{t}. \tag{3.56}$$

The **Mellin convolution theorem** states that

$$(\mathcal{M}(f_1 \star f_2))(s) \;=\; (\mathcal{M}f_1)(s) \cdot (\mathcal{M}f_2)(s), \tag{3.57}$$

which by induction gives

$$(\mathcal{M}(f_1 \star \cdots \star f_n))(s) \;=\; (\mathcal{M}f_n)(s) \cdots (\mathcal{M}f_n)(s). \tag{3.58}$$

We state one last result which is useful for convergence questions. If $g(s)$ is an analytic function for $\Re(s) \in (a, b)$ such that $g(c + iy)$ tends to zero uniformly as $|y| \to \infty$ for any $c \in (a, b)$, then the **inverse Mellin transform**, $(\mathcal{M}^{-1}g)(x)$, is given by

$$(\mathcal{M}^{-1}g)(x) \;=\; \frac{1}{2\pi i} \int_{c-i\infty}^{c+i\infty} g(s) x^{-s} ds \tag{3.59}$$

(provided that the integral converges absolutely); this is known as the Mellin Inversion Theorem. See Appendix 2 of [Pat] for an enumeration of properties of the Mellin transform.

3.4.3 Benford Behavior of Chains and Products

The following theorem give conditions on when chained random variables are Benford. We also isolate the corresponding statement for an associated product of random variables, which is equivalent by our analysis above.

Theorem 3.4.1 (Jang, Kang, Kruckman, Kudo, Miller [JaKKKM]). *Let $\{\mathcal{D}_i(\theta)\}_{i \in I}$ be a collection of one-parameter distributions with associated densities $f_{\mathcal{D}_i(\theta)}$ which vanish outside of $[0, \infty)$. Let $p : \mathbb{N} \to I$, $X_1 \sim \mathcal{D}_{p(1)}(1)$, $X_m \sim \mathcal{D}_{p(m)}(X_{m-1})$, and assume*

1. *for each $m \geq 2$,*

$$f_m(x_m) = \int_0^\infty f_{\mathcal{D}_{p(m)}(1)}\left(\frac{x_m}{x_{m-1}}\right) f_{m-1}(x_{m-1}) \frac{dx_{m-1}}{x_{m-1}} \quad (3.60)$$

   where f_m is the density of the random variable X_m (this condition is always satisfied if $f_{\mathcal{D}_{p(m)}(\theta)}(x) = \theta^{-1} f_{p(m)}(x/\theta)$ for some $f_{p(m)}$);

2. *we have*

$$\lim_{n \to \infty} \sum_{\substack{\ell=-\infty \\ \ell \neq 0}}^{\infty} \prod_{m=1}^{n} (\mathcal{M} f_{\mathcal{D}_{p(m)}(1)}) \left(1 - \frac{2\pi i \ell}{\log B}\right) = 0. \quad (3.61)$$

Then as $n \to \infty$ the distribution of the leading digits of X_n tends to Benford's Law. Further, the error is a nice function of the Mellin transforms. Explicitly, if $Y_n = \log_B X_n$, then

$$|\text{Prob}(Y_n \bmod 1 \in [a, b]) - (b - a)|$$

$$\leq (b - a) \cdot \left| \sum_{\substack{\ell=-\infty \\ \ell \neq 0}}^{\infty} \prod_{m=1}^{n} (\mathcal{M} f_{\mathcal{D}_{p(m)}(1)}) \left(1 - \frac{2\pi i \ell}{\log B}\right) \right|. \quad (3.62)$$

If I is finite and all densities are continuous, then the second condition holds. If W_i has density $f_{\mathcal{D}_i}(1)$ and the W_i's are independent random variables, then $W_1 \cdots W_n$ has the same density as X_n, and the above statements are therefore true for this product.

The above results are stated in terms of the Mellin transform of the densities of the chained random variable X_n (or the density of the product). It's possible to restate the result with the Fourier transform. After some algebra we find

$$(\mathcal{M}f)\left(1 - \frac{2\pi i \xi}{\log B}\right) = \int_0^\infty f(t) e^{-2\pi i \xi \log_B t} dt, \quad (3.63)$$

which implies that $(\mathcal{M}f)\left(1 - \frac{2\pi i \xi}{\log B}\right)$ is the Fourier transform of $\mathfrak{f}(u) = f(e^u)e^u$ at $\xi/\log B$. A simple change of variables shows that if a random variable W_i : $[0, \infty)$ has density $f_i(w_i)$, then the random variable $V_i = \log W_i$ has density $\mathfrak{f}_i(v_i) = f_i(e^{v_i})e^{v_i}$. Thus we can measure the deviation of a product of independent random variables W_i either in terms of the Mellin transform of the densities f_i or the Fourier transforms of the $\mathfrak{f}_i$'s, the densities of the logarithms of the W_i's. We isolate this important result.

Corollary 3.4.2. *Let* $W_1, \ldots, W_n$ *be independent random variables with* $\mathfrak{f}_i$ *the density of* $\log W_i$. *If* $Y_n = W_1 \cdots W_n$ *then*

$$\left| \mathrm{Prob}(Y_n \bmod 1 \in [a, b]) - (b - a) \right| \leq (b - a) \cdot \left| \sum_{\substack{\ell = -\infty \\ \ell \neq 0}}^{\infty} \prod_{m=1}^{n} \widehat{\mathfrak{f}}_m \left(\ell / \log B \right) \right|.$$

(3.64)

Before giving the proof, we discuss the convergence, and refer the reader to [JaKKKM] for generalizations and comments on the weakness of the conditions on our chained random variables. If f is a continuous density function, then $(\mathcal{M}f)\left(1 - \frac{2\pi i \xi}{\log B}\right) < 1$ if $\xi \neq 0$. This is because $f(x)$ is non-negative and

$$(\mathcal{M}f)\left(1 - \frac{2\pi i \xi}{\log B}\right) = \int_0^{\infty} f(t) e^{-2\pi i \xi \log_B t} dt;$$

(3.65)

note the integral is clearly at most $\int_0^{\infty} f(t) dt = 1$ (since f is a density), and in fact is less than this because of the oscillation due to the exponential factor. As $|\xi|$ grows this integral tends to zero rapidly. This follows from our assumption that the Mellin transform is a nice function, and indicates that we have rapid convergence if all the distributions in the chain are equal. An alternate proof of the decay in $|\xi|$ is to note that $(\mathcal{M}f)\left(1 - \frac{2\pi i \xi}{\log B}\right)$ is the Fourier transform of $g(u) = f(e^u)e^u$ at $\xi / \log B$, and this tends to zero by the Riemann–Lebesgue lemma.

Proof. We first calculate f_n, the density of X_n. The basis case is clear, and for the inductive step we note

$$f_n(x_n) = \int_0^{\infty} f_{\mathcal{D}_{p(n)}(1)}\left(\frac{x_n}{x_{n-1}}\right) f_{n-1}(x_{n-1}) \frac{dx_{n-1}}{x_{n-1}}$$
$$= (f_{\mathcal{D}_{p(n)}(1)} \star f_{n-1})(x_n).$$

(3.66)

By the Mellin convolution theorem and induction we have

$$(\mathcal{M}f_n)(s) = (\mathcal{M}(f_{\mathcal{D}_{p(n)}(1)} \star f_{n-1}))(s)$$
$$= (\mathcal{M}f_{\mathcal{D}_{p(n)}(1)})(s) \cdot (\mathcal{M}f_{n-1})(s)$$
$$= \prod_{m=1}^{n} (\mathcal{M}f_{\mathcal{D}_{p(m)}(1)})(s).$$

(3.67)

By the **Mellin inversion theorem** we find

$$f_n(x_n) = \left(\mathcal{M}^{-1}\left(\prod_{m=1}^{n} (\mathcal{M}f_{\mathcal{D}_{p(m)}(1)}(\cdot)) \right) \right)(x_n).$$

(3.68)

To investigate the distribution of the digits of X_n (base B), it's convenient to make a logarithmic change of variables. Thus set $Y_n = \log_B X_n$. We have

$$\mathrm{Prob}(Y_n \leq y) = \mathrm{Prob}(X_n \leq B^y) = F_n(B^y).$$

(3.69)

Taking the derivative gives the density of Y_n, which we denote by $g_n(y)$:

$$g_n(y) = f_n(B^y) B^y \log B.$$

(3.70)

We again use the Uniform Distribution Characterization (Theorem 2.4.2) to prove Benford behavior. The key ingredient here is Poisson Summation. While the argument is similar to ones earlier in the chapter, the resulting expressions are not in the form considered there. Fortunately a trivial modification suffices. Let $h_{n,y}(t) = g_n(y+t)$. Then

$$\sum_{\ell=-\infty}^{\infty} g_n(y+\ell) = \sum_{\ell=-\infty}^{\infty} h_{n,y}(\ell) = \sum_{\ell=-\infty}^{\infty} \widehat{h}_{n,y}(\ell) = \sum_{\ell=-\infty}^{\infty} e^{2\pi i y\ell} \widehat{g}_n(\ell),$$

(3.71)

where $\widehat{f}$ denotes the Fourier transform of f:

$$\widehat{f}(\xi) = \int_{-\infty}^{\infty} f(x) e^{-2\pi i x\xi} dx.$$

(3.72)

Letting $[a,b] \subset [0,1]$, we see that

$$\text{Prob}(Y_n \bmod 1 \in [a,b]) = \sum_{\ell=-\infty}^{\infty} \int_{a+\ell}^{b+\ell} g_n(y) dy$$

$$= \int_a^b \sum_{\ell=-\infty}^{\infty} g_n(y+\ell) dy$$

$$= \int_a^b \sum_{\ell=-\infty}^{\infty} e^{2\pi i y\ell} \widehat{g}_n(\ell) dy,$$

$$|\text{Prob}(Y_n \bmod 1 \in [a,b]) - (b-a)| \le |b-a| \sum_{\ell \neq 0} |\widehat{g}_n(\ell)|,$$

(3.73)

as $\widehat{g}_n(0) = 1$ since g_n is a probability density. We thus need to compute $\widehat{g}_n(\ell)$:

$$\widehat{g}_n(\xi) = \int_{-\infty}^{\infty} g_n(y) e^{-2\pi i y\xi} dy$$

$$= \int_{-\infty}^{\infty} f_n(B^y) B^y \log B \cdot e^{-2\pi i y\xi} dy$$

$$= \int_0^{\infty} f_n(t) t^{-2\pi i\xi/\log B} dt$$

$$= (\mathcal{M}f_n)\left(1 - \frac{2\pi i\xi}{\log B}\right)$$

$$= \prod_{m=1}^{n} (\mathcal{M}f_{\mathcal{D}_{p(m)}(1)})\left(1 - \frac{2\pi i\xi}{\log B}\right).$$

(3.74)

Substituting completes the proof. $\square$

3.4.4 Examples

We give two explicit examples of the types of rapidly converging error estimates easily obtainable from Theorem 3.4.1 and Corollary 3.4.2 (see [JaKKKM] for the calculations). The first example is chaining exponential distributions. Many processes have wait times governed by a Poisson or exponential distribution; thus applications of these results could be to more involved processes where the wait time parameter depends on another process. For our second example we consider chaining uniform distributions. Our street example gives one instance where this could arise, namely when we choose uniformly among options of varying size.

3.4.4.1 Chains (or Products) of the Exponential Distribution

Let $X_1, W_1, \ldots, W_n \sim \text{Exp}(1)$ (the standard exponential distribution) and $X_m \sim \text{Exp}(X_{m-1})$, and set $Y_m = \log_B X_m$. We know that as $n \to \infty$ the distribution of digits of X_n or $W_1 \cdots W_n$ tends to Benford's Law, and we have

$$|\text{Prob}(Y_n \bmod 1 \in [a, b]) - (b - a)| \leq |b - a| \sum_{\ell=1}^{\infty} \left(\frac{2\pi^2 \ell / \log B}{\sinh(2\pi^2 \ell / \log B)} \right)^{n/2},$$

(3.75)

or, equivalently, the probability that the mantissa of X_n (or $W_1 \cdots W_n$) is in $[1, s]$ is within $\log_B s \sum_{\ell=1}^{\infty} \left(\frac{2\pi^2 \ell / \log B}{\sinh(2\pi^2 \ell / \log B)} \right)^{n/2}$ of $\log_B s$. As $\sinh(x)$ grows exponentially in x, we see the above sum converges rapidly (i.e., the large ℓ terms are immaterial), and the error term decreases rapidly with n.

If we take $B = 10$ we find the difference between the probability of observing the mantissa of X_n in $[1, s]$ and the Benford probability of $\log_B s$ is at most $0.0033 \log_B s$ if $n = 2$, $0.00019 \log_B s$ if $n = 3$, $0.000011 \log_B s$ if $n = 5$ and $3.6 \cdot 10^{-13} \log_B s$ if $n = 10$. If $B = 10$ then for all $\ell \geq 1$ we have $\exp(2\pi^2 \ell / \log 10) - \exp(-2\pi^2 \ell / \log 10) \geq \frac{10000}{10001} \exp(2\pi^2 \ell / \log 10)$. Thus the error term is bounded by

$$\log_{10} s \sum_{\ell=1}^{\infty} \left(\frac{17.148\ell}{\exp(8.5726\ell)} \right)^{n/2} \leq 0.057^n \log_{10} s.$$

(3.76)

3.4.4.2 Chains (or Products) of the Uniform Distribution

Let $X_1 \sim \text{Unif}(0, k)$ (without loss of generality we may assume $k \in [1, 10)$) and set $X_m \sim \text{Unif}(0, X_{m-1})$; equivalently, we could study $W_1 \cdots W_n$ where the $\{W_i\}_{i=1}^{n}$ denote independent $\text{Unif}(0, k)$ random variables. We know that as $n \to \infty$, X_n and $W_1 \cdots W_n$ converge to Benford behavior. Explicitly, if $P_n(s)$ is the probability that the base 10 mantissa of X_n (or $W_1 \cdots W_n$) is at most s, then for $n \geq 4$ we have

$$|P_n(s) - \log_{10} s| \leq \frac{k}{s} \frac{(\log k)^{n-1}}{\Gamma(n)} + \left(\frac{1}{2.9^n} + \frac{\zeta(n) - 1}{2.7^n} \right) 2 \log_{10} s,$$

(3.77)

where Γ is the Gamma function and $\zeta(s)$ is the Riemann zeta function ($1 < \zeta(n) < 2$ for $n \geq 2$ and $\zeta(n) = 1 + O(1/2^n)$).

3.4.5 Discussion

We end with a brief discussion of some consequences. Returning to our street example, we see we may reformulate it in terms of a Bayesian model (see [Berg] for more details). In **Bayesian models** we have some data (say x) whose values depend on a parameter (say β, called the prior). Thus there are two densities: that of the data (which depends on β) and that of the prior. In our situation, x would be the street address, drawn from a uniform distribution on say $[1, \beta]$, and then β would be drawn from some distribution modeling how street lengths are distributed. One can of course consider more involved models where the prior depends on a hyperparameter drawn from a different distribution (and so on). These are called hierarchical Bayesian models, and in this setting we again encounter chains of distribution, where the number of chains is basically the number of levels.

One of the major problems in Bayesian theory is to justify the choice of the prior. Many ideas have been proposed (for example, Jeffrey's prior, conjugate priors, empirical Bayes, hierarchical models). In putting priors on hyperparameters, we often make our prior more "diffuse," so to speak, or less informative. Our main result says that, in many cases, a non-informative prior in this hierarchical sense leads to sample data closely approximating Benford's Law; further, in many situations a Benford prior might be the true non-informative prior, rather than classic approaches which are essentially variants on the uniform distribution. Our results can thus be used as a data integrity check in this situation.

Our results immediately apply to the situation of hierarchical Bayesian models with each variable depending on just one other variable, establishing a connection between this field and Benford's Law. In particular, we see that when there are many levels then the observed sample values should approximately follow Benford's Law, and thus these simple digit frequency tests can be used to test some detailed assumptions about hierarchical Bayesian models. In practice there is excellent agreement with Benford's Law even when there are few levels (see the examples with explicit bounds from uniform and exponential chains).

Finally, a major goal of [JaKKKM] was to demonstrate the ease of using the Mellin transform to obtain rapidly converging estimates on deviations from Benford's Law. For many distributions, the associated **Markov chains** rapidly converge to a stationary distribution. It is of great interest to obtain estimates on the rate of convergence. Our results allow us to deduce such bounds in terms of the Mellin transform. Further, our results hold for a large class of underlying distributions; in particular, it is not necessary that the Markov chain converge to a stationary distribution. For more on Markov chains and convergence, see [DiaS-C, Has, MaRa, MeRRTT, Ran, Sinc1, Sinc2].

3.5 WEIBULL RANDOM VARIABLES, SURVIVAL DISTRIBUTIONS AND ORDER STATISTICS

3.5.1 Introduction

This section, written jointly with Victoria Cuff and Allison Lewis, is motivated by two observations. First, since Benford's seminal paper, many investigations have shown that amalgamating data from different sources leads to Benford behavior; second, many standard probability distributions are close to Benford behavior. These observations have, in fact, been used to detect fraud in many data sets (see Chapter 8).

Particularly important and interesting distributions to study are exponential random variables. The distribution of digits of an random variable drawn from the standard exponential distribution has been shown [EngLeu, LeScEv, MiNi2] to be close to Benford's Law base 10 (Leemis, Schmeiser and Evans [LeScEv] observed that the standard exponential is quite close to Benford's Law; this was proved by Engel and Leuenberger [EngLeu], who showed that the maximum difference in the cumulative distribution function from Benford's Law (base 10) is at least 0.029 and at most 0.03). In addition to being of interest in its own right, this result is important because of its applications to **order statistics**. (Recall the order statistics of a set of values $x_1, \dots, x_n$ are $y_i - y_{i-1}$, where the y_i's are the x_i's in increasing order.) For many probability distributions, the resulting order statistics of a sample of N independent random variables has spacings between the adjacent observations (the order statistics) approximately exponentially distributed, and thus the almost Benford behavior of the standard exponential density translates to almost Benford behavior in order statistics. For detailed statements and proofs, see [MiNi2], and see [NiMi2] for a nice application of these results in devising a new test for data integrity.

We provide an alternate proof, due to Miller–Nigrini [MiNi2], to [EngLeu] of the closeness of the standard exponential to Benford behavior. While both proofs apply Fourier analysis to periodic functions, in [EngLeu] the main step (their equation (5)) is interchanging an integration and a limit, whereas here the proof is based on applying Poisson Summation to the derivative of the cumulative distribution function of the logarithms modulo 1, F_B. Benford's Law is equivalent to $F_B(b) = b$, which by calculus is the same as $F'_B(b) = 1$ and $F_B(0) = 0$. Thus studying the deviation of $F'_B(b)$ from 1 is a natural way to investigate the deviations from Benford behavior.

The arguments from [MiNi2] generalize to a variety of other distributions. One particularly important generalization is to the **Weibull distribution** (see Theorem 3.5.3), whose density is

$$f(x; \alpha, \beta, \gamma) = \begin{cases} \frac{\gamma}{\alpha} \left(\frac{x-\beta}{\alpha} \right)^{(\gamma-1)} \exp\left(-\left(\frac{x-\beta}{\alpha} \right)^{\gamma} \right) & \text{if } x \geq \beta, \\ 0 & \text{otherwise,} \end{cases} \tag{3.78}$$

where $\alpha, \gamma > 0$. Note that α adjusts the scale of the data, β translates the input and only γ affects the shape of the distribution. The **exponential distribu-**

tion is just a special case of the Weibull ($\gamma = 1$), as is the **Rayleigh distribution** ($\gamma = 2$). The most common use of the Weibull is in **survival analysis**, where a random variable X (modeled by the Weibull) represents the "time-to-failure," resulting in a distribution where the failure rate is modeled relative to a power of time. The Weibull distribution arises in problems in such diverse fields as food contents, engineering, medical data, politics, pollution and sabermetrics [An, Carr, CoBr, CrRe, Fry, McSABF, Miko, Mi2, TeKaDu, We] to name just a few; see [ArJa] for connections between Weibull distributions and Benford's Law for Internet traffic. Thus the almost Benford behavior of Weibull random variables provides another explanation for the prevalence of Benford behavior in numerous systems.

3.5.2 Exponential Random Variables and Benford's Law

This section closely follows [MiNi2].

Theorem 3.5.1 (Miller–Nigrini [MiNi2]). *Let ζ have the standard (unit) exponential distribution:*

$$\text{Prob}\,(\zeta \in [\alpha, \beta]) \subset \int_\alpha^\beta e^{-t}dt, \quad [\alpha, \beta] \in [0, \infty). \qquad (3.79)$$

For $b \in [0, 1]$, let $F_B(b)$ be the cumulative distribution function of $\log_B \zeta \bmod 1$; thus $F_B(b) := \text{Prob}(\log_B \zeta \bmod 1 \in [0, b])$. Then for all $M \geq 2$,

$$F_B'(b) = 1 + 2 \sum_{m=1}^{\infty} \text{Re}\left(e^{-2\pi imb}\Gamma\left(1 + \frac{2\pi im}{\log B}\right)\right)$$

$$= 1 + 2 \sum_{m=1}^{M-1} \text{Re}\left(e^{-2\pi imb}\Gamma\left(1 + \frac{2\pi im}{\log B}\right)\right)$$

$$+ \mathcal{E}\left(4\sqrt{2}\pi c_1(B)e^{-(\pi^2 - c_2(B))M/\log B}\right), \qquad (3.80)$$

where $c_1(B), c_2(B)$ are constants such that for all $m \geq M \geq 2$ we have

$$e^{2\pi^2 m/\log B} - e^{-2\pi^2 m/\log B} \geq e^{2\pi^2 m/\log B}/c_1^2(B),$$

$$m/\log B \leq e^{2c_2(B)m/\log B},$$

$$1 - e^{-(\pi^2 - c_2(B))M/\log B} \geq 1/\sqrt{2}. \qquad (3.81)$$

For $B \in [e, 10]$ we may take $c_1(B) = \sqrt{2}$ and $c_2(B) = 1/5$, which give

$$\text{Prob}(\log \zeta \bmod 1 \in [a, b]) = b - a + \frac{2r}{\pi} \cdot \sin(\pi(b+a) + \theta) \cdot \sin(\pi(b-a))$$

$$+ \mathcal{E}\left(6.32 \cdot 10^{-7}\right), \qquad (3.82)$$

with $r \approx 0.000324986$, $\theta \approx 1.32427186$, and

$$\text{Prob}(\log_{10} \zeta \bmod 1 \in [a, b]) = b - a + \frac{2r_1}{\pi} \sin(\pi(b+a) - \theta_1) \sin(\pi(b-a))$$

$$- \frac{r_2}{\pi} \sin(2\pi(b+a) + \theta_2) \cdot \sin(2\pi(b-a)) + \mathcal{E}(8.5 \cdot 10^{-5}), \ (3.83)$$

with

$$r_1 \approx 0.0569573, \quad \theta_1 \approx 0.8055888,$$
$$r_2 \approx 0.0011080, \quad \theta_2 \approx 0.1384410. \tag{3.84}$$

Proof. To prove Theorem 3.5.1, it suffices to study the distribution of $\log_B \zeta$ mod 1 when ζ has the standard exponential distribution; see (3.79). The analysis is aided by the fact that the cumulative distribution function for the standard exponential random variable has a nice closed form expression: $\mathcal{F}_B(x) = 1 - \exp(-x)$. We have the following useful chain of equalities. Let $[a, b] \subset [0, 1]$. Then

$$\mathrm{Prob}(\log_B \zeta \text{ mod } 1 \in [a, b]) = \sum_{k=-\infty}^{\infty} \mathrm{Prob}(\log_B \zeta \in [a + k, b + k])$$

$$= \sum_{k=-\infty}^{\infty} \mathrm{Prob}(\zeta \in [B^{a+k}, B^{b+k}])$$

$$= \sum_{k=-\infty}^{\infty} \left(e^{-B^{a+k}} - e^{-B^{b+k}} \right). \tag{3.85}$$

It suffices to investigate (3.85) in the special case when $a = 0$, as the probability of any interval $[\alpha, \beta]$ can always be found by subtracting the probability of $[0, \alpha]$ from $[0, \beta]$. We are therefore led to studying, for $b \in [0, 1]$, the cumulative distribution function of $\log_B \zeta$ mod 1:

$$F_B(b) := \mathrm{Prob}(\log_B \zeta \text{ mod } 1 \in [0, b]) = \sum_{k=-\infty}^{\infty} \left(e^{-B^k} - e^{-B^{b+k}} \right). \tag{3.86}$$

This series expansion converges rapidly, and Benford behavior for ζ is equivalent to the rapidly converging series in (3.86) equalling b for all b.

As Benford behavior is equivalent to $F_B(b)$ equals b for all $b \in [0, 1]$, it is natural to compare $F_B'(b)$ to 1. If the derivative were identically 1 then $F_B(b)$ would equal b plus some constant. However, (3.86) is zero when $b = 0$, which implies that this constant would be zero. It is hard to analyze the infinite sum for $F_B(b)$ directly. By studying the derivative $F_B'(b)$ we find a function with an easier Fourier transform than the Fourier transform of $e^{-B^u} - e^{-B^{b+u}}$, which we then analyze by applying Poisson Summation.

We use the fact that the derivative of the infinite sum $F_B(b)$ is the sum of the derivatives of the individual summands. This is justified by the rapid decay of the summands; see, for example, Corollary 7.3 of [Lang1]. We find

$$F_B'(b) = \sum_{k=-\infty}^{\infty} e^{-B^{b+k}} B^{b+k} \log B = \sum_{k=-\infty}^{\infty} e^{-\beta B^k} \beta B^k \log B, \tag{3.87}$$

where for $b \in [0, 1]$ we set $\beta = B^b$.

Let $H(t) = e^{-\beta B^t} \beta B^t \log B$; note $\beta \geq 1$. As $H(t)$ is of rapid decay in t, we may apply Poisson Summation (Theorem 3.1.1). Thus

$$\sum_{k=-\infty}^{\infty} H(k) = \sum_{k=-\infty}^{\infty} \widehat{H}(k),$$

where $\widehat{H}$ is the Fourier transform of H: $\widehat{H}(u) = \int_{-\infty}^{\infty} H(t)e^{-2\pi itu}dt$. Therefore

$$F_B'(b) = \sum_{k=-\infty}^{\infty} H(k) = \sum_{k=-\infty}^{\infty} \widehat{H}(k) = \sum_{k=-\infty}^{\infty} \int_{-\infty}^{\infty} e^{-\beta B^t}\beta B^t \log B \cdot e^{-2\pi itk}dt.$$

(3.88)

We change variables by taking $w = B^t$. Thus $dw = B^t \log B \, dt$ or $\frac{dw}{w} = \log B \, dt$. As $e^{-2\pi itk} = (B^{t/\log B})^{-2\pi ik} = w^{-2\pi ik/\log B}$ we have

$$\begin{aligned} F_B'(b) &= \sum_{k=-\infty}^{\infty} \int_0^{\infty} e^{-\beta w}\beta w \cdot w^{-2\pi ik/\log B} \frac{dw}{w} \\ &= \sum_{k=-\infty}^{\infty} \beta^{2\pi ik/\log B} \int_0^{\infty} e^{-u}u^{-2\pi ik/\log B}du \\ &= \sum_{k=-\infty}^{\infty} \beta^{2\pi ik/\log B}\Gamma\left(1 - \frac{2\pi ik}{\log B}\right), \end{aligned}$$

(3.89)

where we have used the definition of the Γ-function,

$$\Gamma(s) = \int_0^{\infty} e^{-u}u^{s-1}\, du, \quad \operatorname{Re}(s) > 0.$$

(3.90)

As $\Gamma(1) = 1$ we have the following rapidly converging series expansion:

$$F_B'(b) = 1 + \sum_{m=1}^{\infty}\left[\beta^{2\pi im/\log B}\Gamma\left(1 - \frac{2\pi im}{\log B}\right) + \beta^{-2\pi im/\log B}\Gamma\left(1 + \frac{2\pi im}{\log B}\right)\right].$$

(3.91)

We can improve (3.91) by using additional properties of the Γ-function. If $y \in \mathbb{R}$ then from (3.90) we have $\Gamma(1 - iy) = \overline{\Gamma(1 + iy)}$ (where the bar denotes complex conjugation). Thus the mth summand in (3.91) is the sum of a number and its complex conjugate, which is simply twice the real part. We have formulas for the absolute value of the Γ-function for large argument. We use (see (8.332) on page 946 of [GrRy]) that

$$|\Gamma(1 + ix)|^2 = \frac{\pi x}{\sinh(\pi x)} = \frac{2\pi x}{e^{\pi x} - e^{-\pi x}}.$$

(3.92)

Writing the summands in (3.91) as $2\operatorname{Re}\left(e^{-2\pi imb}\Gamma\left(1 + \frac{2\pi im}{\log B}\right)\right)$, (3.91) becomes

$$\begin{aligned} F_B'(b) = 1 &+ 2\sum_{m=1}^{M-1}\operatorname{Re}\left(e^{-2\pi imb}\Gamma\left(1 + \frac{2\pi im}{\log B}\right)\right) \\ &+ 2\sum_{m=M}^{\infty}\operatorname{Re}\left(e^{-2\pi imb}\Gamma\left(1 + \frac{2\pi im}{\log B}\right)\right). \end{aligned}$$

(3.93)

The rest of the claims of Theorem 3.5.1 follow from simple estimation, algebra and trigonometry. $\qquad \square$

With constants as in the theorem, if we take $M = 1$ and $B = e$ (resp., $B = 10$) the error is at most 0.00499 (resp., 0.378), while if $M = 2$ and $B = e$ (resp.,

$B = 10$) the error is at most $3.16 \cdot 10^{-7}$ (resp., 0.006). Thus just *one* term is enough to get approximately five digits of accuracy base e, and two terms give three digits of accuracy base 10! For many bases we have reduced the problem to evaluating $\mathrm{Re}\left(e^{-2\pi i b}\Gamma\left(1 + \frac{2\pi i}{\log B}\right)\right)$. This example illustrates the power of Poisson Summation, taking a slowly convergent series expansion and replacing it with a rapidly converging one.

Corollary 3.5.2. *Let ζ have the standard exponential distribution. There is no base $B > 1$ such that ζ is Benford base B.*

Proof. Consider the infinite series expansion in (3.80). As $e^{-2\pi i m b}$ is a sum of a cosine and a sine term, (3.80) gives a rapidly convergent Fourier series expansion. If ζ were Benford base B, then $F'_B(b)$ must be identically 1; however, $\Gamma\left(1 + \frac{2\pi i m}{\log B}\right)$ is never zero for m a positive integer because its modulus is non-zero (see (3.92)). As there is a unique rapidly convergent Fourier series equal to 1 (namely, $g(b) = 1$; see [StSh1] for a proof), our $F'_B(b)$ cannot identically equal 1. $\qquad\square$

3.5.3 Weibull Random Variables and Benford's Law

Theorem 3.5.1 can be generalized; see [CuLeMi] for details as the proof is similar to the result for exponential random variables.

Theorem 3.5.3 (Cuff–Lewis–Miller [CuLeMi]). *Let $Z_{\alpha,0,\gamma}$ be a random variable whose density is a Weibull with parameters $\beta = 0$ and $\alpha, \gamma > 0$ arbitrary. For $z \in [0, 1]$, let $F_B(z)$ be the cumulative distribution function of $\log_B Z_{\alpha,0,\gamma} \bmod 1$; thus $F_B(z) := \mathrm{Prob}(\log_B Z_{\alpha,0,\gamma} \bmod 1 \in [0, z))$. Then the following conditions hold.*

1. *The density of $Z_{\alpha,0,\gamma}$, $F'_B(z)$, is given by*

$$F'_B(z) = 1 + 2\sum_{m=1}^{\infty} \mathrm{Re}\left[e^{-2\pi i m\left(z - \frac{\log\alpha}{\log B}\right)} \cdot \Gamma\left(1 + \frac{2\pi i m}{\gamma\log B}\right)\right]. \tag{3.94}$$

   In particular, the densities of $\log_B Z_{\alpha,0,\gamma} \bmod 1$ and $\log_B Z_{\alpha B,0,\gamma} \bmod 1$ are equal, and thus it suffices to consider only α in an interval of the form $[a, aB)$ for any $a > 0$.

2. *For $M \geq \frac{\gamma\log B\log 2}{4\pi^2}$, the error in $F'_B(z)$ from dropping the terms with $m \geq M$ is at most*

$$\frac{1}{\pi^3}2\sqrt{2}M(40 + \pi^2)\sqrt{\gamma\log B} \cdot e^{-\pi^2 M/\gamma\log B}. \tag{3.95}$$

3. *In order to have an error of at most ϵ in evaluating $F'_B(z)$, it suffices to take the first M terms, where*

$$M = \frac{k + \ln k + \frac{1}{2}}{a}, \tag{3.96}$$

   with $k \geq 6$ and

$$k = -\ln\left(\frac{a\epsilon}{C}\right), \quad a = \frac{\pi^2}{\gamma\log B}, \quad C = \frac{2\sqrt{2}(40 + \pi^2)\sqrt{\gamma\log B}}{\pi^3}. \tag{3.97}$$

To see the quality of the fit, we compared the series expansion for the derivative to the uniform distribution through a **Kolmogorov–Smirnov** test; see Figure 3.2 for a contour plot of the discrepancy. Note the good fit observed between the two distributions when $\gamma = 1$ (representing the exponential distributions), which we showed earlier is well fit by the Benford distribution.

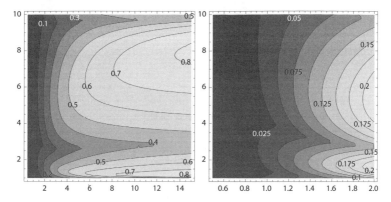

Figure 3.2 Kolmogorov–Smirnov Test. Left: $\gamma \in [0, 15]$. Right: $\gamma \in [0, 2]$. As γ (the shape parameter on the x-axis) increases, the Weibull distribution is no longer a good fit compared to the uniform. Note that α (the scale parameter on the y-axis) has less of an effect on the overall conformance.

The Kolmogorov–Smirnov metric gives a good comparison as it allows us to compare the distributions in terms of both parameters, γ and α. We also look at two other measures of closeness, the L_1- and the L_2-norms, which test the differences between (3.94) and the uniform distribution; see Figure 3.3. The L_1-norm of $f - g$ is $\int_0^1 |f(t) - g(t)| dt$, which puts equal weights on the deviations, while the L_2-norm is $\int_0^1 |f(t) - g(t)|^2 dt$, which gives more weight to larger distances.

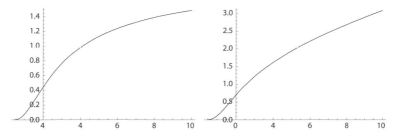

Figure 3.3 Left: L_1-norm of $F_B'(z) - 1$ for $\gamma \in [0.5, 10]$. Right: L_2-norm of $F_B'(z) - 1$ for $\gamma \in [0.5, 10]$. As γ increases the cumulative Weibull distribution is no longer a good fit compared to 1. The L_1- and L_2-norms are independent of α.

The combination of the Kolmogorov–Smirnov tests and the L_1- and L_2 norms shows us that Weibull distributions are almost Benford when γ is modest; as γ

increases the Weibull no longer conforms to the expected leading digit probabilities. The scale parameter α has a small effect on the conformance, but not nearly to the same extent as the shape parameter γ. Fortunately in many applications the shape parameter γ is not too large (it is frequently less than 2 in the Weibull references cited earlier), and thus these results provide additional support for the prevalence of Benford behavior.

3.6 BENFORDNESS OF CAUCHY DISTRIBUTIONS

This section, written jointly with Xixi Edelsbrunner, Karen Huan, Blake Mackall, Jasmine Powell, Caroline Turnage-Butterbaugh and Madeleine Weinstein, is a continuation of the analysis of exponential and Weibull random variables earlier in the chapter. Thus we do not give the most general arguments but rather concentrate on some of the interesting differences arising from Cauchy random variables, our object of study. The interested readers can adapt the previous methods to obtain similarly good estimates of tail probabilities, or see [EdH..].

We say X is a **Cauchy random variable** if its probability density function is

$$p\left(x\right) \;=\; \frac{1}{\pi\left(1+x^2\right)}. \tag{3.98}$$

The function $p(x)$ appears in many areas; in physics, for example, it is the solution to a differential equation that arises in the analysis of resonance. It has also been found to be a good predictor of the distribution of prices of certain stocks (see [ManHu], where it is argued that the Cauchy distribution provides a better fit than the standard random walk model as it leads to more days of large fluctuations). More generally, we can consider

$$p_{a,r}\left(x\right) \;=\; \frac{2r\sin(\pi/r)}{2\pi a}\,\frac{1}{1+|x/a|^r}, \quad a>0,\; r>1,\; x\geq 0 \tag{3.99}$$

(the normalization constant is readily determined by contour integration). For convenience, we adjusted the definition of our random variable to be non-zero only for $x \geq 0$, which avoids absolute values propagating throughout the equations below. Thus $p_{1,2}(x)$ is essentially the density of the standard Cauchy random variable.

The purpose of this section is to explain the observation that samples of data drawn from the Cauchy distribution are very close to Benford (see Table 3.3); as there is a large body of evidence of the applicability of the Cauchy distribution to financial data, this is another explanation for the prevalence of Benford's Law.

As before, we fix a base $B > 1$ and set $Y_{a,r;B} = \log_B X_{a,r}$, where $X_{a,r}$ is a random variable with density $p_{a,r}(x)$. The Uniform Distribution Characterization (Theorem 2.4.2) states that X is Benford base B if and only if $Y_{a,r;B}$ is uniformly distributed modulo 1. If $F_{a,r;B}(y)$ is $Y_{a,r;B}$'s cumulative distribution function (cdf) and $\mathcal{F}_{a,r;B}(y)$ is the cdf of $Y_{a,r;B} \bmod 1$, this is equivalent to showing $\mathcal{F}_{a,r;B}(b) = b$ or $\mathcal{F}'_{a,r;B}(b) = 1$ for all $b \in [0, 1]$. We show that this is almost true, and quantify

Digit	Observed	Predicted (Benford)	Predicted (Cauchy)
1	30908	30103	30930
2	17130	17609	16905
3	11819	12494	11852
4	9340	9691	9382
5	7924	7918	7881
6	6857	6695	6829
7	6065	5799	6022
8	5252	5115	5370
9	4705	4576	4827

Table 3.3 Chi-square test for Cauchy distribution; the critical thresholds (for 8 degrees of freedom) are approximately 15.5% (at the 95% level) and 20.1% (at the 99%). We sampled 100,000 points randomly and independently from the standard Cauchy distribution. The chi-square value comparing to the observed digit frequencies and Benford's Law is 107.169, indicating a bad fit; the chi-square value comparing to the digit frequencies of a Cauchy random variable is 9.64, which (not surprisingly) signifies a good fit with the digit frequencies of a standard Cauchy random variable.

how far $F'_{a,r;B}(b)$ is from 1 (which allows us to quantify how close $X_{a,r}$ is to being Benford base B). We have

$$F_{a,r;B}(y) = \text{Prob}(Y_{a,r;B} \leq y) = \text{Prob}(\log_B X_{a,r} \leq y)$$
$$= \text{Prob}(X_{a,r} \leq B^y) = P_{a,r}(B^y), \tag{3.100}$$

where $P_{a,r}$ is the cdf of $X_{a,r}$. As the pdf is the derivative of the cdf, we find

$$F'_{a,r;B}(y) = \frac{d}{dy}[P_{a,r}(B^y)] = p_{a,r}(B^y)B^y \log B. \tag{3.101}$$

We are reduced to studying how close

$$\mathcal{F}_{a,r;B}(b) = \sum_{n=-\infty}^{\infty} \int_n^{n+b} F'_{a,r;B}(y) dy$$

$$= 2 \sum_{n=-\infty}^{\infty} \int_n^{n+b} p_{a,r}(B^y) B^y \log B \, dy \tag{3.102}$$

is to b; the closer this is to b, the closer $X_{a,r}$ is to being Benford.

Rather than studying this probability directly, it is easier to study the derivative $\mathcal{F}'_{a,r;B}(b)$ and quantify how close this is to 1. The rapid convergence allows us to differentiate under the integral sign, and then we apply Poisson Summation to the sum over n (the Fourier transform of the density follows from standard analysis), yielding

$$\mathcal{F}'_{a,r;B}(b) = 2 \sum_{n=-\infty}^{\infty} p_{a,r}(B^{b+n}) B^{b+n} \log B$$

$$= \sin\left(\frac{\pi}{r}\right) \sum_{n=-\infty}^{\infty} e^{2\pi i \frac{b \log B - \log a}{\log B} n} \csc\left(\frac{\pi}{r}\left(1 - 2\pi i \frac{n}{\log B}\right)\right). \tag{3.103}$$

Note that the $n = 0$ term contributes 1 to the sum in (3.103). The fluctuation from 1 is the deviation of the sum from Benford, given by

$$E_{a,r;B}(b) = \sin\left(\frac{\pi}{r}\right) \sum_{|n|\geq 1} e^{2\pi i \frac{b\log B - \log a}{\log B} n} \csc\left(\frac{\pi}{r}\left(1 - 2\pi i \frac{n}{\log B}\right)\right).$$

(3.104)

Then we can split E_r into

$$E_{a,r;B}(b) = \sin\left(\frac{\pi}{r}\right) \sum_{\substack{n=-(M-1)\\ n\neq 0}}^{M-1} e^{2\pi i \frac{b\log B - \log a}{\log B} n} \csc\left(\frac{\pi}{r}\left(1 - 2\pi i \frac{n}{\log B}\right)\right)$$

$$+ \sin\left(\frac{\pi}{r}\right) \sum_{|n|\geq M} e^{2\pi i \frac{b\log B - \log a}{\log B} n} \csc\left(\frac{\pi}{r}\left(1 - 2\pi i \frac{n}{\log B}\right)\right). \quad (3.105)$$

We have expressed the fluctuation as a rapidly converging sum of trigonometric polynomials. We bound $E_{a,r;B,M}(b)$, the sum over all $|n| \geq M$. A good bound for this tail tells us how well the terms with $|n| \leq M - 1$ approximate the true answer, and we can then compare this to Benford behavior. We find

$$E_{a,r;B,M}(b) =$$

$$4\sin\left(\frac{\pi}{r}\right) \left| \sum_{n=M}^{\infty} \left(\frac{-\sin\left(\frac{\pi}{r}\right)\cosh\left(\frac{2\pi^2 n}{r\log B}\right)}{\cos\left(\frac{2\pi}{r}\right) - \cosh\left(\frac{4\pi^2 n}{r\log B}\right)} + \frac{i\cos\left(\frac{\pi}{r}\right)\sinh\left(\frac{2\pi^2 n}{r\log B}\right)}{\cos\left(\frac{2\pi}{r}\right) - \cosh\left(\frac{4\pi^2 n}{r\log B}\right)} \right) \right|.$$

(3.106)

Further simplification shows

$$E_{a,r;B,M}(b)$$

$$\leq 4\sin\left(\frac{\pi}{r}\right) \sum_{n=M}^{\infty} \left(\left| \frac{\sin\left(\frac{\pi}{r}\right)\cosh\left(\frac{2\pi^2 n}{r\log B}\right)}{1 - \cosh\left(\frac{4\pi^2 n}{r\log B}\right)} \right| + \left| \frac{\cos\left(\frac{\pi}{r}\right)\cosh\left(\frac{2\pi^2 n}{r\log B}\right)}{1 - \cosh\left(\frac{4\pi^2 n}{r\log B}\right)} \right| \right)$$

$$\leq 4\sqrt{2}\sin\left(\frac{\pi}{r}\right) \sum_{n=M}^{\infty} \left| \frac{\cosh\left(\frac{2\pi^2 n}{r\log B}\right)}{2\left(1 - \cosh^2\left(\frac{2\pi^2 n}{r\log B}\right)\right)} \right|. \quad (3.107)$$

As long as B and r satisfy $\cosh^2\left(\frac{2\pi^2}{r\log B}\right) \geq 2$ (which for $B = 10$ happens for $r < 9.726$), we have

$$E_{a,r;B,M}(b) \leq 4\sqrt{2}\sin\left(\frac{\pi}{r}\right) \sum_{n=M}^{\infty} \operatorname{sech}\left(\frac{2\pi^2 n}{r\log B}\right)$$

$$\leq 8\sqrt{2}\sin\left(\frac{\pi}{r}\right) \sum_{n=M}^{\infty} e^{-\left(\frac{2\pi^2 n}{r\log B}\right)} = 8\sqrt{2}\sin\left(\frac{\pi}{r}\right) \frac{e^{-\left(\frac{2\pi^2 M}{r\log B}\right)}}{1 - e^{-\left(\frac{2\pi^2}{r\log B}\right)}}. \quad (3.108)$$

We see that for $B = 10$ the contribution from $|n| \geq 2$ is already small, bounded by about 0.002 (if we took instead $M = 3$ it is bounded by about 0.00003). Thus just keeping the first or first two terms already suffices to essentially see the behavior, which we plot in Figure 3.4.

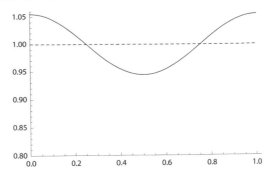

Figure 3.4 The fluctuation in the density of the digit distribution from a standard Cauchy random variable about 1. The maximum fluctuations are at $b = 0, 0.5$ and 1.

Digit	Cauchy	Benford
1	0.30930	0.30103
2	0.16906	0.17609
3	0.11852	0.12494
4	0.09382	0.09691
5	0.07882	0.07918
6	0.06829	0.06695
7	0.06022	0.05799
8	0.05370	0.05115
9	0.04827	0.04576

Table 3.4 Probabilities of first digits (standard Cauchy vs. Benford's Law).

Our analysis allows us to determine the effect of the parameters. When the $n = \pm 1$ terms contribute essentially all of the fluctuation, all a does is shift the plot of the derivative. A similar analysis to the previous section (i.e., integrating the density function) gives the digit probabilities. For the standard Cauchy, these are close to but different from Benford's Law. The probabilities are given in Table 3.4.

Additionally, our analysis shows that as $r \to 1$ the digit probabilities converge to Benford's Law. We briefly remark on why $X_{1,r}$ converge to Benford as $r \to 1$ from above. Base 10, a system is Benford if the probability of a significand at most s is $\log_{10} s$. Consider a function of the form $1/(1 + x)$ for $x \geq 0$ (note that this is not a pdf because its integral over the real line does not converge; however, as $r \to 1$ this is what our densities approach). Letting $t = x + 1$ yields the function $1/t$ on $[1, \infty)$ with antiderivative $\log t$. For any integer $k \geq 0$, the fraction of the range $[10^k, 10^{k+1})$ for which $1/t$ has significand at most s is

$$\frac{\log\left(10^k s\right) - \log\left(10^k\right)}{\log(10^{k+1}) - \log(10^k)} = \frac{\log s}{\log 10} = \log_{10} s, \tag{3.109}$$

which are the Benford probabilities.

PART II

General Theory II: Distributions and Rates of Convergence

Chapter Four

Benford's Law Geometry

Lawrence Leemis[1]

The original discovery of Benford's Law by Simon Newcomb was based on the uneven wear in the pages of logarithm tables. The subsequent independent discovery by Frank Benford was based on conformance of the first digit from a diverse set of data to Benford's Law; both of these discoveries were driven by empirical data. Although Benford's Law applies to a wide variety of data sets, none of the popular parametric distributions, such as the exponential and normal distributions, agree exactly with Benford's Law. After highlighting the failures of several well-known probability distributions in conforming to Benford's Law, we consider what types of probability distributions might produce data that obey Benford's Law, and look at some of the geometry associated with these probability distributions.

4.1 INTRODUCTION

Simon Newcomb (1835–1909) was a largely self-taught American who immigrated from Canada with professional interests in astronomy and mathematics. During his lifetime, calculations were typically performed using logarithm tables. Newcomb noticed that the pages of tables of logarithms had more wear at the beginning of the tables than at the end. The argument to a logarithm table ranges from 1.0 to 10.0 and is arranged in a linear fashion (for example, the arguments between 1.0 and 2.0 take exactly $1/9$ of the pages), yet the tables showed more wear on the earlier pages. Newcomb postulated that those using the logarithm tables tended to look up the largest fraction of values beginning with the digit 1 and the smallest fraction of values beginning with the digit 9. In what can be considered an astoundingly insightful conclusion, particularly considering that his data set could be viewed only by worn pages, he postulated that the distribution of the leading digit X of numbers accessed in the logarithm tables followed a discrete probability distribution with probability mass function

$$P(X = x) = \log_{10}(1 + 1/x), \qquad x = 1, 2, \ldots, 9.$$

Newcomb published what was known as the "logarithm law" in the *American Journal of Mathematics* in 1881. Considering just the extreme values, this law indicates that over 30% of the arguments to a logarithm table will have a leading digit of 1

---

[1]Department of Mathematics, The College of William & Mary, Williamsburg, VA 23187.

because $P(X = 1) = \log_{10}(2) \cong 0.301$, and less than 5% of the arguments to a logarithm table will have a leading digit of 9 because $P(X = 1) = \log_{10}(10/9) \cong 0.0458$.

Frank Benford, Jr. (1883–1948) was an electrical engineer and physicist who spent his career working for General Electric. He apparently independently arrived at the same conclusion as Newcomb concerning the distribution of the leading digit. His rediscovery of what has been named "Benford's Law" came from his collection of "data from as many fields as possible" to determine whether natural and sociological data sets would obey the logarithm law ([Ben]). In 1938 Benford analyzed the leading digits of $20, 229$ data values that he had gathered from a divergent set of sources (for example, populations of counties, American League baseball statistics, numbers appearing in *Reader's Digest*, areas of rivers, physical constants, death rates, drainage rates of rivers, atomic weights). The proportions associated with each of the leading digits are given in Table 4.1, which are a very close fit to Benford's Law.

Digits	1	2	3	4	5	6	7	8	9
Benford's Law	0.301	0.176	0.125	0.097	0.079	0.067	0.058	0.051	0.046
Data	0.306	0.185	0.124	0.094	0.080	0.064	0.051	0.049	0.047

Table 4.1 Benford's leading digit frequencies.

In hindsight, we know part of the explanation of why Benford's data came so close to the proposed distribution of leading digits. First of all, the data set contained observations that spanned several orders of magnitude. This is not a requirement for conformity to Benford's Law, but it seems to help. As shown later in this chapter, a probability distribution can satisfy Benford's Law and span only a single order of magnitude. Second, by choosing such a wide array of data values, Benford was effectively mixing several probability distributions together, and it has been seen that this also enhances conformance to Benford's Law.

Since certain data sets seem to approximate Benford's Law with regularity, a reasonable next step is to search for probability distributions that give rise to data that conforms to Benford's Law. Hill ([Hi4]) framed the question well: "An interesting open problem is to determine which common distributions (or mixtures thereof) satisfy Benford's Law ...." This chapter switches from the traditional analysis of Benford's Law using *data sets* to a search for *probability distributions* that obey Benford's Law.

The analogous search occurred in the early days of probability theory when analysts found so many measurements that produced data that was bell shaped that they named the associated probability distribution the "normal" distribution. (Perhaps any non-bell-shaped distribution was considered to be "abnormal" at the time.) Most of what is known as classical statistics emerged from the derivation of the probability density function of the normal, or Gaussian, distribution.

In order to limit the focus of this chapter, the following assumptions will be made.

- The focus of the analysis is on probability distributions rather than data.

- A probability distribution that might obey Benford's Law is associated with a continuous random variable.

- The probability distribution has support on the positive real numbers or some subset thereof.

- Base 10 is used to represent random variables associated with probability distributions.

- Only the leading digit is of interest. All digits to the right of the leading digit are ignored.

Other chapters in this book (especially Chapter 2) concern mathematical results associated with relaxation of these assumptions.

The next section considers some popular parametric distributions and assesses their conformity to Benford's Law. The following section considers probability distributions that obey Benford's Law exactly and the geometric and algebraic properties that they possess. The last section contains conclusions.

4.2 COMMON PROBABILITY DISTRIBUTIONS

As given in the previous section, let X be a discrete random variable whose support is the integers $1, 2, \ldots, 9$ with probability mass function

$$f_X(x) = P(X = x) = \log_{10}(1 + 1/x), \qquad x = 1, 2, \ldots, 9.$$

In this chapter, the term "Benford distribution" is used to describe this probability distribution. The random variable X has the following associated cumulative distribution function on its support values:

$$F_X(x) = P(X \le x) = \log_{10}(1 + x), \qquad x = 1, 2, \ldots, 9.$$

Using the probability integral transformation, random variates having the Benford distribution are generated via

$$X \leftarrow \lfloor 10^U \rfloor,$$

where U is a uniform random variable on $[0, 1]$, denoted by $U \sim U(0, 1)$.

We now define the probability distribution from which the data will be drawn. Let the continuous random variable T have positive support and cumulative distribution function $F_T(t) = P(T \le t)$. We are interested in the leading digit of a realization of T, which we obtain through the **significand function**. (Recall we may write any positive number x uniquely as $S(x) \cdot 10^{k(x)}$, where $S(x) \in [1, 10)$ and $k(x)$ is an integer; S is called the significand function.) For example, $S(e) = S(10e) = S(e/100) = 2.71828\ldots$. Using the significand function, the leading digit of T can be expressed as

$$Y = \lfloor S(T) \rfloor.$$

The next calculation that is necessary is to determine the probabilities associated with the nine potential leading digits of T having an arbitrary probability distribution. The probability mass function of Y is

$$f_Y(y) = P(Y = y) = P(\lfloor S(T) \rfloor = y)$$

$$= \sum_{i=-\infty}^{\infty} \left[F_T \left((y+1) \cdot 10^i \right) - F_T \left(y \cdot 10^i \right) \right] \qquad (4.1)$$

for $y = 1, 2, \ldots, 9$.

Since Benford's Law seems to apply to a variety of data sets, one would assume that several of the popular parametric models, such as the exponential or Weibull distributions, would provide a close fit to Benford's Law. For certain choices of the parameters of some of these distributions this is in fact the case. The probability mass function of Y was calculated for the $U(1, 10)$, unit exponential, and unit Rayleigh distributions, which respectively have cumulative distribution functions

$$F_T(t) = \frac{t-1}{9}, \qquad 1 < t < 10,$$

$$F_T(t) = 1 - e^{-t}, \qquad t > 0,$$

and

$$F_T(t) = 1 - e^{-t^2}, \qquad t > 0.$$

The following R code, with a carefully selected lower bound `lo` and a carefully selected upper bound `hi` in order to ensure that nearly all of the probability density is captured, calculates the probability mass function of Y for the unit Rayleigh distribution.

```
cdf = function(x) 1 - exp(-x ^ 2)
digits = rep(0, 9)
for (y in 1:9) {
  for (i in lo:hi) {
    digits[y] = digits[y] + cdf((y + 1) * 10 ^ i) - cdf(y * 10 ^ i)
  }
}
print(digits)
```

The results of these calculations for all three probability distributions are shown in Table 4.2. The $U(1, 10)$ distribution provides the worst fit of the three probability distributions because each leading digit is equally likely to occur. The unit exponential probability mass function is monotone like the Benford probability mass function, but it gives too many 1s, 8s, and 9s relative to the Benford distribution. Finally, the unit Rayleigh distribution deviates even further from Benford's Law. Although the unit exponential distribution is the best of the three in terms of proximity to Benford's distribution, none of these perform even as well as Benford's original data set.

Having failed to find a distribution that closely approximates Benford's Law, the search widens for probability distributions that provide a closer approximation. A

Leading digit	1	2	3	4	5	6	7	8	9
Benford's Law	0.301	0.176	0.125	0.097	0.079	0.067	0.058	0.051	0.046
$U(1, 10)$ distr.	0.111	0.111	0.111	0.111	0.111	0.111	0.111	0.111	0.111
Unit exponential distr.	0.330	0.174	0.113	0.086	0.073	0.064	0.058	0.053	0.049
Unit Rayleigh distr.	0.379	0.066	0.063	0.074	0.082	0.087	0.087	0.084	0.079

Table 4.2 Leading digit frequencies for common probability distributions.

probability distribution that provides a surprisingly close approximation to Benford's Law is the **log-normal distribution** with cumulative distribution function

$$F_T(t) = \Phi\left(\frac{\ln t - \mu}{\sigma}\right), \qquad t > 0,$$

where μ is a real-valued parameter, and σ is a positive real-valued parameter. We arbitrarily set $\mu = 0$ and gradually increase σ. Since the approximation to Benford's Law is very close for the log-normal distribution, we use a measure similar to the Kolmogorov–Smirnov goodness-of-fit test statistic to assess the fit:

$$d = \max_{x=1,2,\ldots,9} \{|\Pr(Y = x) - \Pr(X = x)|\}.$$

Table 4.3 gives the value of d for several values of σ.

σ	1/4	1/2	1	2	3
d	1.96×10^{-1}	1.17×10^{-1}	7.30×10^{-3}	1.03×10^{-7}	1.03×10^{-15}

Table 4.3 Assessing conformance to Benford's Law for the log-normal distribution.

The log-normal distribution appears empirically to be approaching Benford's Law as σ increases (see Theorem 3.2.2 and Corollary 5.4.7 for additional theoretical support). What is it about the log-normal distribution that makes this occur? The geometry behind why certain distributions conform well to Benford's Law is taken up in the next section.

4.3 PROBABILITY DISTRIBUTIONS SATISFYING BENFORD'S LAW

Rather than looking at the well-known probability distributions considered by probabilists for modeling or by statisticians for statistical inference, we try to construct probability distributions that satisfy Benford's Law exactly. One of the key insights gleaned from the last section is that random variables whose logarithm provides a symmetric distribution have a good chance of satisfying Benford's Law.

We initially look past the obvious continuous probability distribution that satisfies Benford's Law by brute force, that is,

$$f_T(t) = \log_{10}(1 + 1/\lfloor t \rfloor), \qquad 1 < t < 10.$$

This distribution spans simply one order of magnitude, dividing the nine leading digits between 1 and 9 into cells with probabilities that match Benford's distribution exactly.

In order to find a nontrivial distribution with exact conformance to Benford's Law, we define another random variable: $W = \log_{10} T$. It is easier to construct distributions that conform to Benford's Law by working with W rather than T. For example, we let $W \sim U(0, 1)$, which has probability density function

$$f_W(w) = 1, \qquad 0 < w < 1.$$

By using the transformation technique (see for example [HogMC]), the distribution of $T = 10^W$ has probability density function

$$f_T(t) = \frac{1}{t \ln 10}, \qquad 1 < t < 10.$$

This probability distribution is, from one point of view, the primary continuous distribution whose leading digit satisfies Benford's Law because (a) its base 10 logarithm is $U(0, 1)$, (b) its support spans a single order of magnitude, and (c) its probability density function is a continuous function (unlike the probability density function described in the previous paragraph). Figure 4.1 shows the probability density function of W on the left-hand graph and the probability density function of T on the right-hand graph. The shaded areas on the graphs correspond to the probability that the leading digit is 4 (the digit 4 was an arbitrary choice).

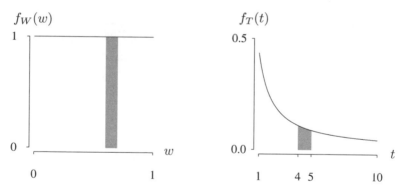

Figure 4.1 Probability density functions of $W \sim U(0, 1)$ and $T = 10^W$ with $P(Y = 4)$ shaded.

This example can be extended to cover two orders of magnitude simply by letting $W \sim U(0, 2)$. This distribution also satisfies Benford's Law exactly. In this case, the probability density function of W is

$$f_W(w) = \frac{1}{2}, \qquad 0 < w < 2.$$

The distribution of $T = 10^W$ has probability density function

$$f_T(t) = \frac{1}{2t \ln 10}, \qquad 1 < t < 100.$$

Figure 4.2 shows the probability density function of W on the left-hand graph and the probability density function of T on the right-hand graph. The shaded areas on the graphs correspond to the probability that the leading digit is 4. Since two orders of magnitude are spanned by the support of T, there are two ranges ($4 \leq T < 5$ and $40 \leq T < 50$) that result in having $Y = 4$ as a leading digit.

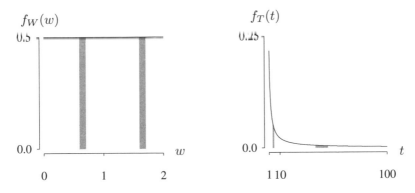

Figure 4.2 Probability density functions of $W \sim U(0, 2)$ and $T = 10^W$ with $P(Y = 4)$ shaded.

The previous two distributions that satisfy Benford's Law exactly can be generalized to cover all uniform distributions for W that cover an integer number of orders of magnitude. Let $W \sim U(a, b)$. As long as $b - a$ is a positive integer, then the support of T is $10^a < T < 10^b$, which covers $b - a$ orders of magnitude. Benford's Law is satisfied exactly because the effect of picking off the leading digit shifts all W values into the interval $(0, 1)$ and shifts all corresponding T values into the interval $(1, 10)$. When $b - a$ is an integer, the support of W that falls outside of $(0, 1)$ that is shifted into the unit interval does so in a fashion that results in T following Benford's Law, as will be seen geometrically in the next paragraph. For example, if $W \sim U(3.507, 6.507)$, then the support of T spans exactly three orders of magnitude and it obeys Benford's Law exactly.

There are non-uniform distributions for W that also satisfy Benford's Law. One simple example is to allow W to have the triangular distribution with minimum 0, mode 1, and maximum 2. In this case, the probability density function of W is

$$f_W(w) = \begin{cases} w, & 0 < w < 1, \\ 2 - w, & 1 \leq w < 2. \end{cases}$$

The distribution of $T = 10^W$ has probability density function

$$f_T(t) = \begin{cases} \frac{\ln t}{t(\ln 10)^2}, & 1 < t < 10, \\ \frac{2\ln 10 - \ln t}{t(\ln 10)^2}, & 10 \leq w < 100. \end{cases}$$

Figure 4.3 shows the probability density function of W on the left-hand graph and the probability density function of T on the right-hand graph. The shaded areas on the graphs correspond to the probability that the leading digit is 4. Since two orders of magnitude are again spanned by the support of T, there are two ranges

$(4 \leq T < 5$ and $40 \leq T < 50)$ that result in having $Y = 4$ as a leading digit. The geometry associated with what is happening by picking off the leading digit is most easily seen by considering the support of W. The probability density function on

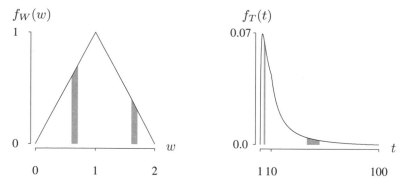

Figure 4.3 Probability density functions of $W \sim \text{triangular}(0, 1, 2)$ and $T = 10^W$ with $P(Y = 4)$ shaded.

the range $1 < w < 2$ is being shifted to the left by one unit, as seen in Figure 4.4. The two shaded bars from Figure 4.3 are stacked on top of one another which reach to the dashed line in Figure 4.4 that provide the basis for the conformance to Benford's Law.

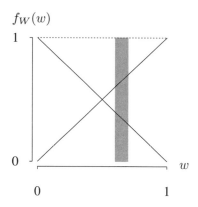

Figure 4.4 Probability density function of $W \sim \text{triangular}(0, 1, 2)$ with $P(Y = 4)$ shaded, shifted onto $(0, 1)$.

All of the examples of random variables W that satisfy Benford's Law have been symmetric distributions. We now consider the case of a nonsymmetric distribution of W that satisfies Benford's Law exactly. Let $W \sim \text{triangular}(0, 1, 3)$. In this case, the probability density function of W is

$$f_W(w) = \begin{cases} 2w/3, & 0 < w < 1, \\ 1 - w/3, & 1 \leq w < 3. \end{cases}$$

The distribution of $T = 10^W$ has probability density function

$$f_T(t) = \begin{cases} \frac{2\ln t}{3t(\ln 10)^2}, & 1 < t < 10, \\ \frac{\ln 10 - \ln t^{1/3}}{t(\ln 10)^2}, & 10 \leq w < 1000. \end{cases}$$

Figure 4.5 shows the probability density function of W on the left-hand graph and the probability density function of T on the right-hand graph. The shaded areas on the graphs correspond to the probability that the leading digit is 4. Since three orders of magnitude are spanned by the support of T, there are three ranges $(4 \leq T < 5, 40 \leq T < 50, \text{ and } 400 \leq T < 500)$ that result in having $Y = 4$ as a leading digit. The geometry associated with what is happening by picking off the leading digit is most easily seen by considering the support of W. The probability

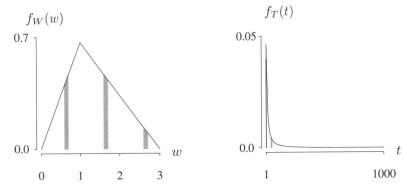

Figure 4.5 Probability density functions of $W \sim \text{triangular}(0, 1, 3)$ and T with $P(Y = 4)$ shaded.

density function on the range $1 < w < 2$ is being shifted to the left by one unit, and the probability density function on the range $2 < w < 3$ is being shifted to the left by two units, as seen in Figure 4.6. The probabilities associated with a leading digit of $Y = 4$, or any other leading digit for that matter, correspond to the rectangle of height 1 in Figure 4.6.

All of the examples in this section can be viewed through a different lens. Instead of shifting orders of magnitude associated with W onto the unit interval, the shifting can be considered to be a finite mixture model (see for instance [McLPe]). Consider the $W \sim \text{triangular}(0, 1, 2)$ example. This is equivalent to two probability density functions, namely

$$f_{W_1}(w) = 2w, \qquad 0 < w < 1,$$

and

$$f_{W_2}(w) = 2(1 - w), \qquad 0 < w < 1,$$

which are mixed together with equal probabilities. In this case

$$f_W(w) = pf_{W_1}(w) + (1-p)f_{W_2}(w) = \frac{1}{2} \cdot 2w + \frac{1}{2} \cdot 2(1-w) = 1, \quad 0 < w < 1.$$

Since $W \sim U(0, 1)$, this implies that W corresponds to a random variable $T = 10^W$ that satisfies Benford's Law.

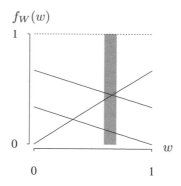

Figure 4.6 Probability density function of $W \sim \text{triangular}(0, 1, 3)$ with $P(Y = 4)$ shaded, shifted onto $(0, 1)$.

4.4 CONCLUSIONS

Benford's Law is approximated to varying degrees for common parametric distributions. An infinite array of probability distributions can be constructed, however, that satisfy Benford's Law exactly. The geometry discussed in this chapter works with $W = \log_{10} T$, which must be $U(0, 1)$ after shifting to account for the various orders of magnitude, to satisfy Benford's Law. The distribution of W, with odd-numbered leading digits shaded, is shown in Figure 4.7.

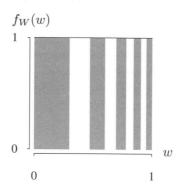

Figure 4.7 Probability density function of $W \sim U(0, 1)$ with odd leading digits of $T = 10^W$ shaded.

Chapter Five

Explicit Error Bounds via Total Variation

Lutz Dümbgen and Christoph Leuenberger[1]

Similar to earlier work in the book, this chapter is concerned with obtaining explicit error estimates for convergence to Benford's law. The analysis is done through the total variation of the densities. This yields reasonable estimates for Benford's law in many cases, and is often simpler to calculate and more elementary than Fourier methods.

5.1 INTRODUCTION

Let us recall a common heuristic explanation of Benford's first digit law: "If a random variable $X > 0$ has a distribution ranging over several orders of magnitude, its first leading digit is likely to follow Benford's law." This is closely related to another phenomenon, applied to $Y := \log_B(X)$ for some integer base $B \geq 2$:

Phenomenon: *If the distribution of a real random variable Y is **diffuse**, its **rounding error***

$$U := Y - \lfloor Y \rfloor \in [0, 1)$$

follows approximately a uniform distribution on $[0, 1)$.

Applying this to $Y = \log_B(X)$, we may conclude the following: For arbitrary integers $\ell \geq 0$ and digits $d_0, \ldots, d_\ell \in \{0, 1, \ldots, b-1\}$ with $d_0 \neq 0$,

$$P\big(X \text{ has leading digits } d_0, \ldots, d_\ell \text{ base } B\big)$$
$$= P\big(\log_B(d) \leq U < \log_B(d + B^{-\ell})\big), \qquad (5.1)$$

where $d := \sum_{j=0}^{\ell} d_j B^{-j}$. The latter probability is approximately equal to

$$\log_B(d + B^{-\ell}) - \log_B(d) = \log_B(1 + B^{-\ell}/d). \qquad (5.2)$$

To formulate the phenomenon above rigorously, one has to specify the meaning of "diffuse" of course. One could think of the distribution of Y ranging over several units or the standard deviation of Y being large. In fact, Feller [Fel, 1971] is arguing along these lines, but Berger and Hill [BerH3, 2011] report fundamental flaws in this type of argument. In the present chapter we show that the heuristic above

---

[1]University of Bern, Bern, Switzerland and University of Fribourg, Fribourg, Switzerland respectively.

does work remarkably well if one quantifies "diffuseness" in terms of the total variation of a probability density or its derivatives. We present explicit bounds due to Dümbgen and Leuenberger [DueLeu, 2008], extending results of Pinkham [Pin, 1961] and Kemperman [Kemp, 1975]. The latter reference is an abstract of a conference talk, which the present authors learned about only recently. (Apparently, J. Kemperman never published proofs of his results.) A particular consequence of the bounds presented here is that the distribution of the remainder U is very close to the uniform distribution in the case of Y being normally distributed with standard deviation one or more.

The remainder of this chapter is organized as follows: In Section 5.2 we provide the distribution of the remainder U in the case of Y having a Lebesgue density f, define our measures of non-uniformity of this distribution and collect some basic facts about the total variation of functions. The main results and examples are then presented in Sections 5.3 and 5.4. All proofs are deferred to Section 5.5.

5.2 PRELIMINARIES

Throughout this chapter we assume that Y is a real random variable with cumulative distribution function (c.d.f.) F and Lebesgue density f.

5.2.1 The Distribution of the Remainder U

The probability of U falling into a Borel set $S \subset [0,1)$ equals $P(U \in S) = \sum_{n \in \mathbb{Z}} P(Y \in n + S)$. This entails that the c.d.f. G of U is given by

$$G(x) := P(U \le x) = \sum_{n \in \mathbb{Z}} (F(n+x) - F(n)) \quad \text{for } 0 \le x \le 1,$$

while the corresponding density g equals

$$g(x) := \sum_{n \in \mathbb{Z}} f(n+x).$$

Note the latter equation defines a periodic function $g : \mathbb{R} \to [0, \infty]$, i.e., $g(x + n) = g(x)$ for arbitrary $x \in \mathbb{R}$ and $n \in \mathbb{Z}$. Strictly speaking, a density of U is given by $\mathbb{I}_{\{0 \le x < 1\}} g(x)$.

5.2.2 Measuring Non-Uniformity of U

We shall quantify the distance between the distribution of U and $\text{Unif}[0, 1)$ by means of the **range of** g,

$$R(g) := \sup_{x,y \in \mathbb{R}} |g(y) - g(x)| \ge \sup_{u \in [0,1]} |g(u) - 1|.$$

The latter inequality follows from $\sup_{x \in \mathbb{R}} g(x) \ge \int_0^1 g(x)\, dx = 1 \ge \inf_{x \in \mathbb{R}} g(x)$. In addition we shall consider the **Kuiper distance** between the distribution of U

and $\mathrm{Unif}[0, 1)$,

$$
\begin{aligned}
\mathrm{KD}(G) &:= \sup_{0 \le x < y \le 1} \big| G(y) - G(x) - (y - x) \big| \\
&= \sup_{0 \le x < y \le 1} \big| P(x \le U < y) - (y - x) \big|,
\end{aligned}
$$

and the **maximal relative approximation error**,

$$
\begin{aligned}
\mathrm{MRAE}(G) &:= \sup_{0 \le x < y \le 1} \left| \frac{G(y) - G(x)}{y - x} - 1 \right| \\
&= \sup_{0 \le x < y \le 1} \left| \frac{P(x \le U < y)}{y - x} - 1 \right|.
\end{aligned}
$$

Expressions (5.1–5.2) show that these distance measures are canonical in connection with Benford's law. Note that $\mathrm{KD}(G)$ is bounded from below by the more standard **Kolmogorov–Smirnov distance**,

$$
\sup_{x \in [0,1]} |G(x) - x|,
$$

and it is not greater than twice the Kolmogorov–Smirnov distance.

Any bound on $\mathrm{R}(g)$ entails bounds on $\mathrm{KD}(G)$ and $\mathrm{MRAE}(G)$:

Proposition 5.2.1. *For arbitrary* $0 \le x \le y \le 1$,

$$
\big| G(y) - G(x) - (y - x) \big| \le (y - x)\big(1 - (y - x)\big) \mathrm{R}(g).
$$

In particular,

$$
\mathrm{KD}(G) \le \mathrm{R}(g)/4 \quad \text{and} \quad \mathrm{MRAE}(G) \le \mathrm{R}(g).
$$

In connection with smooth densities f and g we shall utilize a refinement of Proposition 5.2.1.

Proposition 5.2.2. *Suppose that for some* $L(g) > 0$ *and arbitrary* $x, y \in [0, 1]$,

$$
\big| g(x) - g(y) \big| \le |x - y|(1 - |x - y|)\, L(g).
$$

Then for all $0 \le x \le y \le 1$,

$$
\big| G(y) - G(x) - (y - x) \big| \le (y - x)\big(1 - (y - x)\big) L(g)/6.
$$

In particular,

$$
\mathrm{KD}(G) \le L(g)/24 \quad \text{and} \quad \mathrm{MRAE}(G) \le L(g)/6.
$$

5.2.3 Total Variation of Functions

Let us recall the definition of **total variation** (cf. Royden [Roy, chapter 5]): For any interval $\mathbb{J} \subset \mathbb{R}$ and a function $h : \mathbb{J} \to \mathbb{R}$, the total variation of h on $\mathbb{J}$ is defined as

$$
\mathrm{TV}(h, \mathbb{J}) := \sup_{m \in \mathbb{N};\ t_0 < t_1 < \cdots < t_m;\ t_0, \dots, t_m \in \mathbb{J}} \sum_{i=1}^{m} \big| h(t_i) - h(t_{i-1}) \big|.
$$

In the case of $\mathbb{J} = \mathbb{R}$ we just write $\mathrm{TV}(h) := \mathrm{TV}(h, \mathbb{R})$.

Example 5.2.3. *Suppose that h is absolutely continuous with locally integrable derivative h', i.e.,*

$$h(b) - h(a) = \int_a^b h'(x) \, dx \quad \text{for real numbers } a < b.$$

Then

$$\text{TV}(h) = \int_{\mathbb{R}} |h'(x)| \, dx.$$

Example 5.2.4. *Another important special case is **unimodal probability densities** f on the real line, i.e., f is non-decreasing on $(-\infty, \mu]$ and non-increasing on $[\mu, \infty)$ for some real number μ. Here*

$$\text{TV}(f) = 2f(\mu).$$

Note that total variation may be decomposed as

$$\text{TV}(h, \mathbb{J}) = \text{TV}^+(h, \mathbb{J}) + \text{TV}^-(h, \mathbb{J})$$

with

$$\text{TV}^{\pm}(h, \mathbb{J}) := \sup_{m \in \mathbb{N}; \; t_0 < t_1 < \cdots < t_m; \; t_0, \ldots, t_m \in \mathbb{J}} \sum_{i=1}^m \left(h(t_i) - h(t_{i-1}) \right)^{\pm}$$

and $a^{\pm} := \max(\pm a, 0)$ for real numbers a. Here is a further useful fact in the case of $\mathbb{J} = \mathbb{R}$.

Lemma 5.2.5. *Let $h : \mathbb{R} \to \mathbb{R}$ with $\text{TV}(h) < \infty$. Then both limits $h(\pm\infty) := \lim_{x \to \pm\infty} h(x)$ exist. Moreover, for arbitrary $x \in \mathbb{R}$,*

$$h(x) = h(-\infty) + \text{TV}^+(h, (-\infty, x]) - \text{TV}^-(h, (-\infty, x]).$$

In particular, if $h(\pm\infty) = 0$, then $\text{TV}^+(h) = \text{TV}^-(h) = \text{TV}(h)/2$.

This standard result implies that any function h on the real line with $\text{TV}(h) < \infty$ is the difference of two non-decreasing and bounded functions. Furthermore, $\lim_{|x| \to \infty} h(x) = 0$ whenever both $\int |h(x)| \, dx$ and $\text{TV}(h)$ are finite.

A function h on the real line is called $k \geq 1$ times absolutely continuous if $h \in \mathcal{C}^{k-1}(\mathbb{R})$, and if its derivative $h^{(k-1)}$ is absolutely continuous. With $h^{(k)}$ we denote some version of the derivative of $h^{(k-1)}$. To be more specific, the derivative of an absolutely continuous function is a class of locally integrable functions that differ pairwise only on nullsets. In particular, the derivative cannot be evaluated pointwise. By the derivative of h we mean a function from the above equivalence class.

Here is an important extension of Lemma 5.2.5 which we will prove in Section 5.5.

Lemma 5.2.6. *Let h be integrable and $k \geq 1$ times absolutely continuous such that $\text{TV}(h^{(k)}) < \infty$ for some version of $h^{(k)}$. Then*

$$\lim_{|x| \to \infty} h^{(j)}(x) = 0 \quad \text{for } j = 0, 1, \ldots, k.$$

5.3 ERROR BOUNDS IN TERMS OF $\mathrm{TV}(F)$

Pinkham [Pin, corollary to Theorem 2] proved the inequality

$$\sup_{0 \leq x \leq 1} |G(x) - x| \leq \mathrm{TV}(f)/6$$

via Fourier techniques. Kemperman [Kemp] formulated the refined inequality

$$|G(x) - x| \leq x(1-x)\mathrm{TV}(f)/2 \leq \mathrm{TV}(f)/8 \quad \text{for } 0 \leq x \leq 1,$$

which is also a consequence of Corollary 5.3.2 below.

Theorem 5.3.1. *Suppose that* $\mathrm{TV}(f) < \infty$. *Then* g *is real valued with*

$$\mathrm{TV}(g, [0,1]) \leq \mathrm{TV}(f) \quad \text{and} \quad \mathrm{R}(g) \leq \mathrm{TV}(f)/2.$$

Combining this result with Proposition 5.2.1 yields error bounds for G:

Corollary 5.3.2. *Under the conditions of Theorem 5.3.1, for* $0 \leq x < y \leq 1$,

$$\big|G(y) - G(x) - (y - x)\big| \leq (y-x)(1 - (y-x))\mathrm{TV}(f)/2.$$

In particular,

$$\mathrm{KD}(G) \leq \mathrm{TV}(f)/8 \quad \text{and} \quad \mathrm{MRAE}(G) \leq \mathrm{TV}(f)/2.$$

Remark 5.3.3. *The inequalities in Theorem 5.3.1 are sharp in the sense that for each number* $\tau > 0$ *there exists a density* f *such that the corresponding density* g *satisfies*

$$\mathrm{TV}(g, [0,1]) = \mathrm{TV}(f) = 2\tau \quad \text{and} \quad \max_{0 \leq x < y \leq 1} |g(x) - g(y)| = \tau. \quad (5.3)$$

A first class of examples, with continuous densities f *and* g, *is as follows: Let* f *be a probability density with* $f(0) = \tau \in (0, 2]$ *such that for all integers* $n \geq 0$,

$$f \text{ is } \begin{cases} \text{linear and non-increasing on } [n, n + 1/2], \\ \text{constant on } [n + 1/2, n + 1], \end{cases}$$

whereas for all integers $n < 0$,

$$f \text{ is } \begin{cases} \text{constant on } [n, n + 1/2], \\ \text{linear and non-decreasing on } [n + 1/2, n + 1]. \end{cases}$$

Then f *is unimodal with mode at zero, whence* $\mathrm{TV}(f) = 2f(0) = 2\tau$. *Moreover, one verifies easily that* g *is linear on both* $[0, 1/2]$ *and* $[1/2, 1]$ *with* $g(0) - g(1/2) = g(1) - g(1/2) = \tau$. *Thus* $\mathrm{TV}(g, [0,1]) = 2\tau$ *as well. Figure 5.1 illustrates this construction. The left panel shows a density* f *with* $f(0) = 0.4 = \mathrm{TV}(f)/2$, *and the right panel shows the resulting function* g *with* $\mathrm{TV}(g, [0,1]) = \mathrm{TV}(f) = 0.8$.

Example 5.3.4 (Uniform distributions). *Another simple example is the uniform density* $f(x) = \mathbb{I}_{\{0 \leq x < \tau^{-1}\}}\tau$. *Writing* $\tau^{-1} = m + a$ *for some integer* $m \geq 0$ *and* $a \in (0, 1]$, *one can easily verify that*

$$g(x) = m\tau + \tau\mathbb{I}_{\{0 \leq x < a\}},$$

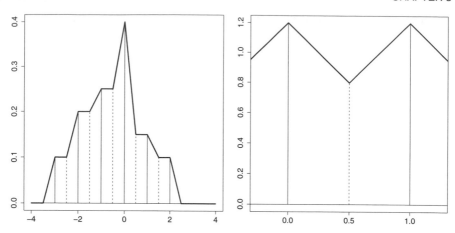

Figure 5.1 A density f (left) and the corresponding g (right) such that $\text{TV}(f) = \text{TV}(g)$.

and this implies (5.3), unless $a = 1$. Moreover,

$$G(x) = \begin{cases} (m+1)x\tau & \text{for } 0 \le x \le a, \\ (mx+a)\tau & \text{for } a \le x \le 1, \end{cases}$$

and elementary calculations reveal that

$$\text{KD}(G) = a(1-a)\,\tau = 4a(1-a)\text{TV}(f)/8,$$
$$\text{MRAE}(G) = \max(a, 1-a)\,\tau = \max(a, 1-a)\text{TV}(f)/2 \quad \text{if } 0 < a < 1.$$

Setting $a = 1/2$ (i.e., $\tau = (m + 1/2)^{-1}$ for some integer $m \ge 0$) yields $\text{KD}(G) = \text{TV}(f)/8$, whereas $\text{TV}(f)/4 \le \text{MRAE}(G) \le \text{TV}(f)/2$ whenever $0 < a < 1$.

Example 5.3.5 (Pareto and exponential distributions). *We recall that if a positive random variable X has density h then $Y = \log_B(X)$ has density $f(y) = B^y \log(B)\, h(B^y)$. Now let X be **Pareto** distributed, i.e., $h(x) = ax^{-a-1}\mathbb{I}_{\{x \ge 1\}}$ for some parameter $a > 0$. Then Y has density*

$$f(y) = c\exp(-cy)\mathbb{I}_{\{y \ge 0\}}$$

with $c := a \log B$, i.e., Y is exponentially distributed with mean c. Since f is unimodal with maximum at $y = 0$, its total variation is given by

$$\text{TV}(f) = 2f(0) = 2c.$$

Corollary 5.3.2 thus implies that

$$\text{KD}(G) \le c/4 \quad \text{and} \quad \text{MRAE}(G) \le c.$$

Hence the Pareto distributed r.v. X follows Benford's law approximatively, provided that $c = a \log B$ is sufficiently small. Intuitively, this means that the realizations of Y are likely to be spread over a range of several powers of B.

Here the c.d.f. G and density g of $U = Y - \lfloor Y \rfloor$ can be determined explicitly: For $x \in [0, 1]$ and $y \in [0, 1)$,

$$G(x) = \sum_{n \in \mathbb{Z}} (F(n + x) - F(z)) = (1 - e^{-cx}) \sum_{k=0}^{\infty} e^{-ck} = \frac{1 - e^{-cx}}{1 - e^{-c}}, \quad (5.4)$$

$$g(y) = \frac{1 - e^{-cy}}{1 - e^{-c}}. \quad (5.5)$$

Since g is decreasing on $[0, 1)$, the maximal relative approximation error equals

$$\mathrm{MRAE}(G) = \sup_{0 \leq x < y \leq 1} \left| \frac{G(y) - G(x)}{y - x} - 1 \right| = \sup_{0 \leq \xi < 1} |g(\xi) - 1|$$

$$= \max\{g(0) - 1, 1 - g(1-)\} = \frac{e^{-c} - 1 + c}{1 - e^{-c}}$$

$$= \frac{c}{2} + \frac{c^2}{12} + O(c^4).$$

Hence for small c, our general bound for $\mathrm{MRAE}(G)$ is sharp up to a factor close to 2.

Example 5.3.6 (Danish fire insurance data). *Let us illustrate Example 5.3.5 with a famous data set consisting of 2167 losses of over one million Danish crowns in 1985. The loss figure is a total loss including damage to buildings, furniture, and personal properties as well as loss of profits. These data were studied by several authors and seem to follow a Pareto distribution, see e.g. Mikosch [Mik, 2004]. We estimated the parameter $c = a \log(10)$ by its maximum likelihood estimator $\hat{c} = 1/\bar{Y} \approx 2.926$. With this value we computed the theoretical frequencies of the leading digits using formula (5.4). Figure 5.2 shows (from left to right) the empirical, the (estimated) theoretical, and the Benford frequencies of the first leading digits $d = 1, \ldots, 9$. Obviously, the empirical and (estimated) theoretical frequencies match very well. The Benford frequencies are quite different, which is not surprising taking into account that $\hat{c}$ is rather large.*

5.4 ERROR BOUNDS IN TERMS OF $\mathrm{TV}(F^{(K)})$

The previous results are for the case of $\mathrm{TV}(f)$ being finite. Kemperman [Kemp] realized already that there exist stronger bounds for smooth densities f. In particular, he proposed the inequality

$$|G(x) - x| \leq \left(1 - (2x - 1)^4\right) \mathrm{TV}(f^{(1)})/64 \quad \text{for } 0 \leq x \leq 1, \quad (5.6)$$

and promised other bounds in terms of higher-order derivatives. We derive similar inequalities which apply, for instance, to normal distributions.

Theorem 5.4.1. *Suppose that f is $k \geq 1$ times absolutely continuous such that $\mathrm{TV}(f^{(k)}) < \infty$ for some version of $f^{(k)}$. Then g is Lipschitz-continuous on $\mathbb{R}$. Precisely, for $x, y \in \mathbb{R}$ with $|x - y| \leq 1$,*

$$|g(x) - g(y)| \leq |x - y|(1 - |x - y|) \frac{\mathrm{TV}(f^{(k)})}{2 \cdot 6^{k-1}} \leq \frac{\mathrm{TV}(f^{(k)})}{8 \cdot 6^{k-1}}.$$

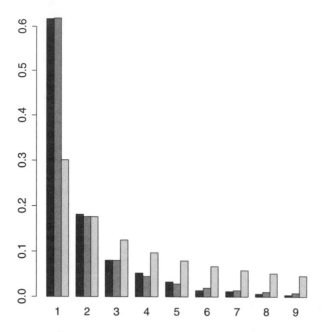

Figure 5.2 Danish Fire Data: Bars from left to right indicate the empirical, the (estimated) theoretical, and the Benford frequencies of the leading digits $d = 1, 2, \ldots, 9$.

Combining this result with Proposition 5.2.2 leads to the following error bounds.

Corollary 5.4.2. *Under the conditions of Theorem 5.4.1, for $0 \le x \le y \le 1$,*

$$\left| G(y) - G(x) - (y - x) \right| \le (y - x)(1 - (y - x))\frac{\mathrm{TV}(f^{(k)})}{2 \cdot 6^k}.$$

In particular,

$$\mathrm{KD}(G) \le \frac{\mathrm{TV}(f^{(k)})}{8 \cdot 6^k} \quad and \quad \mathrm{MRAE}(G) \le \frac{\mathrm{TV}(f^{(k)})}{2 \cdot 6^k}.$$

Remark 5.4.3 (Comparison with Kemperman's bound). *Since $\mathrm{TV}(f^{(1)})$ remains unchanged when replacing f with $f(\cdot - x)$ for any $x \in \mathbb{R}$, one can deduce from (5.6) that*

$$\left| G(y) - G(x) - (y - x) \right| \le \left(1 - (2(y - x) - 1)^4\right) \mathrm{TV}(f^{(1)})/64$$

for $0 \le x < y \le 1$. This entails that

$$\mathrm{KD}(G) \le \mathrm{TV}(f^{(1)})/64 \quad and \quad \mathrm{MRAE}(G) \le \mathrm{TV}(f^{(1)})/8,$$

because $\sup_{0 < u \le 1}(1 - (2u - 1)^4)/u = 8$. Corollary 5.4.2 yields the inequalities

$$\mathrm{KD}(G) \le \mathrm{TV}(f^{(1)})/48 \quad and \quad \mathrm{MRAE}(G) \le \mathrm{TV}(f^{(1)})/12,$$

so (5.6) is stronger in terms of Kuiper distance but weaker in terms of relative approximation errors.

Remark 5.4.4 (Location-scale families). *For the subsequent examples, the following general consideration is useful. Let f_o be a probability density on the real line such that $\mathrm{TV}(f_o^{(k)}) < \infty$ for some integer $k \geq 0$. For $\mu \in \mathbb{R}$ and $\sigma > 0$ let*

$$f(x) = f_{\mu,\sigma}(x) := \sigma^{-1} f_o(\sigma^{-1}(x - \mu)).$$

Then one verifies easily that

$$\mathrm{TV}(f^{(k)}) = \mathrm{TV}(f_o^{(k)})/\sigma^{k+1}.$$

Example 5.4.5 (Log-normal and normal distributions). *For the **standard normal density** $\phi(x) := (2\pi)^{-1/2} \exp(-x^2/2)$, elementary calculations reveal that*

$$\begin{aligned}
\mathrm{TV}(\phi) &= 2\phi(0) &\approx 0.7979, \\
\mathrm{TV}(\phi^{(1)}) &= 4\phi(1) &\approx 0.9679, \\
\mathrm{TV}(\phi^{(2)}) &= 8\phi(\sqrt{3}) + 2\phi(0) &\approx 1.5100.
\end{aligned}$$

In general,

$$\phi^{(k)}(x) = H_k(x)\phi(x)$$

with the Hermite-type polynomial

$$H_k(x) = \exp(x^2/2) \frac{d^k}{dx^k} \exp(-x^2/2)$$

of degree k. Via partial integration and induction one may show that

$$\int H_j(x) H_k(x) \phi(x)\, dx = \mathbb{I}_{\{j=k\}} k!$$

for arbitrary integers $j, k \geq 0$ (see for example [AbrSteg, 1964]). Hence the Cauchy–Schwarz inequality entails that

$$\begin{aligned}
\mathrm{TV}(\phi^{(k)}) &= \int |\phi^{(k+1)}(x)|\, dx \\
&= \int |H_{k+1}(x)| \phi(x)\, dx \\
&\leq \left(\int H_{k+1}(x)^2 \phi(x)\, dx \right)^{1/2} \cdot \left(\int \phi(x)\, dx \right)^{1/2} \\
&= \sqrt{(k+1)!}.
\end{aligned}$$

These bounds yield the following results.

Theorem 5.4.6. *Let $f(x) = f_{\mu,\sigma}(x) = \phi((x - \mu)/\sigma)/\sigma$ for $\mu \in \mathbb{R}$ and $\sigma \geq 1/6$. Then the corresponding functions $g = g_{\mu,\sigma}$ and $G = G_{\mu,\sigma}$ satisfy the inequalities*

$$\begin{aligned}
\mathrm{R}(g_{\mu,\sigma}) &\leq 4.5 \cdot h(\lfloor 36\,\sigma^2 \rfloor), \\
\mathrm{KD}(G_{\mu,\sigma}) &\leq 0.75 \cdot h(\lfloor 36\,\sigma^2 \rfloor), \\
\mathrm{MRAE}(G_{\mu,\sigma}) &\leq 3 \cdot h(\lfloor 36\,\sigma^2 \rfloor),
\end{aligned}$$

where $h(m) := \sqrt{m!/m^m}$ for integers $m \geq 1$.

σ	$h(\lfloor 36\,\sigma^2 \rfloor)$	Upper bounds for		
		$\mathrm{R}(g_{\mu,\sigma})$	$\mathrm{KD}(G_{\mu,\sigma})$	$\mathrm{MRAE}(G_{\mu,\sigma})$
0.5	0.03061	0.13773	0.02296	0.09182
0.6	0.00733	0.03299	0.00550	0.02199
0.7	$6.5572 \cdot 10^{-4}$	$2.9508 \cdot 10^{-3}$	$4.9179 \cdot 10^{-4}$	$1.9672 \cdot 10^{-3}$
0.8	$3.5187 \cdot 10^{-5}$	$1.5834 \cdot 10^{-4}$	$2.6390 \cdot 10^{-5}$	$1.0556 \cdot 10^{-4}$
0.9	$1.8557 \cdot 10^{-6}$	$8.3505 \cdot 10^{-6}$	$1.3918 \cdot 10^{-6}$	$5.5670 \cdot 10^{-6}$
1.0	$5.9132 \cdot 10^{-8}$	$2.6610 \cdot 10^{-7}$	$4.4349 \cdot 10^{-8}$	$1.774 \cdot 10^{-7}$

Table 5.1 Some bounds for $X = B^Y$ with $Y \sim \mathcal{N}(\mu, \sigma^2)$ (the normal distribution with mean μ and variance σ^2).

It follows from Stirling's formula that $h(m) = c_m m^{1/4} e^{-m/2}$ with $c_m \to (2\pi)^{1/4}$ as $m \to \infty$. In particular,

$$\lim_{m \to \infty} \frac{\log h(m)}{m} = -\frac{1}{2},$$

so the bounds in Theorem 5.4.6 decrease exponentially in σ^2. For σ between 0.5 and 1.0 we obtain already quite remarkable bounds; see Table 5.1.

Corollary 5.4.7. *For an integer base $B \geq 2$ let $X = B^Y$ for some random variable $Y \sim \mathcal{N}(\mu, \sigma^2)$ with $\sigma \geq 1/6$. Then for arbitrary digits $d_0, d_1, d_2, \ldots$ in $\{0, 1, \ldots, B-1\}$ with $d_0 \geq 1$ and integers $\ell \geq 0$,*

$$\left| \frac{P(X \text{ has leading digits } d_0, \ldots, d_\ell)}{\log_B(1 + B^{-\ell}/d_{(\ell)})} - 1 \right| \leq 3 \cdot h(\lfloor 36\,\sigma^2 \rfloor),$$

where $d_{(\ell)} = \sum_{i=1}^{\ell} d_i \cdot B^{-i}$ and $h(m) := \sqrt{m!/m^m}$ for integers $m \geq 1$.

Note that Corollary 5.4.7 provides an alternative proof of Theorem 3.2.2. Moreover, the quantity d introduced in Section 4.2 may be bounded by

$$0.75 \cdot h\left(\lfloor 36\sigma^2/\log(10) \rfloor\right)$$

and thus decreases exponentially as $\sigma \to \infty$.

Example 5.4.8 (Weibull and Gumbel distributions). *Let $X > 0$ be a random variable with **Weibull distribution**, i.e., for some parameters $\gamma, \tau > 0$,*

$$P(X \leq r) = 1 - \exp(-(r/\gamma)^\tau) \quad \text{for } r \geq 0.$$

Then the standardized random variable $Y_o := \tau \log(X/\gamma)$ satisfies

$$F_o(y) := P(Y_o \leq y) = 1 - \exp(-e^y) \quad \text{for } y \in \mathbb{R}$$

and has density function

$$f_o(y) = e^y \exp(-e^y), \tag{5.7}$$

*i.e., $-Y_o$ has a **Gumbel distribution**. Thus $Y := \log_B(X)$ may be written as $Y = \mu + \sigma Y_o$ with $\mu := \log_B(\gamma)$ and $\sigma = (\tau \log B)^{-1}$.*

Elementary calculations reveal that for any integer $n \geq 1$,

$$f_o^{(n-1)}(y) \; = \; p_n(e^y) \exp(-e^y)$$

with $p_n(t)$ being a polynomial in t of degree n. Precisely, $p_1(t) = t$, and

$$p_{n+1}(t) \; = \; t(p_n'(t) - p_n(t)) \tag{5.8}$$

for $n = 1, 2, 3, \ldots$. In particular, $p_2(t) = t(1 - t)$ and $p_3(t) = t(1 - 3t + t^2)$. These considerations lead already to the following conclusion:

Corollary 5.4.9. *Let $X > 0$ have Weibull distribution with parameters $\gamma, \tau > 0$ as above. Then for the density function f_o given by (5.7) we have $\mathrm{TV}(f_o^{(k)}) < \infty$ and*

$$\left| \frac{P(X \text{ has leading digits } d_0, \ldots, d_\ell)}{\log_B(1 + B^{-\ell}/d_{(\ell)})} - 1 \right| \; \leq \; 3 \cdot \mathrm{TV}(f_o^{(k)}) \left(\frac{\tau \log B}{6} \right)^{k+1}$$

for arbitrary integers $k, \ell \geq 0$ and digits $d_0, d_1, d_2 \ldots$ as in Corollary 5.4.7.

Explicit inequalities as in the Gaussian case seem to be out of reach. Nevertheless some numerical bounds can be obtained. Table 5.2 contains numerical approximations for $\mathrm{TV}(f_o^{(k)})$ and the resulting upper bounds

$$\beta_\tau(k) \; := \; 3 \cdot \mathrm{TV}(f_o^{(k)}) \left(\frac{\tau \log 10}{6} \right)^{k+1}$$

for the maximal relative approximation error in Benford's law with decimal expansions, where $\tau = 1.0, 0.5, 0.3$. Note that $\tau = 1.0$ corresponds to the standard exponential distribution. For a detailed analysis of this special case we refer to Engel and Leuenberger [EngLeu, 2003] and Miller and Nigrini [MiNi2, 2008].

A final remark on the polynomials p_n in this example: Writing

$$p_n(t) \; = \; \sum_{k=1}^{n} (-1)^{k-1} S_{n,k} \, t^k,$$

it follows from the recursion (5.8) that the coefficients can be calculated inductively via

$$S_{1,1} \; = \; 1, \qquad S_{n,k} = S_{n-1,k-1} + k S_{n-1,k}.$$

*Hence the $S_{n,k}$ are **Stirling numbers of the second kind**; see Graham et al. [GrKnP, 1994, Chapter 6.1].*

k	$\mathrm{TV}(f_o^{(k)})$	$\beta_{1.0}(k)$	$\beta_{0.5}(k)$	$\beta_{0.3}(k)$
0	$7.3576 \cdot 10^{-1}$	$8.4707 \cdot 10^{-1}$	$4.2354 \cdot 10^{-1}$	$2.5412 \cdot 10^{-1}$
1	$9.4025 \cdot 10^{-1}$	$4.1543 \cdot 10^{-1}$	$1.0386 \cdot 10^{-1}$	$3.7388 \cdot 10^{-2}$
2	1.7830	$3.0232 \cdot 10^{-1}$	$3.7790 \cdot 10^{-2}$	$8.1627 \cdot 10^{-3}$
3	4.5103	$\mathbf{2.9348 \cdot 10^{-1}}$	$1.8343 \cdot 10^{-2}$	$2.3772 \cdot 10^{-3}$
4	$1.4278 \cdot 10$	$3.5653 \cdot 10^{-1}$	$1.1142 \cdot 10^{-2}$	$8.6638 \cdot 10^{-4}$
5	$5.4301 \cdot 10$	$5.2038 \cdot 10^{-1}$	$8.1309 \cdot 10^{-3}$	$3.7936 \cdot 10^{-4}$
6	$2.4118 \cdot 10^2$	$8.8699 \cdot 10^{-1}$	$6.9296 \cdot 10^{-3}$	$1.9399 \cdot 10^{-4}$
7	$1.2252 \cdot 10^3$	1.7292	$\mathbf{6.7546 \cdot 10^{-3}}$	$1.1345 \cdot 10^{-4}$
8	$7.0056 \cdot 10^3$	3.7944	$7.4110 \cdot 10^{-3}$	$7.4686 \cdot 10^{-5}$
9	$4.4527 \cdot 10^4$	9.2552	$9.0383 \cdot 10^{-3}$	$5.4651 \cdot 10^{-5}$
10	$3.1140 \cdot 10^5$	$2.4840 \cdot 10$	$1.2129 \cdot 10^{-2}$	$4.4003 \cdot 10^{-5}$
11	$2.3763 \cdot 10^6$	$7.2744 \cdot 10$	$1.7760 \cdot 10^{-2}$	$3.8659 \cdot 10^{-5}$
12	$1.9648 \cdot 10^7$	$2.3083 \cdot 10^2$	$2.8177 \cdot 10^{-2}$	$\mathbf{3.6801 \cdot 10^{-5}}$
13	$1.7498 \cdot 10^8$	$7.8888 \cdot 10^2$	$4.8150 \cdot 10^{-2}$	$3.7732 \cdot 10^{-5}$
14	$1.6698 \cdot 10^9$	$2.8890 \cdot 10^3$	$8.8166 \cdot 10^{-2}$	$4.1454 \cdot 10^{-5}$

Table 5.2 Some bounds for Weibull-distributed X with $\tau \leq 1.0, 0.5, 0.3$. Minimal values are indicated in boldface.

5.5 PROOFS

Proof of Propositions 5.2.1 and 5.2.2. Let $0 \leq x < y \leq 1$ and $\delta := y - x \in (0, 1]$. Then

$$
\begin{aligned}
\left| G(y) - G(x) - (y - x) \right| &= \left| \int_x^y g(u)\,du - \delta \int_{y-1}^y g(u)\,du \right| \\
&= \left| (1 - \delta) \int_x^y g(u)\,du - \delta \int_{y-1}^x g(u)\,du \right| \\
&= \left| \delta(1 - \delta) \int_0^1 \big(g(x + \delta t) - g(x - (1 - \delta)t) \big)\,dt \right| \\
&\leq \delta(1 - \delta) \int_0^1 \left| g(x + \delta t) - g(x + \delta t - t) \right|\,dt \\
&\leq \delta(1 - \delta)\,\mathrm{R}(g).
\end{aligned}
$$

If $|g(u) - g(v)| \leq |u - v|(1 - |u - v|)L(g)$ for arbitrary $u, v \in [0, 1]$, then

$$
\begin{aligned}
\left| G(y) - G(x) - (y - x) \right| &\leq \delta(1 - \delta) \int_0^1 \left| g(x + \delta t) - g(x + \delta t - t) \right|\,dt \\
&\leq \delta(1 - \delta) \int_0^1 t(1 - t)L(g)\,dt \\
&= \delta(1 - \delta)L(g)/6. \qquad \square
\end{aligned}
$$

Proof of Lemma 5.2.6. Define $h^{(k)}(\pm\infty) := \lim_{x\to\pm\infty} h^{(k)}(x)$. If $h^{(k)}(+\infty) \neq 0$, then one can show inductively for $j = k-1, k-2, \ldots, 0$ that $\lim_{x\to\infty} h^{(j)}(x) = \text{sign}(h^{(k)}(+\infty)) \cdot \infty$. Similarly, if $h^{(k)}(-\infty) \neq 0$, then $\lim_{x\to-\infty} h^{(j)}(x) = (-1)^{k-j}\text{sign}(h^{(k)}(-\infty)) \cdot \infty$ for $0 \leq j < k$. In both cases we would get a contradiction to $h^{(0)} = h$ being integrable over $\mathbb{R}$.

Now suppose that $\lim_{|x|\to\infty} h^{(k)}(x) = 0$. It follows from Taylor's formula that for $x \in \mathbb{R}$ and $u \in [-1, 1]$,

$$|h(x+u)| = \left| \sum_{j=0}^{k-1} \frac{h^{(j)}(x)}{j!} u^j + \int_0^u \frac{h^{(k)}(x+v)(u-v)^{k-1}}{(k-1)!} \, dv \right|$$

$$\geq \left| \sum_{j=0}^{k-1} \frac{h^{(j)}(x)}{j!} u^j \right| - \sup_{|s|\geq|x|-1} \frac{|h^{(k)}(s)||u|^k}{k!}.$$

Hence

$$\int_{x-1}^{x+1} |h(t)| \, dt \geq \frac{|h^{(j)}(x)|}{j!} A_{j,k-1} - 2 \sup_{|s|\geq|x|-1} \frac{|h^{(k)}(s)|}{(k+1)!}$$

for any $j \in \{0, 1, \ldots, k-1\}$, where for $0 \leq \ell \leq m$,

$$A_{\ell,m} := \min_{a_0,\ldots,a_m \in \mathbb{R} \,:\, a_\ell = 1} \int_{-1}^{1} \left| \sum_{j=0}^{m} a_j u^j \right| du > 0.$$

This shows that

$$|h^{(j)}(x)| \leq \frac{j!}{A_{j,k-1}} \left(\int_{x-1}^{x+1} |h(t)| \, dt + 2 \sup_{|s|\geq|x|-1} \frac{|h^{(k)}(s)|}{(k+1)!} \right) \to 0$$

as $|x| \to \infty$. $\qquad\square$

Proof of Theorem 5.3.1. For arbitrary $m \in \mathbb{N}$ and $0 \leq t_0 < t_1 < \cdots < t_m \leq 1$,

$$\sum_{n\in\mathbb{Z}} \sum_{i=1}^{m} |f(n+t_i) - f(n+t_{i-1})| \leq \text{TV}(f). \tag{5.9}$$

In particular, for two points $x, y \in [0, 1]$ with $\min(g(x), g(y)) < \infty$, the difference $g(x) - g(y)$ is finite. Hence $g < \infty$ everywhere. Now it follows directly from (5.9) that $\text{TV}(g) \leq \text{TV}(f)$. Moreover, for $0 \leq x < y \leq 1$,

$$\left(g(y) - g(x) \right)^{\pm} = \left(\sum_{n\in\mathbb{Z}} \left(f(n+y) - f(n+x) \right) \right)^{\pm}$$

$$\leq \sum_{n\in\mathbb{Z}} \left(f(n+y) - f(n+x) \right)^{\pm}$$

$$\leq \text{TV}^{\pm}(f) = \text{TV}(f)/2,$$

where the latter equality follows from Lemma 5.2.5. $\qquad\square$

Proof of Theorem 5.4.1.   Throughout this proof let $x, y \in \mathbb{R}$ be generic real numbers with $\delta := y - x \in [0, 1]$. For integers $j \in \{0, \ldots, k\}$ and $N \geq 1$ we define

$$g_N^{(j)}(x, y) := \sum_{n=-N}^{N} \left(f^{(j)}(n + y) - f^{(j)}(n + x) \right).$$

Note that $g(y) - g(x) = \lim_{N \to \infty} g_N^{(0)}(x, y)$ whenever $g(x) < \infty$ or $g(y) < \infty$. To establish a relation between $g^{(j)}(\cdot, \cdot)$ and $g^{(j+1)}(\cdot, \cdot)$ note first that for absolutely continuous $h : \mathbb{R} \to \mathbb{R}$,

$$
\begin{aligned}
h(y) - h(x) &= h(y) - h(x) - \delta\big(h(y) - h(y-1)\big) + \delta\big(h(y) - h(y-1)\big) \\
&= \delta(1 - \delta) \int_0^1 \big(h'(x + \delta t) - h'(x + \delta t - t)\big)\, dt \\
&\quad + \delta\big(h(y) - h(y-1)\big);
\end{aligned}
$$

see also the proof of Propositions 5.2.1 and 5.2.2. Hence for $0 < j \leq k$,

$$
\begin{aligned}
g_N^{(j-1)}(x, y) = \delta(1 - \delta) \int_0^1 g_N^{(j)}(x + \delta t, x + \delta t - t)\, dt \\
+ \delta\big(f^{(j-1)}(N + y) - f^{(j-1)}(-N + y - 1)\big).
\end{aligned}
\tag{5.10}
$$

Recall that $\lim_{|z| \to \infty} f^{(j)}(z) = 0$ for $0 \leq j \leq k$ by virtue of Lemma 5.2.6. In particular, $\mathrm{TV}^{\pm}(f^{(k)}) = \mathrm{TV}(f^{(k)})/2$ by Lemma 5.2.5. Hence

$$
\begin{aligned}
g_N^{(k)}(x, y) &= \sum_{n=-N}^{N} \left(f^{(k)}(n + y) - f^{(k)}(n + x) \right)^+ \\
&\quad - \sum_{n=-N}^{N} \left(f^{(k)}(n + y) - f^{(k)}(n + x) \right)^-
\end{aligned}
$$

satisfies the inequality $\left| g_N^{(k)}(x, y) \right| \leq \mathrm{TV}(f^{(k)})/2$ and converges, as $N \to \infty$, to a limit $g^{(k)}(x, y)$. Moreover, it follows from (5.10) that

$$\left| g_N^{(k-1)}(x, y) \right| \leq \delta(1 - \delta) \frac{\mathrm{TV}(f^{(k)})}{2} + 2\| f^{(k-1)} \|_\infty$$

and, via dominated convergence,

$$\lim_{N \to \infty} g_N^{(k-1)}(x, y) = g^{(k-1)}(x, y) := \delta(1 - \delta) \int_0^1 g^{(k)}(x + \delta t, x + \delta t - t)\, dt$$

with

$$
\begin{aligned}
\left| g^{(k-1)}(x, y) \right| &\leq \delta(1 - \delta) \int_0^1 \left| g^{(k)}(x + \delta t, x + \delta t - t) \right| dt \\
&\leq \delta(1 - \delta) \mathrm{TV}(f^{(k)})/2.
\end{aligned}
$$

Now we perform an induction step: Suppose that for some $1 \leq j < k$,

$$\left| g_N^{(j)}(x, y) \right| \leq \alpha^{(j)} < \infty$$

and

$$g^{(j)}(x,y) := \lim_{N\to\infty} g_N^{(j)}(x,y) \quad \text{exists with} \quad \left|g^{(j)}(x,y)\right| \le \delta(1-\delta)\beta^{(j)}.$$

For $j = k - 1$ this is true with $\beta^{(k-1)} := \mathrm{TV}(f^{(k)})/2$. Now it follows from (5.10) and dominated convergence that

$$\left|g_N^{(j-1)}(x,y)\right| \le \alpha^{(j)} + 2\|f^{(j-1)}\|_\infty$$

and

$$\lim_{N\to\infty} g_N^{(j-1)}(x,y) = g^{(j-1)}(x,y) := \delta(1-\delta)\int_0^1 g^{(j)}(x+\delta t, x+\delta t - t)\,dt,$$

where

$$\left|g^{(j-1)}(x,y)\right| \le \delta(1-\delta)\int_0^1 \left|g^{(j)}(x+\delta t, x+\delta t - t)\right|\,dt$$

$$\le \delta(1-\delta)\int_0^1 t(1-t)\beta^{(j)}\,dt$$

$$= \delta(1-\delta)\beta^{(j)}/6.$$

These considerations show that $g^{(0)}(x,y) := \lim_{N\to\infty} g_N^{(0)}(x,y)$ always exists and satisfies the inequality

$$\left|g^{(0)}(x,y)\right| \le \delta(1-\delta)\frac{\mathrm{TV}(f^{(k)})}{2\cdot 6^{k-1}} \le \frac{\mathrm{TV}(f^{(k)})}{8\cdot 6^{k-1}}.$$

In particular, g is everywhere finite with $g(y) - g(x) = g^{(0)}(x,y)$ satisfying the asserted inequalities. $\qquad\square$

Proof of Corollary 5.4.2. For $0 \le x < y \le 1$ and $\delta := y - x \in (0,1]$,

$$\left|G(y) - G(x) - (y-x)\right| = \left|\delta(1-\delta)\int_0^1 \left(g(x+\delta t) - g(x+\delta t - t)\right)dt\right|$$

$$\le \delta(1-\delta)\frac{\mathrm{TV}(f^{(k)})}{2\cdot 6^{k-1}}\int_0^1 t(1-t)\,dt$$

$$= \delta(1-\delta)\frac{\mathrm{TV}(f^{(k)})}{2\cdot 6^k}. \qquad\square$$

Proof of Theorem 5.4.6. According to Theorem 5.3.1,

$$\mathrm{R}(g_{\mu,\sigma}) \le \frac{\mathrm{TV}(f_{\mu,\sigma})}{2} = \frac{\mathrm{TV}(\phi)}{2\sigma} = \frac{\phi(0)}{\sigma},$$

whereas Theorem 5.4.1 and the considerations preceding Theorem 5.4.6 yield the inequalities

$$\mathrm{R}(g_{\mu,\sigma}) \le \frac{\mathrm{TV}(f_{\mu,\sigma}^{(k)})}{8\cdot 6^{k-1}} = \frac{\mathrm{TV}(\phi^{(k)})}{8\cdot 6^{k-1}\sigma^{k+1}} \le \frac{\sqrt{(k+1)!}}{8\cdot 6^{k-1}\sigma^{k+1}}$$

for all $k \geq 1$. Since the right-hand side equals $0.75/\sigma \geq \phi(0)/\sigma$ if we plug in $k = 0$, we may conclude that

$$R(g_{\mu,\sigma}) \leq \frac{\sqrt{(k+1)!}}{8 \cdot 6^{k-1}\sigma^{k+1}} = 4.5 \cdot \sqrt{\frac{(k+1)!}{(36\,\sigma^2)^{k+1}}}$$

for all $k \geq 0$. The latter bound becomes minimal if $k + 1 = \lfloor 36\,\sigma^2 \rfloor \geq 1$, and this value yields the desired bound $4.5 \cdot h\big(\lfloor 36\,\sigma^2 \rfloor\big)$.

Similarly, Corollaries 5.3.2 and 5.4.2 yield the inequalities

$$KD(G_{\mu,\sigma}) \leq \frac{\sqrt{(k+1)!}}{8 \cdot 6^k\sigma^{k+1}} = 0.75 \cdot \sqrt{\frac{(k+1)!}{(36\,\sigma^2)^{k+1}}},$$

$$MRAE(G_{\mu,\sigma}) \leq \frac{\sqrt{(k+1)!}}{2 \cdot 6^k\sigma^{k+1}} = 3 \cdot \sqrt{\frac{(k+1)!}{(36\,\sigma^2)^{k+1}}}$$

for arbitrary $k \geq 0$, and $k + 1 = \lfloor 36\,\sigma^2 \rfloor \geq 1$ leads to the desired bounds. □

ACKNOWLEDGMENT.

We are grateful to Steven J. Miller, an anonymous referee of [DueLeu] and Pieter C. Allaart for constructive comments and additional references.

Chapter Six

Lévy Processes and Benford's Law

Klaus Schürger[1]

Lévy processes (LPs) can be thought of as random walks in continuous time, having independent and stationary increments. We assume throughout that the characteristic function of the state of an LP at time 1 satisfies a certain "Standard Condition" (SC). A certain subclass of stable LPs have the property that at all positive times their states have infinite variance. If their Lévy measure is non-zero, LPs perform jumps. If an LP X (satisfying (SC)) does not have a Gaussian component, then for each $c > 0$ the expected number of non-zero jumps of size less than c performed by X on any time interval of positive length is infinite! Hence exponential Lévy processes (ELPs) are attractive for modeling many phenomena, such as the evolution of the price per share of a certain stock. (A special case of an ELP is given by a geometric Brownian motion (also called a Black–Scholes process).) It was already known to Benford that often the frequencies of leading digits of stock market indices are well approximated by the corresponding probabilities given by Benford's law. This corresponds to a certain theoretical result in §6.3, which says that within any ELP model the empirical frequencies of a given initial block of digits converge a.s. to the probability of that block given by Benford's law. An analogous result holds in the continuous-time case, with the frequencies replaced by normalized Lebesgue measures.

Using a certain variant of the Poisson Summation Formula (obtained in §6.6), we arrive in §6.2 at convergence results for the expectations of certain normalized functionals of the significand of an ELP. In §6.3 we obtain, using Azuma's inequality for martingales, large deviation results for the above functionals. Combining these with the results in §6.2 and the Borel–Cantelli Lemma yields the desired a.s. convergence results. On the other hand, the just-mentioned large deviation results allow for the construction of non-parametric tests based on certain leading digits of some strictly positive continuous-time process Z which is observed at discrete time points $n = 1, \ldots, T$ (T being a suitably chosen (finite) time horizon). The null hypothesis being tested says that Z belongs to a specified class of ELPs. In §6.1 we recall some basic notions (including infinite divisibility, the Lévy–Khintchin representation, Lévy measures, martingales), and give examples of Lévy processes. In §6.4 we obtain conditions which are sufficient for (SC).

---

[1]Department of Economics, University of Bonn, Adenauerallee 24-42, 53113 Bonn, Germany.

6.1 OVERVIEW, BASIC DEFINITIONS, AND EXAMPLES

A **geometric Brownian motion** (also called a **Black–Scholes process**) being of the form

$$Z_t = Z_0 \exp\left(\mu t + c W_t - c^2 t/2\right), \quad t \in \mathbb{R}^+ \tag{6.1}$$

(here, $Z_0 > 0, c > 0$, and $\mu \in \mathbb{R}$ are constants, and (W_t) is a (standard) Brownian motion (or Wiener process) starting at 0) was introduced in [BlS, Mer] to model the evolution of the price per share of a certain stock. The limited ability of the process (Z_t) (having continuous sample paths!) to reflect financial reality has become clearer over the years. The (convincing) thesis of a recent book by Cont and Tankov [ConT] is that **jumps** are needed in a model. On the other hand, the logarithms of financial asset returns often show high variability which results in **heavy tails** in the empirical distribution of returns (see, e.g., [ConT], page 6). Thus, Merton [Mer2] was led to replace (W_t) in (6.1) by a certain Lévy process of jump-diffusion type (see Example 6.1.22) to model prices exhibiting discontinuities (see also [Lan]). The prototype of jump processes is the Poisson process (see Example 6.1.20), ubiquitous in the modeling of actuarial and insurance problems. Poisson processes and Brownian motion stand at opposite ends of a spectrum of processes forming the class of **Lévy processes** (see Definition 6.1.11) which are discontinuous if their Lévy measure is non-zero. On the other hand, α-**stable** $(1 < \alpha < 2)$ **Lévy processes** (X_t) have the property that, for all $t > 0$, X_t has an infinite variance (see Example 6.1.24).

It was already known to Benford that often the frequencies of leading digits of stock market indices are well approximated by the corresponding probabilities given by Benford's law. Similar results are reported in [PiTTV], and (for one-day returns of certain indices) in [Ley] (see also [Hi4]). This corresponds to a certain theoretical result in §6.3. In fact, based on large deviation results, we shall show that if (W_t) in (6.1) is replaced by a more general Lévy process, then, under a certain "standard condition" (see Definition 6.2.9), as $n \to \infty$,

$$\frac{1}{n} \#\{1 \leq j \leq n \colon D_{B,i}(Z_j) = d_i, \ i = 1, \ldots, m\} \to p_B(d_1, \ldots, d_m) \quad \text{a.s.}$$

for any initial block $(d_1, \ldots, d_m)$ of (B-adic) digits such that $1 \leq d_1 \leq B - 1$. Here, $B \ (= 2, 3, \ldots)$ denotes a certain base, and the left-hand side is the relative frequency of all $j \leq n$ such that the ith leading (B-adic) digit of Z_j equals d_i for $i = 1, \ldots, m$. The right-hand side is the probability of occurrence of the block $(d_1, \ldots, d_m)$ under Benford's law (base B); see (6.21), (6.23) below, and eq. (1.9) in [Schür2], page 1221. An analogous result holds in the continuous-time case.

We now give a sketch of the content of this chapter. In this section we recall the basic notions of characteristic functions, infinite divisibility, generating triples, the Lévy–Khintchin representation, Lévy processes, ..., and we give examples of basic Lévy processes such as Brownian motion, Lévy processes of jump-diffusion type, gamma and α-stable processes. Martingales play an important role in the (motivating) Example 6.1.32, and are essential for obtaining large deviation results in §6.3, which are used to get a.s. results for certain normalized functionals. In §6.2 we derive results about the asymptotic behavior of the expectation of certain functionals (including occupation time). In §6.4 we mention some conditions which are either necessary or sufficient for the "domination" or "standard condition" introduced in §6.2.

Using the large deviation results in §6.3, we construct certain **non-parametric tests** in §6.5 which are based on the observation of leading digits of a certain process. The null hypotheses tested are concerning, respectively, Lévy processes of jump-diffusion type, gamma processes and α-stable processes. The derivation of some theoretical results (including a variant of the Poisson Summation Formula) is deferred to the first appendix to this chapter.

Definition 6.1.1 (Characteristic function). *Assume that a (real-valued) random variable ξ has distribution $\widetilde{Q}$ (notation: $\xi \sim \widetilde{Q}$), i.e.,*

$$P(\xi \in A) = \widetilde{Q}(A) \quad \text{for all Borel sets } A \subset \mathbb{R}.$$

*The **characteristic function** of ξ (or $\widetilde{Q}$) is defined by*

$$h(z) := \mathbb{E}\left[\exp\left(iz\xi\right)\right] = \int_{\mathbb{R}} \exp(izx)\, \widetilde{Q}(dx), \quad z \in \mathbb{R}. \tag{6.2}$$

Note that

$$|h(z)| \leq 1 \quad \text{for all } z \in \mathbb{R}. \tag{6.3}$$

One can show that two random variables with the same characteristic function are identically distributed (see [Fel], page 508). Recall that $L^1(\mathbb{R})$ denotes the family of all functions $f \colon \mathbb{R} \longrightarrow \mathbb{C}$ which are (Borel) measurable and (Lebesgue) integrable, i.e.,

$$\int_{\mathbb{R}} |f(x)|\, dx < \infty. \tag{6.4}$$

Example 6.1.2. *It is almost immediate that, for each $p > 0$, $\cos\left(2\pi pz\right)$ is a characteristic function. Using some basic properties of characteristic functions (see, e.g., [Fel], page 504), for arbitrary numbers $\alpha_1 \geq 0, \alpha_2 \geq 0, \ldots$ and $a_1 \geq 0, a_2 \geq 0, \ldots$ such that $a_1 + a_2 + \cdots = 1$, we have*

$$h(z) := \sum_{k=1}^{\infty} a_k \left(\cos\left(2\pi\alpha_k z\right)\right)^2, \quad z \in \mathbb{R} \tag{6.5}$$

is a characteristic function.

Example 6.1.3. *It follows from **Pólya's criterion** (see [Fel], page 509) that*

$$h^*(z) := (1 - |z|)\, \mathbb{1}_{[-1,1]}(z), \quad z \in \mathbb{R} \tag{6.6}$$

is a characteristic function. Hence, by Example 6.1.2,

$$\widetilde{h}(z) := (1 - |z|)\, \mathbb{1}_{[-1,1]}(z) \sum_{k=1}^{\infty} 2^{-k} \left(\cos\left(2\pi 9^k z\right)\right)^2, \quad z \in \mathbb{R} \tag{6.7}$$

is a non-negative characteristic function in $L^1(\mathbb{R})$ which "wildly fluctuates" in the sense that it has infinite total variation; for details see Definition 6.2.2 and Example 6.6.4.

Definition 6.1.4 (Infinite divisibility). *Let ξ be a (real-valued) random variable having distribution $\widetilde{Q}$ and characteristic function h. We say that ξ (or $\widetilde{Q}$) is **infinitely divisible** if, for each $n \geq 2$, there exist random variables $\xi_{n1}, \ldots, \xi_{nn}$ being independent and identically distributed such that $\xi_{n1} + \cdots + \xi_{nn} \sim \widetilde{Q}$. Equivalently, for each $n \geq 2$ there exists a characteristic function h_n such that $h = (h_n)^n$.*

A series of examples of infinitely divisible distributions will be given later (starting with Example 6.1.17).

For the formulation of a famous result due to Lévy and Khintchin (see [Kh, Lév1]) which characterizes infinitely divisible distributions in terms of their characteristic functions (see also [Ber, ConT, Ka, Sat2]) we need

Definition 6.1.5 (Generating triple; Lévy measure). *By definition, a **generating triple** (β, σ^2, Q) consists of*

- *a real number β;*

- *a real number $\sigma^2 \geq 0$;*

- *a measure Q on (the Borel subsets of) $\mathbb{R}$, called **Lévy measure**, such that*

$$Q(\{0\}) = 0 \quad and \quad \int_{\mathbb{R}} \left(x^2 \wedge 1\right) Q(dx) < \infty. \tag{6.8}$$

Remark 6.1.6. *It is easy to show that a Lévy measure Q satisfies*

$$Q\left(\mathbb{R} \setminus (-\alpha, \alpha)\right) < \infty \quad for\ all\ \alpha > 0. \tag{6.9}$$

Theorem 6.1.7 (**Lévy–Khintchin representation**). *Let ξ be a (real-valued) random variable having distribution $\widetilde{Q}$ and characteristic function h. Then we have*

1. *$\widetilde{Q}$ is infinitely divisible iff there exists a (uniquely determined) generating triple (β, σ^2, Q) such that, for all $z \in \mathbb{R}$,*

$$h(z) = \exp\left[i\beta z - \sigma^2 z^2/2 + \int_{\mathbb{R}} \left(e^{izx} - 1 - izx\mathbb{1}_{[-1,1]}(x)\right) Q(dx)\right];$$

$$\tag{6.10}$$

2. *for each generating triple (β, σ^2, Q), the right-hand side in (6.10) defines the characteristic function of an infinitely divisible distribution.*

Definition 6.1.8 (Cadlag function). *By definition, a **cadlag function** $f \colon \mathbb{R}^+ \longrightarrow \mathbb{R}$ is right-continuous and has (finite!) left limits.*

The term "cadlag" is an acronym from the French phrase "continu à droite, limites à gauche," meaning "continuous on the right, limits on the left." If f is cadlag, we denote the left-hand limit of f at $t > 0$ by $f(t-)$, and put $f(0-) := 0$. Then

$$\Delta f(t) := f(t) - f(t-) \text{ is the (finite!) size of a **jump** of } f \text{ at } t \geq 0. \tag{6.11}$$

The following result collects some useful properties of cadlag functions.

Proposition 6.1.9. *Let* $f\colon \mathbb{R}^+ \longrightarrow \mathbb{R}$ *be cadlag. Then, for each bounded non-void set* $A \subset \mathbb{R}^+$ *we have*

1. *f is bounded on A, i.e.,* $\sup_{t \in A} |f(t)| < \infty$.

2. *for each $r > 0$, there exist only finitely many $t \in A$ such that* $|\Delta f(t)| > r$;

Hence, there are at most countably many $t \in \mathbb{R}^+$ such that $\Delta f(t) \neq 0$.

Proposition 6.1.9 is an easy consequence of

Lemma 6.1.10. *Let f be cadlag. For each non-void set $A \subset \mathbb{R}^+$, let* $F_f(A) := \sup\{|f(s) - f(t)| : s, t \in A\}$. *Then, for each $a > 0$ and $\varepsilon > 0$, there are points* $0 = t_0 < t_1 < \cdots < t_n = a$ *such that* $F_f\left([t_{j-1}, t_j]\right) \leq \varepsilon$ *for all* $j = 1, \ldots, n$.

For a proof of Lemma 6.1.10 see [Bi1], page 110. We are now prepared for the central definition of this chapter.

Definition 6.1.11 (Lévy process). *A (real-valued or one-dimensional)* **Lévy process** $X = (X_t)$ $(t \in \mathbb{R}^+)$ *defined on a probability space* $(\Omega, \mathcal{F}, P)$ *has the following properties:*

1. $X_0 \equiv 0$.

2. *X has* **independent increments**, *i.e., for each $n \geq 2$ and $0 = t_0 < t_1 < \cdots < t_n$, the* **increments** $X_{t_m} - X_{t_{m-1}}$ $(m = 1, \ldots, n)$ *are independent.*

3. *X has* **stationary increments**, *i.e., the distribution of the increment $X_t - X_s$ $(0 \leq s < t)$ only depends on $t - s$.*

4. *For each $\omega \in \Omega$, the* **sample path** $t \mapsto X_t(\omega)$ *is cadlag.*

Let (X_t) be a Lévy process. Then, for each $t \geq 0$, X_t is infinitely divisible. If X is a given Lévy process, then, throughout this chapter (except for the appendices), g **denotes the characteristic function of** X_1.

Combining Theorem 6.1.7 and the remark following Definition 6.1.11 shows that there exists a (uniquely determined) generating triple (β, σ^2, Q) such that, for all $z \in \mathbb{R}$,

$$g(z) = \exp\left[i\beta z - \sigma^2 z^2/2 + \int_{\mathbb{R}} \left(e^{izx} - 1 - izx\mathbb{1}_{[-1,1]}(x)\right) Q(dx)\right]. \quad (6.12)$$

Then (β, σ^2, Q) and Q are called, respectively, the **generating triple** and the **Lévy measure** of g (or of X_1 or of X). If $\sigma^2 > 0$, then X is said to have a **Gaussian component**.

Proposition 6.1.12 ([Sat2], Corollary 8.3). *Let X be a Lévy process with generating triple (β, σ^2, Q). Then, the characteristic function g_t of X_t $(t > 0)$ has the generating triple $(t\beta, t\sigma^2, tQ)$, and*

$$|g_t(z)| = |g(z)|^t \quad \text{for all } z \in R \text{ and } t \geq 0. \quad (6.13)$$

Concerning the existence of a Lévy process we have (see [Sat2], Corollary 11.6)

Theorem 6.1.13 (Existence of Lévy processes). *Let $\widetilde{Q}$ be an infinitely divisible distribution. Then there exists a Lévy process X (defined on a suitable probability space) such that $X_1 \sim \widetilde{Q}$. X is uniquely determined up to identity in law, i.e., if (Y_t) is a Lévy process such that $Y_1 \sim \widetilde{Q}$, then, for each $n \geq 1$ and $0 \leq t_1 < \cdots < t_n$, the $\mathbb{R}^n$-valued random vectors $(X_{t_1}, \ldots, X_{t_n})$ and $(Y_{t_1}, \ldots, Y_{t_n})$ have the same distribution.*

The following result shows that the Lévy measure of a Lévy process X has a nice interpretation in terms of the size and the number of jumps of X; see [ConT], page 76. (For the Δ-notation see (6.11).)

Theorem 6.1.14 (Explicit expression for the Lévy measure). *Let X be a Lévy process with Lévy measure Q. Then, for each Borel set $A \subset \mathbb{R}$,*

$$
\begin{aligned}
Q(A) &= \mathbb{E}\left[\#\{0 < t \leq 1 : \Delta X_t \in A \setminus \{0\}\}\right] \\
&= \mathbb{E}\left[\#\{0 \leq t \leq 1 : \Delta X_t \in A \setminus \{0\}\}\right].
\end{aligned}
\tag{6.14}
$$

More generally, for all $c > 0$ and $s \geq 0$,

$$
\begin{aligned}
Q(A) &= c^{-1}\mathbb{E}\left[\#\{s < t \leq s+c : \Delta X_t \in A \setminus \{0\}\}\right] \\
&= c^{-1}\mathbb{E}\left[\#\{s \leq t \leq s+c : \Delta X_t \in A \setminus \{0\}\}\right].
\end{aligned}
\tag{6.15}
$$

(Note that the second "=" in (6.14) follows from the first "=" since $\Delta X_0 \equiv 0$.)

By (6.15), $Q(A)$ is the expected number, per unit time, of the (non-zero) jumps of X, whose size belongs to A. Note that, by (6.15), the mean number of (non-zero) jumps of X with size in A, occurring in a given time interval, is proportional to the length of that interval. For a proof of (6.14) see [ConT]. The second "=" in (6.15) follows from the first "=" in (6.15) by using the following (amazing) result which says that a Lévy process has **no fixed jumps**. (A cadlag process (Y_t) is said to have a **fixed jump** at time $t > 0$ if $P(\Delta Y_t \neq 0) > 0$.)

Proposition 6.1.15. *Let X be a Lévy process. Then*

$$
\Delta X_t = 0 \quad \text{a.s. for each } t \in \mathbb{R}^+.
\tag{6.16}
$$

Proof. Since $\Delta X_0 \equiv 0$, it suffices to consider $t > 0$. Clearly, (6.16) is equivalent to

$$
\mathbb{E}\left[\exp\left(iz\Delta X_t\right)\right] = 1 \quad \text{for all } z \in \mathbb{R}.
\tag{6.17}
$$

Let X have the generating triple (β, σ^2, Q). By dominated convergence, the stationarity of the increments of X, and Proposition 6.1.12, we get that the expectation in (6.17) equals

$$
\lim_{s \uparrow t} \mathbb{E}\left[\exp\left(iz\left(X_t - X_s\right)\right)\right] = \lim_{s \uparrow t} \mathbb{E}\left[\exp\left(izX_{t-s}\right)\right]
$$

$$
= \lim_{s \uparrow t} \exp\left((t-s)\left[i\beta z - \sigma^2 z^2/2 + \int_{\mathbb{R}}\left(e^{izx} - 1 - izx\mathbb{1}_{[-1,1]}(x)\right) Q(dx)\right]\right)
$$

$$
= 1 \quad \text{for all } z \in \mathbb{R}. \qquad \qquad \square
$$

The first "=" in (6.15) can be proved by applying (6.14) to another Lévy process given in the following

Remark 6.1.16. *Let X be a Lévy process having Lévy measure Q. For fixed $c > 0$ and $s \geq 0$, the process X^* given by $X_t^* := X_{ct+s} - X_s$ ($t \in \mathbb{R}^+$) is easily seen to be a Lévy process having Lévy measure $Q^* := cQ$.*

We now give a series of examples of Lévy processes and hence (see the remark following Definition 6.1.11!) of infinitely divisible distributions.

Example 6.1.17. *For fixed $\mu \in \mathbb{R}$ let $X_t \equiv \mu t$ ($t \in \mathbb{R}^+$) which is the most simple Lévy process. Let $\mu \neq 0$. Even in this simple (deterministic!) case, the process*

$$x(t) := a \exp(X_t) = a \exp(\mu t), \quad t \in \mathbb{R}^+ \ (a > 0) \tag{6.18}$$

(which occurs, e.g., when $a > 0$ euros are continuously compounded with rate μ) has interesting convergence properties related to Benford's law; see Example 6.1.32 below. In order to give a precise formulation of the latter, we first introduce some notation which will be used throughout this chapter.

Let $B \ (= 2, 3, \dots)$ be any base. For any strictly positive r, the (base B) **significand** $S_B(r)$ of r and the jth (base B or B-adic) **leading digit** $D_{B,j}(r)$ of r ($j = 1, 2, \dots$) are analogously defined as in the case $B = 10$. (See also eq. (1.9) in [Schür2], page 1221; recall that we prefer to call $S_B(r)$ the significand and not the mantissa!)

For any base $B \geq 2$ and $m \geq 1$, let

$$I_B(m) := \{(d_1, \dots, d_m) : 1 \leq d_1 \leq B - 1, 0 \leq d_j \leq B - 1 \text{ for } j = 2, \dots, m\} \tag{6.19}$$

(here, $d_1, \dots, d_m$ are B-adic digits). For $(d_1, \dots, d_m) \in I_B(m)$ consider the half-open interval

$$J_B(d_1, \dots, d_m) := \left[\sum_{j=1}^{m} d_j B^{-j+1}, \sum_{j=1}^{m} d_j B^{-j+1} + B^{-m+1} \right). \tag{6.20}$$

Note that $J_B(d_1, \dots, d_m) \subset [1, B)$ and, for all $r > 0$,

$$D_{B,j}(r) = d_j \quad \text{for } j = 1, \dots, m \quad \Leftrightarrow \quad S_B(r) \in J_B(d_1, \dots, d_m). \tag{6.21}$$

Let f_B denote the density of (base B) **Benford's law** (denoted by BL(B)), given by

$$f_B(x) = (x \log B)^{-1} \mathbb{1}_{[1,B)}(x), \quad x \in \mathbb{R} \tag{6.22}$$

(log denoting (as usual) the **natural** logarithm) and put, for $(d_1, \dots, d_m) \in I_B(m)$,

$$p_B(d_1, \dots, d_m) := \int_{J_B(d_1, \dots, d_m)} f_B(x) \, dx = \log_B \left[1 + \left(\sum_{j=1}^{m} d_j B^{m-j} \right)^{-1} \right]. \tag{6.23}$$

For $m = 1$ this gives

$$p_B(d_1) = \log_B ((d_1 + 1)/d_1) \quad \text{for } d_1 = 1, \dots, B - 1. \tag{6.24}$$

It follows from results in Example 6.1.32 below that, for all $(d_1, \ldots, d_m) \in I_B(m)$, as $t \to \infty$,

$$\frac{1}{t} \operatorname{Leb}\{0 \le u \le t : D_{B,j}(x(u)) = d_j, \ j = 1, \ldots, m\} \to p_B(d_1, \ldots, d_m)$$
$$(6.25)$$

(Leb denoting Lebesgue measure). For more details see Example 6.1.32 below.

Example 6.1.18 (Wiener process or Brownian motion). *Recall that* $N(\delta, \widetilde{\sigma}^2)$ *denotes the normal distribution with mean* $\delta \in \mathbb{R}$ *and variance* $\widetilde{\sigma}^2 > 0$. *A **Wiener process** or (standard) **Brownian motion** (abbreviated BM), denoted throughout this chapter by* $W = (W_t)$ $(t \in \mathbb{R}_+)$, *has the following properties:*

1. $W_0 \equiv 0$.

2. W *has independent increments.*

3. $W_t - W_s \sim N(0, t - s)$ *for all* $0 \le s < t$.

4. *All sample paths of* W *are continuous.*

The existence of a BM follows easily from a criterion due to Kolmogorov (see, e.g, [RevYo], page 19). Clearly, W *is a Lévy process. Since the characteristic function* g_t *of* W_t *is given by*

$$g_t(z) = \exp\left(-tz^2/2\right), \quad z \in \mathbb{R}, \quad t \ge 0, \qquad (6.26)$$

the generating triple (β, σ^2, Q) *of* W *is given by* $\beta = 0, \sigma^2 = 1$, *and* $Q \equiv 0$.

Lemma 6.1.19. *Let* W *be a BM. Then*

1. *for each* $s > 0$, $W_{t+s} - W_s$ $(t \in \mathbb{R}_+)$ *is a BM (time homogeneity);*

2. $(-W_t)$ *is a BM (symmetry);*

3. *for every* $c > 0$, (cW_{t/c^2}) *is a BM (scaling property).*

For a proof see [KaSh], page 104, or [RevYo], page 21.

A process of the form

$$Z_t = Z_0 \exp\left(\mu t + cW_t - c^2 t/2\right), \quad t \in \mathbb{R}^+ \qquad (6.27)$$

where $Z_0 > 0, \mu \in \mathbb{R}$, and $c > 0$ are constants, is called a **Black–Scholes process** (or **geometric Brownian motion**) with parameters μ and c. It plays an important role in stochastic finance. If Z_t is modeling the price per share of a certain stock at time t, then μ may be interpreted as the mean rate of return for the stock, and c^2 may be interpreted as the variance of the rate of return (c is also called **volatility**).

Example 6.1.20 (Poisson process). *A **Poisson process** $N = (N_t)$ $(t \in \mathbb{R}^+)$ with parameter* $\lambda > 0$, *defined on* $(\Omega, \mathcal{F}, P)$, *has the following properties:*

1. $N_0 \equiv 0$.

2. N *has independent increments.*

3. *For* $0 \leq s < t$, $N_t - N_s$ *has a **Poisson distribution** with parameter* $\lambda(t-s)$, *i.e.,*

$$P(N_t - N_s = k) = \frac{(\lambda(t-s))^k}{k!} e^{-\lambda(t-s)}, \quad k = 0, 1, 2, \ldots. \quad (6.28)$$

4. *All sample paths of* N *are cadlag, piecewise constant, increase by jumps of size 1, perform on each time interval only finitely many jumps, and satisfy*

$$N_t(\omega) \rightarrow \infty \, (t \rightarrow \infty) \quad \text{for all } \omega \in \Omega.$$

For an explicit construction of N *see [ConT], page 48. Clearly,* N *is a Lévy process. It is easy to see that the characteristic function* g_t *of* N_t *is given by*

$$g_t(z) = \exp\left[\lambda t \left(e^{iz} - 1\right)\right], \quad \text{for all } z \in \mathbb{R} \text{ and } t \geq 0, \quad (6.29)$$

which implies that the Lévy measure $Q^{(\lambda)}$ *of* N *is given by*

$$Q^{(\lambda)}(A) = \begin{cases} \lambda & \text{if } 1 \in A, \\ 0 & \text{if } 1 \notin A, \end{cases} \quad \text{for all Borel sets } A \subset \mathbb{R}, \quad (6.30)$$

i.e., the total mass of $Q^{(\lambda)}$ *equals* λ *and is concentrated on* $\{1\}$. *It is easily checked that the generating triple of* N *equals* $\left(\lambda, 0, Q^{(\lambda)}\right)$.

Example 6.1.21 (Compound (CP-)process). *Let* $\zeta_1, \zeta_2, \ldots$ *be independent and identically distributed random variables defined on* $(\Omega, \mathcal{F}, P)$ *such that the distribution* Q_1 *of* ζ_1 *satisfies* $Q_1(\{0\}) = 0$. *We therefore may assume* $\zeta_n(\omega) \neq 0$ *for all* $\omega \in \Omega$ *and* $n \geq 1$. *Let* (N_t) *be a Poisson process with parameter* $\lambda > 0$ *such that the processes*

$$(\zeta_n) \, (n \geq 1) \text{ and } (N_t) \text{ are independent of each other.} \quad (6.31)$$

Then the process

$$T_t := \sum_{j=1}^{N_t} \zeta_j \quad (t \in \mathbb{R}^+) \quad (6.32)$$

*is called a **compound Poisson (or CP-)process** associated with* λ *and* Q_1. *(Interpretation: Imagine an insurance company and suppose that the number of accidents by time* t *equals* N_t. *If, at the time of the* k*th accident, the company has to pay the amount of* ζ_k, *then* T_t *equals the total payment of the company by time* t.) *It can be shown that* (T_t) *is a Lévy process (see [Sat2], page 18) such that the characteristic function* g_t *of* T_t *is given by*

$$g_t(z) = \exp\left[\lambda t \int_{\mathbb{R}} \left(e^{izx} - 1\right) Q_1(dx)\right] \quad \text{for all } z \in \mathbb{R} \text{ and } t \geq 0, \quad (6.33)$$

which implies that the generating triple of (T_t) *is given by*

$$(\beta, \sigma^2, Q) = \left(\lambda \int_{[-1,1]} x \, Q_1(dx), 0, \lambda Q_1\right). \quad (6.34)$$

Example 6.1.22 (Lévy processes of jump-diffusion type). *We can build a new Lévy process from the processes in Examples 6.1.17, 6.1.18, and 6.1.21, assuming that the processes*

$$(W_t), (\zeta_t), \text{ and } (N_t) \text{ are independent of each other.} \tag{6.35}$$

Then, for constants $\alpha \in \mathbb{R}$ and $c > 0$,

$$X_t := \alpha t + c W_t + \sum_{j=1}^{N_t} \zeta_j \quad (t \in \mathbb{R}^+) \tag{6.36}$$

*is called a **Lévy process of jump-diffusion type** associated with $\alpha \in \mathbb{R}$, $c > 0$, $\lambda \geq 0$ (!), and Q_1 (denoting the distribution of ζ_1, satisfying $Q_1(\{0\}) = 0$; note that we assume $\zeta_n(\omega) \neq 0$ for all $\omega \in \Omega$ and $n \geq 1$). In the degenerate case $\lambda = 0$ we put $N_t \equiv 0$ for all $t \geq 0$ which is suggested by (6.28). Combining (6.26) and (6.33) it follows from (6.35) that X_1 has the characteristic function*

$$g(z) = \exp\left[i\alpha z - c^2 z^2/2 + \lambda \int_{\mathbb{R}} \left(e^{izx} - 1\right) Q_1(\mathrm{d}x)\right], \quad z \in \mathbb{R}. \tag{6.37}$$

Lévy processes of jump-diffusion type have been used for modeling (log-)prices on options markets exhibiting discontinuities (see [ConT, Lan, Mer2]). If at (random!) time t_0 (N_t) jumps from k to $k + 1$, X performs at t_0 a jump of size ζ_{k+1}, representing some rare event (e.g. some "crash"). If $Q_1 = \mathrm{N}(\delta, \widetilde{\sigma}^2)$, X is called a **Merton model** (cf. [ConT, Mer2]); if Q_1 is a **double exponential distribution** of the form

$$Q_1(\mathrm{d}x) = \Big[p\lambda_+ \exp(-\lambda_+ x)\mathbb{1}_{(0,\infty)}(x)$$
$$+ (1 - p)\lambda_- \exp(-\lambda_-|x|)\mathbb{1}_{(-\infty,0)}(x)\Big]\mathrm{d}x \tag{6.38}$$

(where $\lambda_+ > 0$, $\lambda_- > 0$, and $0 \leq p \leq 1$ are real constants), X is called a **Kou model** (cf. [ConT, Kou, KouWa]).

Example 6.1.23 (Gamma process). *Let $f(\cdot|\alpha, \Delta)$ be the density of a **gamma distribution** $\mu_{\alpha,\Delta}$ with parameters $\alpha > 0$ and $\Delta > 0$, given by*

$$f(x|\alpha, \Delta) = \alpha^\Delta x^{\Delta-1} \exp(-\alpha x)/\Gamma(\Delta), \quad x > 0 \tag{6.39}$$

and $f(x|\alpha, \Delta) = 0$ for $x \leq 0$. The characteristic function $\hat{\mu}_{\alpha,\Delta}$ of $\mu_{\alpha,\Delta}$ is given by

$$\hat{\mu}_{\alpha,\Delta}(z) = \left(1 - i\alpha^{-1}z\right)^{-\Delta}, \quad z \in \mathbb{R}. \tag{6.40}$$

Since ([Sat2], page 45) $\hat{\mu}_{\alpha,\Delta}$ can also be written in the form

$$\hat{\mu}_{\alpha,\Delta}(z) = \exp\left[\Delta \int_0^\infty (e^{izx} - 1)x^{-1}e^{-\alpha x}\, dx\right], \quad z \in \mathbb{R}, \tag{6.41}$$

$\mu_{\alpha,\Delta}$ is infinitely divisible with Lévy measure Q given by

$$Q(\mathrm{d}x) = \Delta x^{-1} e^{-\alpha x}\mathbb{1}_{(0,\infty)}(x)\mathrm{d}x. \tag{6.42}$$

*A **gamma process** with parameters $\alpha > 0$ and $\Delta > 0$ is a Lévy process X such that $X_1 \sim \mu_{\alpha,\Delta}$. Then $X_t \sim \mu_{\alpha,\Delta t}$ $(t > 0)$.*

Example 6.1.24 (Stable distribution; stable process). *Let X be a Lévy process such that X_1 has an $\boldsymbol{\alpha}$-**stable distribution** $S_\alpha(\gamma, \delta, \tau)$ depending on parameters $0 < \alpha < 2, \gamma \in \mathbb{R}, \delta \geq 0$, and $-1 \leq \tau \leq 1$. The characteristic function h_α of $S_\alpha(\gamma, \delta, \tau)$ is of the form*

$$h_\alpha(z) = \begin{cases} \exp\left[i\gamma z - \delta|z|^\alpha(1 + i\tau\operatorname{sign}(z)\tan(\pi\alpha/2))\right] & \text{if } \alpha \neq 1, \\ \exp\left[i\gamma z - \delta|z|(1 + 2i\tau\pi^{-1}\operatorname{sign}(z)\log|z|)\right] & \text{if } \alpha = 1, \end{cases} \quad (6.43)$$

where

$$\operatorname{sign}(z) = \begin{cases} 1 & \text{if } z > 0, \\ 0 & \text{if } z = 0, \\ -1 & \text{if } z < 0 \end{cases} \quad (6.44)$$

*(see, e.g., [Br], page 204, [SamTa], page 5, or [Shi], page 341). One can show that $S_\alpha(\gamma, \delta, \tau)$ is infinitely divisible. The process X is called an $\boldsymbol{\alpha}$-**stable process**. Furthermore, combining results in [SamTa], page 18, and [Sat2], page 159, yields that (provided $\delta > 0$)*

$$\mathbb{E}\left[|X_t|^p\right] < \infty \quad \text{for all } t > 0 \text{ and } 0 < p < \alpha \quad (6.45)$$

and

$$\mathbb{E}\left[|X_t|^p\right] = \infty \quad \text{for all } t > 0 \text{ and } p \geq \alpha. \quad (6.46)$$

This implies that, for all $t > 0$, X_t has an infinite variance if $1 < \alpha < 2$.

Martingales, being of fundamental importance in modern probability theory, will appear in this section in the (hopefully!) motivating Example 6.1.32 below, and also in §6.3 where they are used to obtain large deviation results for certain functionals.

We first need

Definition 6.1.25 (Filtration). *Let I be an index set which is one of the sets $\mathbb{R}^+, \mathbb{Z}^+$ or $\{0, 1, \ldots, N\}$ for some $N \geq 1$. By a **filtration** on a measurable space $(\Omega, \mathcal{F})$ we mean a family of σ-algebras $(\mathcal{F}_t)$ $(t \in I)$ such that $\mathcal{F}_s \subset \mathcal{F}_t \subset \mathcal{F}$ for all $s, t \in I$ such that $s < t$.*

Definition 6.1.26 (Martingale). *Let $(\mathcal{F}_t)$ $(t \in I)$ be any filtration. A family of random variables (ξ_t) $(t \in I)$ defined on a probability space $(\Omega, \mathcal{F}, P)$ is called a **martingale** with respect to $(\mathcal{F}_t)$ if the following three conditions hold:*

(M1) *Each ξ_t is **integrable**, i.e., $\mathbb{E}\left[|\xi_t|\right] < \infty$.*

(M2) *(ξ_t) is **adapted** to $(\mathcal{F}_t)$, i.e., for each $t \in I, \xi_t$ is $\mathcal{F}_t$-**measurable** meaning that $\{\xi_t \in A\} \in \mathcal{F}_t$ for all Borel sets $A \subset \mathbb{R}$.*

(M3) *For all $s, t \in I$ such that $s < t$,*

$$\mathbb{E}\left[\xi_t | \mathcal{F}_s\right] = \xi_s \quad a.s. \quad (6.47)$$

Note that, by (M3) and (M1),

$$\mathbb{E}\left[\xi_t\right] = \mathbb{E}\left[\xi_0\right] \quad \text{for all } t \in I. \quad (6.48)$$

Example 6.1.27 (Partial sums of random variables as martingales). *Let $\xi_0, \xi_1, \ldots$ be a sequence of independent integrable random variables defined on $(\Omega, \mathcal{F}, P)$ and let*

$$\mathcal{F}_n := \sigma\{\xi_0, \xi_1, \ldots, \xi_n\}, \quad n \in \mathbb{Z}^+ \tag{6.49}$$

*denote the σ-algebra **generated** by $\xi_0, \xi_1, \ldots, \xi_n$, i.e., $\mathcal{F}_n$ is the smallest sub-σ-algebra of $\mathcal{F}$ such that $\xi_0, \xi_1, \ldots, \xi_n$ are all $\mathcal{F}_n$-measurable. Let $\mathbb{E}[\xi_n] = 0$ for all $n \geq 1$. Then, the partial sums $S_n := \xi_0 + \xi_1 + \cdots + \xi_n$ $(n \in \mathbb{Z}^+)$ form a martingale with respect to $(\mathcal{F}_n)$.*

Example 6.1.28 (Exponential Lévy processes as martingales). *Let $X = (X_t)$ be a Lévy process such that $\mathbb{E}[\exp(X_t)] < \infty$ for all t. Let $(\mathcal{F}_t)$ $(t \in \mathbb{R}^+)$ be the **filtration generated by** $\mathbf{X}$, i.e.,*

$$\mathcal{F}_t := \sigma\{X_s : s \leq t\}, \quad t \in \mathbb{R}^+ \tag{6.50}$$

*(the definition of $\mathcal{F}_t$ being similar to that in (6.49)). Then the **exponential Lévy process** $(\exp(X_t))$ $(t \in \mathbb{R}^+)$ is easily shown to be a martingale with respect to $(\mathcal{F}_t)$ iff*

$$\mathbb{E}[\exp(X_t)] = 1 \quad \text{for all } t \in \mathbb{R}^+. \tag{6.51}$$

(In order to prove (M3), write $X_t = (X_t - X_s) + X_s$ and note that $X_t - X_s$ is independent of $\mathcal{F}_s$.)

Example 6.1.29. *Let the random variable ξ (defined on $(\Omega, \mathcal{F}, P)$) be integrable and let $(\mathcal{F}_t)$ $(t \in I)$ be any filtration. Then $\xi_t := \mathbb{E}[\xi|\mathcal{F}_t]$ $(t \in I)$ is a martingale with respect to $(\mathcal{F}_t)$. In order to prove (M3) one uses the "**iteration property**" of **conditional expectations**, which says that for any σ-algebras $\mathcal{G}_1 \subset \mathcal{G}_2 \subset \mathcal{F}$,*

$$\mathbb{E}[\mathbb{E}[\xi|\mathcal{G}_2]\,\mathcal{G}_1] = \mathbb{E}[\xi|\mathcal{G}_1] \quad a.s. \tag{6.52}$$

In the case $I = \{0, 1, \ldots, N\}$ $(N \geq 1)$ this example will be crucial when applying Azuma's Inequality in §6.3 (see proof of Theorem 6.3.4).

Theorem 6.1.30 (Exponential Lévy processes as martingales). *Let (X_t) be a Lévy process with a generating triple (β, σ^2, Q). Assume that*

$$\int_{|x| \geq 1} e^{\alpha|x|}\, Q(dx) < \infty \quad \text{for all } \alpha > 0. \tag{6.53}$$

For each $c > 0$ let

$$\mu(c) := -c\beta - c^2\sigma^2/2 - \int_{\mathbb{R}} \left(e^{cx} - 1 - cx\mathbb{1}_{[-1,1]}(x)\right) Q(dx). \tag{6.54}$$

Then we have that, for all $c > 0$,

$$(\exp(cX_t + \mu t)) \quad (t \in \mathbb{R}^+) \quad \text{is a martingale iff } \mu = \mu(c) \tag{6.55}$$

(the filtration is the one generated by X).

Note that (6.53) holds e.g. if Q has compact support (use (6.9)).

Proof of Theorem 6.1.30. We shall use that

$$\mathbb{E}\left[\exp\left(cX_t\right)\right] < \infty \text{ holds for all } t \in \mathbb{R}^+ \Leftrightarrow \int_{|x| \geq 1} e^{cx}\, Q(dx) < \infty \qquad (6.56)$$

(see [Sat2], page 159, or [Kr]). Let (for fixed $t > 0$) $\psi_1(z) := \mathbb{E}\left[\exp\left(izX_t\right)\right]$ and

$$\psi_2(z) := \exp\left(t\left[i\beta z - \sigma^2 z^2/2 + \int_{\mathbb{R}}\left(e^{izx} - 1 - izx\mathbb{1}_{[-1,1]}(x)\right)Q(dx)\right]\right).$$

By (6.53) and (6.8) one obtains, using the dominated convergence theorem, that ψ_1 and the argument of the exponential function defining ψ_2 are, as functions of $z \in \mathbb{C}(!)$ holomorphic on $\mathbb{C}$. (To prove this, one verifies the Cauchy–Riemann equations and shows that the partial derivatives involved are continuous; see, e.g., [StSh2], page 13.) Since $\psi_1 = \psi_2$ on $\mathbb{R}$, we have $\psi_1 = \psi_2$ on $\mathbb{C}$ (see [StSh2], page 52). Hence, taking $z = c/i$, we get for $c > 0$ and $t \geq 0$,

$$\mathbb{E}\left[\exp\left(cX_t + \mu t\right)\right] = \exp\left(t(\mu - \mu(c))\right). \qquad (6.57)$$

Hence (6.55) follows from the result in Example 6.1.28. $\qquad\square$

Example 6.1.31. *Let* $W - (W_t)$ *be a BM. It follows easily from Theorem 6.1.30 that, for each $c > 0$,*

$$\left(\exp\left(c\,W_t - c^2 t/2\right)\right) \qquad (t \in \mathbb{R}_+) \quad \text{is a martingale,} \qquad (6.58)$$

with the filtration given by (6.50) for $X = W$.

Note that the process in (6.58) is a geometric BM with parameters 0 and c (see Example 6.1.18).

Example 6.1.32 (Motivation). *For fixed $a > 0$ and $\mu \neq 0$ consider the (deterministic!) process $x(t) = a\exp(\mu t)$ $(t \in \mathbb{R}^+)$ as in (6.18). Choose any base $B \geq 2$. We shall show that, for each function $G\colon [1, B] \longrightarrow \mathbb{R}$ which is bounded and measurable, that (as $t \to \infty$)*

$$\frac{1}{t}\int_0^t G(S_B\left(x(u)\right))\, du \to \int_1^B G(x)f_B(x)\, dx \qquad (6.59)$$

where f_B is the density of BL(B), *given by (6.22). Choosing $G := \mathbb{1}_{J_B(d_1,\dots,d_m)}$ (for $(d_1,\dots,d_m) \in I_B(m)$ given by (6.19)), we obtain from (6.59), using (6.21), that (as $t \to \infty$)*

$$\frac{1}{t}\,\mathrm{Leb}\{0 \leq u \leq t : D_{B,j}(x(u)) = d_j, \, j = 1,\dots,m\} \to p_B(d_1,\dots,d_m) \qquad (6.60)$$

($p_B(d_1,\dots,d_m)$ given by (6.23)).

According to the usual approximation scheme it suffices to prove (6.59) for $G = \mathbb{1}_A$ ($A \subset [1, B]$ being any Borel set).

First let $\mu > 0$. Note that, for each integer j and $t \geq 0$,

$$S_B(x(t)) = ae^{\mu t}B^{-j} \Leftrightarrow j = \left\lfloor \frac{\log(a\exp(\mu t))}{\log B} \right\rfloor =: n(t) \qquad (6.61)$$

($\lfloor v \rfloor$ (floor of v) denoting the largest integer $\leq v$). Let $n_0 := 1 + \left\lfloor \frac{\log a}{\log B} \right\rfloor$. Clearly

$$\mu^{-1} \log \left(B^{n_0}/a \right) \geq 0 \quad \text{and} \quad \mu^{-1} \log(B^{n(t)}/a) \leq t. \tag{6.62}$$

In the sequel let $t \geq \mu^{-1} \log B$. Then $n_0 \leq n(t)$. For each set $A_1 \subset \mathbb{R}$ and $\alpha \in \mathbb{R}$, we shall put

$$\alpha A_1 := \{ \alpha \tilde{a} : \tilde{a} \in A_1 \} \tag{6.63}$$

and, if $A_1 \subset (0, \infty)$,

$$\log A_1 := \{ \log \tilde{a} : \tilde{a} \in A_1 \}. \tag{6.64}$$

By (6.61), the left-hand side in (6.59) equals (as $t \to \infty$)

$$\frac{1}{t} \operatorname{Leb} \{ 0 \leq u \leq t : S_B(x(u)) \in A \}$$

$$= o(1) + \frac{1}{t} \sum_{j=n_0}^{n(t)-1} \operatorname{Leb} \{ u : S_B(x(u)) = a e^{\mu u} B^{-j}, \ S_B(x(u)) \in A \}$$

$$= o(1) + \frac{1}{t} \sum_{j=n_0}^{n(t)-1} \operatorname{Leb} \left(\mu^{-1} \log(a^{-1} B^j A) \right)$$

$$= o(1) + \frac{n(t) - n_0}{\mu t} (\log B) \int_A f_B(x) \, dx$$

$$= o(1) + \int_A f_B(x) \, dx.$$

Hence, if $\mu > 0$, (6.59) holds for general G. If $\mu < 0$, the proof of (6.59) for $G = \mathbb{1}_A$ can be based on the validity of (6.59) for $\mu > 0$ and the fact that, for all $\alpha > 0$, either $S_B(\alpha) = S_B(1/\alpha) = 1$ or $S_B(\alpha) \cdot S_B(1/\alpha) = B$. (Note that the first alternative only occurs if $\alpha = B^m$ for some $m \in \mathbb{Z}$.) This finishes the proof of (6.59). Now let X be a Lévy process with generating triple (β, σ^2, Q) and suppose Q satisfies (6.53). Note that the process $Z_t = a \exp\left(cX_t + \mu t\right)$ (for $a > 0, c > 0$, and $\mu \in \mathbb{R}$) can be factorized as $Z_t = Z_t^{(1)} \cdot Z_t^{(2)}$ where $Z_t^{(1)} := a \exp\left((\mu - \mu(c))t\right), Z_t^{(2)} = \exp\left(cX_t + \mu(c)t\right)$ ($\mu(c)$ given by (6.54)). Hence, if $\mu \neq \mu(c)$, Z fluctuates around the (deterministic!) process $Z^{(1)}$ for which (6.59) holds—due to factors forming a mean 1 martingale (recall (6.55) and (6.48)).

This suggests

Conjecture 6.1.33. *Suppose that in (6.59) $(x(t))$ is replaced by a process of the form*

$$Z_t = a \exp\left(cX_t + \mu t\right) \quad (t \in \mathbb{R}^+) \tag{6.65}$$

where $a > 0, c > 0$, and $\mu \in \mathbb{R}$ are constants, and X is a Lévy process. Let $B \geq 2$ be any base and let $G: [1, B] \longrightarrow \mathbb{R}$ be bounded and measurable. Then (under certain assumptions on X), as $t \to \infty$,

$$\mathbb{E}\left[t^{-1} \int_0^t G\left(S_B(Z_u)\right) du \right] \to \int_1^B G(x) f_B(x) \, dx \tag{6.66}$$

or even

$$t^{-1} \int_0^t G\left(S_B(Z_u)\right) du \;\longrightarrow\; \int_1^B G(x) f_B(x)\, dx \quad a.s. \tag{6.67}$$

*Note that by dominated convergence, (6.66) is **necessary** for (6.67).*

We shall prove (6.66) in §6.2, and (6.67) in §6.3—provided the characteristic function g of X_1 satisfies a certain "standard condition" (see Definition 6.2.9 in §6.2). In particular, it follows from our results that (6.67) holds for a geometric Brownian motion given by (6.27).

6.2 EXPECTATIONS OF NORMALIZED FUNCTIONALS

In this section we study the asymptotic behavior of the significand of the process

$$Z_t := a\theta^{c_t X_t + d_t}, \quad t \in \mathbb{R}^+ \tag{6.68}$$

which is slightly more general than the one in (6.65). Here, $a > 0, \theta > 0$ ($\theta \neq 1$), c_t, and d_t ($t \geq 0$) are real numbers, and we shall assume throughout that

$$c_t \geq \widetilde{c} > 0, \quad t \geq 0. \tag{6.69}$$

Furthermore X is a real-valued Lévy process. Recall the convention that throughout this chapter (except for the appendices)

$$g \text{ denotes the characteristic function of } X_1.$$

If a base $B \geq 2$ has (often tacitly!) been chosen, $G\colon [1, B] \longrightarrow \mathbb{R}$ denotes a function which is bounded and measurable. Similarly as in (6.59) we investigate the asymptotic behavior of the **continuous-time functional**

$$L_t := \int_0^t G\left(S_B(Z_u)\right) du, \quad t \in \mathbb{R}^+ \tag{6.70}$$

where c_t and d_t (as functions of t) are assumed to be cadlag.

The **discrete-time functional**

$$\widetilde{L}_n := \sum_{j=1}^n G\left(S_B(Z_j)\right), \quad n \in \mathbb{Z}^+ \tag{6.71}$$

will be important for statistical applications in §6.5. If $G = \mathbb{1}_A$ for some Borel set $A \subset [1, B]$, (L_t) is the **occupation time** of A by the significand process $(S_B(Z_t))$, given by

$$L_t = \mathrm{Leb}\{0 \leq u \leq t : S_B(Z_u) \in A\}, \quad t \in \mathbb{R}^+. \tag{6.72}$$

In particular, for $A = J_B(d_1, \ldots, d_m)$ (given by (6.20)) we obtain, using (6.21), that

$$L_t = \mathrm{Leb}\{0 \leq u \leq t : D_{B,j}(Z_u) = d_j, j = 1, \ldots, m\}, \quad t \in \mathbb{R}^+. \tag{6.73}$$

The desired almost sure convergence results for the normalized functionals given by (6.70) and (6.71), respectively, will be obtained in two steps. In the present section it is shown (assuming the Standard Condition (SC) introduced in Definition 6.2.9) that, given a base B, the *expectations* of the normalized functionals, viz. $\mathbb{E}\left[L_t/t\right]$ and $\mathbb{E}\left[\tilde{L}_n/n\right]$, converge to $\mathbb{E}\left[G(\xi)\right]$ for $\xi \sim \mathrm{BL}(B)$ (see Theorem 6.2.14). Here, the crucial tool is the Poisson Summation Formula (see Theorem 6.6.3 in Appendix 6.6). In §6.3 it will be shown (assuming (SC)) that the *centered* normalized functionals, viz. $L_t/t - \mathbb{E}\left[L_t/t\right]$ and $\tilde{L}_n/n - \mathbb{E}\left[\tilde{L}_n/n\right]$, tend to zero almost surely. (This follows from Theorem 6.3.4 combined with the Borel–Cantelli Lemma.) Here, the basic tool is Azuma's Inequality for discrete-time martingales with bounded increments (see Theorem 6.3.1). Combining the above results, we finally obtain that (L_t/t) and $(\tilde{L}_n/n)$ converge almost surely to $\mathbb{E}\left[G(\xi)\right]$ for $\xi \sim \mathrm{BL}(B)$ (see Theorem 6.3.5).

We shall also consider the **mod p functionals** (M_t^*) $(t \in \mathbb{R}^+)$ and $(\widetilde{M}_n)$ $(n \in \mathbb{Z}^+)$ which, for fixed $p > 0$, are defined as follows. Let $H\colon [0,p] \longrightarrow \mathbb{R}$ be bounded and measurable and put

$$M_t^* := \int_0^t H\left((c_u X_u + d_u) \bmod p\right) du, \quad t \in \mathbb{R}^+ \tag{6.74}$$

and

$$\widetilde{M}_n := \sum_{j=1}^n H\left((c_j X_j + d_j) \bmod p\right), \quad n \in \mathbb{Z}^+. \tag{6.75}$$

Here X is a Lévy process, and c_t, d_t $(t \geq 0)$ are real numbers such that (6.69) holds. In (6.74), c_t and d_t (as functions of t) are assumed to be cadlag. Using a simple transformation (based on results in Lemma 6.2.1 below), large deviation and convergence results for (L_t) and $(\tilde{L}_n)$ carry over to corresponding results for (M_t^*) and $(\widetilde{M}_n)$, respectively. This shows (see Theorem 6.3.6 and Remark 6.3.9 in §6.3) that, under the Standard Conditions (SC) (see Definition 6.2.2 below), Lévy processes have nice equidistribution properties.

The transformation just mentioned is based on the fact that

$$\log_B S_B(r) = (\log_B r) \bmod 1, \quad r > 0 \tag{6.76}$$

and, for $\alpha > 0$ and $p > 0$,

$$\alpha\,(r \bmod p) = (\alpha r) \bmod (\alpha p), \quad r \in \mathbb{R}. \tag{6.77}$$

Using (6.76) and (6.77) one easily obtains

Lemma 6.2.1. *Fix any base $B \geq 2$ and $p > 0$. Let $H\colon [0,p] \longrightarrow \mathbb{R}$ be bounded and measurable. Define $G\colon [1, B] \longrightarrow \mathbb{R}$ by*

$$G(x) := H\left(p \log_B x\right), \quad 1 \leq x \leq B \tag{6.78}$$

and let

$$Z_t := B^{c_t^* X_t + d_t^*}, \quad t \in \mathbb{R}^+ \tag{6.79}$$

where $c_t^* = c_t/p$ and $d_t^* := d_t/p$ $(t \in \mathbb{R}^+)$. Then

$$G\left(S_B(Z_t)\right) = H\left((c_t X_t + d_t) \bmod p\right), \quad t \in \mathbb{R}^+ \tag{6.80}$$

implying that $L_t = M_t^*$ $(t \geq 0)$ *and* $\widetilde{L}_n = \widetilde{M}_n$ $(n \geq 0)$. *Furthermore we have*

$$\int_1^B G(x) f_B(x)\, dx \;=\; p^{-1} \int_0^p H(y)\, dy. \tag{6.81}$$

Throughout we shall assume that the characteristic function g of X_1 satisfies a certain "domination condition" which guarantees that the Poisson Summation Formula holds (see Theorem 6.6.3 in Appendix 6.6).

Definition 6.2.2 (Domination Condition; total variation). *The characteristic function* g *is said to satisfy the* **Domination Condition** *(D) if there exists a function* $\widetilde{g}$ *on* $\mathbb{R}$ *such that*

$$|g(z)| \leq \widetilde{g}(z) \quad \text{for all } z \in \mathbb{R}, \tag{D1}$$

$$\widetilde{g} \in L^1(\mathbb{R}), \tag{D2}$$

and

$$\mathrm{TV}\,(\widetilde{g}) < \infty. \tag{D3}$$

Here, for a function $\psi \colon \mathbb{R} \longrightarrow \mathbb{C}$, $\mathrm{TV}(\psi)$ *denotes the* **total variation** *of* ψ *(on* $\mathbb{R}$*) which, as usual, is defined by*

$$\mathrm{TV}(\psi) \;:=\; \sup\left\{ \sum_{j=1}^m |\psi(z_{j+1}) - \psi(z_j)| \right\}, \tag{6.82}$$

the supremum taken over all $m \geq 1$ *and real numbers* $-\infty < z_1 < \cdots < z_{m+1} < \infty$. *The total variation of* ψ *on an interval is defined similarly.*

Note that

$$\mathrm{TV}(\psi) < \infty \;\Rightarrow\; \mathrm{TV}(|\psi|) < \infty. \tag{6.83}$$

Remark 6.2.3. *There exist characteristic functions* h *satisfying* (D) *such that* $\mathrm{TV}(h) = \infty$ *(see Example 6.6.4).*

Remark 6.2.4. *Assume that* g *satisfies* (D). *By* (6.13) *this implies that the characteristic function* g_t *of* X_t $(t \geq 1)$ *is integrable. Hence, by Fourier inversion (see Appendix 6.6),* X_t $(t \geq 1)$ *has a density* $\widetilde{f}_t$ *which is continuous and bounded.*

Example 6.2.5 (Total variation of Brownian sample paths). *Although the sample paths of a BM* (W_t) *defined on* $(\Omega, \mathcal{F}, P)$ *are continuous, they show some very "pathological" behavior. For example, it can be shown (see, e.g., [KaSh], page 106, or [RevYo], page 29) that, for almost all* $\omega \in \Omega$, *the sample path* $t \mapsto W_t(\omega)$ *is of infinite total variation on each interval* $[a, b]$ $(0 \leq a < b < \infty)$!

In order to obtain convergence results like (6.66) or (6.67), we have to assume a "standard condition" (see Definition 6.2.9 below) being stronger than (D). For its definition we need

Definition 6.2.6. *The characteristic function* g *is said to satisfy condition* (CF) *if, for some* $\tau > 0$,

$$\lim_{z \to \infty} z^\tau |g(z)| \;=\; 0. \tag{CF(τ)}$$

Remark 6.2.7. *Assume that g satisfies* (D), *and let the density f be given by Fourier inversion (see Appendix 6.6). Then g is the characteristic function of f, i.e., $g(z) = \int_{\mathbb{R}} e^{izx} f(x) \, dx$, $z \in \mathbb{R}$. Hence, by the **Riemann–Lebesgue Lemma** which holds for arbitrary functions in $L^1(\mathbb{R})$ (see, e.g., [Bi3], page 345), $g(z) \to 0$ $(z \to \infty)$ which is weaker than* (CF).

Example 6.2.8 (Lévy processes with Gaussian component). *Assume that X has a generating triple (β, σ^2, Q) such that $\sigma^2 > 0$. Then g satisfies* (D) *and* (CF).

Definition 6.2.9 (Standard Condition). *Let X be a Lévy process and let the numbers c_t $(t \in \mathbb{R}^+)$ be as in* (6.69). *We shall say that the **Standard Condition** (SC) is satisfied if at least one of the following three conditions holds:*

   (SC1) *$g \in L^1(\mathbb{R})$, and $|g|$ is decreasing on $\mathbb{R}^+$.*

   (SC2) *g satisfies* (D), *and $c_t = \widetilde{c} > 0$ for all $t \in \mathbb{R}^+$.*

   (SC3) *g satisfies* (D) *and* (CF).

   Note that (SC) $\Rightarrow$ (D).

Example 6.2.10. *Let X be a **gamma process** with parameters $\alpha > 0$ and $\Delta > 1$. Then g satisfies* (D) *and* (CF) *(cf. Example 6.1.23).*

Example 6.2.11 (α-stable distribution). *Let h_α $(0 < \alpha < 2)$ be the characteristic function of an α-stable distribution (see Example 6.1.24). Then h_α satisfies* (D) *and* (CF(τ)) *for all $\tau > 0$.*

Remark 6.2.12. *Let $X = (X_t)$ be a Lévy process with generating triple (β, σ^2, Q) and denote (as above) by g the characteristic function of X_1. The main convergence results (Theorems 6.2.14 and 6.3.5) will be proved assuming the Standard Condition* (SC) *which entails that*

$$g \in L^1(\mathbb{R}). \tag{6.84}$$

It is clear from Theorem 6.1.14 that X is continuous iff $Q = 0$ (see [Sat2], page 135). Then, by (6.84) and (6.134), we must have $\sigma^2 > 0$ which means that X has a Gaussian component. Now assume $\sigma^2 = 0$. Since, by (6.134), $|g(z)| \geq \exp(-2Q(\mathbb{R}))$, $z \in \mathbb{R}$, it follows from (6.84) that $Q(\mathbb{R}) = \infty$ which, by (6.9) is equivalent to

$$Q([-\alpha, \alpha]) = \infty \quad \text{for all } \alpha > 0. \tag{6.85}$$

By Theorem 6.1.14, (6.85) implies that for each $\alpha > 0$, the expected number, per unit time, of (non-zero) jumps of size $\leq \alpha$, performed by X is infinite!

Lemma 6.2.13. *Suppose g satisfies* (D). *Then, for any numbers $a > 0$, $\theta > 0$ $(\theta \neq 1)$, and $c > 0$, $S_B(a\theta^{cX_t})$ $(t \geq 1)$ has a density which, for $1 \leq x \leq B$, is given by*

$$\widetilde{h}_t(x) := f_B(x) \sum_{j=-\infty}^{\infty} \exp\left(-2\pi i j \frac{\log(x/a)}{\log B}\right) g_t\left(\frac{2\pi j c \log \theta}{\log B}\right). \tag{6.86}$$

(Here, f_B is the density of $\mathrm{BL}(B)$, given by (6.22), and g_t is the characteristic function of X_t.)

Proof. Fix $\theta > 1$ and $t \geq 1$. For all $1 \leq s \leq B$ we have

$$S_B(a\theta^{cX_t}) < s \Leftrightarrow X_t \in \bigcup_{j=-\infty}^{\infty} [a_j(1), a_j(s)), \qquad (6.87)$$

where, for $u \geq 1$, $a_j(u) := (c \log \theta)^{-1} \log (a^{-1} B^j u)$. By Remark 6.2.4, X_t has a density $\widetilde{f_t}$ which is continuous and bounded. Hence, by (6.87),

$$P\Big(S_B(a\theta^{cX_t}) < s\Big) = \sum_{j=-\infty}^{\infty} \int_{a_j(1)}^{a_j(s)} \widetilde{f_t}(x)\, dx, \quad 1 \leq s \leq B. \qquad (6.88)$$

Substituting $\log y = cx \log \theta - \log (a^{-1} B^j)$ gives, for $1 \leq s \leq B$,

$$P\Big(S_B(a\theta^{cX_t}) < s\Big) = \frac{1}{c|\log \theta|} \sum_{j=-\infty}^{\infty} \int_1^s \widetilde{f_t}\left(\frac{j \log B}{c \log \theta} + \frac{\log(y/a)}{c \log \theta}\right) \frac{1}{y}\, dy.$$

Since this also holds in the case $0 < \theta < 1$, we conclude that $S_B(a\theta^{cX_t})$ has a density $\widetilde{h_t}$ given by

$$\widetilde{h_t}(x) := \frac{1}{xc|\log \theta|} \sum_{j=-\infty}^{\infty} \widetilde{f_t}\left(\frac{j \log B}{c \log \theta} + \frac{\log(x/a)}{c \log \theta}\right), \quad 1 \leq x \leq B \qquad (6.89)$$

for all $\theta > 0$ $(\theta \neq 1)$. Applying the Poisson Summation Formula (see Theorem 6.6.3) to $\widetilde{f_t}$ and g_t gives that, for all $\lambda \neq 0$ and $\Delta \in \mathbb{R}$,

$$\sum_{j=-\infty}^{\infty} \widetilde{f_t}\left(\frac{j\pi}{\lambda} + \Delta\right) = \frac{|\lambda|}{\pi} \sum_{j=-\infty}^{\infty} \exp\left(-2ij\Delta\lambda\right) g_t(2j\lambda). \qquad (6.90)$$

Choosing $\lambda = (\log B)^{-1} \pi c \log \theta$ and $\Delta = (c \log \theta)^{-1} \log (x/a)$ (for fixed $1 \leq x \leq B$) we obtain (6.86) from (6.89) and (6.90). $\qquad \square$

In the sequel we put

$$\|G\|_\infty := \sup_{1 \leq x \leq B} |G(x)|. \qquad (6.91)$$

Theorem 6.2.14. *Let the process (Z_t) be given by (6.68).*

1. *Suppose g satisfies (D). Then we have, for each base $B \geq 2$ that*

$$\left| \mathbb{E}\left[G\left(S_B(Z_t)\right)\right] - \int_1^B G(x) f_B(x)\, dx \right|$$

$$\leq \left| \int_1^B G(x) f_B(x)\, dx \sum_{j \neq 0} \left| g_t\left(\frac{2\pi j c_t |\log \theta|}{\log B}\right) \right| \right|$$

$$\leq 2\|G\|_\infty \sum_{j=1}^{\infty} \left| g\left(\frac{2\pi j c_t |\log \theta|}{\log B}\right) \right|^t, \quad t \geq 1. \qquad (6.92)$$

2. *Assume the Standard Condition* (SC). *Then we have, for all bases $B \geq 2$, as $t \to \infty$,*

$$\mathbb{E}\left[G\left(S_B(Z_t)\right)\right] \to \int_1^B G(x) f_B(x)\, dx \qquad (6.93)$$

which implies

$$\mathbb{E}\left[L_t/t\right] \to \int_1^B G(x) f_B(x)\, dx \quad (t \to \infty) \qquad (6.94)$$

and

$$\mathbb{E}\left[\widetilde{L}_n/n\right] \to \int_1^B G(x) f_B(x)\, dx \quad (n \to \infty). \qquad (6.95)$$

*In particular ($\xrightarrow{w}$ denoting **weak convergence**),*
$$S_B(Z_t) \xrightarrow{w} \mathrm{BL}(B) \quad (t \to \infty) \qquad (6.96)$$
or (equivalently), for all $m \geq 1$ and $(d_1, \ldots, d_m) \in I_B(m)$,
$$P\left(D_{B,j}(Z_t) = d_j,\, j = 1, \ldots, m\right) \to p_B(d_1, \ldots, d_m) \quad (t \to \infty) \quad (6.97)$$
($p_B(d_1, \ldots, d_m)$ given by (6.23)).

Remark 6.2.15. *Let the random variable ξ_B have density f_B. It follows from Theorem 6.2.14 (assuming* (SC)*) that*
$$G\left(S_B(Z_t)\right) \xrightarrow{w} G\left(\xi_B\right) \quad (t \to \infty) \qquad (6.98)$$
for all functions $G \colon [1, B] \longrightarrow \mathbb{R}$ which are bounded and measurable. Example 2.5 in [Schür2] suggests, however, that, in general, $G\left(S_B(Z_t)\right)$ does not converge in probability (as $t \to \infty$).

Proof of Theorem 6.2.14.

1. Fix $t \geq 1$. Using Lemma 6.2.13 and (6.13), we get, putting $\widetilde{a} := a\theta^{d_t}$,

$$\left| \mathbb{E}\left[G\left(S_B(Z_t)\right)\right] - \int_1^B G(x) f_B(x)\, dx \right|$$

$$= \left| \int_1^B G(x)\widetilde{h}_t(x)\, dx - \int_1^B G(x) f_B(x)\, dx \right|$$

$$\leq \left| \int_1^B G(x) f_B(x)\, dx \right| \left| \sum_{j \neq 0} \exp\left(-2\pi i j \frac{\log(x/\widetilde{a})}{\log B}\right) g_t\left(\frac{2\pi j c_t \log\theta}{\log B}\right) \right|$$

$$\leq \left| \int_1^B G(x) f_B(x)\, dx \right| \sum_{j \neq 0} \left| g_t\left(\frac{2\pi j c_t |\log\theta|}{\log B}\right) \right|$$

$$\leq 2\|G\|_\infty \sum_{j=1}^\infty \left| g\left(\frac{2\pi j c_t |\log\theta|}{\log B}\right) \right|^t .$$

2. The claim (6.93), following from (6.92) and Corollary 6.6.6, entails (6.95). A similar argument yields (6.94). A well-known criterion for weak convergence (see [Bi3], Theorem 25.8) shows that (6.93) implies (6.96). The equivalence of (6.96) and the validity of (6.97) for all $m \geq 1$ and $(d_1, \ldots, d_m) \in I_B(m)$ follows from (6.21) and Theorem 25.8 in [Bi3]. □

6.3 A.S. CONVERGENCE OF NORMALIZED FUNCTIONALS

Recall that if a base $B \geq 2$ and a number $p > 0$ have (often tacitly!) been chosen, $G \colon [1, B] \longrightarrow \mathbb{R}$ and $H \colon [0, p] \longrightarrow \mathbb{R}$ are functions which are bounded and measurable. Recall the definitions of the functionals (L_t) and $(\widetilde{L}_n)$ which are given by

$$L_t := \int_0^t G\left(S_B(Z_u)\right) du, \quad t \in \mathbb{R}^+ \tag{6.99}$$

$((Z_t)$ given by (6.68)), and

$$\widetilde{L}_n := \sum_{j=1}^n G\left(S_B(Z_j)\right), \quad n \in \mathbb{Z}^+. \tag{6.100}$$

We shall also study the **mod p functionals** (M_t^*) and $(\widetilde{M}_n)$ given by

$$M_t^* := \int_0^t H\left((c_u X_u + d_u) \bmod p\right) du, \quad t \in \mathbb{R}^+ \tag{6.101}$$

and

$$\widetilde{M}_n := \sum_{j=1}^n H\left((c_j X_j + d_j) \bmod p\right) du, \quad n \in \mathbb{Z}^+. \tag{6.102}$$

If, in (6.99), $G = \mathbb{1}_A$ for some Borel set $A \subset [1, B]$, then (L_t) is the **occupation time** of A by $(S_B(Z_t))$ (cf. (6.72) and (6.73)).

We shall assume throughout that

$$c_t \geq \widetilde{c} > 0 \quad \text{for all } t \in \mathbb{R}^+. \tag{6.103}$$

In (6.99) and (6.101), c_t and d_t (as functions of t) are assumed to be cadlag.

It turns out that, due to large deviation results (assuming the Standard Condition (SC)), the normalized functionals (L_t/t), $(\widetilde{L}_n/n)$, (M_t^*/t), $(\widetilde{M}_n/n)$ converge almost surely. In order to obtain the desired large deviation results, we shall need (for proofs see [LedTa, McD, Schür1])

Theorem 6.3.1 (Azuma's Inequality). *Let $\xi_0, \xi_1, \ldots, \xi_N$ $(N \geq 1)$ be a martingale such that*

$$\xi_0 = 0 \quad \text{a.s.} \tag{6.104}$$

Assume that there exist constants $\widetilde{c}_1, \ldots, \widetilde{c}_N$ such that

$$|\xi_n - \xi_{n-1}| \leq \widetilde{c}_n \quad \text{a.s.,} \quad n = 1, \ldots, N. \tag{6.105}$$

Then (putting $\exp(-\infty) := 0$)

$$P\left(\max_{0 \leq n \leq N} |\xi_n| \geq v\right) \leq 2\exp\left(-\frac{v^2}{2(\widetilde{c}_1^2 + \cdots + \widetilde{c}_N^2)}\right), \quad v > 0. \tag{6.106}$$

Remark 6.3.2. *Azuma's Inequality can also be applied to obtain large deviation results for certain sequences of integrable random variables* not *being martingales. In order to achieve this, one first tries to construct a suitable martingale to which Theorem 6.3.1 applies (method of bounded differences (increments)). This idea is used e.g. in the proof of Theorem 6.3.4 below (based on Example 6.1.29).*

In the sequel let $(\mathcal{F}_t)$ $(t \in \mathbb{R}^+)$ be the filtration generated by the Lévy process X occurring in (Z_t), i.e.,

$$\mathcal{F}_t := \sigma\{X_s : s \le t\}, \quad t \in \mathbb{R}^+ \tag{6.107}$$

(see Example 6.1.28). Note that $\mathcal{F}_0 = \{\emptyset, \Omega\}$ which follows from Definition 6.1.11. Similarly, we denote by

$$\widetilde{\mathcal{F}}_n := \sigma\{X_m : m \le n\} \quad (n \in \mathbb{Z}^+) \tag{6.108}$$

(see Example 6.1.27) the filtration generated by (X_n) $(n \in \mathbb{Z}^+)$. Note that $\widetilde{\mathcal{F}}_0 = \{\emptyset, \Omega\}$.

The following result is crucial for the applicability of Azuma's inequality in our situation.

Proposition 6.3.3. *Assume the Standard Condition* (SC) *(cf. Definition 6.2.9). Then we have that*

1. *there exists a constant $\mu_1 > 0$ such that, for all $0 \le r < s \le t$,*

$$|\mathbb{E}[L_t|\mathcal{F}_s] - \mathbb{E}[L_t|\mathcal{F}_r]| \le 2(s - r + \mu_1)\|G\|_\infty \quad a.s.; \tag{6.109}$$

2. *there exists a constant $\widetilde{\mu}_1 > 0$ such that, for all $0 \le k < m \le n$,*

$$\left|\mathbb{E}[\widetilde{L}_n|\widetilde{\mathcal{F}}_m] - \mathbb{E}[\widetilde{L}_n|\widetilde{\mathcal{F}}_k]\right| \le 2(m - k + \widetilde{\mu}_1)\|G\|_\infty \quad a.s. \tag{6.110}$$

If G is always non-negative or non-positive, the factor 2 in (6.109) *and* (6.110) *can be omitted. The constants μ_1 and $\widetilde{\mu}_1$ depend on g, B, θ, and $\widetilde{c}$ if* (SC1) *or* (SC2) *holds; if only* (SC3) *holds (g satisfying* (CF(τ))*), they depend also on τ.*

The constants μ_1 and $\widetilde{\mu}_1$ above do not depend on $a > 0$ or d_t $(t \in \mathbb{R}^+)$, and they depend on c_t $(t \in \mathbb{R}^+)$ only through the lower bound $\widetilde{c}$ in (6.103).

Proof of Proposition 6.3.3. We only prove part 2. (This gives us an explicit upper bound (see (6.114) below) which will be important for statistical applications in §6.5. The proof of part 1 is similar to that of part 2 in that sums are replaced by integrals, and Fubini's theorem is used at a certain step.)

Fix $0 \le k < m \le n, N \ge 1$, and let

$$V_r := S_B\left(a\theta^{c_r X_r + d_r}\right), \quad r = 1, 2, \dots.$$

Then, since $\widetilde{L}_j$ is $\widetilde{\mathcal{F}}_s$-measurable when $j \le s$,

$$\mathbb{E}[\widetilde{L}_n|\widetilde{\mathcal{F}}_m] - \mathbb{E}[\widetilde{L}_n|\widetilde{\mathcal{F}}_k]$$

$$= \widetilde{L}_m - \widetilde{L}_k + \sum_{r=m+1}^{n} \mathbb{E}[G(V_r)|\widetilde{\mathcal{F}}_m] - \sum_{r=k+1}^{n} \mathbb{E}[G(V_r)|\widetilde{\mathcal{F}}_k]$$

$$= \sum_{r=k+1}^{m} (G(V_r) - \mathbb{E}[G(V_r)|\widetilde{\mathcal{F}}_k])$$

$$+ \sum_{r=m+1}^{n} (\mathbb{E}[G(V_r)|\widetilde{\mathcal{F}}_m] - \mathbb{E}[G(V_r)|\widetilde{\mathcal{F}}_k]) \quad a.s.$$

implying

$$|E[\widetilde{L}_n|\widetilde{\mathcal{F}}_m] - E[\widetilde{L}_n|\widetilde{\mathcal{F}}_k]|$$

$$\leq 2\|G\|_\infty(m - k + N - 1) + \sum_{r=m+N}^{\infty} |E[G(V_r)|\widetilde{\mathcal{F}}_m] - E[G(V_r)|\widetilde{\mathcal{F}}_k]| \text{ a.s.}$$

(6.111)

(Note that the factor 2 can be omitted if either G is always non-negative or if G is always non-positive.) Using Theorem 6.7.1 gives, for $r \geq m + 1$, that

$$\mathbb{E}[G(V_r)|\widetilde{\mathcal{F}}_m] = G_{r-m,r}(X_m) \quad \text{a.s.}$$

(6.112)

and

$$\mathbb{E}[G(V_r)|\widetilde{\mathcal{F}}_k] = G_{r-k,r}(X_k) \quad \text{a.s.}$$

(6.113)

where, putting $a(x) := a\theta^{xc_r+d_r} \quad (x \in \mathbb{R})$,

$$G_{u,r}(x) := \mathbb{E}\left[G(S_B(a(x)\theta^{c_r X_u}))\right], \quad x \in \mathbb{R}, \ u = 1, 2, \ldots.$$

(Here, we use that X has independent and stationary increments.) By Lemma 6.2.13 (noting that (SC) $\Rightarrow$ (D)) we obtain (g_u denoting the characteristic function of X_u)

$$G_{u,r}(x)$$

$$= \int_1^B G(z)f_B(z)\left[\sum_{j=-\infty}^{\infty} \exp\left(-2\pi i j \frac{\log(z/a(x))}{\log B}\right) g_u\left(\frac{2\pi j c_r \log \theta}{\log B}\right)\right] dz.$$

Using (6.13) we get for $u \geq 1, v \geq 1$, and $x, y \in \mathbb{R}$, noting that $|\exp(iw)| = 1$ for all $w \in \mathbb{R}$, and $g_\beta = g_\alpha g_{\beta-\alpha}$ for all $0 \leq \alpha \leq \beta$,

$$|G_{u,r}(x) - G_{v,r}(y)|$$

$$\leq \|G\|_\infty \sum_{j\neq 0}\left[\left|g_u\left(\frac{2\pi j c_r|\log\theta|}{\log B}\right)\right| + \left|g_v\left(\frac{2\pi j c_r|\log\theta|}{\log B}\right)\right|\right]$$

$$\leq 4\|G\|_\infty \sum_{j=1}^{\infty}\left|g\left(\frac{2\pi j c_r|\log\theta|}{\log B}\right)\right|^{u\wedge v}$$

uniformly with respect to $x, y \in \mathbb{R}$. Combining (6.111)–(6.113) yields (this will also be needed for statistical applications in §6.5)

$$|E[\widetilde{L}_n|\widetilde{\mathcal{F}}_m] - E[\widetilde{L}_n|\widetilde{\mathcal{F}}_k]|$$

$$\leq \|G\|_\infty\left[2(m - k + N - 1) + 4\sum_{j=1}^{\infty}\sum_{r=N}^{\infty}\left|g\left(\frac{2\pi j c_{r+m}|\log\theta|}{\log B}\right)\right|^r\right] \text{ a.s.}$$

(6.114)

(If G is always non-negative or non-positive, $2(m - k + N - 1)$ can be replaced by $m - k + N - 1$.) Now, (6.109) follows from Corollary 6.6.6. □

Combining Proposition 6.3.3 and Azuma's Inequality gives the desired large deviation results for (L_t/t) and $(\widetilde{L}_n/n)$:

Theorem 6.3.4. *Assume* (SC) *(see Definition 6.2.9). Let* $\|G\|_\infty > 0$. *Then we have that*

1. *there exists a constant* $K_1 > 0$ *such that, for all* $v > 0$ *and* $t \geq 1$,

$$P\left(\left\|\frac{1}{t}L_t - \mathbb{E}\left[\frac{1}{t}L_t\right]\right\| \geq v\right) \leq 2\exp\left(-K_1 v^2 t\right);\qquad(6.115)$$

2. *there exists a constant* $K_2 > 0$ *such that, for all* $v > 0$ *and* $n \geq 1$,

$$P\left(\left\|\frac{1}{n}\widetilde{L}_n - \mathbb{E}\left[\frac{1}{n}\widetilde{L}_n\right]\right\| \geq v\right) \leq 2\exp\left(-K_2 v^2 n\right).\qquad(6.116)$$

The constants K_1 *and* K_2 *depend on* g, B, θ, $\widetilde{c}$, *and* $\|G\|_\infty$ *if* (SC1) *or* (SC2) *holds; if only* (SC3) *holds* (g *satisfying* (CF(τ))), *they depend also on* τ.

The constants K_1 and K_2 above do not depend on $a > 0$, d_t ($t \in \mathbb{R}^+$), and they depend on c_t ($t \in \mathbb{R}^+$) only through the lower bound $\widetilde{c}$ in (6.103).

Proof of Theorem 6.3.4.

1. Let $t \geq 1$ be fixed. We apply Azuma's Inequality to the martingale (cf. Example 6.1.29)

$$\xi_m := \mathbb{E}\left[L_t | \mathcal{F}_{t(m)}\right] - \mathbb{E}\left[L_t\right], \quad m = 0, 1, \ldots, N$$

(the filtration $(\mathcal{F}_t)$ given by (6.107)) where $N := \lceil t/\mu_1 \rceil$ and $t(m) := mt/N$, $m = 0, 1, \ldots, N$. Here $\lceil v \rceil$ (ceiling of v) is the smallest integer $\geq v$, and μ_1 occurs in Proposition 6.3.3. Since L_t is $\mathcal{F}_t$-measurable, $\xi_N = L_t - \mathbb{E}[L_t]$ a.s., and (since $\mathcal{F}_0 = \{\emptyset, \Omega\}$) $\xi_0 = 0$ a.s. By Proposition 6.3.3 we have, for $m = 1, \ldots, N$,

$$\begin{aligned} |\xi_m - \xi_{m-1}| &= \left|\mathbb{E}\left[L_t | \mathcal{F}_{t(m)}\right] - \mathbb{E}\left[L_t | \mathcal{F}_{t(m-1)}\right]\right| \\ &\leq 2\left(t/N + \mu_1\right)\|G\|_\infty =: \widetilde{c}_m \quad \text{a.s.} \end{aligned}$$

Now (since $(\alpha + \beta)^2 \leq 2(\alpha^2 + \beta^2)$ for all $\alpha, \beta \in \mathbb{R}$, and since $t \geq 1$),

$$\begin{aligned} \widetilde{c}_1^2 + \cdots + \widetilde{c}_N^2 &= 4N\left(t/N + \mu_1\right)^2 \|G\|_\infty^2 \\ &\leq 8\left(t^2/N + N\mu_1^2\right)\|G\|_\infty^2 \\ &\leq 8\mu_1\left(\mu_1 + 2\right)\|G\|_\infty^2 t =: K_1^{-1} t/2 \end{aligned}$$

which, by Azuma's Inequality, implies (6.115).

2. Let $n \geq 1$ be fixed. We apply Azuma's Inequality to the martingale

$$\xi_m := \mathbb{E}[\widetilde{L}_n | \widetilde{\mathcal{F}}_m] - \mathbb{E}[\widetilde{L}_n], \quad m = 0, 1, \ldots, n$$

(the filtration $(\widetilde{\mathcal{F}}_n)$ given by (6.108)). Clearly, $\xi_n = \widetilde{L}_n - \mathbb{E}[\widetilde{L}_n]$ a.s. and $\xi_0 = 0$ a.s. By Proposition 6.3.3 we have, for $m = 1, \ldots, N$,

$$\begin{aligned} |\xi_m - \xi_{m-1}| &= |\mathbb{E}[\widetilde{L}_n | \widetilde{\mathcal{F}}_m] - \mathbb{E}[\widetilde{L}_n | \widetilde{\mathcal{F}}_{m-1}]| \\ &\leq 2\left(1 + \widetilde{\mu}_1\right)\|G\|_\infty =: \widetilde{c}_m \quad \text{a.s.} \end{aligned}$$

Since $\widetilde{c}_1^2 + \cdots + \widetilde{c}_n^2 = 4(1 + \widetilde{\mu}_1)^2 \|G\|_\infty^2 n =: K_2^{-1} n/2$, (6.116) follows from Azuma's Inequality. $\qquad\square$

Combining Theorems 6.3.4 and 6.2.14 we obtain the following result which (besides Theorem 6.3.4) is our main result about the functionals (L_t) and $(\widetilde{L}_n)$.

Theorem 6.3.5. *Assume* (SC) *(see Definition 6.2.9).*

1. *We have, as $t \to \infty$,*

$$\frac{1}{t}L_t \to \int_1^B G(x)f_B(x)\,dx \quad a.s.\ and\ in\ L^r(1 \leq r < \infty) \qquad (6.117)$$

(the density f_B given by (6.22)). In particular, as $t \to \infty$,

$$\frac{1}{t}\mathrm{Leb}\left\{0 \leq u \leq t : D_{B,j}\left(a\theta^{c_u X_u + d_u}\right) = d_j, j = 1, \dots, m\right\}$$
$$\to p_B(d_1, \dots, d_m) \quad a.s.$$
$$(6.118)$$

holds for each $m \geq 1$ and $(d_1, \dots, d_m) \in I_B(m)$ $(p_B(d_1, \dots, d_m)$ given by (6.23)).

2. *We have, as $n \to \infty$,*

$$\frac{1}{n}\widetilde{L}_n \to \int_1^B G(x)f_B(x)\,dx \quad a.s.\ and\ in\ L^r(1 \leq r < \infty). \qquad (6.119)$$

In particular, for each $m \geq 1$ and $(d_1, \dots, d_m) \in I_B(m)$, as $n \to \infty$,

$$\frac{1}{n}\#\left\{1 \leq j \leq n : D_{B,i}\left(a\theta^{c_j X_j + d_j}\right) = d_i, i = 1, \dots, m\right\}$$
$$\to p_B(d_1, \dots, d_m) \quad a.s. \quad (6.120)$$

(In part 1, we tacitly assume that c_t and d_t are cadlag (as functions of t).)

Proof. By Theorem 6.2.14,

$$\mathbb{E}\left[L_t/t\right] \to \int_1^B G(x)f_B(x)\,dx \quad (t \to \infty).$$

Hence, for each $v > 0$ and $0 < \varepsilon < 1$, there exists a number $t_0(v, \varepsilon) \geq 1$ such that, for all $t \geq t_0(v, \varepsilon)$,

$$\left|\mathbb{E}\left[L_t/t\right] - \int_1^B G(x)f_B(x)\,dx\right| \leq \varepsilon v.$$

By Theorem 6.3.4, this implies that, for all $v > 0, 0 < \varepsilon < 1$, and $t \geq t_0(v, \varepsilon)$,

$$P\left(\left|\frac{1}{t}L_t - \int_1^B G(x)f_B(x)\,dx\right| \geq v\right) \leq 2\exp\left(-K_1(1-\varepsilon)^2 v^2 t\right)$$

which, by the first Borel–Cantelli Lemma, entails

$$L_n/n \to \int_1^B G(x)f_B(x)\,dx \quad a.s.\ (n \to \infty).$$

This, in turn, implies (6.117) since $|L_t - L_s| \le \|G\|_\infty$ for $0 \le s \le t \le s+1$. Applying (6.117) to $G := \mathbb{1}_{J_B(d_1,\dots,d_m)}$ ($J_B(d_1,\dots,d_m)$ given by (6.20)) proves (6.118). Part 2 is proved in a similar fashion. $\square$

Using a simple transformation (see results in Lemma 6.2.1) one obtains from Theorems 6.2.14, 6.3.4, and 6.3.5 without extra work, large deviation and almost sure convergence results for the $\bmod\, p$ functionals (M_t^*) in (6.101) and $(\widetilde{M}_n)$ in (6.102). This reveals that, under (SC), Lévy processes have nice equidistribution properties.

Theorem 6.3.6. *Assume* (SC). *Fix any* $p > 0$ *and* $H\colon [0,p] \longrightarrow \mathbb{R}$ *which is bounded and measurable.*

1. *Then* $\mathbb{E}\,[M_t^*/t] \to p^{-1}\int_0^p H(x)\,dx \quad (t \to \infty)$.   (6.121)

2. *Letting* $\|H\|_\infty > 0$, *then, for all* $v > 0$ *and* $t \ge 1$,

$$P\left(\left| \frac{1}{t}M_t^* - \mathbb{E}\left[\frac{1}{t}M_t^*\right] \right| \ge v \right) \le 2\exp\left(-K_3 v^2 t\right).$$ (6.122)

3. *As* $t \to \infty$, *then*

$$M_t^*/t \to p^{-1}\int_0^p H(x)\,dx \quad \text{a.s. and in } L^r (1 \le r < \infty).$$ (6.123)

In particular, for each Borel set $A \subset [0,p]$,

$$\frac{1}{t}\,\mathrm{Leb}\,\{0 \le u \le t:\, X_u \bmod p \in A\} \to \frac{1}{p}\,\mathrm{Leb}(A) \quad \text{a.s.}$$ (6.124)

The constant $K_3 > 0$ *only depends on* $g, p, \widetilde{c}$, *and* $\|H\|_\infty$ *if* (SC1) *or* (SC2) *holds; if only* (SC3) *holds* (g *satisfying* (CF(τ))), *it also depends on* τ.

Corollary 6.3.7. *Let* (W_t) *be a BM. Then, for each* $p > 0$ *and each Borel set* $A \subset [0,p]$, *as* $t \to \infty$,

$$\frac{1}{t}\,\mathrm{Leb}\,\{0 \le u \le t:\, W_u \bmod p \in A\} \to \frac{1}{p}\,\mathrm{Leb}(A) \quad \text{a.s.}$$ (6.125)

Remark 6.3.8. *Fix* $p > 0$. *It is easily seen from Corollary 6.3.7 that there exists a set* $\Omega_p \in \mathcal{F}$ *satisfying* $P(\Omega_p) = 0$ *such that for all intervals* $A \subset [0,p)$

$$\lim_{t\to\infty} \frac{1}{t}\,\mathrm{Leb}\,\{0 \le u \le t:\, W_u(\omega) \bmod p \in A\} = \frac{1}{p}\,\mathrm{Leb}(A), \quad \omega \notin \Omega_p.$$ (6.126)

Remark 6.3.9. *The arguments which lead to Theorem 6.3.6 show that, under* (SC), (6.121)–(6.123) *remain true for* $(\widetilde{M}_n/n)$ *instead of* (M_t^*/t). *Instead of* (6.124) *we have under* (SC) *that, for each Borel set* $A \subset [0,p]$, *as* $n \to \infty$,

$$\frac{1}{n}\#\,\{1 \le j \le n:\, X_j \bmod p \in A\} \to \frac{1}{p}\,\mathrm{Leb}(A) \quad \text{a.s.}$$ (6.127)

6.4 NECESSARY AND SUFFICIENT CONDITIONS FOR (D) OR (SC)

The following result gives a condition which is sufficient for (CF) (see also [Fel], page 514).

Proposition 6.4.1. *Let h be the characteristic function of some random variable ξ such that $h \in L^1(\mathbb{R})$. Let f be the continuous and bounded density of ξ, which (by Fourier inversion) is given by*

$$f(x) = \frac{1}{2\pi} \int_{\mathbb{R}} e^{-izx} h(z)\, dz, \quad x \in \mathbb{R} \tag{6.128}$$

(see Appendix 6.6). Assume that $f'(x)$ exists and is finite for all but a countable set of $x \in \mathbb{R}$, and let $f' \in L^1(\mathbb{R})$. Then h satisfies (CF(1)).

Proof. Integration by parts yields, for $n \geq 1$ and $z \neq 0$,

$$\int_{-n}^{n} e^{izx} f(x)\, dx = \frac{1}{iz}\left(e^{izn} f(n) - e^{-izn} f(-n)\right) - \frac{1}{iz}\int_{-n}^{n} e^{izx} f'(x)\, dx$$

(cf. [HeSt], page 299). Combining (6.128) and the Riemann–Lebesgue Lemma (see Remark 6.6.2) we obtain for $z \neq 0$, that

$$zh(z) = z \int_{-\infty}^{\infty} e^{izx} f(x)\, dx = i \int_{-\infty}^{\infty} e^{izx} f'(x)\, dx = o(1) \quad \text{as } |z| \to \infty. \;\Box$$

In the sequel, $X = (X_t)$ is a Lévy process such that (β, σ^2, Q) is the generating triple of the characteristic function g of X_1. Recall that

$$Q(\{0\}) = 0 \quad \text{and} \quad \int_{\mathbb{R}} \left(x^2 \wedge 1\right) Q(dx) < \infty. \tag{6.129}$$

Proposition 6.4.2.   *1. Assume $\sigma^2 > 0$ (i.e., X has a Gaussian component). Then g satisfies (D) and (CF(τ)) for all $\tau > 0$.*

2. Let $\sigma^2 = 0$. Suppose there exists a constant $\varepsilon_1 > 0$ such that, for some $z_0 \geq 1$,

$$z^2 \int_{[-1/z,1/z]} x^2 Q(dx) \geq \varepsilon_1 \log z \quad \text{for all } z \geq z_0. \tag{6.130}$$

Then g satisfies (CF(τ)) for all $0 < \tau < 2\varepsilon_1$. If, moreover,

$$\varepsilon_1 > 1/2, \tag{6.131}$$

then g satisfies (D).

3. Let $\sigma^2 = 0$. If g satisfies (D) or (CF), then

$$Q([-\alpha, \alpha]) = \infty \quad \text{for all } \alpha > 0 \tag{6.132}$$

which (by (6.129)) is equivalent to

$$Q([-1, 1]) = \infty. \tag{6.133}$$

In view of Theorem 6.1.14, the intuitive meaning of condition (6.130) is that X performs, per unit time, "sufficiently many" jumps of small size.

Proof of Proposition 6.4.2.

1. By the Lévy–Khintchin representation, for $z > 0$,

$$
\begin{aligned}
|g(z)| &= \exp\left[-\sigma^2 z^2/2 + \int_{\mathbb{R}} (\cos(zx) - 1)\, Q(dx)\right] \\
&\leq \exp\left[-\sigma^2 z^2/2 - \frac{z^2}{2} \int_{[-\pi/z,\pi/z]} \left(\frac{\sin(zx/2)}{zx/2}\right)^2 x^2\, Q(dx)\right]
\end{aligned}
$$

$$(6.134)$$

   which entails part 1.

2. Let $\sigma^2 = 0$ and put, for $\tau > 0$,

$$
\psi_\tau(z) := -\tau \log z + \frac{z^2}{2} \int_{[-\pi/z,\pi/z]} \left(\frac{\sin(zx/2)}{zx/2}\right)^2 x^2 Q(dx), \quad z > 0.
$$

   Clearly, by (6.134),

$$
\psi_\tau(z) \to \infty \ (z \to \infty) \Rightarrow g \text{ satisfies (CF}(\tau)). \tag{6.135}
$$

   Since $\sin x$ is concave on $[0, \pi/2]$, we have

$$
\sin \alpha \geq 2\alpha/\pi \quad \text{for } 0 \leq \alpha \leq \pi/2. \tag{6.136}
$$

   By (6.136) and (6.130), we get for $z \geq \pi z_0$ that

$$
\begin{aligned}
\psi_\tau(z) &\geq -\tau \log z + \frac{2z^2}{\pi^2} \int_{[-\pi/z,\pi/z]} x^2 Q(dx) \\
&\geq (-\tau + 2\varepsilon_1 + o(1)) \log z \quad (\text{as } z \to \infty)
\end{aligned}
$$

   which, by (6.135), proves the first claim. In order to prove the second claim, note that, by (6.134), (6.136), and (6.130), for $z \geq \pi z_0$,

$$
|g(z)| \leq \exp\left[-2(z/\pi)^2 \int_{[-\pi/z,\pi/z]} x^2 Q(dx)\right] \leq (z/\pi)^{-2\varepsilon_1}
$$

   which entails that (D) is satisfied if (6.131) holds.

3. Let $\sigma^2 = 0$. Note that, by (6.9),

$$
Q(\mathbb{R} \setminus (-\alpha, \alpha)) < \infty \quad \text{for all } \alpha > 0. \tag{6.137}
$$

   Hence, if $Q([-\alpha_0, \alpha_0]) < \infty$ for some $\alpha_0 > 0$, then $Q(\mathbb{R}) < \infty$. By (6.134), this implies that $|g(z)| \geq \exp(-2Q(\mathbb{R})) > 0$ $(z \geq 0)$ which shows that g satisfies neither (D) nor (CF). Clearly, by (6.137), (6.133) implies (6.132). $\square$

Example 6.4.3. *Let* $(X_t^{(\alpha)})$ *$(\alpha > 0)$ be a Lévy process such that $X_1^{(\alpha)}$ has Lévy measure $Q^{(\alpha)}$ given by*

$$Q^{(\alpha)}(dx) = \alpha x^{-1} \mathbb{1}_{(0,1]}(x)dx. \tag{6.138}$$

Then, instead of (6.129), $Q^{(\alpha)}$ satisfies the stronger moment condition

$$\int_{\mathbb{R}} |x| Q^{(\alpha)}(dx) < \infty, \quad \alpha > 0. \tag{6.139}$$

Since

$$z^2 \int_{[-1/z,1/z]} x^2 Q^{(\alpha)}(dx) = \alpha/2, \quad \alpha > 0, z \geq 1, \tag{6.140}$$

$Q^{(\alpha)}$ $(\alpha > 0)$ does not satisfy condition (6.130). Let $\sigma^2 = 0$ and let $\widetilde{h}_\alpha$ denote the characteristic function of $X_1^{(\alpha)}$. Then $\widetilde{h}_\alpha \notin L^1(\mathbb{R})$ iff $0 < \alpha \leq 1$. In fact, it is easily seen (using (6.138)) that

$$|\widetilde{h}_\alpha(z)| = \exp\left[\alpha \int_0^z \frac{\cos x - 1}{x} dx\right], \quad z \geq 0. \tag{6.141}$$

It is well known (cf. [GrRy], formula 3.782.1) that

$$\int_0^z \frac{\cos x - 1}{x} dx = -\gamma - \log z - \int_z^\infty \frac{\cos x}{x} dx, \quad z > 0 \tag{6.142}$$

*where $\gamma = 0.5772\ldots$ is **Euler's constant** defined by*

$$\gamma := \lim_{n \to \infty} \left(1 + 1/2 + 1/3 + \cdots + 1/n - \log n\right).$$

Although the integrand on the right-hand side in (6.142) is not Lebesgue-integrable on $[z, \infty)$ $(z > 0)$, we get by partial integration

$$\int_z^\infty \frac{\cos x}{x} dx = -\frac{\sin z}{z} + \int_z^\infty \frac{\sin x}{x^2} dx, \quad z > 0. \tag{6.143}$$

Combining (6.141)–(6.143) yields

$$|\widetilde{h}_\alpha(z)| = (1 + o(1)) \exp(-\alpha\gamma) z^{-\alpha} \quad (z \to \infty) \tag{6.144}$$

which implies that $\widetilde{h}_\alpha \notin L^1(\mathbb{R})$ iff $0 < \alpha \leq 1$. Nevertheless, by (6.144), $\widetilde{h}_\alpha$ satisfies (CF(τ)) for $0 < \tau < \alpha$ $(\alpha > 0)$, and it satisfies (D) iff $\alpha > 1$. One can show (see [FisVa, Sat1, Tuc] and [Sat2], page 177) that, for each $\alpha > 0$ and $t > 0$, $X_t^{(\alpha)}$ has a density.

It is interesting to compare this example with the following one.

Example 6.4.4. *Let the characteristic function g of X_1 have the form*

$$g(z) = \exp\left[\int_{\mathbb{R}} (e^{izx} - 1) Q_r(dx)\right], \quad z \in \mathbb{R} \tag{6.145}$$

where the Lévy measure Q_r $(r > 0)$ is given by

$$Q_r(dx) = \frac{1}{x} \left(\log \frac{1}{x}\right)^r \mathbb{1}_{(0,1]}(x)dx. \tag{6.146}$$

Then $Q_r\left((0,1]\right) = \infty$ and $\int_0^1 x Q_r(dx) < \infty$ $(r > 0)$. It is easily seen that, for each $r > 1$,

$$z^2 \int\limits_{[-1/z,1/z]} x^2 Q_r(dx) \geq \log z \quad \text{for all } z \geq \exp\left(2^{1/(r-1)}\right).$$

(In fact, for $z \geq 1$, the left-hand side equals

$$z^2 \int_0^{1/z} x \left(\log(1/x)\right)^r dx \geq z^2 \left(\log z\right)^r \int_0^{1/z} x\, dx = 2^{-1} \left(\log z\right)^r.)$$

So, by Proposition 6.4.2, g satisfies (D) and (CF) if $r > 1$. It turns out that g satisfies (D) and (CF) for all $r > 0$. In fact, estimating more carefully, one obtains that

$$|g(z)| \leq \exp\left(\frac{1}{2(r+1)}\right) \left(\frac{3\pi}{2z}\right)^{(2(r+1))^{-1}\left(\log\left(2z/(3\pi)\right)\right)^r} , \quad z \geq 4\pi \quad (6.147)$$

which shows that, for each $r > 0$, g satisfies (D) and (CF).

6.5 STATISTICAL APPLICATIONS

Suppose we observe some strictly positive continuous-time process $Z = (Z_t)$ and want to test (based on observing certain leading digits of (Z_t)) some **null hypothesis** $H_0(c^*)$ $(c^* > 0)$ saying that there exist constants $c \geq c^*$, $\mu \in \mathbb{R}$, and $Z_0 > 0$ such that (Z_t) is of the form

$$Z_t = Z_0 \exp\left(\mu t + c X_t\right), \quad t \in \mathbb{R}^+ \quad (6.148)$$

where (X_t) is a process which belongs to a certain class of Lévy processes all satisfying (SC).

Fix any base $B \geq 2$ and any block $(d_1, \ldots, d_m) \in I_B(m)$ of leading digits. The desired **non-parametric test** is then based on the observation of the leading digits $D_{B,1}(Z_n), \ldots, D_{B,m}(Z_n)$ for $n = 1, \ldots, T$, where T is a suitably chosen (finite) time horizon. Let G occurring in the definition of $\widetilde{L}_n$ (see (6.100)) be given by

$$G := \mathbb{1}_{J_B(d_1,\ldots,d_m)} \quad (6.149)$$

(see (6.20) and (6.21)). The simple idea is now to reject $H_0(c^*)$ iff

$$|\widetilde{L}_T/T - p_B(d_1, \ldots, d_m)| \geq v \quad (6.150)$$

for some $0 < v < 1$ which is chosen in advance. Here, $\widetilde{L}_T/T$ is the relative frequency of time points $n \leq T$ such that $(D_{B,1}(Z_n), \ldots, D_{B,m}(Z_n)) = (d_1, \ldots, d_m)$. If $H_0(c^*)$ is true, then (even for moderate values of T), with high probability, $\widetilde{L}_T/T$ should be close to $p_B(d_1, \ldots, d_m)$ given by (6.23) (this is suggested by Theorems 6.3.4 and 6.3.5).

In order to construct the desired test, we shall need the following inequalities (valid under $H_0(c^*)$):

$$\left| \mathbb{E}\left[\frac{1}{n}\widetilde{L}(n)\right] - p_B(d_1, \ldots, d_m) \right|$$

$$\leq \frac{2}{n} p_B(d_1, \ldots, d_m) \sum_{j=1}^{\infty} \sum_{r=1}^{\infty} \left| g\left(\frac{2\pi c r}{\log B}\right) \right|^j, \quad n \geq 1 \quad (6.151)$$

(this follows immediately from (6.92)) and, for $0 \le k < m \le n$,

$$\left| \mathbb{E}[\tilde{L}_n | \tilde{\mathcal{F}}_m] - \mathbb{E}[\tilde{L}_n | \tilde{\mathcal{F}}_k] \right|$$

$$\le m - k + 4 \sum_{j=1}^{\infty} \sum_{r=1}^{\infty} \left| g\left(\frac{2\pi cr}{\log B} \right) \right|^j \quad \text{a.s.} \qquad (6.152)$$

(see (6.114); note that, in (6.151) and (6.152) we have that $c \ge c^* > 0$). In order for (6.151) and (6.152) to become effective, assume that there exists a finite constant Σ^* such that

$$\sum_{j=1}^{\infty} \sum_{r=1}^{\infty} \left| g\left(\frac{2\pi cr}{\log B} \right) \right|^j \le \Sigma^* \qquad (6.153)$$

holds whenever $H_0(c^*)$ is true. By Azuma's Inequality, we get, using (6.152) and (6.153), that, for all $v > 0$ and $n \ge 1$,

$$P\left(\left| \frac{1}{n} \tilde{L}_n - \mathbb{E}\left[\frac{1}{n} \tilde{L}_n \right] \right| \ge v \right) \le 2 \exp\left(-\frac{v^2 n}{2(1 + 4\Sigma^*)^2} \right) \qquad (6.154)$$

holds under $H_0(c^*)$ (see proof of Theorem 6.3.4). Now choose an upper bound $0 < p_0 < 1$ for the probability of a **Type I error** (i.e., rejecting $H_0(c^*)$ although it is true) and specify the **critical region** determined by (6.150) by choosing some $0 < v < 1$. Finally fix any $0 < \varepsilon < 1$. Clearly, under $H_0(c^*)$, by (6.151) and (6.153),

$$\left| \mathbb{E}[\tilde{L}_T / T] - p_B(d_1, \ldots, d_m) \right| \le \varepsilon v \qquad (6.155)$$

provided

$$T \ge \frac{2 p_B(d_1, \ldots, d_m) \Sigma^*}{\varepsilon v} =: \frac{a_1}{\varepsilon}. \qquad (6.156)$$

Hence, if T satisfies (6.156), we obtain from (6.154) and (6.155) that

$$P\left(\left| \frac{1}{T} \tilde{L}_T - p_B(d_1, \ldots, d_m) \right| \ge v \right) \le 2 \exp\left(-\frac{(1 - \varepsilon)^2 v^2 T}{2(1 + 4\Sigma^*)^2} \right). \qquad (6.157)$$

Assuming (6.156) and

$$2 \exp\left(-\frac{(1 - \varepsilon)^2 v^2 T}{2(1 + 4\Sigma^*)^2} \right) \le p_0$$

or (equivalently)

$$T \ge \frac{2(1 + 4\Sigma^*)^2 \log (2/p_0)}{(1 - \varepsilon)^2 v^2} =: \frac{a_2}{(1 - \varepsilon)^2}, \qquad (6.158)$$

the level of significance of the test equals p_0. Clearly, the maximum $(a_1 \varepsilon^{-1}) \vee (a_2(1 - \varepsilon)^{-2})$ as a function of $0 < \varepsilon < 1$ is minimal if $a_1 \varepsilon^{-1} = a_2(1 - \varepsilon)^{-2}$. One of the two solutions of this equation is located in $(0, 1)$, and is given by

$$\varepsilon_0 := 1 + \frac{a_2}{2a_1} - \sqrt{\left(1 + \frac{a_2}{2a_1} \right)^2 - 1}. \qquad (6.159)$$

This proves (extending a result in [Schür2])

Proposition 6.5.1. *Let the time horizon T satisfy*

$$T \geq \frac{a_1}{\varepsilon_0} = a_1 + \frac{a_2}{2} + \sqrt{\left(a_1 + \frac{a_2}{2}\right)^2 - a_1^2}. \tag{6.160}$$

Then the level of significance of the above test equals p_0. In particular, since the right-hand side in (6.160) gets smaller if a_1 is getting smaller,

$$T \geq a_2. \tag{6.161}$$

(Note that a_2 does not depend on $(d_1, \ldots, d_m)$!)

Example 6.5.2 (Lévy processes of jump-diffusion type)**.** *Recall (compare Example 6.1.22) that a **Lévy process of jump-diffusion type** is of the form*

$$X_t = \alpha t + cW_t + \sum_{j=1}^{N_t} \zeta_j, \quad t \in \mathbb{R}^+ \tag{6.162}$$

(see, e.g., [ConT], page 111, or [Lan], page 289). Here, $c > 0$ and α are real numbers, (W_t) is a BM, and (N_t) is a Poisson process with parameter $\lambda \geq 0$. (In the degenerate case $\lambda = 0$ we put $N_t \equiv 0$ for all $t \in \mathbb{R}^+$.) Finally, $\zeta_1, \zeta_2, \ldots$ are independent and identically distributed random variables. Additionally it is assumed that the three processes (W_t), (ζ_n), and (N_t) are independent of each other. Let $\zeta_1 \sim Q_1$ (assuming that $Q_1(\{0\}) = 0$). By Example 6.1.22, (X_t) is a Lévy process such that X_1 has the characteristic function

$$g(z) = \exp\left[i\alpha z - c^2 z^2/2 + \lambda \int_{\mathbb{R}} \left(e^{izx} - 1\right) Q_1(dx)\right], \quad z \in \mathbb{R}. \tag{6.163}$$

By Proposition 6.4.2, g satisfies (D) and (CF(τ)) for all $\tau > 0$.

Suppose we observe the process

$$Z_t = Z_0 \exp(X_t), \quad t \in \mathbb{R}^+ \tag{6.164}$$

where $Z_0 > 0$ (X given by (6.162)), i.e., in (6.148), $\mu = 0$ and $c = 1$. So we have to estimate the double series in (6.153) for $c = 1(!)$. Note that if, for the process in (6.162), $\lambda = 0$ and $\alpha = \mu - c^2/2$, then the process in (6.164) is a **Black–Scholes process** with parameters μ and c. Now we want to test $H_0(c^*)$ $(c^* > 0)$ which says that there exist $\alpha \in \mathbb{R}$, $c \geq c^*$, $\lambda \geq 0$, and a distribution Q_1 on $\mathbb{R}$ (satisfying $Q_1(\{0\}) = 0$) such that X (in (6.164)) is a Lévy process of jump-diffusion type associated with α, c, λ, and Q_1. (Note that, here, $H_0(c^*)$ is a null hypothesis which differs from that at the beginning of this section!) In order to estimate the double series in (6.153) (for $c = 1(!)$), first note that, by (6.163),

$$|g(z)| = \exp\left[-c^2 z^2/2 + \lambda \int_{\mathbb{R}} (\cos(zx) - 1)Q_1(dx)\right]$$
$$\leq \exp\left(-c^2 z^2/2\right).$$

Hence,

$$\sum_{j=1}^{\infty} \sum_{k=1}^{\infty} \left| g\left(\frac{2\pi k}{\log B}\right) \right|^j \leq \sum_{j=1}^{\infty} \sum_{k=1}^{\infty} \exp\left[-2j\left(\frac{c\pi k}{\log B}\right)^2\right].$$

Using inequality (6.165) in Lemma 6.5.4 below gives that, under $H_0(c^*)$, the last double series is not greater than

$$(\pi + 2) \left(\frac{\log B}{2\pi c^*} \right)^2 =: \Sigma^*.$$

By (6.156) and (6.158), this yields

$$a_1 = \frac{(\pi + 2)p_B(d_1, \ldots, d_m)(\log B)^2}{2\pi^2 (c^*)^2 v},$$

$$a_2 = \frac{2 \left[1 + (\pi + 2)(\log B)^2/(\pi c^*)^2 \right]^2 \log (2/p_0)}{v^2}.$$

(In [Schür2] this was obtained for a much stronger null hypothesis.)

For a numerical example choose $B = 10$, $c^* = 1$, $p_0 = v = 0.1$, $m = 1$, and $d_1 = 1$. Then $p_B(d_1) = \log_{10} 2$ (see (6.24)), $a_1 = 4.1573$, and $a_2 = 8479.6677$. Hence, by (6.160), the time horizon T has to satisfy $T \geq 8488$. Even if we choose a much longer block of leading digits, by (6.161), we must use at least 8480 observations!

Remark 6.5.3. *For the test above it turns out that it pays out much less than (perhaps) expected to base the test on a long block $(d_1, \ldots, d_m)$ of leading digits in order to reduce the number of observations. Apparently the reason for this is that the upper bound in (6.154) does not depend on $(d_1, \ldots, d_m)$ which, in turn, is due to the fact that the upper bound in (6.116) depends on G only through $\|G\|_\infty$.*

Lemma 6.5.4. *If $a > 0$ and $\alpha > 1$, then*

$$\sum_{j=1}^{\infty} \sum_{k=1}^{\infty} \exp (-ajk^\alpha) \leq \frac{1}{a} \left[1 + \frac{\pi/\alpha}{\sin (\pi/\alpha)} \right]. \tag{6.165}$$

Proof. Comparing with a suitable Riemann integral gives that the inner series is not greater than

$$\exp (-aj) \left[1 + \alpha^{-1} a^{-1/\alpha} \Gamma(1/\alpha) j^{-1/\alpha} \right].$$

Comparing again with a certain Riemann integral and using that

$$\Gamma(1/\alpha)\Gamma(1 - 1/\alpha) = \frac{\pi}{\sin (\pi/\alpha)}$$

(see [Iw] or [StSh2], page 164) finally yields (6.165). $\qquad\square$

Example 6.5.5 (Gamma processes). *Suppose we want to test the null hypothesis $H_0(c^*, \alpha^*, \Delta^*)$ $(c^* > 0, \alpha^* > 0, \Delta^* > 1)$ which says that there exist numbers $c \geq c^*$, $\mu \in \mathbb{R}$, $0 < \alpha \leq \alpha^*$, and $\Delta \geq \Delta^*$ such that the observed process (Z_t) is as in (6.148) where (X_t) is a **gamma process** with parameters α and Δ (see Example 6.1.23). Then the characteristic function g of X_1 satisfies (D) and (CF) (see Example 6.2.10), and*

$$|g(z)| = (1 + (z/\alpha)^2)^{-\Delta/2}, \quad z \in \mathbb{R}.$$

In the sequel we put $\lambda(B) := [2\pi c^*/(\alpha^* \log B)]^2$. *Under* $H_0(c^*, \alpha^*, \Delta^*)$, *the double series in* (6.153) *is not greater than*

$$\sum_{j=1}^{\infty} \sum_{k=1}^{\infty} (1 + \lambda(B)k^2)^{-\Delta^* j/2} \leq \sum_{k=1}^{\infty} \frac{1}{(1 + \lambda(B)k^2)^{\Delta^*/2} - 1}$$

$$\leq \frac{1}{(1 + \lambda(B))^{\Delta^*/2} - 1} + \int_1^{\infty} \frac{dx}{(1 + \lambda(B)x^2)^{\Delta^*/2} - 1} =: \Sigma^*.$$

For a numerical example choose $B = 10$, $c^* = \alpha^* = 1$, $\Delta^* = 2$, $p_0 = v = 0.1$, $m = 1$, *and* $d_1 = 1$. *Then* $\lambda(10) = 7.4461$, $\Sigma^* = 0.2686$, $a_1 = 1.6171$, *and* $a_2 = 2578.2083$. *Hence, by* (6.160), *the time horizon* T *has to satisfy* $T \geq 2582$.

Example 6.5.6 (α-stable Lévy processes). *Suppose we want to test* $H_0(c^*, \alpha^*, \delta^*)$ $(c^* > 0, 1 < \alpha^* < 2, \delta^* > 0)$ *which says that the observed strictly positive continuous-time process* (Z_t) *is of the form*

$$Z_t = Z_0 \exp(\mu t + cX_t), \quad t \in \mathbb{R}^+$$

for constants $c \geq c^*$, $\mu \in \mathbb{R}$, *and* $Z_0 > 0$, *and, furthermore,* (X_t) *is an* **α-stable Lévy process** *for some* $\alpha^* \leq \alpha < 2$ *and certain parameter values* $\gamma \in \mathbb{R}$, $\delta \geq \delta^*$, *and* $-1 \leq \tau \leq 1$ *such that*

$$X_1 \sim S_\alpha(\gamma, \delta, \tau). \tag{6.166}$$

Here, $S_\alpha(\gamma, \delta, \tau)$ *is an* **α-stable distribution** *with parameters* γ, δ, *and* τ *(see Example 6.1.24). Let* h_α *(as in* (6.43)) *denote the characteristic function of* $S_\alpha(\gamma, \delta, \tau)$. *It follows from* (6.43) *that, under* $H_0(c^*, \alpha^*, \delta^*)$, *the double series in* (6.153) *is not greater than*

$$\sum_{j=1}^{\infty} \sum_{k=1}^{\infty} \exp\left(-\delta^* j \left(\frac{2\pi c^* k}{\log B}\right)^\alpha\right) = \sum_{j=1}^{\infty} \sum_{k=1}^{\infty} \exp(-ajk^\alpha)$$

where $a := \delta^* (\frac{2\pi c^*}{\log B})^\alpha$. *By Lemma* 6.5.4, *the last double series is not greater than*

$$\frac{1}{a}\left[1 + \frac{\pi/\alpha}{\sin(\pi/\alpha)}\right] \leq \frac{1}{\delta^*}\left(\frac{\log B}{2\pi c^*}\right)^\alpha \left[1 + \frac{\pi/\alpha^*}{\sin(\pi/\alpha^*)}\right]$$

since $x(\sin x)^{-1}$ *is increasing on* $(\pi/2, \pi)$. *Assume*

$$\log B \leq 2\pi c^*. \tag{6.167}$$

Then, under $H_0(c^*, \alpha^*, \delta^*)$, *we can choose in* (6.153)

$$\Sigma^* := \frac{1}{\delta^*}\left(\frac{\log B}{2\pi c^*}\right)^{\alpha^*} \left[1 + \frac{\pi/\alpha^*}{\sin(\pi/\alpha^*)}\right].$$

For a numerical example, choose $B = 10$, $c^* = \delta^* = 1$, $\alpha^* = 1.5$, $p_0 = v = 0.1$, $m = 1$, *and* $d_1 = 1$. *Then* (6.167) *holds. We calculate* $\Sigma^* = 0.7584$, $a_1 = 4.5660$, $a_2 = 9748.0703$ *which, by* (6.160), *gives* $T \geq 9758$.

6.6 APPENDIX 1: ANOTHER VARIANT OF POISSON SUMMATION

We shall obtain the Poisson Summation Formula under sufficient conditions formulated in terms of characteristic functions. This case is particularly useful when dealing with an infinitely divisible distribution which, by the Lévy–Khintchin representation, is given by the generating triple of its characteristic function.

Let g be the characteristic function of some random variable ξ. Suppose $g \in L^1(\mathbb{R})$. By **Fourier inversion**, this implies that ξ has a density f given by

$$f(x) = \frac{1}{2\pi} \int_{-\infty}^{\infty} e^{-ixz} g(z)\, dz, \quad x \in \mathbb{R} \tag{6.168}$$

which is continuous and bounded (cf. [Fel], page 509, or [Kat], page 158). In [Fel], page 630, the following variant of the **Poisson Summation Formula** is obtained:

Theorem 6.6.1. *Let g be a characteristic function such that $g \in L^1(\mathbb{R})$, and let the density f be given by* (6.168). *Let $\lambda \neq 0$ be a real number such that the function*

$$H_1(s) := \sum_{k=-\infty}^{\infty} g(s + 2\lambda k), \quad s \in \mathbb{R} \tag{6.169}$$

is continuous. Then

$$H_1(s) = \frac{\pi}{|\lambda|} \sum_{k=-\infty}^{\infty} f\left(\frac{k\pi}{\lambda}\right) \exp\left(\frac{i\pi ks}{\lambda}\right), \quad s \in \mathbb{R} \tag{6.170}$$

where the sums in (6.169) *and* (6.170) *are defined by*

$$\sum_{k=-\infty}^{\infty} := \lim_{N \to \infty} \sum_{|k| \le N}.$$

Remark 6.6.2.

1. *In [Fel] only the case $\lambda > 0$ is considered. Clearly* (6.170) *also holds in the case $\lambda < 0$.*

2. *Note that the function H_1 in* (6.169) *has period $2|\lambda|$. This follows from the Riemann–Lebesgue Lemma which holds for arbitrary functions in $L^1(\mathbb{R})$ (cf. [Bi3], page 345, or [Kat], page 155.*

3. *For other conditions which are sufficient for the Poisson Summation Formula see [StSh1], page 154, or [Zy], Vol. I, page 68.*

We have repeatedly used the following variant of the **Poisson Summation Formula** which is a straightforward consequence of Theorem 6.6.1 (f^{n*} denoting the n-fold convolution of the density f):

Theorem 6.6.3 (Poisson Summation Formula)**.** *Let g be a characteristic function satisfying the Domination Condition* (D) *(see Definition 6.2.2), and let f be the* *density given by (6.168). Then, for all real numbers $\lambda \neq 0$, Δ, and integers $n \geq 1$,*

$$\sum_{k=-\infty}^{\infty} \exp\left(-i\Delta(s+2\lambda k)\right) g^n(s+2\lambda k)$$

$$= \frac{\pi}{|\lambda|} \sum_{k=-\infty}^{\infty} \exp\left(\frac{i\pi ks}{\lambda}\right) f^{n*}\left(\frac{k\pi}{\lambda} + \Delta\right), \quad s \in \mathbb{R}. \quad (6.171)$$

The series on both sides in (6.171) are absolutely and uniformly convergent with respect to $s \in \mathbb{R}$.

Example 6.6.4. *Note that in Theorem 6.6.3 it is not assumed that $\mathrm{TV}(g) < \infty$. The following example shows that there are characteristic functions $g \in L^1(\mathbb{R})$ having compact support such that $\mathrm{TV}(g) = \infty$. In fact, define g by*

$$g(t) = (1 - |t|) \, \mathbb{1}_{[-1,1]}(t) \sum_{k=1}^{\infty} 2^{-k} \left(\cos\left(2\pi 9^k t\right)\right)^2, \quad t \in \mathbb{R}. \quad (6.172)$$

Then g is a characteristic function such that $g \in L^1(\mathbb{R})$ (see Example 6.1.3) and $\mathrm{TV}(g) = \infty$.
(Hint: For $k \geq 1$ consider the increments $g(t_{2j}(k)) - g(t_{2j-1}(k))$ where $t_{2j-1}(k) = (j - 1/4)9^{-k}$ and $t_{2j}(k) = j \cdot 9^{-k}$ $(j = 1, 2, \ldots, \lfloor 2^{-1} 9^k \rfloor =: n(k))$. Using the inequality $|(\cos \beta)^2 - (\cos \alpha)^2| \leq 2|\beta - \alpha|$, finally yields an inequality of the form

$$\sum_{j=1}^{n(k)} |g(t_{2j}(k)) - g(t_{2j-1}(k))| \geq \left(\frac{1}{2} - \frac{\pi}{7}\right) \left(\frac{9}{2}\right)^k + const.)$$

Proof of Theorem 6.6.3. It suffices to prove (6.171) for fixed $\lambda > 0$ and Δ. By (D2), there exists some $s_0 \in [0, 2\lambda]$ such that

$$\widetilde{H}(s_0) := \sum_{k=-\infty}^{\infty} \widetilde{g}(s_0 + 2\lambda k) < \infty. \quad (6.173)$$

Here, $\widetilde{g}$ occurs in Definition 6.2.2 (see also [Zy], Vol. I, page 68). In fact, by (D2),

$$\int_0^{2\lambda} \left\{ \sum_{k=-\infty}^{\infty} \widetilde{g}(s+2\lambda k) \right\} ds = \sum_{k=-\infty}^{\infty} \int_0^{2\lambda} \widetilde{g}(s+2\lambda k) \, ds$$

$$= \int_{\mathbb{R}} \widetilde{g}(x) \, dx < \infty$$

implying that

$$\sum_{k=-\infty}^{\infty} \widetilde{g}(s+2\lambda k) < \infty \quad \text{for almost all } s \in [0, 2\lambda]. \quad (6.174)$$

Let $v_k(\lambda)$ denote the total variation of $\widetilde{g}$ on $[2\lambda k, 2\lambda(k+1)]$. Then, for each $s \in$

$[0, 2\lambda]$,

$$\sum_{k=-\infty}^{\infty} \widetilde{g}(s + 2\lambda k) \leq \sum_{k=-\infty}^{\infty} |\widetilde{g}(s + 2\lambda k) - \widetilde{g}(s_0 + 2\lambda k)|$$

$$+ \sum_{k=-\infty}^{\infty} \widetilde{g}(s_0 + 2\lambda k)$$

$$\leq \sum_{k=-\infty}^{\infty} v_k(\lambda) + \widetilde{H}(s_0)$$

$$= \mathrm{TV}(\widetilde{g}) + \widetilde{H}(s_0) < \infty. \tag{6.175}$$

(Note that the series in (6.174) has period 2λ.) Let $n \geq 1$ be fixed. Clearly, $\exp(-i\Delta z)g^n(z)$ (as a function of z) is the characteristic function of the density $f^{n*}(x + \Delta)$. The estimates in (6.175) show that the series on the left-hand side in (6.171) is a uniform limit of partial sums depending continuously on s. Hence (6.171) follows from Theorem 6.6.1. $\square$

In the proofs of some results in this chapter some consequences of Theorem 6.6.3 were used (see Corollary 6.6.6 below). For their proofs we need

Lemma 6.6.5. *Let g be a characteristic function such that $g \in L^1(\mathbb{R})$. Then*

$$\varphi(t) := \sup_{|z| \geq t} |g(z)| < 1 \quad \text{for all } t > 0. \tag{6.176}$$

Proof. Assuming $g \in L^1(\mathbb{R})$ implies $|g(z)| < 1$ for all $z \neq 0$ (see [Fel], page 501). Since g is continuous and, by the Riemann–Lebesgue Lemma, $g(z) \to 0$ as $|z| \to \infty$, (6.176) follows. $\square$

As a consequence of Theorem 6.6.3 we obtain

Corollary 6.6.6. *Let g be a characteristic function and let the numbers c_t $(t \in \mathbb{R}^+)$ satisfy*

$$c_t \geq \widetilde{c} > 0 \quad \text{for all } t \in \mathbb{R}^+. \tag{6.177}$$

Suppose the Standard Condition (SC) (cf. Definition 6.2.9). Then, for each $\alpha > 0$ and $\beta > 0$, we have that

$$\lim_{t \to \infty} \sum_{j=1}^{\infty} |g(\alpha c_t j)|^{\beta t} = 0, \tag{6.178}$$

$$\sup_{m \geq 1} \sum_{j=1}^{\infty} \sum_{r=\widetilde{r}_0}^{\infty} |g(\alpha c_{r+m} j)|^r =: \widetilde{\mu} < \infty, \tag{6.179}$$

and (assuming c_t (as a function of t) to be measurable)

$$\sup_{s \geq 0} \sum_{j=1}^{\infty} \int_{r_0}^{\infty} |g(\alpha c_{u+s} j)|^u \, du =: \mu < \infty. \tag{6.180}$$

Here, we put $\tilde{r}_0 := 2$ *and* $r_0 := 1$ *if* (SC1) *or* (SC2) *holds; if only* (SC3) *holds* (g *satisfying condition* (CF(τ)) *in Definition 6.2.6*), $\tilde{r}_0 = r_0 := \lceil 2/\tau \rceil$. *Furthermore,* $\tilde{\mu}$ *and* μ *are constants which depend on* g, α, *and* $\tilde{c}$ *if* (SC1) *or* (SC2) *holds; if only* (SC3) *holds, they also depend on* τ.

Proof. We only prove (6.179). If (SC1) or (SC2) holds, then the double series in (6.179) is not greater than

$$\sum_{j=1}^{\infty} \sum_{r=1}^{\infty} |g(\alpha \tilde{c} j)|^r \leq \sum_{r=1}^{\infty} (\varphi(\alpha \tilde{c}))^{r-1} \sum_{j=1}^{\infty} |g(\alpha \tilde{c} j)|$$

($\varphi(\cdot)$ defined by (6.176)) which is finite by (6.176) and Theorem 6.6.3. Now let g satisfy (D) and (CF(τ)). Put $\tilde{r}_0 := \lceil 2/\tau \rceil$. First note that, by (6.177) and (6.176), for each $j \geq 1$,

$$\sum_{r=1}^{\infty} |g(\alpha c_{r+m} j)|^r \leq \sum_{r=1}^{\infty} (\varphi(\alpha \tilde{c}))^r =: K_0 < \infty. \tag{6.181}$$

By (CF(τ)), there exists some $t_0 \geq 1$ such that

$$t^{\tau} |g(t)| \leq 2^{-1} (\alpha \tilde{c})^{\tau} \quad \text{for all } t \geq t_0. \tag{6.182}$$

Putting $j_0 := \lceil t_0/(\alpha \tilde{c}) \rceil$, it follows from (6.181) and (6.182) (noting that $\alpha c_{r+m} j \geq t_0$ holds for all $j \geq j_0$, $r \geq 1$, and $m \geq 1$) that the double series in (6.179) is bounded from above by

$$j_0 K_0 + \sum_{r=\tilde{r}_0}^{\infty} \sum_{j=j_0+1}^{\infty} \left[(\alpha c_{r+m} j)^{\tau} |g(\alpha c_{r+m} j)| (\alpha c_{r+m} j)^{-\tau} \right]^r$$

$$\leq j_0 K_0 + \sum_{r=\tilde{r}_0}^{\infty} 2^{-r} \sum_{j=j_0+1}^{\infty} j^{-\tau r}$$

$$\leq j_0 K_0 + \sum_{r=\tilde{r}_0}^{\infty} 2^{-r} \sum_{j=j_0+1}^{\infty} j^{-2}$$

$$\leq j_0 K_0 + 1$$

which proves (6.179). $\qquad \qquad \square$

The proof of (6.178) uses Theorem 6.6.3, Lemma 6.6.5, and the following simple analytical result (its proof being left to the reader):

Lemma 6.6.7. *Let* $a_1, a_2, \ldots$ *be real numbers such that* $0 \leq a_n < 1$ ($n = 1, 2, \ldots$) *and* $\sum_{n=1}^{\infty} a_n < \infty$. *Then*

$$\lim_{t \to \infty} \sum_{n=1}^{\infty} a_n^t = 0.$$

6.7 APPENDIX 2: AN ELEMENTARY PROPERTY OF CONDITIONAL EXPECTATIONS

In §6.3 we used the following result on conditional expectations:

Theorem 6.7.1. *Let* $h\colon \mathbb{R}^{m+n} \longrightarrow \mathbb{R}^+$ *be Borel measurable. Let* ξ *and* η *be, respectively,* $\mathbb{R}^m$- *and* $\mathbb{R}^n$-*valued random vectors which are defined on some probability space* $(\Omega, \mathcal{F}, P)$. *Let, for some sub-σ-algebra* $\mathcal{G}$ *of* $\mathcal{F}$, ξ *be* $\mathcal{G}$-*measurable and let* η *be independent of* $\mathcal{G}$, *i.e., for all Borel sets* $A_1 \subset \mathbb{R}^n$ *and sets* $A_2 \in \mathcal{G}$,

$$P\left(\{\eta \in A_1\} \cap A_2\right) = P\left(\eta \in A_1\right) P(A_2).$$

Then, if $\varphi\colon \mathbb{R}^m \longrightarrow \mathbb{R}^+ \cup \{\infty\}$ *is defined by*

$$\varphi(x) := \mathbb{E}\left[h(x, \eta)\right], \quad x \in \mathbb{R}^m, \tag{6.183}$$

we have

$$\mathbb{E}\left[h(\xi, \eta)|\mathcal{G}\right] = \varphi(\xi) \quad a.s. \tag{6.184}$$

Proof. Since both sides in (6.184) are $\mathcal{G}$-measurable, it suffices to show that

$$\int_A h(\xi, \eta)\, dP = \int_A \varphi(\xi)\, dP \quad \text{for all } A \in \mathcal{G}. \tag{6.185}$$

First assume that h is bounded. To prove (6.185), let Q_1 be the distribution of the $\mathbb{R}^{m+1}$-valued random vector $(\mathbb{1}_A, \xi)$ (for fixed $A \in \mathcal{G}$) and let $\widetilde{Q}_1$ be the distribution of η. Since $(\mathbb{1}_A, \xi)$ (being $\mathcal{G}$-measurable) is independent of η, the product measure $Q_1 \otimes \widetilde{Q}_1$ is the distribution of the $\mathbb{R}^{m+n+1}$-valued random vector $(\mathbb{1}_A, \xi, \eta)$. Hence, by Fubini's theorem,

$$
\begin{aligned}
\mathbb{E}\left[\mathbb{1}_A h(\xi, \eta)\right] &= \int_{\mathbb{R}^{m+n+1}} z\, h(x, y)\, Q_1 \otimes \widetilde{Q}_1(dz, dx, dy) \\
&= \int_{\mathbb{R}^{m+1}} z \left(\int_{\mathbb{R}^n} h(x, y)\, \widetilde{Q}_1(dy)\right) Q_1(dz, dx) \\
&= \int_{\mathbb{R}^{m+1}} z\, \mathbb{E}\left[h(x, \eta)\right] Q_1(dz, dx) \\
&= \int_{\mathbb{R}^{m+1}} z\, \varphi(x)\, Q_1(dz, dx) \\
&= \mathbb{E}\left[\mathbb{1}_A \varphi(\xi)\right]
\end{aligned}
$$

which is equal to the right-hand side in (6.185). For general h let $h_N := h \wedge N$ $(N = 1, 2, \ldots)$ which is bounded. From what has already been proved, we get

$$\mathbb{E}\left[h_N(\xi, \eta)|\mathcal{G}\right] = \varphi_N(\xi) \quad \text{a.s.} \tag{6.186}$$

if φ_N is given by

$$\varphi_N(x) := \mathbb{E}\left[h_N(x, \eta)\right], \quad x \in \mathbb{R}^m.$$

As $N \to \infty$, by monotone convergence,

$$\mathbb{E}\left[h_N(\xi, \eta)|\mathcal{G}\right] \to \mathbb{E}\left[h(\xi, \eta)|\mathcal{G}\right] \quad \text{a.s.}$$

and the limit $\varphi(x) := \lim_{N \to \infty} \varphi_N(x) = \mathbb{E}\left[h(x, \eta)\right]$ exists which, by (6.186), implies

$$\varphi(\xi) = \mathbb{E}\left[h(\xi, \eta)|\mathcal{G}\right] \quad \text{a.s.} \qquad \square$$

PART III
Applications I: Accounting and Vote Fraud

Chapter Seven

Benford's Law as a Bridge between Statistics and Accounting

Richard J. Cleary and Jay C. Thibodeau[1]

Undergraduate accounting students almost universally take a course in elementary business statistics as part of their program, but only rarely are applications of the material seen in future accounting courses. Benford's Law provides a natural way to tie the topics together so that the principles of statistical thinking are reinforced in the accounting curriculum. Conversely, for students primarily interested in statistics, applications of Benford's Law provide easy to understand, real-world examples. In this chapter we explore these connections and consider the questions of when and how to effectively deliver this material in such a way that Benford's Law is more than just a conversation piece but rather a tool that helps accountants make stronger and more efficient decisions using sound statistical practice.

We begin by looking at the current state of statistics education for accounting students, and we present some of the ways in which accounting practice, particularly in auditing, can benefit from a statistical point of view. We then demonstrate how Benford's Law can be used to reinforce the key concepts that appear at the intersection of ideas from statistics and accounting. We conclude with some suggestions for how to effectively incorporate Benford's Law into the curriculum as an example of an analytical procedure in the introductory financial statement auditing class that is required in all accounting programs.

7.1 THE CASE FOR ACCOUNTANTS TO LEARN STATISTICS

There is ample evidence that the accounting profession, or at least the academic wing of the accounting profession, sees learning how to think about variation and uncertainty as an essential part of the training for undergraduate accounting majors. A quick glance at ten highly regarded accounting programs reveals that all of these programs require at least one course in probability and/or statistics. Typically these courses are not limited to accounting majors, but also attract students from a variety of the business disciplines. Interestingly, we also noted that statistics courses

---

[1]Mathematics & Science Division, Babson College, and Department of Accountancy, Bentley University.

tend to appear early in the curriculum (sophomore year), when little if any of the technical content of accounting has been introduced.

Although most auditing professionals are likely to have had a first course in statistics, such courses tend to provide an overview of key statistical principles with examples from generic contexts and rarely, if ever, utilize audit examples. In addition, most statistics courses taken during undergraduate general education are likely to emphasize calculation rather than "big picture" decision making. It comes as no surprise, then, that as a discipline, accounting gets shortchanged in the number of problems and cases devoted to it compared with the other business disciplines.

As a result, we believe that auditing students could benefit from more statistical education and related critical thinking demonstrated in real-life auditing contexts. What should future auditors learn in a statistics course? If the students have had limited exposure to the vocabulary and practice of accounting when they take the course, then it makes sense to consider this question at a conceptual level rather than as a listing of specific topics and techniques. This concept-oriented approach appeals to the statisticians and mathematicians who primarily teach the first course in business statistics. These instructors may believe that an introduction to statistics should be an opportunity for students to learn and appreciate some of the intellectual dimensions of chance and uncertainty.

The American Statistical Association's 2005 report *Guidelines for Assessment and Instruction in Statistics Education* (GAISE) is an excellent document that presents both the rationale and practical advice on teaching this sort of high-level critical thinking. The GAISE report [GAISE] includes 22 specific recommendations for what students should know after the first statistics course. Here we choose the four examples from that list that we think are particularly important goals of a first statistics course for future auditors.

Students should know

- Goal 1: That variability is natural, predictable, and quantifiable;

- Goal 2: That random sampling allows results of surveys and experiments to be extended to the population from which the sample was taken;

- Goal 3: How to interpret statistical results in context;

- Goal 4: How to critique news stories and journal articles that include statistical information, including identifying what's missing in the presentation and the flaws in the studies or methods used to generate the information.

The authors of this chapter, one a professor specializing in statistics and the other a professor specializing in auditing, have worked together for several years trying to encourage students and professionals in both fields to recognize the opportunities in the links between them. In the rest of this chapter our goal is to demonstrate how Benford's Law can illustrate these four concepts in the accounting context for both students and practitioners. To make the examples concrete, we will focus our applications on the auditing process in which the financial statements of a company or institution are reviewed by an independent CPA firm to ensure that they are not misleading. The next section of the chapter reviews the ways in which auditors work and introduces the necessary vocabulary.

7.2 THE FINANCIAL STATEMENT AUDITOR'S WORK ENVIRONMENT

7.2.1 Regulation

The objective of a **financial statement audit** is to provide reasonable assurance that the financial statements prepared by a company have been presented in accordance with the set of accounting rules that are in place for the company in a particular jurisdiction. For publicly traded companies in the United States, an annual audit of the financial statements and related footnotes is required by the **Securities and Exchange Commission** (**SEC**). The SEC was established by Congress to enforce the Securities and Exchange Act of 1934. The Act requires publicly held companies to file annual audited financial statements (Form 10-K) with the SEC. While an audit is not required for privately held companies in the United States, many of these companies do have their annual financial statements audited for other reasons. For instance, many banking relationships require that an audit is conducted on an annual basis to add credibility to the company's financial statements [AbdTh]. Benford's Law is one of many possible analytical techniques that auditors might use in looking for fraud in the financial statements of a company.

In recent years, the work of financial statement auditors has received increased scrutiny. Specifically, a significant number of high profile financial statement frauds (e.g., Enron and WorldCom) that occurred around the year 2000 reduced confidence in the financial statements being reported by publicly traded companies. As a result, the US Congress passed the **Sarbanes–Oxley Act** (**SOX**) of 2002 in an effort to restore confidence in the capital markets. One of most dramatic changes mandated by SOX has been governmental regulation of the audit profession. Specifically, Section 103 of SOX established the **Public Company Accounting Oversight Board** (**PCAOB**).

Under the law, the PCAOB is required to perform detailed inspections of the audit process employed by each auditing firm. In addition, the PCAOB has assumed responsibility for establishing all standards pertaining to audits of publicly traded companies. Overall, in executing its responsibilities related to inspections and standard-setting, the PCAOB has made it clear that the interests of the investing public will always come first. Importantly, for a profession that previously had responsibility to formulate its own standards and was subject only to evaluation by its peers, the change has been dramatic. Indeed, the financial statement auditor now operates in a highly regulated work environment and all of their professional judgments are being scrutinized. While there are many benefits from these changes, it has almost certainly reduced the willingness of auditors to use data interrogation techniques like Benford's Law. Without statistical expertise, practicing auditors fear having to explain and justify choices they make on the basis of a statistical or mathematical test.

7.2.2 Financial Statement Auditing Process

When completing the financial statement audit, an auditor is required to gather enough evidence to allow a determination of whether the financial statements have been recorded in accordance with the relevant financial accounting rules. Once enough evidence is gathered, an auditor will reach a conclusion about the financial statements and prepare a written report that communicates the auditors' opinion about the correspondence of the financial statements with the prescribed rules. Stated simply, auditors add credibility to financial statements, which provides more informational value to investors.

The audit report expresses the independent auditor's professional opinion regarding the fairness of presentation of the financial statements. If the financial statements present fairly, in all material respects, an entity's financial position (i.e., the balance sheet), results of operations (i.e., the income statement), and cash flows (i.e., the statement of cash flows) in conformity with Generally Accepted Accounting Principles (GAAP); and the audit was performed in accordance with the appropriate standards (Generally Accepted Auditing Standards for privately held companies or PCAOB Standards for publicly held companies), then a standard unqualified opinion can be issued; see [AbdTh, LoRSST]. The typical wording for such a report follows.

Privately Held Company Standard Unqualified Report

Independent Auditor's Report
 We have audited the accompanying balance sheets of the XYZ Company as of December 31, 20X2 and 20X1, and the related statements of income, retained earnings, and cash flows for the years then ended. These financial statements are the responsibility of the Company's management. Our responsibility is to express an opinion on these financial statements based on our audits.

 We conducted our audits in accordance with auditing standards generally accepted in the United States of America. Those standards require that we plan and perform the audit to obtain reasonable assurance about whether the financial statements are free of material misstatement. An audit includes examining, on a test basis, evidence supporting the amounts and disclosures in the financial statements. An audit also includes assessing the accounting principles used and significant estimates made by management, as well as evaluating the overall financial statement presentation. We believe that our audits provide a reasonable basis for our opinion.

 In our opinion, the financial statements referred to above present fairly, in all material respects, the financial position of the XYZ Company as of [at] December 31, 20X2 and 20X1, and the results of its operations and its cash flows for the years then ended in conformity with accounting principles generally accepted in the United States of America.

The wording in the Independent Auditor's Report is important because it makes clear that auditors provide reasonable and not absolute assurance that the financial statements are free of material misstatement. Note that the auditing vocabulary can be readily translated into statistical terms. The term "reasonable assurance" for an auditor is essentially the notion of "confidence" for a statistician, and what auditors call "material" difference is what a statistics text would call "practically significant." In order for an auditor to provide absolute assurance that the financial statements are free of all misstatements, an auditor would have to examine one hundred percent of the evidence, which would be far too costly. This is where auditors (whether they know it or not) are applying the first two goals mentioned from the GAISE report [GAISE] in Section I: understanding and quantifying variability and the importance of random sampling. The third goal from the GAISE report, putting data in context, is also vital. An auditor knows that it would be far too costly to design his/her work to target small monetary dollar or other immaterial mistakes that would not impact an investor's decision-making process. The fourth goal, careful reading of the statistical work of others, is vital for senior managers and partners who spend less time in the field conducting audits, but must read and approve the work of their colleagues as well as provide guidance in difficult cases.

The most vital overall objective for auditors is to efficiently use the available evidential matter to protect against the possibility of a "blown" opinion, that is, the possibility that audited financial statements that were given an unqualified opinion actually did in fact contain a material misstatement. As a result, the process formulated by auditors needs to be designed to ensure that auditors have gathered enough evidence to reach a conclusion about the financial statements. Perhaps not surprisingly, auditors have long struggled with the issue of how to keep audit costs under control, while also gathering enough evidence to ensure that the risk of a "blown" opinion has been mitigated. Ultimately, this is a matter of professional judgment. This judgment can be enhanced with appropriate application of statistical thinking and tools such as Benford's Law.

7.2.3 An Auditor's Professional Judgment Context

An auditor's professional judgment is on display throughout the entire financial statement audit. Because of the use of samples to gather evidence about various populations, the process of gathering and evaluating audit evidence relies substantially on statistical theory. Unfortunately, auditors have limited knowledge and experience applying relevant statistical thinking. In addition, there are few, if any, useful quantitative guidelines to assist auditors during the execution phase of the audit process. Our work in this area has only reinforced our belief that more effective statistical practice can help improve auditor's professional judgments and ultimately improve the effectiveness and efficiency of the financial statement auditing process.

Despite the importance of statistical thinking, many auditors have received very little training on statistical theory during their undergraduate education. In addition, it is quite possible that the auditor has not had any further training on using statistics to make good decisions. We therefore believe that an opportunity currently exists to help auditors become far better consumers of statistical information and better producers of audit evidence based on established statistical practice. In our view, such grounding will also help the defensibility of audit work papers, a very important consideration for auditing firms that seek to mitigate their litigation risk.

Given the importance of helping auditors improve their statistical thinking, we believe Benford's Law can be used by professors as a motivating example to help students make the connection between auditing and statistical theory. In fact, during their undergraduate education, we believe that there is a wonderful opportunity to help future auditors (i.e., our students) understand the world of statistics and learn how the audit process can be improved as a result.

7.2.4 The Importance of Statistical Thinking on the Audit

As auditors now respond to the new regulatory environment, the profession is further challenged to find a practical way to address and document the thought process of planning our engagements. As a profession, we must identify the right level of audit effort that will result in effective and efficient audits being completed. Interestingly, in recent years, there has been escalating rhetoric surrounding the substantial cost to public corporations of implementing the Sarbanes–Oxley Act (SOX) of 2002. Corporate managers in particular are expressing a great deal of frustration with onerous compliance costs, particularly since they cannot envision commensurate returns. It has become clear that auditors need to think hard about how to be more efficient in executing the financial statement auditing process, while always maintaining effectiveness.

The bottom line is that, as Goal 1 from Section 7.1 should make clear, risk assessments for an auditor are necessarily imprecise because of the nature of sampling error. Thus the selection of which fraud risk factors to follow up on, along with the procedures used and the extent of work that is planned for this endeavor, is a difficult judgment for auditors. Understanding the benefits and costs of applying an analytical technique, if properly considered as an aid in decision making, should help auditors understand and balance their audit firm's exposure to error risks [ClTh]. In addition, it may provide a concrete basis to help auditors provide the documentation necessary to justify the extent of work performed. This is always important in a highly regulated work environment such as public company auditing.

Of course, the application of statistical thinking to decision problems like these is quite difficult. While it would be ideal to develop a statistical formula or rule-of-thumb, it is hard to mathematically quantify different steps in an auditor's judgment and decision-making process. For example, we continue to believe that tools like Benford's Law can be useful in helping auditors detect fraud. In particular, in situations where fraudsters use small dollar amount accounting entries that are designed to "fall" under the financial statement auditor's materiality thresholds, Benford's Law would be a great tool to help identify anomalous patterns of data that may indicate fraud.

However, it is important to remember that there are substantial costs to the auditing firm for the extra work that is completed on identified risk factors that result in no audit findings. Without a doubt, for auditors, the false positives have become a strong deterrent to the use of Benford's Law in recent years. The work required on these "blind alleys" typically necessitates an additional site visit by the auditing firm to follow up on the possible fraud. In addition, it would include costs to travel to and from the site, along with the professional service time of the auditors assigned to follow up on the discrepancy. Stated simply, the cost can be substantial.

As a result, the judicious use of Benford's Law is likely to be the only way the tool's use will be kept from extinction for practicing auditors. So, auditors must think hard about the situations where it should be employed. For example, an auditor should only use Benford's Law when the risk of misstatement is higher, due to the industry and/or financial health of the company being examined. In such circumstances, the auditor may wish to use Benford's Law on those accounts that represent the greatest risk of material misstatement. In addition, an auditor needs to know enough about statistics to minimize the number of false positives that occur, and when they do occur, to quickly root out the explanation. The next sections detail some of the statistical ideas and practices that would help prepare auditors for this work.

7.3 PRACTICAL AND STATISTICAL HYPOTHESES

Thanks to the efforts of Mark J. Nigrini and others who demonstrated the power of Benford's Law as a fraud detection tool, use of first digit and/or first two digit analysis became a regularly discussed, if rarely applied, technique. Benford's Law has been shown to be applicable in a number of auditing contexts, including external, internal, and governmental auditing contexts. For example, Nigrini & Mittermaier [NiMit] show how external auditors can use Benford's Law as an analytical procedure to help discover surprising patterns in transaction activity. Nigrini [Nig6, Nig7] was the first to highlight the potential of Benford's Law as an effective fraud detection process. He outlined a number of practical applications where a fraud auditor could effectively employ digital analysis using Benford's Law including accounts payable data, general ledger estimations, duplicate payments, and customer refunds.

Importantly, Benford's Law has also been shown to be an effective tool for internal and governmental auditors as well. Nigrini [Nig6, Nig7] outlines a number of different contexts where a digital analysis can add value for internal auditors, including the revenue, canceled checks, inventory, and disbursement areas. Nigrini [Nig4] also demonstrated the applicability of Benford's Law in a taxpayer compliance context, raising the possibility of its effectiveness as a tool for IRS (governmental) auditors.

By around 2004, Benford's Law was incorporated into leading auditing software packages to be available for practicing auditors, although, as pointed out by Cleary and Thibodeau [ClTh], this was not always done in a statistically sound way. In particular, the difference between testing the entire distribution of first digits as a whole, as opposed to a "digit-by-digit" analysis, made applying Benford's Law in practice too risky for auditors. The well-documented successes in which such an analysis uncovered a significant fraud were evidence of the technique's potential, but added to audit cost. The complexity of pursuing a "false positive" indication of fraud was generally deemed too expensive to make use of the technique routine. As discussed in Section 7.2, this cost concern became even more important after the passage of Sarbanes–Oxley.

When we look at the realities of the business statistics education and audit practice environments as described in Sections 7.1 and 7.2, it is easy to see how an audit manager might be reluctant to apply Benford's Law, or any data interrogation technique. Students in a first statistics course, typically the only course practicing auditors have taken, gain some experience in using hypothesis tests to make decisions. A big part of the difficulty in implementation of Benford's Law results from the fact that these courses do not usually discuss the ways in which those decisions have to be implemented in the field, as described in Goal 3 of Section 7.1. While this lack of alignment between the business curriculum and field practice has resulted in a reluctance to use analytic techniques, it also raises an important question. How can the statistics and accounting curricula be integrated to provide students with the experience they need to translate a statistical result to an effective decision in the field?

Let us consider how the audit process using Benford's Law as a test of hypothesis looks when translated into the terms of statistical hypothesis testing. Note that we present the tests themselves using the vocabulary and notation from what statisticians would call the "frequentist," or "data-centric" point of view. It would not be any more difficult to align the terminology with the Bayesian approach, in which the auditor's prior estimate of the probability of fraud is explicitly part of the statistical process. Certainly an auditor's prior estimate of the probability of fraud would be a factor in choosing whether to apply Benford's Law, so the Bayesian approach does seem like a good fit. However, we choose the frequentist paradigm because, for better or worse, changing a first course in statistics to a Bayesian viewpoint is unlikely to happen any time soon.

When an auditor begins an engagement with a particular client, we assume that they have a presumption that their client's financial statements are free of material misstatement. This means that the null hypothesis, the statement we are testing, is one that suggests the accounts are in order. The practical hypothesis of interest, which we denote PH_0 is

PH_0: There is no material misstatement.

The auditor's job is to carefully examine both the actual accounts, and the procedures and controls that the client has in place, to look for evidence the null hypothesis is incorrect. If such evidence is found and verified by the auditor, then the auditor must reach the alternate conclusion:

PH_a: There is a material misstatement.

The statements PH_0 and PH_a above are not actually statistical hypotheses. A statistical hypothesis is a statement about a particular statistical distribution or its parameters, the numerical summaries of the distribution. The statements that the auditor is concerned with are much more general and are not typically interpreted with a single number or result. However, the usual descriptions of **Type I and Type II errors** as covered in a first statistics course are still of value. The Type II error, failing to reject PH_0 when it is actually false, leads to the "blown" opinion that is the auditor's worst fear. A Type I error, rejecting PH_0 when it is actually true is almost as serious. This results in challenging a client's financial statements that are materially correct, which can result in a substantial increase in the costs to complete an audit and ultimately the loss of a client. A handful of errors of either type could critically damage the reputation and credibility of an auditor.

Choosing appropriate analytical tests to help in this decision-making process is clearly of vital importance. If there is concern that fraud might be present the auditor might choose to use any of a number of statistical tests that indicate whether the data entries of interest conform to Benford's Law. This leads to a genuine statistical hypothesis and alternative:

SH_0: The data entries in question follow Benford's Law.

SH_a: The data entries in question do not follow Benford's Law.

It is the moving back and forth between the practical hypotheses about a business operation and the statistical outcomes of a particular analytical procedure that our current statistics and accounting curricula fail to experience. An auditor who performs a Benford analysis and rejects SH_0, a procedure that uses our first two goals from the statistics course, need not immediately reject PH_0 as well. Further investigation may reveal any of a number of reasons (which we detail below) to explain the discrepancy between the data and Benford's distribution. This is the critical step using Goal 3, putting the statistical results in context of the field work.

In Table 7.1 we review the interaction between these related but different pairs of hypotheses.

Recall that the auditor is always on the lookout to reduce the risk of a "blown" opinion, so when SH_0 is rejected this "red flag" mandates a detailed (and potentially expensive) follow-up. In the next section, we address the ways in which our courses can assist students as they learn how to proceed in the critical practice step of moving from rejecting the statistical hypothesis to making a decision on the practical hypotheses. As an example we consider how to teach our students to use the information in Table 7.1 as they make the connection between their courses in statistics and auditing.

7.4 FROM STATISTICAL HYPOTHESIS TO DECISION MAKING

Imagine that an auditor in the field has a run a Benford's Law analysis on the first, or first and second digits, in a particular data field of interest, and has discovered that the statistical hypothesis that the data conform to Benford's Law is rejected.

	SH$_0$ rejected – possible evidence of fraud.	SH$_0$ accepted – no evidence of fraud.
PH$_0$ false: Material misstatement exists.	Follow-up by auditors may • Lead to detection of fraud. • Fail to find evidence of the fraud; perhaps resulting in a "blown" opinion.	Procedure included in work notes, no follow-up. High risk of a "blown" opinion.
PH$_0$ true: Statements materially correct.	Follow up by auditors may • Produce reasonable explanation. • Lead to false charge of fraud; added costs and potential loss of client.	Procedure included in work notes, no follow-up needed. Correct decision likely to be reached.

Table 7.1 Interaction between practical and statistical hypotheses in Benford's Law.

(In the language of the previous section, SH$_0$ has been rejected.) This is a possible indication of fraud, but fabricated data is one of just several possible reasons for this result. In the following paragraphs we list four possible reasons that the test may have turned out this way, and we suggest how follow-up testing and documentation can help the auditor choose among them. We also detail which skills from the first course in statistics come into play at each step.

Reason 1: The data in this field should not have been expected to conform to Benford's Law.

It is well known that for Benford's Law to apply, often the values in a data set range over several orders of magnitude. In teaching about Benford's Law in a first statistics course, we have found that pulse rates, measured in beats per minute, are an excellent and easily understood example of a variable for which there is no hope for Benford's Law to apply to the first digit, as the pulse rates of students almost all begin with five through eight. Our pulse rates are simply too close to the same size for Benford's Law to apply.

Evaluating whether this is the case for a particular variable is a classic example of the statistical skill of diagnostic checking, a staple of a good first statistics course. Just as statisticians learn to verify that a set of data is not wildly skewed before applying a classical t-test, auditors should learn to look at the size of values in a data set and not expect Benford's Law to apply to those fields in which entries are very similar in size. An auditor applying good statistical methodology would, we hope, have evaluated the relative size of the entries prior to applying a test for Benford's Law. In any event, the first step for an auditor after rejecting SH$_0$ should be to graph the data and check the size of the entries to make sure that expecting Benford's Law to hold was reasonable. Ideally, the data should be plotted prior to carrying out a Benford test to rule out this possibility ahead of time.

Reason 2: There is an easily explained feature of the data that causes a first digit or pair of digits to repeat surprisingly often.

For example, the firm being audited may have a long-standing relationship with a vendor who is paid for a weekly delivery that is essentially constant from week to week. The vendor receives the same amount each week and the leading digit of that amount shows up far too often. This is another explanation that should typically be found by looking at the data in a summary way prior to applying analytical testing, but is also easy to discover after the fact.

Reason 3: A Type I error has occurred. That is, SH_0 *is actually true and the data in this field do follow a Benford's Law distribution, but this particular sample was not representative.*

Much to the dismay of practitioners in auditing and other fields, errors in testing are unavoidable. Goal 1, understanding the nature of random variation, informs students that no method is completely reliable. The unfortunate case in which a Type I error occurs is likely to be expensive for auditors. At a minimum, a follow-up test should be done on related fields to look for a broader pattern of fraud. Individual entries in this field will have to be scrutinized. This is the area where an auditor's experience in knowing how to look for fraudulent entries would be vital. As detailed in [ClTh], early versions of Benford's tests in auditing software made this outcome far too likely, probably leading to additional costs for auditors.

Reason 4: Some of the data was, in fact, fraudulent.

Here is where the promise of Benford's Law as an analytical technique actually pays off. The data reveals an anomaly, an analysis of the data set shows that it should conform to Benford's Law, and then a careful examination of individual entries reveals that at least some are not correctly entered. This is a case where the auditor can, if the amounts involved in the improper entries are material, move with some confidence from rejecting SH_0 to rejecting PH_0 and uncovering a material misstatement.

Note that the auditor who steps through this list of reasons would be following guidelines for good data analysis as taught in a successful first statistics course that achieved our goals from Section 1. Data would be previewed using summary statistics and graphics, the appropriateness of a particular procedure ascertained, a statistical test carried out, and the results interpreted in context. Due to the nature of randomness, carrying out these steps with care does not guarantee success, but it would minimize the risks of failure.

7.5 EXAMPLE FOR CLASSROOM USE

Here is an example of how instructors in either statistics or auditing can use Benford's Law with real data in a classroom setting to illustrate both statistical and auditing principles.

Repeated trials in a variety of settings have revealed that the first digits of street addresses tend to follow Benford's Law in most communities. While this sort of data would not typically be part of a financial statement audit at a public company, there are instances where such data is collected and there would be a strong motivation for fraud. One such case would be petitions collecting signatures to get political candidates eligible for a ballot. This work is often done under time pressure by volunteers or temporary workers who might be tempted to invent data, presuming that a careful check of records is unlikely. Another case could be a small business that needs a certain number of customers in order to qualify to participate in a potentially lucrative government program.

To carry out the experiment, each student in the class is provided with a piece of a page from the white pages of a phone book. Experience shows that about one quarter of a page will be adequate for the class experiment. The students are asked to randomly select a number of residential customers from the page and to record the first digits of each customer's street address; for instance somebody living at 353 Broad Street would be recorded as a "3." Data from all students are then collected and recorded in a spreadsheet.

Instructors should ensure that the total number of residences selected across all students is at least 150 to be sure that the sample size is large enough to justify a chi-squared goodness of fit test. Upon collection, the data are summarized in a table like the one below. This particular table represents actual data from the Boston, Massachusetts phone book collected at a seminar on Benford's Law at Harvard University, given by the authors in 2008.

When the appropriate chi-squared goodness of fit test is carried out, we get the following result:

First digit	Observed	Expected	$(\text{Obs} - \text{Exp})^2/\text{Exp}$
1	128	117.4017	0.956749
2	63	68.67559	0.469051
3	45	48.72611	0.284937
4	42	37.79491	0.467863
5	21	30.88069	3.161457
6	25	26.10925	0.047126
7	28	22.61686	1.281266
8	17	19.94948	0.436074
9	21	17.84542	0.557643
Totals	390	390	7.662165 $\longleftarrow$ χ^2-stat
			0.467147 $\longleftarrow$ p-value

The instructor then leads a facilitated discussion with students about the meaning of this result. We suggest that questions such as the following would be useful:

What is the null hypothesis? What does the p-value mean? Does this result support the notion that street addresses in Boston follow Benford's Law?

A useful variation of this experiment is then to have some students act as a control group. They are asked to make up first digits of street addresses without looking at the phone book pages. Not surprisingly, their distribution tends to be much closer to a uniform distribution of first digits. The purpose of this discussion is to highlight the difference between the real data and the artificial data.

Note that this example is consistent with all of the theory discussed in this chapter. Using real data in an active classroom-learning environment aligns perfectly with the recommendations in the GAISE report and is considered a best practice in statistics education. Further, the discussion of why some sets of data, in this case the control group discussed above, fail to follow Benford's Law is key to the decision making, as discussed in Section 7.4.

7.6 CONCLUSION AND RECOMMENDATIONS

In the regulated public company auditing market, the auditor's responsibility to detect fraud during a financial statement audit has increased dramatically. Such added responsibility has placed a great demand on auditors to devise more effective and efficient processes to detect incidents of financial statement fraud. We believe that auditors can make greater use of Benford's Law and other statistical techniques that have been greatly facilitated by the use of Computer Assisted Auditing Techniques (CAATs) like the ACL auditing software package [LoRSST].

For example, an auditor seeking to apply the Benford command using ACL (the market's leading audit software product) need only identify the appropriate data field (e.g., invoice amount) within the appropriate data file (e.g., client accounts receivable file) in order to successfully run the command. The auditor would then have to consider whether additional audit testing should be completed on any field that did not conform to the Benford's Law probability distribution, and eventually issue their audit opinion using this review as one of the components. Empowering students to have the ability to make good decisions with Benford's Law is a challenging but important task for those of us working in the area of business education.

One final advantage of incorporating a unit on Benford's Law into the first course in statistics for business majors is the extent to which this will help differentiate this course from the standard liberal arts statistics course. Many business statistics books focus far too much on techniques and computation and not enough on the big picture notions of how to think in a statistical manner. Sometimes there are very few examples from disciplines outside of business; in fact, some statistics educators have referred to business statistics texts as "the liberal arts books with the interesting material removed." Including Benford's Law helps to counteract such complaints. Indeed, since Benford's Law has intellectual merit and applications in so many areas, we believe it should be an important part of the statistics and auditing curriculum.

How can we improve the teaching of a first course in business statistics, and upper-level courses for accounting majors, so that practicing auditors can use analytical procedures with confidence? Given the real-world environment for auditors spelled out in Section 7.2 and the disconnect between statistical decisions and practical decisions pointed out in Section 7.3, the task is quite daunting. We believe, however, that it is important to try to encourage sensible application of Benford's Law by students in the undergraduate and possibly the graduate and continuing education curriculum.

We conclude with three recommendations for the undergraduate business curriculum. These are steps that we have initiated at our own school that require relatively little in the way of resources. Getting faculty teaching both disciplines to cooperate and implement these recommendations may take some time, but we believe the payoff could be substantial.

Recommendation 1: The first course in business statistics should be taught using the GAISE guidelines, with a particular focus on learning about data in real-life contexts.

The first course in business statistics should be, and is typically advertised as, a place to learn to make good decisions with data. Projects, exams, and assignments should be sure to include the important step of connecting the statistical hypothesis to the practical decision.

Recommendation 2: Intermediate-level auditing courses should include material from the prerequisite statistics course in examples and cases.

Using real data and cases is common in the financial statement auditing course. Recognizing that these cases can be better understood with statistical techniques, or just statistical thinking, is much less common. Instructors in this important course should take advantage of their students' statistical experience. Inviting a statistician to visit for a case analysis or guest lecture, or even team teaching the course, would be worthwhile.

Recommendation 3: Benford's Law should be the topic of choice to help students understand the connection between statistical and practical hypotheses as applied to auditing.

No other analytical procedure combines so many appealing features. Benford's Law is intellectually interesting, is well understood statistically and mathematically, and has been incorporated into leading software. The technique has a well-documented literature with both successful cases and cautionary tales.

Finally, the coauthors have enjoyed working together on educating auditors to become more effective decision makers with statistics, and we hope our suggestions encourage other collaborations between statisticians and auditors.

Chapter Eight

Detecting Fraud and Errors Using Benford's Law

Mark Nigrini[1]

8.1 INTRODUCTION

This chapter will review Benford's Law as it relates to detecting fraud and errors. We start with an introduction and a review of selected parts of Benford's original 1938 paper [Ben]. Thereafter, several complaint data sets will be reviewed. The next section will discuss several Benford analyses of fraudulent data. The concluding discussion will discuss the findings and shed some light on when Benford's Law might, or might not, detect fraud or errors.

I first came across Benford's Law in a decision theory course taught by Professor Marty Lévy in 1989. In the 600-page textbook, Berger ([Berg], 1985), in two paragraphs, notes that if a statistician's prior expectation was that the digits should be equally distributed, then this prior distribution was both an improper and a non-informative prior, and according to Benford's Law, the correct prior distribution was the logarithmic distribution shown in equation 8.1 below:

$$P(D_1 = d_1) = \log\left(1 + \frac{1}{d_1}\right); \quad d_1 \in \{1, 2, \ldots, 9\}, \quad (8.1)$$

where P indicates the probability of observing the event in parentheses and log refers to the log to the base 10. Berger's discussion used about two-thirds of a page in the textbook and his discussion included words describing the digit patterns as *interesting and intriguing*. After this, Berger returned to his business of non-informative priors, and prior and posterior probabilities. I remember walking to the library to find a copy of Benford's paper and I also remember being very happy at finding a paper copy of the *Proceedings of the American Philosophical Society* on the shelves, waiting to be photocopied. That night I read the paper over several times and was very excited by the prospect that *if Benford's Law were true then perhaps auditors could use these expected digit frequencies to test whether a data set was authentic or not.* I only had a few ways to test whether Benford's Law was *true*. One way would have been to have tested the digit frequencies of many data sets and then compared the results to Benford's Law. This approach was easier said than done. At that time the internet as we know it now didn't exist. Obtaining data was very difficult and analyzing it was just as difficult. I remember my 286

---

[1]Department of Accounting, West Virginia University, Morgantown, West Virginia 26506.

computer straining away to analyze just 1,000 records in Lotus 1-2-3. A more practical approach was to read and evaluate everything that had been written on the topic and to consult with published authors in the field. Thus began what has turned out to be a long-time relationship with Benford's Law.

8.2 BENFORD'S ORIGINAL PAPER

Benford ([Ben], 1938) shows the results of his analysis of 20,229 records from a total of 20 sets of data. Back in the 1930s this would have been a very time-consuming activity because everything would have needed to be done by hand. Benford analyzes his own data by using a metric that can be described as the **Total Absolute Deviation**. This metric simply sums the absolute differences between the actual percentages and what he called the "theoretical frequencies" (in his own paper he doesn't call the expected frequencies "Benford's Law"). He concludes that the best fit occurs for "outlaw numbers that are without known relationship" as opposed to those numbers that "follow an orderly course." This conclusion of Benford's is correct when related to his observed digit patterns, but it isn't true in general. There are many examples of recursive data series that have an excellent fit to Benford's Law, the best known of which is the Fibonacci sequence.

Benford continues, noting that "In natural events, and in events of which man considers himself an originator, there are plenty of examples of geometric or logarithmic progressions." Benford believed that the basis of the expected digit patterns was that the used numbers in the world generally formed a geometric series. He cites many examples of geometric progressions in the real world. These examples include psychological reactions to stimuli, our sense of loudness, music scales, examples from medicine, engineering phenomena, and examples from physics. Benford then claims, "Nature counts e^0, e^x, e^{2x}, e^{3x}, ... and builds and functions accordingly." According to Benford, it is this geometric basis of events and matter that give us the expected frequencies of Benford's Law. In his conclusion, Benford claims that the expected digit patterns are really a theory about phenomena and events and not really a theory about numbers, which are essentially just symbols used by human beings.

Benford makes no claim that his theory could be used in any constructive way. He essentially tests some empirical data, develops the equations for the expected digit frequencies, and provides some discussion points. In essence, Benford's paper tells us something about how the world works, but he doesn't claim that this knowledge could be terribly useful in any application. This is not surprising, given how much time would have been necessary to analyze his data. My thoughts on reading Benford's paper were, "If there were indeed predictable patterns to the digits in tabulated data, then perhaps auditors could use these expected patterns to test whether data was authentic or fraudulent."

8.3 CASE STUDIES WITH AUTHENTIC DATA

This section reviews four studies of authentic data, and in each case the data follows Benford's Law. The first analysis is of interest received amounts from tax return data. The second application is of census results from the 2000 United States census. The third application shows the results of an analysis of streamflow data, and the final analysis shows the results of analysis of accounts payable data.

In a data analysis setting it is important to evaluate beforehand whether the data is expected to conform to Benford's Law. No valid conclusion can be reached by analyzing the conformity of data that isn't expected to follow Benford's Law (for conformity to Benford). As such, it was necessary to "convert" the mathematical underpinnings of Benford's Law into some workable requirements for conformity. These requirements are set out in Nigrini ([Nig9], 2011) and are summarized below.

1. The data must represent the sizes of facts or events. Examples of such data would include the populations of towns and cities, the flow rates of rivers, or the sizes of heavenly bodies. Financial examples include the market values of companies on the major stock exchanges, the revenues of these companies, or their daily sales volumes.

2. There should be no built-in minimum or maximum values in the data set. An example of a minimum would be a stockbroker that has a minimum commission charge of $50 for a buy or sell transaction. The broker would then have many people whose small trades are charged the $50 minimum. A data table of these commission charges for a month would have an excess of first digit 5s and second digit 0s. A built-in minimum of zero is acceptable.

3. The data set should not represent numbers used as identification numbers or labels. These are numbers that we have given to events and individual entities, objects and items in place of words. Examples of these include social security numbers, bank account numbers, county numbers, highway numbers, car license plate numbers, flight numbers, or telephone numbers. These numbers have digit patterns that have some meaning to the persons that developed the sequence.

8.3.1 Analysis of Taxpayer Data

Nigrini ([Nig4], 1996) is an early example of an analysis of several large financial data sets. The paper flowed from my dissertation, which was a comprehensive review of using Benford's Law in a taxpayer compliance setting. The dissertation analyzed, amongst others, interest paid and interest received numbers from 1985 and 1988 individual tax returns. The tests were done on the first and second digits of the numbers. The data is no longer available to me and since that time I have moved on from analyzing the first and second digits as two separate graphs to analyzing

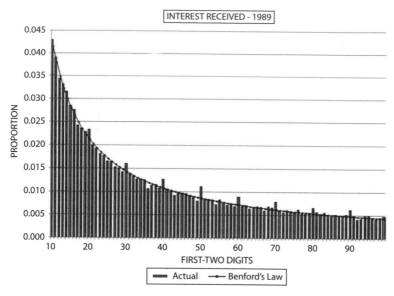

Figure 8.1   The first-two digits of the interest received amounts reported by taxpayers on
1989 tax returns.

the first-two digits in a single graph. That analysis is therefore reproduced on data
that is still available to me (the 1989 Individual Tax Model Files).

The data on the Individual Tax Model Files (ITMFs) are compiled by the Internal
Revenue Service (IRS) from a stratified sample of unaudited individual income tax
returns (Forms 1040, 1040A, and 1040EZ) filed by U.S. citizens and residents (see
[IRS]). The 1989 sample of 95,713 returns was selected from the population of
109.7 million returns filed in 1990. The population included all returns processed
except for tentative and amended returns. All 1989 returns were assigned to sample
strata based on amounts related to the larger of total income or total loss amounts
and the size of business receipts. In addition, the 1989 strata were based on the
presence or absence of various forms and schedules (such as Schedule C or F).
Returns were randomly selected from the strata at rates ranging from 0.02 percent
to 100 percent. A review of the sampling method, methods of disguising taxpayer
data to preserve privacy, and a listing of the indicator and amount fields is presented
in [IRS].

Figure 8.1 shows the first-two digits of the 77,685 tax returns that reported interest received amounts. These amounts follow the three requirements listed above.
The first-two digits (10, 11, 12, . . . , 99) are shown on the x-axis while the y-axis
shows the actual and expected proportions. The line shows the expected proportions of Benford's Law while the bars show the actual proportions. The results
indicate that the conformity to Benford's Law is good, as expected. The data conforms to the requirements stated above and because of third party reporting (banks
and brokerages that report interest amounts paid to taxpayers to both the taxpayer

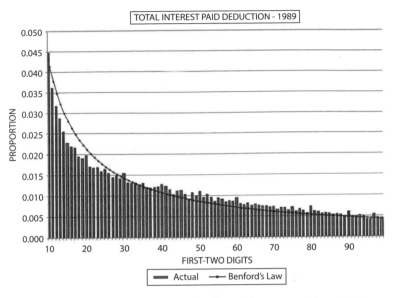

Figure 8.2   The first two digits of the total interest paid amounts claimed by taxpayers as deductions on 1989 tax returns.

and the IRS) this field is subject to minimal tax evasion (underreporting of interest income). By way of a contrast, the first-two digits of total interest paid (a deduction allowed to taxpayers in 1989) are shown in Figure 8.2 for the 48,407 tax returns claiming interest payments.

The graph follows the same format as the graph in Figure 8.1. The graph shows that the conformity to Benford's Law is acceptable, but weaker than the graph for the interest received amounts. The data also conforms to the requirements stated above. The conclusion drawn is that the fit to Benford's Law is weaker than for interest received, possibly because the interest paid amounts are subject to less strict third party reporting requirements. This field is consequently open to more tax evasion (by overstating total interest paid) and the number invention that goes along with making up overstated amounts, which may explain the weaker fit to Benford's Law. The weaker fit is not proof of tax evasion but rather the results (a weaker fit to Benford) are consistent with more manipulation in this field.

8.3.2 Analysis of Census Data

The United States (U.S.) conducts a census every 10 years, as required by the U.S. Constitution. The results of the census are used to allocate congressional seats (the more people that live in an area, the more representatives they will garner in Congress), electoral votes (relevant in elections for the U.S. president), and various types of federal funding and state aid. Therefore, the census is taken very seri-

	SUMLEV	REGION	DIVISION	STATE	COUNTY	STNAME	CTYNAME	CENSUS2000POP	ESTIMATESBASE2000	POPESTIMATE2000	POPESTIMATE2001
2	40	3	6	1	0	Alabama	Alabama	4447100	4447382	4451849	4464034
3	50	3	6	1	1	Alabama	Autauga County	43671	43671	43872	44434
4	50	3	6	1	3	Alabama	Baldwin County	140415	140415	141358	144988
5	50	3	6	1	5	Alabama	Barbour County	29038	29038	29035	74273
6	50	3	6	1	7	Alabama	Bibb County	20826	19889	19936	20942
7	50	3	6	1	9	Alabama	Blount County	51024	51022	51181	51999
8	50	3	6	1	11	Alabama	Bullock County	11714	11626	11604	11381
9	50	3	6	1	13	Alabama	Butler County	21399	21399	21313	21036
10	50	3	6	1	15	Alabama	Calhoun County	112249	112243	111342	111019
11	50	3	6	1	17	Alabama	Chambers County	36583	36614	36593	36281
12	50	3	6	1	19	Alabama	Cherokee County	23988	23986	24053	24078
13	50	3	6	1	21	Alabama	Chilton County	39593	39593	39781	40030
14	50	3	6	1	23	Alabama	Choctaw County	15922	15922	15850	15652
15	50	3	6	1	25	Alabama	Clarke County	27867	27873	27833	27634
16	50	3	6	1	27	Alabama	Clay County	14254	14254	14264	14257
17	50	3	6	1	29	Alabama	Cleburne County	14123	14123	14156	14196
18	50	3	6	1	31	Alabama	Coffee County	43615	43620	43540	43581
19	50	3	6	1	33	Alabama	Colbert County	54984	54984	55024	54841
20	50	3	6	1	35	Alabama	Conecuh County	14089	14089	14040	13894
21	50	3	6	1	37	Alabama	Coosa County	12202	11855	11843	11746
22	50	3	6	1	39	Alabama	Covington County	37631	37633	37490	36939
23	50	3	6	1	41	Alabama	Crenshaw County	13665	13665	13696	13700
24	50	3	6	1	43	Alabama	Cullman County	77483	77483	77580	77661
25	50	3	6	1	45	Alabama	Dale County	49129	49128	49068	49001
26	50	3	6	1	47	Alabama	Dallas County	46365	46379	46181	45910
27	50	3	6	1	49	Alabama	DeKalb County	64452	64454	64638	65519
28	50	3	6	1	51	Alabama	Elmore County	65874	65874	66230	67566
29	50	3	6	1	53	Alabama	Escambia County	38440	38440	38402	38430
30	50	3	6	1	55	Alabama	Etowah County	103459	103450	103313	102914

Figure 8.3   An extract from the Excel file downloaded from the U.S. Census Bureau's website.

ously, and the total cost of the 2010 census is estimated by the U.S. Census Bureau (Census Bureau) to come in at around $14.5 billion. Many people have a stake in the accuracy and integrity of the census. Given the importance of these numbers, and their cost, we should expect the numbers to be as accurate as possible within the inherent limitations of counting people, some of whom might not wish to be counted. Aggregate census data is available from the 2000 census from the website of the Census Bureau, http://www.census.gov.

The website of the Census Bureau gives county-by-county census results from the 2000 census. In any analysis, it is important to understand the data (where the data is from, which numbers were directly measured and which were computed, and so on). In this case, a file layout document is available and should be read before any analysis is undertaken. The data was downloaded from the Census Bureau's website and an extract from the Excel file is shown in Figure 8.3.

It is usual to perform some data cleansing before an analysis can be done. In this case, all the records with county numbers of "0" need to be deleted because these are actually the state totals as opposed to county counts. Also, it was necessary to determine exactly which of the fields with "2000" in the field name contained the official census counts. After the data cleansing steps we are left with a data set of 3,143 records. The sum of the population counts was 281,424,602 and the minimum amount was 67 people for Loving County, Texas, and the maximum amount was 9,519,331 people for Los Angeles County, California. The census data satisfied the three criteria mentioned above and the data was expected to conform to Benford's Law.

Figure 8.4 shows the results of the census data. The results show an acceptable level of conformity to Benford's Law. The conformity seems better if we look at the first digits only (not shown). The conclusion is that the data conforms to Benford's Law at an acceptable level given that there are only 3,143 records. Note that smaller

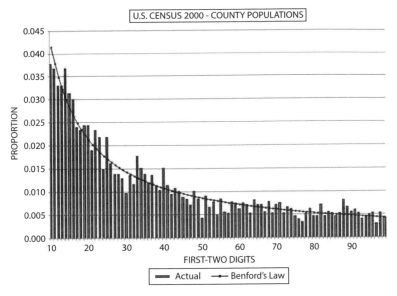

Figure 8.4   The results of an analysis of the first-two digits of the 2000 U.S. Census.

data sets are expected to have a weaker level of conformity due simply to the fact that they are small. This ties in with sampling theory in statistics, which states that as the sample size increases, the sample statistics should better approximate the population parameters.

8.3.3 Analysis of Streamflow Data

Nigrini and Miller ([NiMi1], 2007) analyze a large amount of earth science data (especially hydrology data). The results showed that streamflow data conformed closely to Benford's Law and that deviations from the Benford proportions in other earth science data could be indicators of either (a) an incomplete data set, (b) the sample not being representative of the population, (c) excessive rounding of the data, (d) data errors, inconsistencies, or anomalies, or (e) conformity to a power law with a large exponent.

Streamflow data was obtained from the U.S. Geological Survey (USGS) website, http://www.usgs.gov. The agency's website lists many programs, including the National Water Information System (NWIS). Under this program the USGS operates and maintains approximately 7,300 stream gages, which provide data for many diverse users. There are six main reasons for the collection of accurate, regular, and dependable streamflow data:

- *Interstate and International Waters:* Interstate compacts, court decrees, and international treaties may require long-term, accurate, and unbiased streamflow data at key points in a river.

- *Streamflow Forecasts:* Upstream flow data is used for flood and drought forecasting in order to improve estimates of risk and impacts for better hazard response and mitigation.

- *Sentinel Watersheds:* Accurate streamflow data is needed to describe the changes in the watersheds due to changes in climate, land, and water use.

- *Water Quality:* Streamflow data is a component of the water quality program of the USGS.

- *Design of Bridges and Other Structures:* Streamflow data is required to estimate water level and discharge during flood conditions.

- *Endangered species:* Data is required for an assessment of survivability in times of low flows.

The methods used for measuring flow at most stream gages are almost identical to those used 100 years ago. Acoustic Doppler technology can widen the range of conditions for which accurate flow measurements are possible, but is not yet seen as providing enhanced efficiency or accuracy at most locations. No new technology has yet been found to provide accurate data over a wide range of hydrologic conditions more cost-effectively than the traditional current meter methods.

The data used in the study was obtained from the *USGS Water Data for the Nation*. The data used was the annual data *Calendar Year Streamflow Statistics for the Nation*. To obtain a large data set, the only condition that was imposed was that the period of record included calendar year 1950 or later. The data therefore consisted of all the annual average readings for any site that had an annual average recorded in any of the years from 1950 to 2005. The only sites that were *excluded* were sites that only had data for the pre-1950 period. After some data cleansing to remove duplicates and obvious errors (such as negative numbers and zeros) there were 457,440 records. Each record is an average annual streamflow from a station from 1874 to 2004 as measured in cubic feet per second. The annual flows ranged from 0.001 to 980,900 cubic feet per second. The data was highly skewed and the average annual flow was 2,199 cubic feet per second.

This data set is particularly interesting because (a) the period covered is 130 years and it is rare for any data set to cover such an extended period, (b) at the time, the data set was the largest analyzed in the Benford's Law literature to date, (c) the range in streamflows indicates that the sites covered everything from the smallest streams to the largest waterways, (d) the measurement technology has been unchanged over the entire period, which suggests that there are no distortions due to technological changes, and (e) the data set is used for a variety of important purposes. The results of an analysis of the digits of the streamflow numbers is shown in Figure 8.5.

As usual, the expected proportions of Benford's Law are shown by the smooth monotonically decreasing line from 0.41 to 0.044. The actual proportions are shown as vertical bars. The visual fit to Benford's Law is excellent with a Mean

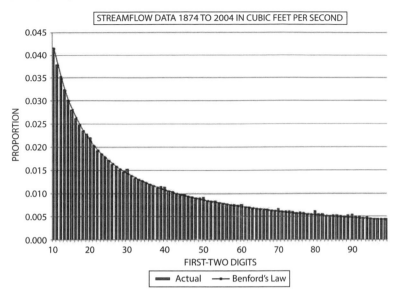

Figure 8.5   The first-two digit proportions of the streamflow data and the expected proportions of Benford's Law.

Absolute Deviation (average of |Actual − Benford's Law|) of 0.00013. The low
Mean Absolute Deviation means that, on average, the deviation of the actual percentage from that of Benford's Law was one-tenth of one-tenth of one percent. A
visual review of the graph shows no sign of the "overs" or "unders" being clustered
in certain parts of the graph, nor are any of the "overs" or "unders" systematic by
occurring (say) at multiples of 10 (10, 20, 30, . . . , 90). The near-perfect visual fit
to Benford's Law suggests that the data is consistent with the geometric pattern (or
a combination of interweaving geometric series) assumed by Benford's Law.

8.3.4 Analysis of Accounts Payable Data

The preceding case studies all related to taxation and other non-financial data.
The taxation data was a data set formed by combining the same type of income and
expense across about 100,000 taxpayers. The final data set relates to combining
many different kinds of expenses for a single company for a single year. The data
set, reviewed in Drake and Nigrini ([DrNi], 2000), is made up of the dollar amounts
of the invoices processed by a single company. There are a few general issues to
consider before using Benford's Law tests on corporate data. First, the data should
be from a reporting time period such as a month, a quarter, or a fiscal year so
that there is a sufficiently large set. Investigators should be able to reconcile the
data total to the ledger accounts. Second, the data should be for an identifiable
business entity. If data from two or more unrelated divisions are combined, then

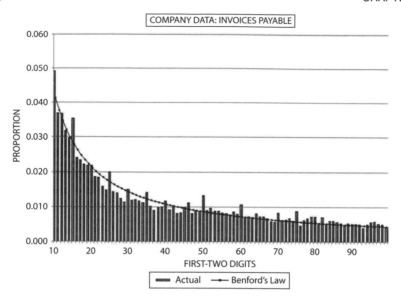

Figure 8.6   Invoice data for a company.

abnormal digit and number duplications existing in the data from one division may be lost when merged with the data from another division. That is, combined data might conform to Benford's Law whereas individual data sets might show only a weak conformity or a lack of conformity. Third, data should be analyzed at a transactional level if possible. For example, disbursements should be analyzed on an invoice-by-invoice basis and employee travel expense claims should be analyzed (if possible) on a line-by-line basis. If financial numbers are summed, then non-Benford patterns arising from invented numbers may be lost. For example, if only the totals of travel and expense amounts are analyzed, then the totals could hide round line-item numbers such as $250 or $1,000.

The data cleansing steps for corporate data usually include the deletion of numbers that are (1) positive numbers less than $10, and (2) zeros and negative numbers. Small (under $10) positive numbers are deleted because they are usually immaterial and it seems appropriate to ignore these small numbers when conducting investigative procedures that are designed to find frauds or large dollar errors. Negative and positive numbers should be analyzed separately because they are usually subject to different types of errors and misstatements.

The data analyzed in this section is from a NASDAQ-listed company with offices in Ohio. The data was analyzed by the author together with internal auditors to detect inefficiencies, possible errors, and also possible fraud. A Benford's Law test was appropriate on the 36,515 invoices totaling about $90 million.

The invoice data results are shown in Figure 8.6. The objective of the first-two digits test is to look for spikes (first-two digits where the actual proportion exceeds the expected proportion by a significant margin). Investigators would then look into the actual numbers causing the spikes.

Investigators also check for spikes corresponding to numbers forming psychological boundaries. The usual psychological boundaries are $500, $1,000, $5,000, $10,000, and $100,000. Spikes (excesses) of (say) 48 and 49, and 98 and 99 would therefore be of interest. Such spikes might signal that managers are intentionally using numbers that are just below psychological boundaries because they believe that numbers at or above the boundaries could be subject to audit.

The first-two digits graph should also be analyzed for spikes just below first-two digits corresponding to internal corporate limits. For example, claims adjusters at an insurance company might fall into two groups with authorization limits at $3,000 and $10,000. An analysis of claims amounts would include checking for abnormal excesses at or just below 29 and 99. Administrative managers at a company may have expense approval limits of $5,000. A look for invoice amount spikes at 48 and 49 could provide evidence that managers might be splitting their purchases into "approvable" parts.

Investigator targets are the numbers that correspond to significant positive differences (spikes) on the graph. These numbers are overused and the excesses could be the result of processing inefficiencies, errors, biases, or fraud. Investigators usually ignore first-two digits where the actual proportion is less than the expected proportion since these occur primarily because the sum of the actual proportions must equal 1.00.

In this example, the review of the summary statistics showed that 14.0 percent of all invoices were for $50 or less. These low-value invoices were costly to process and were mainly for courier charges. Some of the invoices were employee claims for the monthly cost of $22.18 for an additional telephone line so that they could use the head office computer from home. The processing of these claims (signatures, documentation, and reimbursement) was costly, and the auditors suggested that the company make direct payments to the local telephone company to reduce processing costs.

The spike (abnormal excess) of first-two digits 10 was due to an excessive number of invoices for $10 and $100. The $10 invoices were mainly freight charges and the $100 invoices were mainly travel advances for sales staff. The processing of both these amounts was costly and the auditors recommended ways to reduce these costs. The controls for travel advances were not error proof, and it was possible that sales staff were reimbursed for the full costs of a travel trip without the deduction for the advance.

Many credit memos and adjustments (not shown in Figure 8.6 because only positive numbers greater than or equal to 10 are included on the graph) were for a vendor used for moving expenses incurred when employees were recruited from out of state. The auditors started an investigation of the excessive number of corrections.

The spike at 15 was due to an excessive number of invoices for $15 and $1,500. The $15 invoices were for freight, and the $1,500 invoices were programmer bonuses for meeting deadlines. The programmer bonuses payments controls were judged to be valid. The spike at 25 was due mainly to many $25 charges for airfreight. The recommendation was that these invoices should be summary billed (one bill for all the invoices for a week) by the airfreight vendor.

The spike at 46 was due to $46.17 payments to department workers for being employee of the month. The suggestion was made that the monthly bonuses be increased to an amount that gave the employees $50 after tax since $50 seems a lot more than $46.17, but the actual incremental amount is marginal. The spike at 79 was due to 26 payments of $794.98. This was the after-tax amount paid as a bonus to current employees who recruited a new employee. This amount was low compared to industry norms and it was suggested that it be increased to pay at least $1,000 after tax.

There were many payments for amounts of round thousand dollars. The investigation showed that these were mainly payments to directors as director's fees and to researchers at universities. The payments were properly authorized. In summary, the analysis did not reveal any fraud, but it showed several characteristics of the data that were not easily detectable by the usual auditing methods.

8.4 CASE STUDIES WITH FRAUDULENT DATA

Bolton and Hand ([BolHa], 2002) state that the statistical tools for fraud detection all have a common theme in that the observed data is usually compared with a set of expected values. Depending on the context, these expected values can be derived in various ways and could vary on a continuum from single numerical or graphical summaries all the way to complex multivariate behavior profiles. They contrast supervised methods of fraud detection which uses samples of both fraudulent and non-fraudulent records, or unsupervised methods which identify transactions or customers that are most dissimilar to some norm (i.e., outliers). They note that we can seldom be certain by statistical analysis alone that a fraud has been perpetrated. Rather, the analysis should be regarded as an alert that an observation is anomalous, or more likely to be fraudulent than others, so that it can be investigated in more detail. They advocate the concept of a **suspicion score** where higher scores are correlated with observations that are more unusual or more like previous fraudulent values. Their review of detection tools includes a review of Benford's Law and expected digit patterns. This section reviews and gives examples of fraudulent data related to accounts payable amounts, payroll data, and reported corporate numbers.

8.4.1 Fictitious Vendor Fraud

Nigrini ([Nig7], 1999) reviews an interesting case, *State of Arizona v. Wayne James Nelson* (CV92-18841), where Nelson was found guilty of trying to defraud the state of $2 million. Nelson, a manager in the office of the Arizona State Treasurer, argued that he had diverted funds to a bogus vendor to show the absence of safeguards in a new computer system. The amounts of the 23 checks are shown in Table 8.1.

Date	Amount (dollars)
October 9th	1,927.48
	27,902.31
October 14th	86,241.90
	72,117.46
	81,321.75
	97,473.96
October 19th	93,249.11
	89,658.17
	87,776.89
	92,105.83
	79,949.16
	87,602.93
	96,879.27
	91,806.47
	84,991.67
	90,831.83
	93,766.67
	88,338.72
	94,639.49
	83,709.28
	96,412.21
	88,432.86
	71,552.16
Total	1,878,687.58

Table 8.1 The checks that a treasurer for the State of Arizona wrote to a fictitious vendor called Advanced Consulting. The funds were diverted to his own use.

Since no services were ever delivered, Nelson must have invented all the numbers in his scheme, and because people are not random, invented numbers are unlikely to follow Benford's Law. There are several indications that the data is made up of invented numbers. First, as is often the case in fraud, he started small and then increased dollar amounts. The increase was geometric in nature, at least to the threshold of $100,000. Most of the dollar amounts were just below $100,000. It's possible that $100,000+ amounts would receive additional scrutiny or that checks above that amount required human signatures instead of automated check writing. By keeping the dollar amounts just below a control threshold, the manager tried to conceal the fraud. The digit patterns of the check amounts are almost opposite to those of Benford's Law. Over 90% have a 7, 8, or 9 as a first digit. Had each vendor been tested against Benford's Law, this set of numbers also would have had a low conformity to Benford's Law, signaling possible irregularities.

The numbers seem to have been chosen to give the appearance of randomness. None of the check amounts was duplicated; there were no round numbers; and all the amounts included cents. Benford's Law is quite counterintuitive; people do not naturally assume that some digits occur more frequently. Subconsciously though, the manager repeated some digits and digit combinations. Among the first-two digits of the invented amounts, 87 , 88 , 93, and 96 were all used twice. For the last-two digits, 16 , 67, and 83 were duplicated. There was a tendency towards the higher digits; note that 7 through 9 were the most frequently used digits, in contrast to Benford's Law. A total of 160 digits were used in the 23 numbers. The counts for the ten digits from 0 to 9 were 7, 19, 16, 14, 12, 5, 17, 22, 22, and 26, respectively. An investigator familiar with Benford's Law could have easily spotted the fact that these numbers—invented to seem random by someone ignorant of Benford's Law—fall outside of the expected patterns and so merit a closer investigation.

8.4.2 Payroll Fraud

An early application of digital analysis was published in Nigrini ([Nig2], 1994). A large metropolitan housing authority used off-duty policemen to patrol its housing complexes for the 8,000 housing units that it managed. From 1981 to 1991 the head of security managed to embezzle about $500,000 by submitting phony time records and pay claims for work done by police officers. The policeman named on each timesheet was a real person that worked for the authority, but the purported work done and hours worked were phony.

Each payday the security chief would go to the bank to cash checks for policemen that had worked, but were now back on regular duties in the city. There were usually one or two checks drawn for work done that were cashed, and the cash kept by the security chief for his own use.

Since the security chief had to invent a fictitious work schedule, the dollar amounts of the fraudulent checks lent themselves to an interesting application of Benford's Law. The time period of the fraud was divided into two five year periods. The Ben-

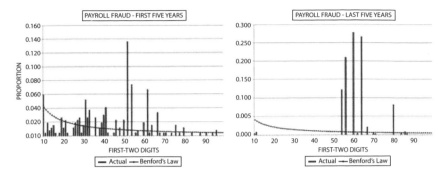

Figure 8.7   The fictitious payroll amounts that the head of security embezzled over a period
of ten years. The left panel shows the analysis of the first five years of the fraud
and the results of the last five years are shown in the right-hand panel.

ford tests were designed to also see whether the security chief's number invention
patterns changed over time or whether he was consistent over time (in academic
terms it was a test for intertemporal consistency).

The first-two digits of the 273 fraudulent checks for the first five years of the
fraud (1981 to mid-1986) are shown in the left panel of Figure 8.7. Some large
positive spikes are evident and we can see that many different combinations were
used. The fraudster used 52 of the possible 90 first-two digit frequencies. His most
frequently used numbers were $520, $540, $624, $312, $416, and $100. The check
amounts ranged from $50 to $1,352 and totaled approximately $125,000. Recall
from the preceding vendor fraud case that Nelson started off small and then quickly
increased the fraudulent check amounts. This was also the case here because the
dollar amount tripled in the last five years.

The first-two digits of the 600 fraudulent checks for the last five years of the fraud
(mid-1986 to 1991) are shown in the right panel of Figure 8.7. Not only do we get a
larger number of significant spikes, but the significant spikes are larger. The Mean
Absolute Deviation is larger for the right-hand graph (the numbers for the last five
years). Only 14 of the possible 90 first-two digit combinations were used, telling
us that the security chief was gravitating towards using the same numbers over and
over again. The most frequently used numbers were $600, $640, $560, $540, and
$800. The check amounts ranged from $540 (much higher than the previous lows)
to $1,120 and totaled approximately $375,000. The total amount embezzled tripled
in the last five years.

Some interesting conclusions can be drawn from the case. First, as time passed
the security chief gravitated towards reusing the same set of numbers. We also note
that the quantity and amounts of the checks increased. In addition, the security
chief used the names of valid policemen. An audit might have showed that these
policemen often worked 40 hour weeks, yet there were no arrest or activity records
for these energetic policemen who were able to work two full-time physically demanding jobs for that week. Finally, given the size of the spikes on the 1986–1991
graph, it is almost certain that these digit combinations would have spiked dur-

ing an analysis of the general disbursements account (the account from which the policemen were paid).

This fraud would still be in progress had it not been that one Friday in 1991, the security chief entered the bank to cash his usual package of checks. The teller happened to know one of the "prior-week" policeman whose check was cashed and who happened to be on duty in the bank at the time. Later that afternoon she told the policeman that the security chief had cashed his check and would probably have the cash at the station on Central Parkway soon. He was rather surprised by the statement since he had spent his off time that week working in the bank. The security chief was probably more surprised when he was arrested (probably not by the bank policeman) but none were as surprised as the management of the bank who were successfully sued for $100,000 for negligence by the housing authority.

8.4.3 Accounting Fraud

On November 11, 2001, Enron Corporation filed a Form 8-R with the Securities and Exchange Commission (SEC) in which it revised its results for 1997 to 2000 (all years inclusive). On December 2, 2001, Enron filed for bankruptcy. This set off a chain of events that culminated in the **Sarbanes–Oxley Act** (**SOX**) in July 2002 which, among other things, requires chief executives and chief financial officers (CFOs) to certify at regular intervals that both their reports and their financial statements contain no untruths and omit no material facts (*Reforming Corporate Governance: In Search of Honesty*, The Economist, August 17, 2002).

Starting with the March 14th indictment of Arthur Andersen in 2002, the topics of corporate fraud and accounting were covered by the financial press and financial programming on television on a regular basis until the June 15th conviction of Andersen on obstruction of justice charges. In addition to the Andersen case, there were also other developments within the audit community as a whole. This high visibility of accounting in a negative vein (with articles such as *Special Report: The Trouble with Accounting: When the Numbers Don't Add Up*, The Economist, February 9, 2002, expressing the sentiment of the time) gave rise to the research question as to whether the level of earnings management around this time period was more or less than "normal." Did the high visibility of corporate accounting impact upon management's actions with regard to earnings management? The approach taken in Nigrini ([Nig8], 2005) was to look at the digit patterns of reported earnings and at selected numbers, as reported by Enron.

The Wall Street Journal (WSJ) includes a daily *Digest of Corporate Earnings Reports* that reports and summarizes the earnings releases of the previous day. The information reported includes

- Company name, Ticker symbol, and the Stock Exchange that the company is listed on;

- Reporting Period (e.g., Q3/31 would indicate quarter ending 3/31);

	2001	2002
Quarter ended March 31	5,483	4869
Quarter ended 12/31, 1/31, 2/28, or 4/30	624	547
New York Stock Exchange listing	1,747	1,633
AMEX or NASDAQ listing	4,360	3,783
Total Number of Records	6,107	5,416

Table 8.2 A summary of all the Earnings Reports published in the *Digest of Corporate Earnings Reports* from April 1 to May 31, 2001, and from April 1 to May 31, 2002.

- Revenue (in millions, with a percentage change);

- Income from continuing operations (in millions, with a percentage change);

- Net Income (in millions, with a percentage change);

- Earnings Per Share (in dollars, with comparison to year-earlier period and percentage change).

The information is timely and, given a standard format for each company, a database of the earnings announcements was created. The period studied included all the Earnings Reports published in the *Digest of Corporate Earnings Reports* from April 1 to May 31, 2001 and from April 1 to May 31, 2002. A summary of the Earnings Reports is given in Table 8.2.

Most of the Earnings Reports (about 90 percent) were for the quarter ended March 31, 2001 or March 31, 2002. About 30 percent of the companies were **New York Stock Exchange (NYSE)** listings, with the remainder of the companies being listed on the **American Stock Exchange (AMEX)** or **NASDAQ** exchanges. Toronto and Foreign listings were omitted because the original numbers were not denominated in U.S. dollars. Also omitted were companies listed on the NYSE that were foreign and that did not report an **EPS** (earnings per share) number.

Benford's Law was used to detect the manipulation of revenues. Companies with less than $1 million in revenues were excluded from the analysis to avoid the situation where a company reported (say) $798,000 and this number was shown by the WSJ as $0.80 million (revenue numbers were shown in millions to two decimal places). Such a company would have a true second digit nine that would be analyzed as if it were a second digit zero. In 2001 there were 182 companies with revenues under $1 million and in 2002 there were 186 such companies. The numbers of companies reporting revenues of $1 million or more were 4,792 companies in 2001 and 4,196 companies in 2002. Benford's Law is still a valid expected distribution when all numbers below an integer power of 10 (10^1, 10^2, 10^3, ...) are deleted.

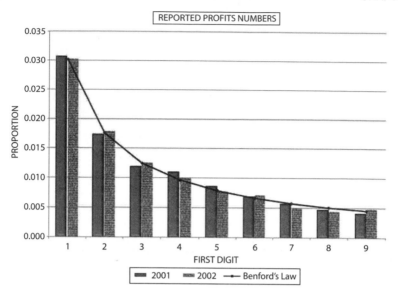

Figure 8.8   Comparison of first digit of revenue numbers and Benford's law.

The first digits of the revenue numbers were tested for conformity as a preliminary test to check that Benford's Law was a valid expectation for the second digits.
The results are shown in Figure 8.8.

The first digits show a close conformity to Benford's Law for both 2001 and
2002 for all practical purposes. If the first digits had shown a bad fit to Benford's
Law it would have been a stretch to attribute any second digit findings to earnings
manipulation, as opposed to nonconformity being due to the data not having the
attributes assumed by Benford's Law. The second digits of the revenue numbers
are shown in Figure 8.9.

The second digits, as seen as a whole, have a close conformity to Benford's Law
for each of the years in question. What is of interest is that for both 2001 and 2002
the 0 is overstated by an average of 0.9 percent while again for both 2001 and 2002,
there is a shortage of 8s and 9s as compared to Benford's Law. These results are
consistent with the hypothesis that when corporate net incomes are just below psychological boundaries, managers would tend to round these numbers up. Numbers
such as $798,000 and $19.97 million might be rounded up to numbers just above
$800,000 and $20 million respectively, possibly because the latter numbers con-
vey a larger measure of size despite the fact that in percentage terms they are just
marginally higher. A sign that this rounding-up was occurring would be an excess
of second digit 0s and a shortage of second digit 9s in reported net income numbers. The direction of the deviations in Figure 8.9 is consistent with a rounding-up
hypothesis and consistent with an upward management of revenue numbers where,
for example, a number with first-two digits of 29 or 99 is managed upwards to have
first-two digits of 30 or 10, respectively. The actual percentage of second digit 0s is

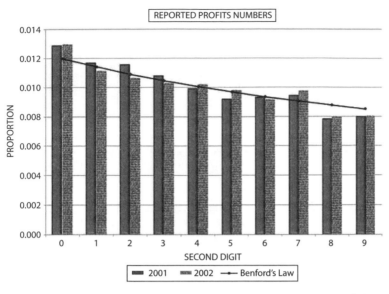

Figure 8.9   Comprison of second digit of revenue numbers and Benford's law.

13.0 percent in 2002 as opposed to 12.8 percent in 2001 and this by itself suggests that rounding upwards was more prevalent in 2002 than it was in 2001. It is a little puzzling that the proportion of second digit 9s was not lower in 2002 than what it was in 2001.

The preceding analysis focused on an analysis of corporate numbers around psychological thresholds. The collapse of Enron and the indictment and collapse of Arthur Andersen has been well covered by the news media and documented in the literature. It was interesting to assess whether the Enron numbers show any signs of attaining thresholds. Note that second digit zeros are seen to be consistent with rounding upwards. On November 11, 2001 Enron filed a Form 8-R in which it revised its results for 1997 to 2000 (four years) inclusive. Table 8.3 shows the original numbers as reported by Enron for 1997 to 2000 inclusive.

From the Enron numbers in Table 8.3 it can be seen that for three of the four years the company reported revenues that just exceeded a multiple of $10 billion, giving the revenue numbers a second digit 0 for three of the four years. Also, for three of the four years, Enron reported Net Income before the cumulative effect of accounting changes that just exceeded a multiple of $100 million, giving the income numbers a second digit 0 for three of the four years. The company seemed to make it clear that they had no control over the Cumulative effect of accounting changes and reported two EPS numbers, one for before, and one for after the effects of accounting changes. For the one year that Enron's revenue numbers did not marginally exceed a multiple of $10 billion it can be seen that the EPS number is $1.01. This number has a second digit zero and just makes a threshold of $1.00.

Total Revenues (in $ millions)	100,789	40,112	31,260	20,273
Net Income before cumulative effect of accounting changes (in $ millions)	979	1,024	703	105
Cumulative effect of accounting changes, net of tax (in $ millions)	–	(131)	–	–
Net Income (in $ millions)	979	893	703	105
Earnings Per Share of Common	1.12	1.10	1.01	0.16

Table 8.3   The fraudulent numbers reported by Enron over the period 1997 to 2000.

In Table 8.3 there are 12 "headline" reported numbers. Of these 12 numbers, seven numbers have a second digit zero. Using the Benford's Law proportion of 11.968 percent the probability of seven zeros occurring just by chance can be calculated using the binomial probability distribution. Using this distribution, the probability of seven or more second digit zeros occurring is 0.000013. So, we would only expect seven second digit zeros in 12 numbers occurring 13 times out of a million. These results have an excess of second digit zeros related to numbers just making psychological thresholds.

8.5 DISCUSSION

The logic underlying using Benford's Law for fraud detection is that authentic data should conform to Benford's Law, whereas fraudulent data would not conform to Benford. Without a careful consideration of the data and the research approach, this logic could have problems. First, some data is not expected to conform to Benford in the first place. For example, salary data is too clustered (there isn't enough of a spread) for Benford's Law to apply to a single set of data from a single location in a single period. Second, most data sets have some small anomaly that manifests itself as a spike on a first-two digits graph. A carefully executed analysis would first ask how fraud might occur, and how it might affect the digit patterns. Then the challenge would be to find exactly those signs of fraud.

This chapter reviewed several data sets that were judged to be authentic. These data sets included selected taxpayer data, census data, streamflow data, and accounts payable data. The fraudulent data included a fictitious vendor created by a state employee, some payroll fraud at a housing authority, and a review of published accounting numbers, including the numbers reported by Enron before its demise due to accounting fraud. The use of Benford's Law as a forensic investigation tool of data can detect anomalies that are not easily detectable by other means. Nigrini and Miller ([NiMi2], 2009) believe that deviations from the Benford proportions could be indicators of either (a) an incomplete data set, (b) the sample not being representative of the population, (c) excessive rounding of the data, (d) data errors, inconsistencies, or anomalies, or (e) conformity to a power law with a large exponent. Also, the procedure of cleansing the data and performing a Benford analysis will help an investigator to understand exactly what it is that they are dealing with.

Chapter Nine

Can Vote Counts' Digits and Benford's Law Diagnose Elections?

Walter R. Mebane, Jr.[1]

Abstract: The digits in vote counts can help diagnose both the strategies voters use in elections and nonstrategic special mobilizations affecting votes for some candidates. The digits can also sometimes help diagnose some kinds of election fraud. The claim that deviations in vote counts' second digits from the distribution implied by Benford's Law (2BL) is an indicator for election fraud, generally fails for precinct vote counts. I show that such tests routinely fail in data from elections in the United States, Germany, Canada and Mexico, countries where it is usually thought that there is negligible fraud. I illustrate how the conditional mean of precinct vote counts' second digits can respond to strategic behavior by voters in response to the presence of a coalition among political parties.

9.1 INTRODUCTION

The diagnostic use of the second digits of vote counts in connection with Benford's Law seems to have first been suggested by Pericchi and Torres [PerTo1], which met a skeptical response (see [Cart] for observations that the digits in vote counts do not follow Benford's Law). The theme was taken up by [Meb1, Meb6], and [Meb4] used the so-called second-digit Benford's Law (2BL) to diagnose likely fraud in Iran's 2009 election. [PerTo2] claim that Benford's Law applied to vote counts' second digits provides a sufficient standard for diagnosing election fraud, and here again there are skeptical voices [ShiMa, Lop, DeMyOr, Meb5]. Cantu and Saleigh [CanSe] find that Benford's Law approximately describes the first digits in some district-level election returns in some Argentine elections.

[Meb7] emphasizes how patterns in the **conditional mean of the second significant digits** of precinct vote counts—a statistic he denotes $\hat{j}_x$ or $\hat{j}_{xy}$—help diagnose the strategies voters are using in elections in the United States, Germany, Canada, Mexico and other places. The digits also respond to other kinds of mobilization that affect voters. It is best to think of precinct vote counts as following not Ben-

---

[1]Department of Political Science and Department of Statistics, University of Michigan, Ann Arbor, MI. The author thanks Jake Gatof, Joe Klaver, William Macmillan and Matthew Weiss for their assistance.

ford's Law but rather distributions in families of Benford-like distributions. Vote counts are mixtures of several distinct kinds of processes: some that determine the number of eligible voters in each precinct; some for how many eligible voters actually vote; some for which candidate each voter chooses; some for how the voter's choice is recorded. Such mixtures can produce numbers that follow Benford-like distributions but not Benford's Law [Rod, GrJuSc]. The following tests have been described as 2BL tests [Meb1], but it is more precise [Meb3] to use 2BL to refer to second-digit Benford-like tests.

Tests for the second digits of vote counts come in two forms. One uses a **Pearson chi-square statistic**: $X_{2BL}^2 = \sum_{j=0}^{9}(n_j - Nr_j)^2/(Nr_j)$, where N is the number of vote counts of 10 or greater (so there is a second digit), n_j is the number having second digit $j \in \{0, 1, \ldots, 9\}$ and $r_j = \sum_{k=1}^{9} \log_{10}(1 + (10k + j)^{-1})$ is given by the Benford's Law formula. For independent vote counts, this statistic should be compared to the chi-square distribution with nine degrees of freedom. To make this comparison [PerTo2] advocate using the significance probability $\alpha = (1 + [-ep \log(p)]^{-1})^{-1}$ [SelBaBe, 62, equation 3], where p is the p-value of X_{2BL}^2.[2] The second statistic, inspired by [GrJuSc], is the mean of the second digits, denoted $\hat{j}$. If the distribution of the counts' second-digits has frequencies r_j as given by Benford's Law, then the second digits' expectation is $\bar{j} = \sum_{j=0}^{9} jr_j = 4.187$.

No formal theory exists to support interpretations of the patterns in the second significant digits of precinct vote counts. [Meb7] finds that in several countries— with both plurality and mixed electoral systems—the second significant digits of precinct or polling station vote counts behave in regular ways that match the strategies voters are using. But mobilization by strategy is merely one type of mobilization that leaves regular traces in vote counts' digits. [Meb7]'s interpretations are partly supported by a simulation exercise [Meb3, Meb6].

9.2 2BL AND PRECINCT VOTE COUNTS

The claim by [PerTo2] that failure of vote counts' second digits to match the distribution implied by Benford's Law provides a sufficient standard for diagnosing election fraud is almost certainly false, at least when precinct or polling station vote counts are examined. Consider the following examples of 2BL test statistics computed using precinct (or polling station) vote counts from the United States, Germany, Canada and Mexico. The data show extensive deviations from what Benford's Law implies. [Meb7] argues at length that the deviations are caused by district imbalance, voters' strategies and other kinds of mobilization that affect vote counts in normal elections. In these cases fraud has little to do with it (although the Mexican case is complicated due to vote buying).

The hypothesis that precinct vote counts follow the 2BL distribution is rejected when the hypothesis is tested using data from American federal and state legislative

---

[2][PerTo2] use a modified statistic that adjusts for the maximum number of possible votes. An argument against their modification is that often the maximum is unknown or, as when there is no voter registration or when registration occurs on election day, the maximum is random and endogenous to voting decisions and consequently it is unreasonable to condition on it.

elections of 1984–1990.[3] For candidates affiliated with the Democratic and Republican parties, Table 9.1 reports χ^2_{2BL}, the corresponding significance probability α and $\hat{j}$. The hypothesis is rejected for all of the 28 test statistics shown in Table 9.1. In all but a few instances $\hat{j}$ differs significantly from $\bar{j}$.

Year	Office	Party	N	X^2_{2BL}	α	$\hat{j}$	$\hat{j}_{lo}$	$\hat{j}_{hi}$
1984	President	Democrat	152,286	135.40	.00	4.21	4.20	4.22
		Republican	152,373	148.34	.00	4.27	4.26	4.29
	U.S. Rep	Democrat	143,659	87.84	.00	4.22	4.21	4.24
		Republican	133,359	112.34	.00	4.24	4.23	4.26
	State House	Democrat	146,221	104.88	.00	4.22	4.20	4.23
		Republican	134,682	98.36	.00	4.23	4.21	4.24
	State Senate	Democrat	73,952	28.50	.02	4.19	4.17	4.21
		Republican	66,066	87.57	.00	4.27	4.25	4.29
1986	U.S. Rep	Democrat	142,660	117.90	.00	4.20	4.19	4.22
		Republican	134,650	101.73	.00	4.20	4.18	4.21
	State House	Democrat	151,116	112.56	.00	4.18	4.16	4.19
		Republican	139,161	68.54	.00	4.20	4.19	4.22
	State Senate	Democrat	82,621	91.37	.00	4.16	4.14	4.18
		Republican	79,993	29.48	.01	4.22	4.20	4.24
1988	President	Democrat	153,330	184.70	0	4.23	4.22	4.24
		Republican	153,353	79.44	0	4.23	4.22	4.25
	U.S. Rep	Democrat	140,013	90.22	0	4.23	4.21	4.24
		Republican	131,817	37.04	0	4.21	4.19	4.22
	State House	Democrat	137,145	68.99	0	4.21	4.20	4.23
		Republican	124,800	63.84	0	4.24	4.22	4.25
	State Senate	Democrat	74,800	73.13	0	4.23	4.21	4.25
		Republican	69,565	50.92	0	4.25	4.23	4.27
1990	U.S. Rep	Democrat	140,976	132.74	0	4.17	4.15	4.18
		Republican	136,928	119.33	0	4.15	4.13	4.16
	State House	Democrat	152,878	162.62	0	4.15	4.14	4.17
		Republican	140,680	95.72	0	4.17	4.16	4.19
	State Senate	Democrat	87,014	104.54	0	4.14	4.12	4.16
		Republican	81,878	53.34	0	4.16	4.14	4.18

Table 9.1 Second-Digit Tests, United States Federal and State Elections, 1984–1990. Note: Statistics for precinct vote counts. N denotes the number of precincts with ten or more votes for the candidate. $\alpha = (1 + [-ep\log(p)]^{-1})^{-1}$ where p is the p-value of X^2_{2BL}, $\hat{j}_{lo}$ and $\hat{j}_{hi}$ are the lower and upper bounds of the 95% confidence interval for $\hat{j}$. Data source: [Ki..].

The hypothesis that precinct votes counts are 2BL-distributed is also often rejected when the hypothesis is tested using data from American elections during the

---

[3]I have precinct data from presidential, U.S. House and state legislative elections in 1984, 1986, 1988 and 1990. The precinct data come from the Record of American Democracy (ROAD) [Ki..]. The data include every state except California.

2000s.[4] Table 9.2 reports χ^2_{2BL}, α and $\hat{j}$ for candidates affiliated with the Democratic and Republican parties. The hypothesis is rejected in six out of seven instances for Democrats but is never rejected for Republicans. We find $\hat{j}$ differs significantly from $\bar{j}$ in every instance for the Democrats but never for Republicans. [MebKe] and [Meb7] argue that differences between Tables 9.1 and 9.2 trace to differences across the decades in the patterns of electoral mobilization by the political parties.

Year	Office	Party	N	X^2_{2BL}	α	$\hat{j}$	$\hat{j}_{lo}$	$\hat{j}_{hi}$
2006	U.S. Rep	Democrat	121,516	29.71	.01	4.22	4.20	4.23
		Republican	109,183	4.35	.22	4.18	4.16	4.20
	State House	Democrat	99,689	41.25	.00	4.22	4.20	4.24
		Republican	95,963	14.23	.40	4.18	4.17	4.20
	State Senate	Democrat	61,419	21.20	.12	4.24	4.21	4.26
		Republican	54,138	15.18	.36	4.18	4.16	4.21
2008	President	Democrat	137,427	77.57	.00	4.25	4.23	4.27
		Republican	134,519	20.04	.16	4.20	4.19	4.22
	U.S. Rep	Democrat	135,878	84.75	.00	4.25	4.24	4.27
		Republican	126,228	5.05	.30	4.19	4.17	4.20
	State House	Democrat	120,226	74.69	.00	4.25	4.24	4.27
		Republican	111,637	7.81	.47	4.17	4.16	4.19
	State Senate	Democrat	65,023	77.68	.00	4.28	4.26	4.30
		Republican	61,385	11.83	.48	4.21	4.19	4.23

Table 9.2   Second-Digit Tests, United States Federal and State Elections, 2006–2008. Note: Statistics for precinct vote counts. N denotes the number of precincts with ten or more votes for the candidate. $\alpha = (1 + [-ep\log(p)]^{-1})^{-1}$ where p is the p-value of X^2_{2BL}, $\hat{j}_{lo}$ and $\hat{j}_{hi}$ are the lower and upper bounds of the 95% confidence interval for $\hat{j}$. Data source: 33 states in 2006 and 41 states in 2008 and 29 states in 2010; collected by the author.

In German Bundestag elections each voter casts two votes. *Erststimmen* (first votes) determine the winner of each *Wahlkreis* (district) through a plurality voting rule, and *Zweitstimmen* (second votes) determine the overall share of the seats each party has in the Bundestag through proportional representation (PR) rules.[5]

The hypothesis that polling station vote counts are 2BL-distributed is usually rejected when the hypothesis is tested using data from the German Bundestag elec-

---

[4]For several states I have precinct vote count data (collected by the author) for the presidential and U.S. House elections of 2006 and 2008, as well as precinct data for state legislative elections. Data come from 33 states in 2006 and 41 states in 2008. The states with data in 2006 are AL, AK, AZ, AR, CA, DE, FL, GA, HI, ID, IA, KS, LA, ME, MD, MI, MN, MS, NE, NH, NY, NC, ND, OH, PA, RI, SC, TN, TX, VT, VA, WI and WY. The states with data in 2008 are AL, AK, AZ, AR, CA, CT, DC, DE, FL, GA, HI, ID, IL, IN, IA, KS, LA, ME, MD, MI, MN, MS, NH, NM, NY, NC, ND, OH, OK, PA, RI, SC, SD, TN, TX, VT, VA, WA, WV, WI and WY. Data are not available for every precinct in some states.

[5]To receive seats through the PR process, a party must receive more than 5 percent of the valid *Zweitstimmen* or win three *Wahlkreise* based on *Erststimmen* [Bun3, Section 6].

tions of 2002, 2005 and 2009 (see also [ShiMa]).[6] The χ^2_{2BL}, α and $\hat{\jmath}$ statistics do not vary substantially over the three years, so Table 9.3 reports statistics for the SPD, CDU/CSU, PDS/Linke and Green parties pooled over years.[7] The hypothesis fails to be rejected only for the Green *Erststimmen*. We have $\hat{\jmath}$ differing significantly from $\bar{\jmath}$ in all but one case for *Erststimmen* but for only two in five instances for *Zweitstimmen*. [Meb6, Meb7] argues that these deviations from 2BL can be explained by the effects on the digits of district imbalance, rolloff and strategic voting with the latter involving a combination of "wasted vote" reasoning and "threshold insurance" calculations [Gsch, HePa, ShiHeTh].

Type	Party	N	X^2_{2BL}	α	$\hat{\jmath}$	$\hat{\jmath}_{lo}$	$\hat{\jmath}_{hi}$
Erststimmen	SPD	264,929	158.45	0	4.24	4.23	4.25
	CDU/CSU	266,731	337.43	0	4.27	4.26	4.29
	FDP	234,416	217.94	0	4.27	4.26	4.28
	PDS/Linke	182,193	158.27	0	4.11	4.09	4.12
	Green	216,109	4.71	.26	4.19	4.18	4.20
Zweitstimmen	SPD	264,529	65.07	0	4.18	4.17	4.19
	CDU/CSU	266,627	175.01	0	4.21	4.20	4.22
	FDP	250,433	69.78	0	4.18	4.17	4.19
	PDS/Linke	190,590	129.52	0	4.11	4.10	4.13
	Green	233,480	35.54	0	4.17	4.16	4.19

Table 9.3   Second-Digit Tests, German Federal Elections, 2002–2009. Note: Statistics for polling station vote counts. N denotes the number of polling stations with ten or more votes for the candidate. $\alpha = (1 + [-ep \log(p)]^{-1})^{-1}$ where p is the p-value of X^2_{2BL}, $\hat{\jmath}_{lo}$ and $\hat{\jmath}_{hi}$ are the lower and upper bounds of the 95% confidence interval for $\hat{\jmath}$. Data source: [Bun2, Bun1, Bun4].

The hypothesis that the polling station counts are 2BL-distributed is rejected for most parties most of the time when the hypothesis is tested using data from the Canadian federal elections of 2004–2011.[8] Of the 20 test instances shown in Table 9.4, the hypothesis is not rejected in only three instances: for NDP in 2004 and 2006, and for the Liberal party in 2011. $\hat{\jmath}$ usually differs significantly from $\bar{\jmath}$. [Meb7] argues that this pattern traces to the fact that Canadian voters usually act strategically—somewhat in accord with instrumental rationality [BlNa, BlNaGN, ChKo] but without any nationally oriented coalition awareness [BlGs]. The varying results for the Liberal party and for NDP reflect the former's decline and the latter's rise in 2011 to Official Opposition status [LeDuc1, LeDuc2, LeDuc3, LeDuc4].

Federal elections in Mexico since 1994 have been closely contested with both volatility in outcomes and frequent charges that election fraud was widespread, so it is controversial whether there is significant fraud in any recent elections. Fraud

---

[6]Data come from [Bun2, Bun1, Bun4].

[7]Here "Green" refers to *Bündnis 90/Die Grünen*.

[8]Data are from [Ele1, Ele2, Ele3, Ele4].

Year	Party	N	X^2_{2BL}	α	$\hat{\jmath}$	$\hat{\jmath}_{lo}$	$\hat{\jmath}_{hi}$
2004	Liberal	59,165	163.4	.00	4.06	4.03	4.08
	Conservative	55,105	222.6	.00	4.03	4.00	4.05
	NDP	48,383	7.3	.45	4.18	4.15	4.20
	Bloc Québécois	14,554	247.1	.00	3.83	3.78	3.88
	Green Party	27,864	1700.5	.00	3.50	3.46	3.53
2006	Liberal	59,200	149.5	.00	4.06	4.03	4.08
	Conservative	60,834	176.1	.00	4.06	4.03	4.08
	NDP	55,635	12.2	.47	4.18	4.15	4.20
	Bloc Québécois	15,084	241.1	.00	3.84	3.79	3.88
	Green Party	27,864	1700.5	.00	3.48	3.45	3.51
2008	Liberal	61,541	96.4	.00	4.10	4.07	4.12
	Conservative	64,267	181.6	.00	4.05	4.03	4.07
	NDP	59,944	38.1	.00	4.25	4.23	4.28
	Bloc Québécois	15,719	225.6	.00	3.85	3.81	3.90
	Green Party	37,736	655.3	.00	3.81	3.78	3.84
2011	Liberal	57,377	18.97	.20	4.14	4.12	4.17
	Conservative	66,307	404.73	.00	3.96	3.94	3.99
	NDP	66,791	60.13	.00	4.13	4.11	4.15
	Bloc Québécois	15,717	33.14	.00	4.31	4.26	4.35
	Green Party	19,081	2248.08	.00	3.21	3.17	3.25

Table 9.4   Second-Digit Tests, Canadian Federal Elections, 2004–2011. Notes: Statistics for polling station vote counts. N is the number of polling stations with a vote count > 9. $\alpha = (1 + [-ep\log(p)]^{-1})^{-1}$ where p is the p-value of X^2_{2BL}, $\hat{\jmath}_{lo}$ and $\hat{\jmath}_{hi}$ are the lower and upper bounds of the 95% confidence interval for $\hat{\jmath}$. Data source: [Ele1, Ele2, Ele3, Ele4].

occurred in the presidential election of 1988 ([Cast, 80–87, 199]; [Mag, 5]). Allegations of fraud and postelection protests followed the elections especially of 1994 [McCDo], 2006 [Kles, Lop] and 2012 [Sala1, Sala2, Sande], although in these cases it is less clear whether substantial fraud actually occurred.

While it may be less a matter of consensus that Mexican elections are largely free of fraud than the elections we have examined from the United States, Germany or Canada, test results regarding the hypothesis that Mexican polling station counts follow the 2BL distribution are comparable to the results from those countries. Using for example data from the Mexican federal elections for *Presidente* and for *Diputados Federales* of 2006 and 2012, the hypothesis is rejected for most parties most of the time.[9] Following the point made by [Meb1] that the *casilla* (ballot box) is too low a level of aggregation for 2BL tests to give meaningful results, I consider each of these counts aggregated to the *sección*, a small administrative unit usually containing several *casillas*. Of the 20 test statistics shown in Table 9.5, the

---

[9]Data are from [Ins1, Ins5]. Results from 1994 and 2000 are similar.

hypothesis is not rejected in only seven instances.[10] In 2012 these non-rejections include the parties or coalitions that finished in second (MP) and in third (PAN) place in the presidential election. For the winning party the 2BL hypothesis is always rejected, and $\hat{\jmath}$ usually differs significantly from $\bar{\jmath}$.

Year	Office	Party	N	X^2_{2BL}	α	$\hat{\jmath}$	$\hat{\jmath}_{lo}$	$\hat{\jmath}_{hi}$
2006	*Presidente*	PAN	62,490	48.11	.00	4.25	4.23	4.28
		APM	63,915	108.96	.00	4.08	4.05	4.10
		PBT	63,143	25.30	.04	4.24	4.21	4.26
		NA	12,303	1254.78	.00	3.29	3.24	3.34
		ASDC	35,364	16.64	.30	4.15	4.12	4.18
	Diputados	PAN	62,621	34.56	.00	4.23	4.21	4.26
		APM	64,424	36.22	.00	4.14	4.12	4.17
		PBT	62,718	11.35	.49	4.19	4.17	4.21
		NA	43,295	17.57	.26	4.14	4.12	4.17
		ASDC	27,229	234.43	.00	3.92	3.89	3.96
2012	*Presidente*	PAN	65,114	19.36	.19	4.17	4.15	4.19
		CM	66,658	99.24	.00	4.29	4.27	4.32
		MP	64,869	11.34	.49	4.22	4.19	4.24
		NA	38,244	223.56	.00	3.98	3.95	4.00
	Diputados	PAN	64,503	6.46	.41	4.18	4.16	4.20
		PRI	27,361	8.09	.48	4.17	4.13	4.20
		PVEM	15,855	119.23	.00	3.95	3.90	3.99
		NA	47,431	31.31	.01	4.13	4.10	4.15
		CM	39,001	149.04	.00	4.36	4.33	4.39
		MP	63,963	7.57	.46	4.21	4.18	4.23

Table 9.5   Second-Digit Tests, Mexican Federal Elections, 2006 and 2012. Notes: Statistics for *sección* vote counts. N is the number of *secciones* with a vote count > 9. $\alpha = (1 + [-ep\log(p)]^{-1})^{-1}$ where p is the p-value of X^2_{2BL}, $\hat{\jmath}_{lo}$ and $\hat{\jmath}_{hi}$ are the lower and upper bounds of the 95% confidence interval for $\hat{\jmath}$. Data source: [Ins1, Ins5].

9.3 AN EXAMPLE OF STRATEGIC BEHAVIOR BY VOTERS

The frequent rejections of the 2BL hypothesis do not imply that election fraud is present in all these elections. Instead [Meb7] argues that the deviations from 2BL

---

[10]The parties and coalitions shown in Table 9.5 are as follows: APM, coalición Alianza por México (PRI, PVEM); ASDC, Partido Alternativa Social Democrática y Campesina; CM, coalición Compromiso por México (PRI, PVEM); MP, coalición Movimiento Progresista (PRD, PT, Movimiento Ciudadano); NA, Partido Nueva Alianza; PAN, Partido Acción Nacional; PBT, coalición Por el Bien de Todos (PRD, PT, Convergencia); PCD, Partido Centro Democrático; PRD, Partido de la Revolución Democrática; PRI, Partido Revolucionario Institucional; PT, Partido del Trabajo; PVEM, Partido Verde Ecologista de México.

are caused by district imbalance, voters' strategies and other kinds of mobilization that affect vote counts in normal politics.

To illustrate how normal political activity can affect the distribution of digits in precinct vote counts and how statistics based on the digits can help diagnose the political behavior, I present one case drawn from Mexican data. The statistic of interest is the **conditional digit mean**, $\hat{j}_x$, which is estimated using non-parametric regression [BowAz][11]; $\hat{j}_x$ shows how the mean of the second digit of the vote counts for the candidates affiliated with a party or coalition varies with covariates defined by the margins between the candidates in each legislative district. In districts where a party won, the margin is the difference between the proportion of votes received by the winner and the third-place candidate (denoted $\mathfrak{M}_{13}$), and in districts where a party finished in second place the margin is the difference between the proportion of votes received by the second-place candidate and the third-place candidate (denoted $\mathfrak{M}_{23}$). [Meb6, Meb7] motivates these choices for covariates in terms of basic "wasted vote" instrumental behavior: the key quantity given such strategic behavior is the difference between each of the top two parties and the party that comes in third [Cox1, Cox2].

Figures 9.1(a,b) are examples of estimating $\hat{j}_x$ using *Diputados Federales* election *sección* data in 2012. The figures shows results for districts in which CM and MP finished in first and second place. Figure 9.1(a) displays $\hat{j}_x$ (conditional district means) based on the second digits of *sección* votes for CM. $\hat{j}_x$ for districts where CM won and MP was second is plotted for positive values of the x-axis, and for these estimates the covariate is $\mathfrak{M}_{13}$. $\hat{j}_x$ for districts where CM was second and MP won is plotted for negative values of the x-axis, and for these estimates the covariate is $-\mathfrak{M}_{23}$. $\hat{j}_x$ is shown surrounded by 95 percent confidence bounds, and $\bar{j}$ is indicated by a horizontal dotted line. Figure 9.1(b) displays the analogously estimated $\hat{j}_x$ based on votes for MP.

Elections for *Diputados Federales* in 2012 are especially interesting because of the complex mix of coalitions that competed. In particular the PRI and PVEM parties formed the partial coalition CM: CM had candidates in 199 districts, but PRI and PVEM had separate candidates in the other 101 districts ([Ins4, 4–11]; [Ins3, 23–30]). Several other parties (PRD, PT and Movimiento Ciudadano) formed the coalition MP, which unified on all candidates [Ins2]. Estimating $\hat{j}_x$ suggests that voters' strategies are different when the coalitions are present than when they are absent. When CM together runs a candidate there is evidence of more strategic vote switching than when PRI and PVEM sponsor candidates separately. When CM together runs a candidate there seems to be more strategic vote switching not only for the CM candidate but also for candidates supported by opposing parties and coalitions.

The sensitivity of strategic behavior to the presence of a coalition as opposed to merely the parties comprising the coalition is apparent in races in which MP and CM or PRI were in the lead. In districts where CM and MP finished in first and second place, both coalitions have $\hat{j}_x$ values that are never significantly less than 4.35 (Figures 9.1(a,b)). Based in part on simulations, [Meb6, Meb7] identi-

---

[11]Nonparametric regressions are computed using the sm package of **R** [R].

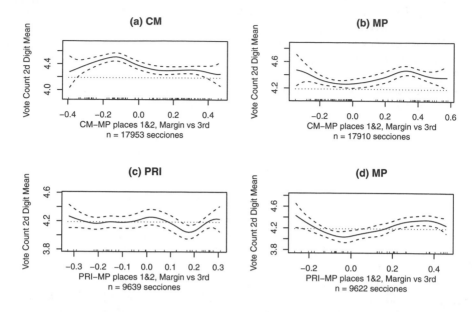

Figure 9.1   Mexico 2012: *Diputados, Sección* Count Second-Digit Mean (Districts) by
$-\mathfrak{M}_{23}$ and $\mathfrak{M}_{13}$. Note: Nonparametric regression of *Mayoría Relativa* vote
counts' second digits based on *sección* data. Rug plots show locations of *sección*
values of $-\mathfrak{M}_{23}$ and $\mathfrak{M}_{13}$.

fies $\hat{j} \approx 4.35$ as a key indicator of strategic vote switching behavior. In districts
where MP won and CM was second, $\hat{j}_x$ is slightly greater than 4.35 for MP when
$.27 < \mathfrak{M}_{13} < .42$ and for CM when $-.25 < -\mathfrak{M}_{23} < -.04$. These values are
evidence of strategic vote switching adding to the vote totals of both winners and
second-place finishers in all the districts where the two coalitions led. In contrast
consider the districts in which PRI and MP finished in first and second (Figures
9.1(c,d)). When PRI is sponsoring candidates not as part of the CM coalition, $\hat{j}_x$
for PRI is never significantly greater than $\bar{j}$ and indeed is somewhat less than $\bar{j}$ in
some of the districts where PRI won.[12] Note $\hat{j}_x$ for MP is frequently less than $\bar{j}$ in
districts where MP was second behind a PRI candidate, a condition that never occurred when an MP candidate finished second behind a CM candidate.[13] In districts
where the MP candidate defeated the second-place PRI candidates, $\hat{j}_x$ for MP rises
above 4.35 only when $.21 < \mathfrak{M}_{13}$. Indeed $\hat{j}_x$ for MP in the MP-winning districts
resembles $\hat{j}_x$ for the advantaged candidate in a simulation [Meb6, Meb7] with no
strategic vote switching and no turnout decline. In any case, evidence based on $\hat{j}_x$
suggesting there is strategic vote switching in favor of the MP candidate when such
a candidate is running against a strong PRI candidate is much less than the evidence
when an MP candidate is running against a strong CM candidate.

---

[12] $\hat{j}_x$ for PRI in Figure 9.1(c) is significantly less than $\bar{j}$ when $.15 < \mathfrak{M}_{13} < .22$.
[13] $\hat{j}_x$ for MP in Figure 9.1(d) is significantly less than $\bar{j}$ when $-.11 < -\mathfrak{M}_{23}$.

The sensitivity of strategic behavior to the presence of coalitions is apparent as well in races in which PAN is one of the leading parties. Looking at districts where PAN and CM or PRI led shows clearly how strategic behavior varies with the presence of a coalition candidate. In districts where PAN and CM were first and second (Figures 9.2(c,d)), $\hat{j}_x$ for CM is never meaningfully different from 4.35,[14] and $\hat{j}_x$ for PAN is never significantly different from 4.35 for winning PRI candidates. Evidence of strategically switched vote gains is strong in these cases. When PAN finished second behind CM, $\hat{j}_x$ for PAN does not differ significantly from $\bar{j}$ when $-.11 < -\mathfrak{M}_{23}$ and is significantly less than 4.35 when $-.2 < -\mathfrak{M}_{23}$. The mean of the point estimates of $\hat{j}_x$ when $-.2 < -\mathfrak{M}_{23}$ is $\hat{j}_x \approx 4.25$. The estimate of $\hat{j}_x$ is bounded significantly below the value of $\hat{j}_x \approx 4.35$: probably strategic vote switching can be ruled out for this party in these districts. As $-\mathfrak{M}_{23}$ decreases below $-\mathfrak{M}_{23} = -.2$, $\hat{j}_x$ increases to the point that when $-\mathfrak{M}_{23} < -.34$, $\hat{j}_x$ becomes significantly greater than 4.35. Perhaps these values point to some but not all PAN candidates receiving strategically switched votes when finishing behind a CM candidate.

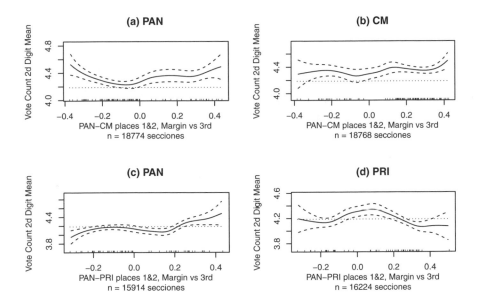

Figure 9.2   Mexico 2012: *Diputados*, *Sección* Count Second-Digit Mean (Districts) by $-\mathfrak{M}_{23}$ and $\mathfrak{M}_{13}$. Note: Nonparametric regression of *Mayoría Relativa* vote counts' second digits based on *secciòn* data. Rug plots show locations of *secciòn* values of $-\mathfrak{M}_{23}$ and $\mathfrak{M}_{13}$.

---

[14]$\hat{j}_x$ in Figure 9.2(d) is significantly greater than 4.35 when $.4 < \mathfrak{M}_{13}$: the lower bound of the confidence interval for $\hat{j}_x$ at $\mathfrak{M}_{13} = .42$ is 4.36 and the point estimate for $\hat{j}_x$ is $\hat{j}_x \approx 4.5$.

9.4 DISCUSSION

Are statistics based on the second significant digits of precinct vote counts meaningful? [Meb7] argues that not only can digit tests help diagnose strategic voting [Meb6] but also they are sensitive to other aspects of normal politics such as kinds of mobilization that go well beyond the scope of strategic voting. The claim by [PerTo2] that violation of Benford's Law provides a sufficient standard for diagnosing election fraud is almost certainly false, but Benford's Law is not irrelevant for the forensic examination of elections. While precinct vote counts often do not match the distribution implied by 2BL, often they do. The simulation that [Meb7] relies on takes an election with vote counts that satisfy 2BL as a point of departure and shows that familiar kinds of political manipulation produce regular patterns of departure from 2BL. [Meb7] tries to show that the simulated patterns very often match the patterns found in real data from elections in several countries.

When considered against the background of more complicated patterns that occur in various electoral settings, tests based on the second significant digits of precinct vote counts may be useful for detecting election fraud. [Meb4] uses such tests to diagnose likely fraud in Iran's 2009 election. [Meb7] compares the Iran 2009 findings to similar statistics for federal elections in Mexico. The Mexican results strongly suggest effects of vote buying, which some (e.g. [Sande]) consider election fraud. The analysis becomes intricate, involving covariates such as the mayoral party affiliation. As [DeMyOr] and [Meb5] argue, it is unlikely that tests based on precinct vote counts' second digits will support simple rules of thumb to diagnose election fraud. Even in Russian elections where tests based on the last digits of turnout figures diagnose fraud [MyOrSh, MebKa1, KaMeb], the second digits of polling station vote counts provide plausible strategic diagnostics along with some hints of fraud [MebKa2].

The fact that digit tests are sensitive to many normal aspects of politics may be good for general political science interests, but it at least complicates the potential for using the tests to diagnose election fraud. The question of whether the patterns in digits produced by fraud differ sharply from the patterns produced by normal politics is not an easy one to answer. [Meb7] gives some cases where likely fraud produces very distinctive patterns (such as the Iranian election of 2009 [Meb4], and other elections), but also cases where natural political shocks produce patterns that would otherwise be interpreted as political coercion. Except in exceptional, flagrant cases, there is no reason to think that forensically diagnosing elections should be any simpler than forensic examinations are in the face of a sophisticated adversary in any other realm.

Chapter Ten

Complementing Benford's Law for Small N: A Local Bootstrap

Boudewijn F. Roukema[1]

In analyzing the 2009 Iranian presidential election initially published results, this chapter echoes a theme common throughout this book. Apparent anomalies revealed by the initial application of Benford's Law (BL) need to be followed up by more detailed analyses. The local bootstrap model described below provides a non-parametric method of simulating electoral data—using the data itself. The minimalist assumption is made that the proportions of people voting in electoral regions of a given population size can be simulated by randomly drawing (bootstrap resampling) the voting rates per candidate from the claimed true data for those regions. This can be thought of as giving a very small nudge to the data, without introducing any external assumptions of what statistical distributions the voting follows. If an apparent BL signal disappears when the data are nudged just slightly, and if no similar signal reappears by chance among 10,000 statistically independent nudges (local bootstrap simulations on a computer), then the signal is statistically significant.

Not only was the dramatic BL spike for one of the four presidential candidates found to be highly significant, but, as in the case studies by Nigrini in Chapter 8, another method of follow-up was also used. The spike was used as a tracer to select a suspect subset of the full data set. This subset has several statistically unusual properties, rendering the null hypothesis of an unaltered data set extremely unlikely. A similar local bootstrap plus BL approach is most likely applicable to other settings.

10.1 THE 2009 IRANIAN PRESIDENTIAL ELECTION

The 2009 **Iranian presidential election** first round, held on 12 June 2009, was a locally and internationally important geopolitical event. Since it took place in a country with a high rate of internet usage, it was unsurprising that the ministry responsible for organizing the election published initial results on the world wide web. Fortunately for those interested in the first-digit Benford's Law (see Definition

[1]Toruń Centre for Astronomy, Faculty of Physics, Astronomy and Informatics, Nicolaus Copernicus University, ul. Gagarina 11, 87-100 Toruń, Poland.

1.4.1), these initial data were published for 366 electoral regions (**shahrestans**) varying by well over an order of magnitude in electoral size. A quick analysis [Rou, v1] was circulated four days later, showing what appeared to be an obvious first-digit anomaly for K, one of the four candidates. Benford's Law using non-primary digits was also applied to the data [Meb2, BebSca, BerRin].

The three conditions suggested by Nigrini (see Chapter 8) for practical applications of Benford's Law (to detect **vote fraud**) are mostly satisfied for this data set, with the exception that the total voting population bounds the full set of vote counts from above. However, some skepticism about the application of BL to this case was expressed.

It is useful to apply a small N statistical first-digit frequency test that is as non-parametric as possible in a way that leaves no doubt regarding "how close" the observed system should be to a Benford's Law limit. One approach is a **local bootstrap model**, designed to closely mimic the data in a way that should statistically reproduce its first-digit distributions, given some simple hypotheses about the general behavior of the system. This method was calibrated on several presidential-election first rounds from before 2009 and applied to the 2009 Iranian election, confirming the improbability of candidate K's first-digit anomaly to high significance [Rou]. The computing power required for the test is higher than for a direct application of Benford's Law, but practical on a personal computer.

10.2 APPLICABILITY OF BENFORD'S LAW AND THE K7 ANOMALY

First let us consider the election data. For a given election, let us write v_{ij} for the vote count for jth candidate in the ith electoral region. The latter has $x_i \geq \sum_j v_{ij}$ total votes (inequality occurs due to invalid votes). Let the voting rate for a given candidate in a given electoral region be

$$w_{ij} := v_{ij}/x_i. \tag{10.1}$$

For BL to be applicable, the standard deviation of the intrinsic distribution of potential voters per voting area for a given election, estimated using the actual (official) total votes, $\sigma(\log_{10} x_i)$, and the standard deviation in the voting rates for a candidate j, $\sigma(\log_{10} w_{ij})$, should both be high in order that the resulting v_{ij} first-digit distribution approaches BL.

Figure 10.1 shows these parameters for the Iranian 2009 and five prior presidential-election first rounds (the data are public; see [Rou, Table 1 and Section 2.3]). The Iranian 2005 and 2009 elections have the broadest distributions in both parameters. Thus, these are the elections for which BL is most likely to apply. Moreover, of the four 2009 candidates, K has the highest voting rate spread $\sigma(\log_{10} w_{ij}) = 0.42$, and should be closest to the BL limit. How uniform is the **folded logarithmic distribution** of the total votes? Figure 10.2 shows this distribution. It is a little noisy. Multiplying x_i by w_{ij} should smooth this out and reduce the chance of deviation from the BL limit. Yet, Figure 10.3 shows what appears to be a dramatic deviation from BL, considering just the thick line (BL model) for the moment.

Figure 10.4 shows the folded logarithmic distribution of K's vote counts. As

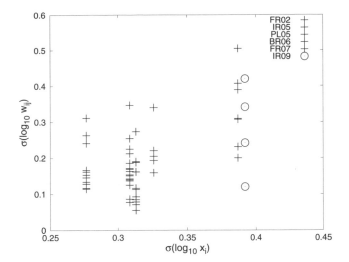

Figure 10.1 Logarithmic spread of voting rates $\sigma(\log_{10} w_{ij})$ as a function of the logarithmic spread of the underlying voting populations $\sigma(\log_{10} x_i)$ for elections in (left to right) Poland 2005, France 2002, France 2007, Brazil 2006, Iran 2005, and Iran 2009. For the four candidates in the latter, $\sigma(\log_{10} w_{ij}) = 0.12, 0.34, 0.42,$ and 0.24 for candidates A, R, K, and M, respectively.

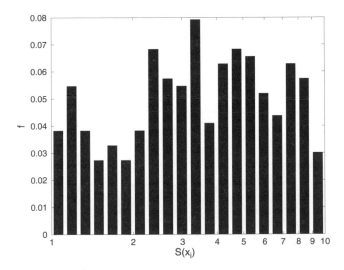

Figure 10.2 Folded logarithmic distribution of the total vote count in the 2009 Iranian presidential-election first round, shown as a frequency histogram over the significand $S(x_i) = 10^{\log_{10} x_i - \lfloor \log_{10} x_i \rfloor}$. This implicitly shows what the first digit distribution would be for a candidate getting 100% of the votes in every shahrestan.

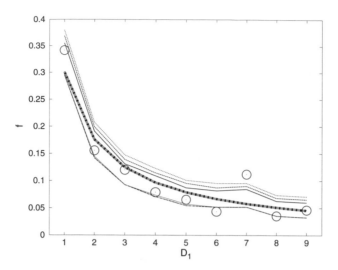

Figure 10.3 First-digit frequency distribution $f(D_1)$ for candidate K ($\odot$) in the Iranian 2009 election. The Benford's Law limit is shown as a thick line (passing by chance through the $f(9)$ data point). The excess in $f(7)$ is obvious. A local bootstrap model (Section 10.3), calibrated on earlier, similar elections, gives a conservative estimate of confidence levels. Lower and upper confidence levels of $c_\theta = 0.05\%, 0.5\%, 2.5\%, 97.5\%, 99.5\%, 99.95\%$, are shown. The upper three confidence levels can be distinguished; the lower three are almost indistinguishable in this plot.

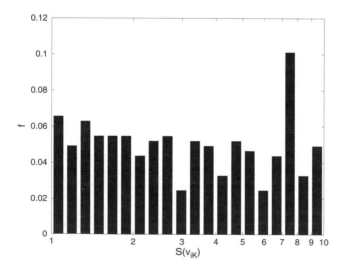

Figure 10.4 Folded logarithmic distribution of votes for candidate K, as for Figure 10.2, over $10^{\log_{10} v_{iK} - \lfloor \log_{10} v_{iK} \rfloor}$.

expected, this is generally flatter and smoother than the total vote count distribution shown in Figure 10.2, with the exception of the spike which approximately corresponds to the significand $7 \leq S < 8$. The "K7" spike appears to be highly discrepant. Is this a "smoking gun," revealing artificial interference in the data?

10.3 A CONSERVATIVE ALTERNATIVE TO BENFORD'S LAW: A SMALL N, EMPIRICAL, LOCAL BOOTSTRAP MODEL

The **local bootstrap model**, defined in [Rou, Section 2.2] and summarized here, is conservative; i.e., it is of high specificity (the chance of falsely rejecting the hypothesis that the data are fully authentic is low), but weak power (the method is likely to fail to detect some artifacts in the data). The aim is to use a bootstrap of the data itself to simulate the voting rates for each shahrestan given its approximate total vote size. This allows the modeling of the possibility that a candidate tends to be more popular in big cities, for example, without needing to make any specific hypothesis on the type of distribution (e.g. normal, log-normal, unimodal, or bimodal). Using the above notation, let us define a realization of the model ([Rou, Definition 2.1]).

Definition 10.3.1 (Local bootstrap realization). *Sort the set of* x_i *such that the sequence* $(x_{i_1}, x_{i_2}, \ldots, x_{i_m})$ *is in ascending order. A local bootstrap realization for candidate* j *is a set of simulated votes*

$$\{w_{i_{k'}j} \, x_{i_k}\}_{k=1,\ldots,m} \tag{10.2}$$

where $\{k'(k)\}$ *are drawn from a random realization* G *of a Gaussian distribution of width* $m\Delta$, *where* $\Delta := \log_{10}(10/9) \approx 0.0458$ *and centered at* k, *truncated at the limits of* $\{k\}$, *i.e.,*

$$k'(k) := \max(1, \min(m, \lfloor G(k, m\Delta) + 0.5 \rfloor)). \tag{10.3}$$

The value of Δ and the Gaussian smoothing are designed in order that the data are randomly reselected (allowing repeats) within approximately the logarithmic width of the narrowest first-digit interval, i.e., from 9 to 10. Varying Δ with respect to the value defined here helps to understand it. In the limit $\Delta \to 0$, the realization approaches an exact copy of the data itself: the real data will be considered to be highly probable, no matter what its real nature is. For $\Delta \gg \log_{10}(10/9)$, the realization would tend towards the assumption that the voting rate is independent of the size of the total voting population, and possibly imply a rejection of the real data because of an oversimplified model. The word "local" refers to Δ being small.

An example of a **local bootstrap realization** is shown in Fig. 10.5. Apart from the general similarity of the distributions evident in the figure, one conservative property of this approach that can be seen is that the bootstrap realization has a similar number of outliers (in this case, extremely low voting rate tails) to the original distribution.

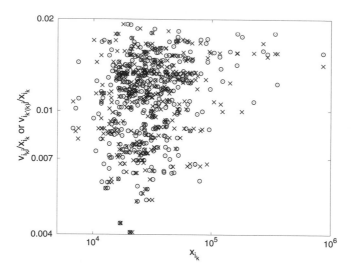

Figure 10.5 Example local bootstrap realization for candidate 1 in the Polish 2005 presidential-election first round. Real voting rates w_{ij} (⊙) and bootstrapped voting rates $w_{i_{k'}j}$ (×) are shown. The latter statistically imitate the former.

Since **bootstrap methods** tend to overestimate the variance in a data set and require bias and skewness corrections, the method of Definition 10.3.1 can be calibrated using the earlier five elections [Rou, Table 1]. Let us assume that the earlier election results were all fully valid. In this case, we have both empirical distributions of first-digit frequencies, $c_{ej}(d_1)$ for digit d_1 for votes $\{v_{ij}\}_{i=1,...,m}$ for candidate j in a given election, and bootstrap distributions $c_{bj}(d_1)$ for simulated votes $\{w_{i_{k'}j}\}_{i=1,...,m}$. An almost-everywhere-smooth, continuous, piecewise fit to the five elections' real and local bootstrap data gave the correction from bootstrap to "empirical" confidence levels [Rou, Eqs (8), (9)]:

$$c_e(c_b) = \begin{cases} 0.5 \left(\frac{c_b}{c_b^*}\right)^{\beta^-} & \text{if } c_b < c_b^*, \\ 1 - 0.5 \left(\frac{1-c_b}{1-c_b^*}\right)^{\beta^+} & \text{if } c_b \geq c_b^*, \end{cases} \quad (10.4)$$

where

$$\beta^- = 1.824, \ \beta^+ = 1.487, \ c_b^* = 0.566. \quad (10.5)$$

Since $\beta^- > 1$ and $\beta^+ > 1$, this confirms that direct use of ensembles of **local bootstrap realization**s leads to conservative confidence levels near the extremes: first-digit frequencies are rejected less strongly than they should be, given the calibration on the previous elections.

The direct bootstrap confidence level (one-sided) for rejecting the K7 spike is $c_{bK7} > 99.924\%$. Applying Eqs (10.4) and (10.5), the calibrated confidence level is $c_{eK7} > 99.9960\%$, i.e., $1 - c_{eK7} \approx 4 \times 10^{-5}$. So the **local bootstrap method** provides a conservative complement to applying BL directly. Of course, as with any statistical analysis, meta-questions regarding the number of statistical tests applied need to be considered. A Šidàk–Bonferonni correction factor [Abd] of 36, for considering all of the digits for the four candidates, increases the latter probability to $p < 1.5 \times 10^{-3}$, which is still low.

We can now return to Figure 10.3, which shows the confidence levels c_e in addition to the BL limit. By the bootstrap nature of the model, the confidence levels show a positive bump for $D_1 = 7$. For $\Delta \to 0$, the model would approach the official data exactly, making it impossible to reject the data in comparison to the model. For $\Delta := \log_{10}(10/9)$ as defined in Definition 10.3.1, i.e., just nudging the official data very slightly, a Gaussian spread in voting rates w_{ij} with a width of the narrowest interval (for $D_1 = 9$) shifts the model $f(7)$ distribution down towards the BL model. Sociologically, this represents the assumption that among the shahrestans of approximately a given voting population x, the voting rates w_{ij} for K among those shahrestans are drawn from a random probability density function identical to that in the official results for shahrestans with $x_i \approx x$, where "$\approx$" is defined by the Gaussian distribution of width $m\Delta$ in the ordered sequence $(x_{i_1}, x_{i_2}, \ldots, x_{i_m})$.

For example, typically (apart from truncation at the lower and upper limits), an official voting rate of 0.95% for K is resampled with a 68% chance of lying among the voting rates for K in the 17 shahrestans with slightly lower populations and 17 shahrestans with slightly higher populations. The shahrestans of those sizes might have some peculiar characteristics (e.g. a sharply bimodal distribution because of cities in regions of different ethnic mixes), and these are (statistically) reflected in the model. Intuitively, this could be thought of as giving a very gentle "nudge" to the official data set. Figure 10.3 shows that for this data set, the official data are fragile, and become extremely unlikely when nudged just slightly.

10.4 USING A SUSPECTED ANOMALY TO SELECT SUBSETS OF THE DATA

As recommended by Nigrini for accounting data (Section 8.3.4), the data corresponding to anomalous spikes compared to BL can be usefully investigated further, possibly revealing innocent explanations that help understand the data rather than leading to further statistical anomalies. In the 2009 Iranian election, the K7 spike can be considered as a **tracer**. If it is just a 1 in 670 fluke, then the shahrestans identified by the K7 spike should not be expected to also have unusual statistical properties.

Figure 10.6 shows that among the most populous shahrestans, the K7 spike does indeed imply another anomaly: of the six most populous shahrestans, the K7 spike selects those three that voted the most for candidate A. This is clearly unusual (e.g. $p < 5 \times 10^{-4}$; see [Rou, Section 4.4.1]) and correlates with the most sociologically likely possibility of interference, that someone from the Ministry altered the data in support of the incumbent, candidate A. Table 10.1 gives more details, revealing an unusual **second-digit distribution**: the big shahrestan $d_1 = 7$ vote counts have $d_2 = 0$ in every case. Nigrini (Section 8.4.3) argues in favor of a psychological bias

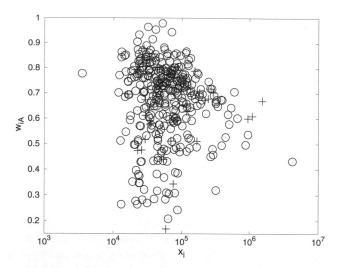

Figure 10.6 Voting rates w_{ij} for candidate A as a function of shahrestan total voting population x_i. Shahrestans corresponding to the K7 spike are shown as plus symbols;
all other shahrestans are shown with circles. The K7 spike selects those of the
six most populous shahrestans who (officially) voted the most for candidate A.

Shahrestan	x_i	v_{iK}	w_{iA}
Tabriz	876919	3513	0.497
Shiraz	947168	7078	0.600
Karaj	950243	8057	0.537
Isfahan	1095399	7002	0.609
Mashhad	1536106	7098	0.669
Tehran	4179188	43073	0.433

Table 10.1   Vote counts for K and voting rates for A for the six most populous voting areas
(as in [Rou, Table 8]).

	Candidate j			
Shahrestan subset	A	R	K	M
$N(v_{ij}$ even $\mid d_1(v_{iK}) = 7)$	18	15	18	16
$N(v_{ij}$ odd $\mid d_1(v_{iK}) = 7)$	23	26	23	25
p	0.266	0.059	0.266	0.106

Table 10.2   Number of K7-selected shahrestans N with odd or even vote counts v_{ij}, and
cumulative binomial tail probabilities p (as in [Rou, Table 10]).

for selecting zero as a second digit, though in a different sociological setting (Enron Corporation) and with a different inferred psychological explanation. Even if we neglect this, the probability that all three second digits are equal, but not necessarily zero, can be calculated using BL, giving $p \approx 0.01037$ [Rou, Section 4.4.1]. The coincidences continue. Table 10.2 shows that the K7-selected shahrestans mostly have odd vote counts, no matter who the candidate is. Considering the cumulative binomial tail probabilities p in the table to be independent, the probability that all four candidates have their K7-selected subset as dominated by odd votes as they are in the official results is $p \approx 4.4 \times 10^{-4}$ [Rou, Section 4.4.2]. Thus, not only does the local bootstrap method confirm the low probability of the K7 spike revealed by the Benford's Law limit, but the K7 spike leads to several improbable patterns in subsets of the data that it selects.

10.5 WHEN LOCAL BOOTSTRAPS COMPLEMENT BENFORD'S LAW

The implicit sociological model used in the local bootstrap model here is

$$v_{ij} = w_{ij}x_i \qquad (10.6)$$

for each candidate j in shahrestan i, where $\{x_i\}$ is a known realization of an underlying sociological process, and $\{w_{ij}\}$ is considered to be a random process (popularity) characteristic of candidate j and described by an implicitly known probability distribution at each shahrestan population size $\approx x$. When the aim is to have a test of high specificity and low power is acceptable, and the number of data points is small, this model is generic enough to infer "raw" bootstrap confidence intervals for first-digit frequencies in a wide variety of situations. These will in general be conservative, and more work—calibration using comparable data sets—would be required for calculating "empirical" confidence levels as described above.

Figures 10.1 and 10.2 show that the 2009 Iranian presidential-election first round is better than several other recent elections for applying a first-digit BL analysis. The other elections have narrower logarithmic widths in x_i and w_{ij}. Thus, local bootstraps are more likely to be necessary as a complement to direct application of the first-digit BL to provide conservative estimates of confidence intervals for similar, post-2009 presidential-election first rounds. Countries with populations smaller than about 40 million are not spanned by the populations of the five countries used in the calibration above, so a recalibration would be preferable if extending to lower population countries.

Of course, given more credible knowledge about a system, a more detailed, parametric statistical model could be built, and then first-digit frequencies could be compared between model and observation. But Benford's Law has the special appeal of being trivial to apply to a given data set without requiring the adoption of possibly wrong assumptions about the processes and properties of the system. A local bootstrap model follows the spirit of being essentially **non-parametric** (apart from the choice of Δ, but BL also depends on the choice of a base), so although it is less trivial (but still straightforward) to calculate, it has the same advantage of bypassing a need to make detailed assumptions about the system, and the additional advantage of not needing to assume that the system is near the Benford's Law limit.

PART IV
Applications II: Economics

Chapter Eleven

Measuring the Quality of European Statistics

Bernhard Rauch, Max Göttsche, Gernot Brähler, Stefan Engel[1]

Because of the Stability and Growth Pact criteria the euro countries have an in-centive to manipulate their macroeconomic statistics. In a recent study we showed a significant deviation from Benford's Law of the first digits distribution of Greek financial statistics. This result supports the effectiveness of Benford's Law in de-tecting fraud, as Greece has been convicted of data manipulation. In this chapter we use a different approach: we analyze Greek statistics which are not relevant to government deficit spending, and compare the findings with the results of our prior research. Our hypothesis was that the social data set should conform better with Benford's Law than the financial data set, as the incentive for manipulation is lower. Our results show that in contrast to their financial data, the Greek social statistics data have a good fit with Benford's Law. Once again, we interpret our outcome as a sign for the effectiveness of the Benford test.

11.1 INTRODUCTION

For a company, an excessively high level of debt can lead to bankruptcy, whereas a country is able to use inflation to reduce public debt denominated in domestic currency. However, if a country is a member of a monetary union, an unsustainable level of debt may endanger the monetary system in the entire union. For this reason, current and prospective members of the European Monetary Union are obliged by the Stability and Growth Pact to restrict public deficit to 3 percent and public debt to 60 percent of **Gross Domestic Product (GDP)**.

If a euro country does not meet the criteria of the Stability and Growth Pact, the European Commission can apply **Excessive Deficit Procedures (EDP)** to this country. EDP include restrictions for the country's government's policies and ex-panded fiscal monitoring of the deficit.

However, EDP have not yet been successful in restricting public debt to a sus-tainable level in all euro states. Some of the euro countries reach ratios of govern-ment debt to GDP of more than 100 percent ([EUC1], 2012): These countries are

---

[1]Rauch: University of Regensburg, Department of Economics, Universitätsstraße 31, 93053 Re-gensburg, Germany; Göttsche and Engel: Catholic University of Eichstätt-Ingolstadt, Department of Auditing and Controlling, Auf der Schanz 49, 85049, Ingolstadt, Germany; Brähler: Ilmenau Univer-sity of Technology, Department of Taxation Theory and Auditing, Helmholtzplatz 3, 98693 Ilmenau, Germany.

Greece (165.3%), Italy (120.1%), Ireland (108,2%) and Portugal (107,8%). Since the beginning of the euro crisis in 2010, tremendous efforts have been necessary to stabilize the monetary union.

Given the high pressure from both the financial markets and the European Commission for the euro countries to meet the criteria of the Stability and Growth Pact, there is an incentive for governments to manipulate their data related to fiscal monitoring. Therefore, the quality of the statistics in the European Union (**EU**) is an important prerequisite for effective debt monitoring. To ensure the quality of the statistics, effective auditing methods are necessary. Benford's Law has been successfully applied to detect manipulation and "cosmetic earnings management" in the accounting data of companies. Since macroeconomic data are similar in nature to accounting data, Benford's law should be applicable to them as well.

In this chapter, we pose the question of whether there is a difference in the quality of data for statistics which are related to fiscal monitoring and for social statistics, which have no relation to fiscal monitoring. Our hypothesis is that the quality of EU government statistics related to fiscal monitoring could be affected by the pressure of the financial markets and the European Commission. For both types of statistics, we compare the distribution of the first digits with the distribution of the first digits generated by Benford's Law.

11.2 MACROECONOMIC STATISTICS IN THE EU

European Government Finance Statistics are currently provided on the basis of the **European System of Accounts** (**ESA** 95). These data are the "basis for fiscal monitoring in the EU." The data for the statistics are reported by the different member states. The member states and their statistical authorities are responsible for the compliance of the reported data with legal provisions. On the level of the European Union, Eurostat, the EU statistical authority, is responsible for the statistical methodology and for the quality of assessment of the data provided by the member states, including data provided in the context of the EDP (cf. [EUC2], 2011).

EDP can be applied by the European Commission and the European Council to countries which do not fulfill the criteria of the Stability and Growth Pact. The Stability and Growth Pact limits the permitted government debt for the countries of the eurozone to three percent of GDP (cf. [EUC2], 2011).

In the last few years, there have been discussions concerning the quality of the statistics reported to Eurostat, especially in the case of those reported by Greece. In a report on the Greek EDP data and statistics, the European Commission (cf. [EUC3], 2010) pointed out that "These most recent revisions are an illustration of the lack of quality of Greek fiscal statistics (and of macroeconomic statistics in general) and show that the progress in the compilation of fiscal statistics in Greece and the intense scrutiny of the Greek fiscal data by Eurostat since 2004 (including 10 EDP visits and 5 reservations on the notified data), have not sufficed to bring the quality of Greek fiscal data to the level reached by other EU Member States."

As a consequence of the lack of quality in the EU statistics, Eurostat (cf. [EUC2], 2011) was given more power in 2010. This includes a "system of regular monitoring and verification of upstream public financial data." The extended competence of Eurostat also includes more "in-depth, methodological visits to the Member States." For the future, the European Commission plans to "support the implementation of public accounting standards" to provide the ESA-based information.

11.3 BENFORD'S LAW AND MACROECONOMIC DATA

Similarly to the accounting data of companies, unmanipulated macroeconomic data from different sources with different distributions can be expected to be Benford distributed. This is confirmed in Nye and Moul ([NyM], 2007), Gonzales-Garcia and Pastor ([GonPa], 2009) and Rauch et al. ([RauGBE], 2011). However, one has to be careful when interpreting a deviation from Benford's Law in a certain data set. It cannot be considered as conclusive proof of poor data quality, since it could be based on e.g. structural shifts in the data set, as argued by Gonzales-Garcia and Pastor ([GonPa], 2009). Nevertheless, in our opinion a deviation from Benford's Law should be regarded as a "red flag," indicating data that need closer inspection and further testing.

Consequently, we do not use a hypothesis framework to investigate the conformity of a data set with Benford's Law. Rather, we compare data sets according to the extent of their deviation from Benford's Law. This relation is used to establish a ranking of data sets, i.e., of countries. The position of each country in this ranking can be used to indicate the probability of manipulation in its data and determines the order in which further auditing procedures should be carried out.

We restrict the analysis to the first valid digit. The evaluation of the data sets' conformity with Benford's Law is based on χ^2 test statistics,

$$\chi^2 = n \sum_{i=1}^{9} \frac{(h_i - p_i)^2}{p_i} \tag{11.1}$$

where n denotes sample size, p_i expected and h_i observed relative frequencies.

Furthermore, to ensure that the ranking induced by χ^2 test statistics is not a result of variation in sample size between the countries or the choice of a particular measure, we use three measures of the distance between the actual data distribution and the Benford distribution which are insensitive to sample size. The first measure is the χ^2 statistics divided by the sample size n as in Leemis et al. ([LeScEv], 2000):

$$c = \chi^2/n. \tag{11.2}$$

The other measures are the normalized Euclidian distance d^* as in Cho and Gaines ([ChGa], 2007),

$$d^* = \frac{\sqrt{\sum_{i=1}^{9} (h_i - p_i)^2}}{\sqrt{\sum_{i=1}^{8} p_i^2 + (1 - p_9)^2}}, \tag{11.3}$$

and the distance measure a^* used by Judge and Schechter ([JuSc], 2009),

$$a^* = \frac{|\mu_e - \mu_b|}{9 - \mu_b}, \tag{11.4}$$

where μ_e denotes the mean of the data set and μ_b denotes the mean of the Benford distribution of first digits.

11.3.1 EDP-Related Statistics

In a recent study (cf. [RauGBE], 2011) we analyze EU macroeconomic data published in the Eurostat database.[1] The aim of the study is to analyze macroeconomic data related to the EDP and to provide a ranking of the EU member states based on the deviation of their data from Benford's Law. This ranking could be useful as an indicator for manipulated data. The data set consists of the following categories:

1. Government statistics / government deficit and debt: government deficit / surplus, debt and associated data.

2. Government statistics / annual government finance statistics: government revenue, expenditure and main aggregates.

3. National accounts / annual national accounts: GDP and main components—current prices

4. Financial accounts: balance sheets, assets and liabilities consolidated.

5. Financial accounts: financial transactions, assets and liabilities consolidated.

Our study analyzes the first digits of 156 single positions per country and per year covering the period from 1999 to 2009. The analysis consists of two steps. First, we analyze the aggregated data set. For this set, the results of the three distance measures independent of sample size show a good fit of the first digit distribution with Benford's Law. Second, investigating each country individually, we calculate the mean of the χ^2 test statistics for each country and rank the countries according to this mean value.

On the individual level, among all euro countries Greece shows the highest deviation from Benford's Law with a mean value of 17.74 for the χ^2 test statistics, followed by Belgium with a value of 17.21 and Austria with a value of 15.25. In contrast, we calculate the lowest deviation for the Netherlands with 7.83. The introduced measures independent of sample size, χ^2/n, d^* and a^*, support the results of the χ^2 test statistics.

The so-called PIIGS countries, an acronym for Portugal, Italy, Ireland, Greece and Spain, are strongly affected by a high level of debt and might therefore be expected to manipulate data. However, as far as the PIIGS states are concerned, apart from Greece only the Irish data indicates a substantial deviation from Benford's Law in our study. Potentially poor data was indicated only by a^* and d^* for Italy; for Spain and Portugal, we could not find any such indication. Furthermore, Portugal shows the second-lowest mean of the χ^2 test statistics. Our results do not support the common assumption that data reported by the "PIIGS" are generally of lower quality than those reported by other euro states.

---

[1] http://epp.eurostat.ec.europa.eu/portal/page/portal/statistics/search_database.

For all measures, Greece shows the highest deviation from Benford's Law among all euro countries. As mentioned above, the European Commission concluded in a report that the quality of Greek EDP data and macroeconomic statistics is insufficient (cf. [EUC3], 2010). At least in the case of Greece, the high deviation from Benford's Law could be interpreted as an indicator for low data quality.

11.3.2 Social Statistics

As mentioned before, there is high pressure for the euro countries to comply with the Stability and Growth Pact criteria. This in turn gives countries an incentive to falsify statistics related to the fiscal monitoring of euro countries and EDP. Our hypothesis is that countries are more willing to falsify their statistics if they have an incentive to do so. On the contrary, if there is no incentive for data manipulation for the countries, the data quality will be higher.

The aim of our study is to use Benford's Law to investigate the quality of macroeconomic data which is not related to fiscal monitoring and EDP provided by EU member states. If there is no data manipulation, one would expect that the data will conform well with Benford's Law. As the Greek data produced the highest deviation from Benford's Law in our recent study (cf. [RauGBE], 2011) and the manipulation of Greek data was criticized by the European Commission (cf. [EUC3], 2010), we decided to investigate our hypothesis using Greek statistics.

We choose social statistics, which are not related to fiscal monitoring and the EDP process, as these are more likely to be subject to fluctuations than population statistics or area data. We assume that there is no pressure or incentive for Greece to falsify these data and we would therefore expect no significant deviation for this data from Benford's Law. Our data set contains Greek data taken from the database of Eurostat[1] in June 2012. We select the data from two different subsections of the Eurostat database. The first part of the data is from the group "social protection" under the theme "living conditions and welfare." The other part of the data set is from the group "labour market policy" under the theme "population and social conditions." Our sample includes the following categories:

1. Social Protection Expenditure

2. Social Protection Receipts

3. Public expenditure on labor market policy interventions

The data set contains 1,322 observations for the period from 1999 to 2009 with a total of 267 possible observations per year and an average of 120.18 observations per year. The main reason for the difference between possible observations and average observations is that data from the third group include a considerable number of entries with missing values, especially for the years 1999 to 2005 and the year 2009. Data are expressed in absolute values in millions of euros; currency conversions were calculated by Eurostat.

Again we calculate the χ^2 test statistics and the three measures independent of sample size, χ^2, χ^2/n, d^* and a^*, for the whole data set and per year. Results are presented in Figure 11.1.

Our results indicate a good fit of the investigated data for the whole sample as well as for the single years. We could not find a significant deviation for any one of the single years or for the sample as a whole. The results for the three measures independent of sample size support the results of the χ^2 test statistics.

Digit	Benford	1999	2000	2001	2002	2003	2004	2005	2006	2007	2008	2009
1	0,30	0,28	0,26	0,30	0,29	0,34	0,34	0,27	0,30	0,34	0,27	0,33
2	0,18	0,20	0,18	0,19	0,17	0,17	0,18	0,18	0,20	0,16	0,17	0,18
3	0,12	0,11	0,11	0,10	0,15	0,11	0,08	0,11	0,06	0,11	0,16	0,08
4	0,10	0,13	0,10	0,12	0,06	0,10	0,16	0,12	0,13	0,09	0,07	0,07
5	0,08	0,08	0,11	0,07	0,11	0,10	0,05	0,09	0,07	0,09	0,06	0,07
6	0,07	0,08	0,09	0,08	0,08	0,07	0,07	0,08	0,08	0,06	0,10	0,07
7	0,06	0,03	0,02	0,07	0,09	0,04	0,06	0,03	0,07	0,07	0,07	0,10
8	0,05	0,06	0,06	0,01	0,02	0,05	0,04	0,06	0,05	0,06	0,05	0,05
9	0,05	0,05	0,09	0,07	0,02	0,02	0,02	0,05	0,04	0,03	0,05	0,05
χ^2		3,50	9,01	6,29	8,78	3,99	11,26	3,91	7,20	3,74	7,03	6,31
χ^2/n		0,03	0,09	0,06	0,09	0,04	0,10	0,04	0,05	0,02	0,04	0,06
d*		0,06	0,08	0,06	0,08	0,06	0,09	0,05	0,07	0,05	0,06	0,07
a*		0,00	0,06	0,00	0,01	0,05	0,05	0,02	0,01	0,01	0,03	0,01
Size		104	104	101	98	106	116	110	142	173	161	107

Notes: Table includes test statistics from the χ^2 goodness of fit test.
Values for χ^2/n, d*,a* are provided for comparision reasons.
5% significance level 15.51.

Figure 11.1   Distribution of first digits for Greek social statistics.

11.3.3 EDP-Related Statistics versus Social Statistics

Figure 11.2 summarizes the results for Greece from our first study. Figure 11.2 compares the distribution of Greek EDP-related statistics with the distribution generated by Benford's Law. There is a significant deviation of the debt-related statistics from Benford's Law for all digits.

The results of the second study are presented in Figure 11.3. As shown in Section 11.3.2, we identify a good fit for the social statistics with Benford's Law for all digits.

The different results of the two studies are illustrated in Figure 11.4, comparing the χ^2 test statistics for the EDP-related statistics and the social statistics for the period from 1999–2009. The 5 percent significance level for the χ^2 test statistics (with 8 degrees of freedom) is 15.51. We identify significant deviations, for the Greek EDP-related statistics, from Benford's Law for seven of the ten years. In contrast, we find no significant deviations for the social statistics in the same period.

Comparing the results of the first study with the results of the second study, we find a lack of data quality for Greek EDP-related statistics, measured by the deviation of the first digit distribution from the distribution generated by Benford's Law. On the contrary, we find evidence that data not related to EDP and debt do not show a significant deviation from Benford's Law. Our results support the hypothesis that, at least in the case of Greece, data will conform well with Benford's Law if there is no pressure or incentive to falsify statistics (cf. [RauGBK], 2014).

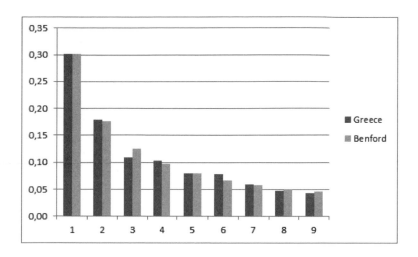

Figure 11.2   First digits of Greek EDP-related statistics.

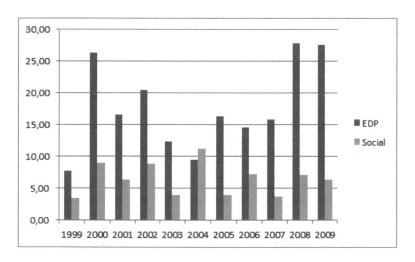

Figure 11.3   First digits of Greek social statistics.

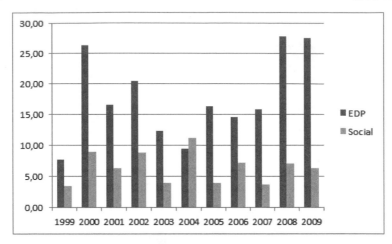

Figure 11.4 χ^2 test statistics for Greece.

11.4 CONCLUSION

For the euro countries, there is high pressure to comply with the Stability and Growth Pact criteria. The compliance of the member states with the Stability and Growth Pact criteria is monitored by the European Commission through their statistical authority Eurostat. If a country's government deficit spending does not comply with the criteria of the Stability and Growth Pact, the European Commission can apply EDP to this country's government. Therefore, governments have an incentive to falsify statistics related to fiscal monitoring and EDP.

The basis for the fiscal monitoring through Eurostat is the statistics provided by the member states. Particularly during the current euro crisis, the European Commission has raised considerable doubts concerning the data quality of the Greek EDP-related statistics (cf. [EUC3], 2010). In view of Greece's current economic situation, there is a strong incentive for Greece to manipulate the EDP-related statistics.

In a recent study (cf. [RauGBE], 2011) we could show that the first digits distribution of Greek EDP-related statistics show a significant deviation from Benford's Law, indicating low data quality. This result is hardly surprising, considering the current Greek situation.

In this chapter, we analyzed statistics which are not relevant to government deficit spending and fiscal monitoring. Our hypothesis was that countries are more willing to falsify data if they have an incentive to do so, whereas if there is no incentive to manipulate data, the quality of the data will be higher. We used Benford's Law to examine the quality of Greek social statistics, which are not likely to be manipulated, as there is no incentive or pressure for Greece to falsify these statistics.

Our results show that the Greek social statistics data have a good fit with Benford's Law, for the whole sample as well as for the single years. This supports our hypothesis that the quality of statistical data will be higher and will therefore conform well with Benford's Law if there is no incentive for data manipulation.

The results of our two studies can thus be interpreted as a sign for the efficiency of Benford's Law as a measurement method to check the quality of macroeconomic statistics.

Chapter Twelve

Benford's Law and Fraud in Economic Research

Karl-Heinz Tödter[1]

Science and academia are not immune to dishonesty and deception. Scientific misconduct appears in various forms. Fabrication, falsification, and plagiarism are as old as science itself. Journal editors and referees have a difficult task when confronted with empirical research results based on large data sets and complex econometric techniques.

Replication is considered the prime strategy against scientific misconduct, but it is seldom performed in empirical economics. Benford's Law provides tools for checking reliability and detecting fraud. The chapter reviews applications of Benford's Law to uncover fraud in macroeconomic data, forecasts, and econometric regression results. Moreover, the potential of Benford's Law to enhance the efficiency of replication as a strategy against fraud in published research is discussed. We conclude that routine applications of Benford tests could uncover data anomalies and provide valuable hints of irregularities in empirical economics research.

12.1 INTRODUCTION

Science and academia are not immune to dishonesty and deception. Scientific misconduct is as old as science itself. Galileo, Kepler, Newton, Mendel, Pasteur, and Freud have all been suspected of fraud in some way [Jud]. In an 1830 text entitled *Reflections on the Decline of Science in England*, mathematician, philosopher, and computer pioneer Charles Babbage documented his concern about scientific misconduct, which he classified as hoaxing, forging, trimming, and cooking [Han]. The US Office of Science and Technology Policy defines **scientific misconduct** as "fabrication, falsification, or plagiarism in proposing, performing, or reviewing research or in reporting research results" [MaAdV]. **Fabrication** is invention of data or cases, **falsification** is intentional distortion of data or results, and **plagiarism** is unquoted copying of data or text.[2]

Increasing specialization of science, its growing social and economic relevance, keen competition for research funding, and strong publication pressure in academia

---

[1] Research Centre, Deutsche Bundesbank, Frankfurt am Main, Germany

[2] Beyond that narrow definition, [MaAdV] report "a range of questionable practices that are striking in their breadth and prevalence," according to a survey conducted among several thousand scientists. They concluded that overall, 33% of the respondents admitted having engaged in at least one of the top ten misbehaviors during the previous three years.

have all raised the temptation to make up research results [GrLaW, Fre, Sta]. The development of the internet and of "text mining" platforms have increased the risk of plain plagiarism,[3] but other forms of research misconduct cannot be uncovered by the same means. The traditional control mechanisms in the publication process are easily overstrained, particularly when editors and referees are confronted with empirical research results based on large data sets and/or complex econometric techniques. Replication is considered as prime strategy against scientific miscon-duct, but it remains ([Ham], p.715) "an activity that most economists applaud but few perform." Since empirical research in economics is rarely replicated by in-dependent experts, complementary methods for checking reliability and detecting fraud are called for.

This chapter reviews applications of Benford's Law to detect fraud in economic data, forecasts, and research. Regression results and quantitative forecasts form a large and important part of published research output in economics. Section 12.2 discusses some issues on Benford's Law. Section 12.3 reports applications of Ben-ford's Law to macroeconomic data and forecasts, including Consensus forecasts and national accounting data of euro member states. In the case of Greece the suspicion of data manipulation raised by tests of Benford's Law was confirmed officially by the European Commission [RauGBE] (for more on Greece, see Chap-ter 11). Section 12.4 reviews work on Benford's Law to check for anomalies in published regression output, and Section 12.5 gauges the potential of Benford's Law to enhance the efficiency of replication as a strategy against fraud in published research. We end with some concluding remarks in Section 12.6

12.2 ON BENFORD'S LAW

The density and distribution functions of Benford's Law for a variable X defined on $[1, 10)$ are

$$f(x) = \frac{1}{x \log 10}, \quad F(x) = \log_{10}(x). \tag{12.1}$$

From (12.1), the following probabilities of first digits (D_1) and conditional proba-bilities of second digits (D_2) are obtained and shown in Table 12.1:

$$\text{Prob}(D_1 = d_1) = \log_{10}\left(1 + \frac{1}{d_1}\right), \quad d_1 \in \{1, 2, \ldots, 9\}, \tag{12.2}$$

$$\text{Prob}(D_2 = d_2 | D_1 = d_1) = \log_{10}\left(1 + \frac{1}{10d_1 + d_2}\right), \quad d_2 \in \{0, 1, 2, \ldots, 9\}. \tag{12.3}$$

---

[3]Recently, two prominent cases of plagiarism in their dissertations led to the demise of the German Minister of Defence Karl-Theodor zu Guttenberg in 2011 and of the President of Hungary Pál Schmitt in 2012. Also in 2012, Annette Schavan, German Federal Minister of Education and Research, was accused of plagiarism in her dissertation from 1980 at the University of Düsseldorf, and resigned in February 2013.

	Benford's Law		Benford's Law for Rounded Figures		
d_1,d_2	$P(d_1)$	$P(d_2)$	$P(d_1)^{\text{a)}}$	$P(d_2)^{\text{b)}}$	$P(d_2)^{\text{c)}}$
0	–	0.120	-	0.103	0.506
1	0.301	0.114	0.198	0.117	–
2	0.176	0.109	0.222	0.111	–
3	0.125	0.104	0.146	0.107	–
4	0.097	0.100	0.109	0.102	–
5	0.079	0.097	0.087	0.098	0.494
6	0.067	0.093	0.073	0.095	–
7	0.058	0.090	0.062	0.092	–
8	0.051	0.088	0.054	0.089	–
9	0.046	0.085	0.048	0.086	–
$\mathbb{E}(D)^{\text{d)}}$	*3.440*	*4.187*	*3.693*	*4.261*	*2.470*
$\text{Var}(D)^{\text{d)}}$	*6.057*	*8.254*	*5.738*	*8.063*	*6.249*

a) Figures rounded to one significant digit. b) Figures rounded to two significant digits. c) Second digits rounded to zero and five.

d) $\mathbb{E}(D)$ is the expected value and $\text{Var}(D)$ is the variance of the distribution.

Table 12.1 Benford's Law for original and rounded figures.

Newspapers, monographs, and scientific journals often publish only one or two significant digits. Benford's Law is scale and base invariant [Hi2, Pin], but it is not invariant to rounding. From (12.1) probabilities are obtained for figures rounded to *one* significant digit, which means that figures 0.5 or more are rounded up and figures less than 0.5 are rounded down:

$$P(D_1 = d_1) = \begin{cases} \log_{10}\left(1 + \frac{11}{19}\right), & d_1 = 1, \\ \log_{10}\left(1 + \frac{1}{d_1 - \frac{1}{2}}\right), & d_1 \in \{2,3,\ldots,9\}. \end{cases} \tag{12.4}$$

If figures are rounded to two significant digits, the probabilities of first digits remain as in (12.2), whereas the conditional probabilities of second digits are

$$P(D_2 = d_2 | D_1 = d_1) = \begin{cases} \log_{10}\left(1 + \frac{1}{20d_1}\right) + \log_{10}\left(1 + \frac{1}{20d_1 + 19}\right), & d_2 = 0, \\ \log_{10}\left(1 + \frac{1}{10d_1 + d_2 - \frac{1}{2}}\right), & d_2 \neq 0. \end{cases} \tag{12.5}$$

In Table 12.1 it can be seen that the unconditional probabilities of rounded first and second digits are not monotonically decreasing. The final column applies in the case when second digits are **bold rounded** to the "half integers" 0 and 5, which means, e.g., that figures in the range $[3.25, 3.75)$ are rounded to 3.5 and figures in the range $[3.75, 4.25)$ to 4.0.

To test leading digits for conformity with Benford's Law, the χ^2–**goodness-of-fit test** is often used; the χ^2-distribution is a special case of the Gamma distribution with one parameter, called the degrees of freedom. The test checks whether the sum of squared deviations between observed relative frequencies of leading digits (h_{d_1}, h_{d_2}) and Benford probabilities $(\text{Prob}(D_1=d_1) = p_{d_1}, \text{Prob}(D_2=d_2) = p_{d_2})$ as given in (12.2) and (12.3) significantly differs from zero:

$$Q_1 = n \sum_{d_1=1}^{9} \frac{(h_{d_1} - p_{d_1})^2}{p_{d_1}}, \quad Q_2 = n \sum_{d_2=0}^{9} \frac{(h_{d_2} - p_{d_2})^2}{p_{d_2}}. \tag{12.6}$$

The sample size is denoted by n, the statistic Q_1 (Q_2) has an approximate χ^2-distribution with 8 (9) degrees of freedom for testing first (second) digits.

The following test checks whether the arithmetic mean of observed leading digits, defined as $\overline{d_1} = \sum_{d_1=1}^{9} d_1 * h_{d_1}$, and $\overline{d_2} = \sum_{d_2=0}^{9} d_2 * h_{d_2}$, deviates from its expectation ($\mathbb{E}(D_1)$, $\mathbb{E}(D_2)$) under Benford's Law. The test statistics

$$M_1 = n \frac{(\overline{d_1} - \mathbb{E}(D_1))^2}{\text{Var}(D_1)}, \quad M_2 = n \frac{(\overline{d_2} - \mathbb{E}(D_2))^2}{\text{Var}(D_2)}, \tag{12.7}$$

have an approximate χ^2-distribution with 1 degree of freedom for testing first and second digits; the expectations ($\mathbb{E}(D_1)$, $\mathbb{E}(D_2)$) and variances ($\text{Var}(D_1)$, $\text{Var}(D_2)$) are shown in Table 12.1. The power of both tests depends on the specific pattern of deviations from Benford's Law. However, since the M-tests "use up" only one degree of freedom, they have potentially greater power than the Q-tests.

The uniform distribution is often regarded as natural and thus most plausible. A tampered data set is likely to have more evenly distributed leading digits than Benford's Law requires because mostly people are unaware that data sets should be Benford. This ignorance suggests, for testing purposes, replacing the observed distribution of first digits by a linear combination ($0 \leq \lambda \leq 1$) of Benford's Law and the uniform law (UL): $h_{d_1} = (1 - \lambda)p_{d_1} + \lambda(1/9)$. For example, the probability of the first digit "1" drops from 0.301 to $0.301 - 0.190\lambda$, and that of the digit "9" increases from 0.046 to $0.046 + 0.065\lambda$. The mean of the mixed distribution increases from $\mathbb{E}(D_1) = 3.44$ to $3.44 + 1.56\lambda$. Thus, both tests have power to detect this type of manipulation. Both statistics turn out as approximately the same functions of λ : $Q_1(\lambda) \approx M_1(\lambda) \approx 0.401n\lambda^2$. Since the critical 5% level of the Q_1-test (=15.51) is 4 times as large as that for the M_1-test (=3.84), the latter is more powerful in detecting violations of Benford's Law, as shown in columns 2 and 3 of Table 12.2 for several sample sizes.

For a second comparison, consider the **Generalized Benford Law (GBL)**, which is a power law that includes Benford's Law as a special case for $\alpha = 1$, as can be seen by invoking L'Hopital's rule to evaluate the limit as α tends to unity:

$$g(x) = \frac{1 - \alpha}{(10^{1-\alpha} - 1)x^\alpha}, \quad -\infty \leq \alpha \leq \infty. \tag{12.8}$$

For $\alpha > 0$, the density is monotonically decreasing. If $\alpha > 1$ ($\alpha < 1$) the density is steeper (flatter) than Benford's Law, approaching the uniform distribution as α tends to zero. With negative α-values, the Generalized Benford Law becomes monotonically increasing, shifting more mass to higher digits. For example, assuming $\alpha = \{2, 1, 0, -1, -2\}$, the probabilities of leading digit "1" are $\{0.56,$ $0.30, 0.11, 0.03, 0.01\}$ and of "9" are $\{0.01, 0.05, 0.11, 0.19, 0.27\}$, respectively. Columns 4 and 5 of Table 12.2 show the range of α-values for which the tests fail to reject the null hypothesis of Benford's Law at the 5% level of significance. At small and moderate sample sizes, fairly large deviations from $\alpha = 1$ are required by

	Linear combination of BL and UL		GBL	
n	Q_1-test [a]	M_1-test [b]	Q_1-test [a]	M_1-test [b]
	requires $\lambda >$	$\lambda > \ldots$	$\alpha \notin \ldots$	$\alpha \notin \ldots$
	$\ldots$			
50	0.88	0.44	(0.13, 1.91)	(0.57, 1.47)
100	0.62	0.31	(0.39, 1.63)	(0.69, 1.32)
200	0.44	0.22	(0.57, 1.44)	(0.78, 1.23)
500	0.28	0.14	(0.73, 1.27)	(0.86, 1.14)
1000	0.20	0.10	(0.81, 1.19)	(0.90, 1.10)
5000	0.09	0.04	(0.91, 1.09)	(0.96, 1.04)

BL = Benford's Law, GBL = Generalized Benford Law, UL = uniform distribution. a) Q_1-test rejects BL at 5% significance level with critical value 15.51.

b) M_1-test rejects BL at 5% significance level with critical value 3.84.

Table 12.2   Sensitivity of the Q_1- and M_1-test against alternative distributions.

both tests to detect violations of Benford's Law. However, the range of insignificant α-values is much wider for the Q_1-test, implying that it is uniformly less powerful against the Generalized Benford Law distributions than the M_1-test.

The M_1-test cannot discriminate between distributions of first digits that have identical means. The **Maximum Entropy (MaxEnt)** distribution of first significant digits discussed by Lee et al. in Chapter 17 of this volume provides an example. For the parameters $\alpha = 0.646$ (GBL) and $\lambda = 0.154$ (MaxEnt) the probabilities of the first digits are different, e.g., $P(D_1 = 1) = 0.22(0.19)$ for GBL (MaxEnt), yet both have identical mean $E(D_1) = 4$. The Q_1-test can discriminate between both distributions but requires a large sample size ($n \geq 1417$ at the 5% level).

12.3 BENFORD'S LAW IN MACROECONOMIC DATA AND FORECASTS

National accounting data play a major role in the European Union (EU) because euro countries have to comply with the stability criteria of the Stability and Growth Pact. The member states must report their national and financial accounts to *Eurostat*, the European statistics agency. To avoid sanctions, governments may polish their economic stance in the official data. Rauch et al. (see [RauGBE] and Chapter 11) applied Benford's Law to investigate macroeconomic data in the EU. The authors checked the quality of data related to public deficit, public debt, and gross national product, all of which are relevant to the calculation of the deficit ratio and the debt ratio. First, aggregate data of the 27 EU member states were checked and yielded a very good fit to Benford's Law.

In a second step, data of individual euro and non-euro countries for the 11 years between 1999 and 2009 were tested. [RauGBE] obtained a robust ranking of the quality of the data set according to the Q_1-statistic and related criteria. For the data reported by Greece, they found that the Q_1-statistic exceeds the 5% significance level in 6 of 11 years. These data showed "the greatest deviation from Benford's Law among all euro countries" ([RauGBE], p.243), followed by Belgium and Austria. Among the non-euro countries, the greatest deviations from Benford's Law were observed for Romania, Latvia, and Estonia. The suspicion of data manipulation in the case of Greece was confirmed officially by the European Commission and thus provides evidence for the effectiveness of Benford's Law in revealing anomalies in official macroeconomic data ([RauGBE], p.243). This article was widely commented on in the press after it appeared in the German Economic Review in August 2011. Meanwhile, the International Monetary Fund (IMF) and Eurostat have declared their interest in Benford procedures.

Quantitative forecasts are research output that plays an important role as an information source for economic agents. In particular, monetary policy of central banks and fiscal policy of governments strongly rely on macroeconomic forecasts. The monthly journal *Consensus Forecasts* publishes forecasts of the annual growth rates of the current and the subsequent year for a broad set of macroeconomic variables and countries, which are widely read and analyzed in the business and financial press. The forecasts are based on monthly surveys among commercial banks and private as well as public research institutes. *Consensus Forecasts* publishes both the individual forecasts and the aggregate (mean) forecasts averaged over the participating forecasters. The public pays the most attention to forecasts of the growth rate of real gross domestic product (GDP) and of the inflation rate of the consumer price index (CPI). Günnel and Tödter [GüTd] investigated forecasts for Germany for data anomalies and information inefficiencies. They collected about 18,000 observations of leading digits from 55 panelists to check the forecasts of annual growth rates of real GDP and CPI inflation from October 1989 to July 2004.

The first digits of *Consensus Forecasts* have a very limited range: most reported growth and inflation rates are between 0 and 4. Therefore, [GüTd] looked at the distribution of "first digits after the decimal point," as these digits are of comparable importance in terms of their economic significance. Another important feature of the data set is rounding. The forecasters have to report figures rounded to one digit after the decimal point, requiring adjustment of Benford's Law to take into account rounding effects. If second digits are rounded, values in the range of, e.g., [1.35, 1.45) are reported as 1.4. The observed second digits strongly violate Benford's Law (adjusted for rounding effects) because there is a massive excess of 0s and 5s: in the four subsamples shown, almost 50% of all forecasts look like $d_1.0$ and $d_1.5$ with any first digit d_1.

[GüTd] also checked the forecasts of individual panelists for excessive rounding. The sample size for individual forecasters ranges between 600 and 700 observations. In the extreme, one institute recorded the "half integers" 0 and 5 in 70% of all observations. Such bold rounding could reflect imitation and herding behavior [Ost]. [GüTd] estimated that on average 30% of the observations come from "bold rounded" figures. If there are leading forecasters, other institutes might resort to bold rounding to conceal "plagiarism of numbers." Reporting "half integers" might also reflect the fact that some forecasters do not use forecasting models but simply make "educated guesses." The strong preference for "half integers" may also be a crude way to express forecast uncertainty. Rounding forecasts to "half integers" is not fraud per se, but it is a data anomaly that seriously compromises the information content of the surveys and distorts the mean of the aggregated growth and

	Regression coefficients		Standard errors	
	1st digits	2nd digits	1st digits	2nd digits
Empirica				
No. of observations	4606	3977	3037	2773
Ave. no. of obs. per article	100	86	66	60
M_1- and M_2-test	1.42	0.56	1.30	0.01
Applied Economics Letters				
No. of observations	5171	4650	2921	2619
Ave. no. of obs. per article	73	65	41	37
M_1- and M_2-test	0.10	3.13*	1.56	0.79

Source: [GüTd]. * denotes significance at the 10% level.

Table 12.3 First and second digits of regression coefficients and standard errors.

inflation forecasts of the pool.

12.4 BENFORD'S LAW IN PUBLISHED ECONOMIC RESEARCH

Although as early as 1972 the US economist Hal Varian [Va] suggested Benford's Law as a diagnostic tool for screening model output and forecasts in economics, only in the last two decades have people recognized that Benford's Law could fruitfully be used to screen data for hints of manipulation and fraud. In particular, Nigrini's work [Nig1, Nig2, Nig4, Nig5, Nig6] was instrumental in introducing Benford's Law in accountancy, auditing, business, and finance. Meanwhile, tax authorities in the US, Germany, Switzerland, and The Netherlands, among other countries, apply Benford's Law to check tax declarations for anomalies.

Leading digits of **regression** results were first Benford tested by Diekmann [Die] in the field of sociology and by Günnel and Tödter [GüTd] in economics. The latter analyzed whether Benford's Law applies to estimated **coefficients** and **standard errors** of **econometric regressions**. Focusing on first and second digits, they screened 46 articles in four volumes of *Empirica* (2003 to 2006) and 71 articles in one volume of *Applied Economics Letters* (2006), comprising about 30,000 observations in total. [GüTd] collected data from a broad range of regression models and estimators, such as Ordinary Least Squares (OLS), Instrumental Variables (IV), Generalized Method of Moments (GMM), Quantile- and Tobin-Regressions. All figures were taken from regressions with empirical data, neglecting those with artificially generated data. To avoid double counting, only results presented in tables were counted, discarding figures scattered in the text. Table 12.3 provides an overview of the data set.

There are fewer observations on second digits than on first because about 10% of the observations show only one significant digit. Often authors just report t-values, i.e., the ratio of estimated coefficients to their estimated standard errors. In these cases [GüTd] recalculated standard errors from the published t-values, although this might induce measurement errors due to rounding in some cases.

| | Regression coefficients | | Standard errors | |
	1st digits	2nd digits	1st digits	2nd digits
No. of observations	9777	8627	5928	5392
Ave. no. of obs. per article	84	74	66	60
Standard deviation	136	123	121	116
No. of articles	117	117	90	90
% doubtful*	26	11	30	14
M-tests (average)	3.65	1.41	3.88	1.91
M-tests (std. dev.)	5.45	2.30	6.60	2.56

Source: [Töd]. *Percentage of articles with M-statistics exceeding the critical 5% value 3.84.

Table 12.4 Benford analysis of leading digits in individual articles.

As Table 12.3 shows, Benford's Law is confirmed by the M-test for first and second digits of published regression coefficients and standard errors. Summing up, [GüTd] conclude that "in economic research Benford's Law applies to regression coefficients and standard errors" (p.284).

A **publication bias** arises from the tendency of researchers, referees, and editors to prefer significant (positive) to insignificant (negative) results [RobSt]. To report statistically significant coefficients, authors have an incentive to engage in questionable activities that might range from extensive data mining to outright manipulation of data and regression output.[4] Thus, published first digits of coefficients and standard errors with t-values exceeding 1.96 in absolute value, usually considered significant, may be more likely to have been engineered than those below ([Han], p.24). To check for inflated t-values, [GüTd] divided their sample of standard errors from *Empirica* (2006) into two groups and found that deviations from Benford's Law occurred more often in the group with t-values above 1.96, which they interpreted as evidence for irregularities.[5]

In a closely related study, using the same data set of published regression output in *Empirica* and *Applied Economics Letters*, Tödter [Töd] investigated data anomalies in individual articles. He classified an article as doubtful, deserving further scrutiny, if the appropriate M-test rejects Benford's Law at the 5% level. As Table 12.4 shows, a surprisingly large proportion of articles had to be classified as doubtful. First digits of regression coefficients (standard errors) were found doubtful in 26% (30%) of the articles, in stark contrast to second digits of regressions coefficients (standard errors) with 11% (14%) doubtful articles.

This result makes sense because manipulation of first digits is "more effective" in achieving a desired result than that of second digits. The evidence on doubtful articles is reflected by the M-test statistics. On average, the M-statistics are substantially larger when testing first digits as compared to second digits. In contrast,

---

[4]Despite occasional warnings in the literature of the dangers of pretesting biases [DM], data mining is widespread in econometrics and regarded as unproblematic, e.g., [S-i-M] proudly announces "I just ran two million regressions."

[5]Auspurg and Hinz [AuHi] investigate t-values near 1.96 as an indicator of publication bias and of data manipulation.

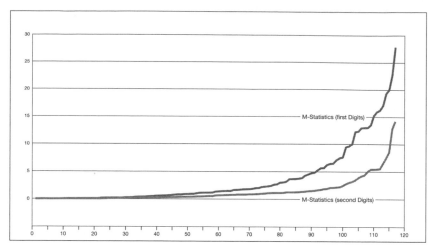

Figure 12.1 Mean test for first and second digits of regression coefficients.

a test for differences in mean [HogMC] between the first digits of regression coefficients and standard errors is statistically insignificant, as is a test for differences
in means of second digits.[6] But the differences between first and second digits are
statistically different from zero, for both regression coefficients and standard errors.
Figure 12.1 displays the values of the M-statistics in increasing order for first and
second digits of regression coefficients in the 117 articles investigated (for standard errors a similar figure is obtained). Figure 12.1 reveals the striking difference
between first and second digits, suggesting that manipulation of regression results
mostly pertains to first digits because second digits largely confirm Benford's Law.

List et al. [LiBEM] conducted a survey at an American Economic Association
(AEA) meeting and report that between 4 and 5% of the respondents self-reported
falsification of data and about 5 to 7% of research in the top 30 journals is believed
to be falsified. What does the evidence reported in [Töd] reveal about data anomalies in the population of research articles? Let Ω ($0 \leq \Omega \leq 1$) denote the unknown
proportion of Benford's Law violations in a population of articles with regression
output. Then, the null hypothesis of Benford's Law for a randomly selected article
is rejected with probability

$$\theta = \alpha(1 - \Omega) + (1 - \beta)\Omega, \tag{12.9}$$

where $\alpha(1 - \beta)$ denotes the size (power) of the test. Provided the test is unbiased
$(1 - \beta > \alpha)$, the rejection probability increases with Ω as it should. If $\Omega = 0(1)$,
the rejection probability is equal to the size (power) of the test. Now, let $\hat{\theta}$ be the
proportion of Benford violations in a sample of articles. Equation (12.9) suggests
the estimate $\hat{\Omega} = (\hat{\theta} - \alpha)/(1 - \alpha - \beta)$. Though the preselected significance

---

[6]Tests for differences between the two journals, *Empirica* and *Applied Economics Letters*, were
insignificant as well. Moreover, the M-statistics were not correlated with the number of observations in
the articles.

level is known, the formula is not applicable since the power is unknown, both for individual articles and on average for all articles sampled. Power increases with the sample size (n), but it also depends on the specific pattern of manipulation, which is unknown and likely varies from article to article. However, in large and heavily contaminated samples power would be close to one, allowing us to provide a lower bound for the ratio of doubtful articles in the population:

$$\hat{\Omega} \geq \frac{\hat{\theta} - \alpha}{1 - \alpha}. \tag{12.10}$$

If the population is clean, the expected share of rejections is equal to the size of the test ($\hat{\theta} = \alpha$), and (12.10) yields the trivial estimation $\hat{\Omega} \geq 0$.

From Table 12.4, the estimated range for the proportion of contaminated first digits of regression coefficients (standard errors) is $\hat{\Omega} \geq 22\%$ ($\geq 26\%$). For second digits of regression coefficients (standard errors) a substantially smaller estimate of the contamination ratio is obtained: $\hat{\Omega} \geq 6\%$ ($\geq 9\%$). This evidence is not proof of falsification, but anomalies revealed by Benford tests provide useful signals to initiate closer investigations. Since resources are scarce, these signals are very valuable. Benford tests are a simple and practical tool to obtain first hints of anomalies, manipulations, and falsifications in empirical research. Tödter ([Töd], p.349) concludes, "Benford tests do not provide conclusive evidence, but they can help identify papers that need closer inspection and thus complement the control mechanisms already in place."

Recently, Diekmann and Jann ([DieJa], p.397) questioned Benford's Law as a good instrument to reveal fraud in statistical and scientific data. They argued that for a valid test "the probability of 'false positives' and 'false negatives' has to be low." This probability of a false decision (ϕ) is the sum of falsely rejecting the null hypothesis and falsely accepting it,

$$\phi = \alpha(1 - \Omega) + \beta\Omega, \tag{12.11}$$

where Ω is the probability of fraud in the population and α (β) is the probability of a Type I (II) error. If a small Type I error is chosen, $\beta > \alpha$ can safely be assumed, implying that the probability of a false decision lies somewhere in the interval $\alpha < \phi \leq \beta$. This is typical for statistical tests in general and not a characteristic of Benford tests. Nonetheless, [DieJa] (p.400) claim that replication "is the most promising remedy to reduce erroneous results in science." In the next section it is argued that replication and Benford tests are best seen as complementary strategies: Benford tests increase the effectiveness of replication studies.

12.5 REPLICATION AND BENFORD'S LAW

Independent review of research results is a cornerstone of science, and **replication** is widely considered as the prime strategy against scientific misconduct. However, in contrast to natural sciences, social sciences lack a distinct tradition of replication.[7] In their "call for replication studies," Burman et al. ([BurRA], p.787) argue,

---

[7]The electrochemists Stanley Pons and Martin Fleischman reported in 1989 that a small tabletop nuclear fusion experiment at room temperature (cold fusion) had produced excess heat. The result was

"A basic requirement for **scientific integrity** is the ability to replicate the results of research, and yet, with some occasional historical exceptions, replication has never been an important part of economic research. The absence of replication studies is particularly problematic because empirical economic research is often prone to error."

In economics, few professional journals request from authors the filing of data, programs, and documentation. Even if they do, attempts to replicate studies mostly fail. McCullough et al. [McCMH] analyzed 150 articles from the *Journal of Money, Credit, and Banking* but failed to replicate the results in more than 90% of the cases. But, as [McCVi] (p.888) point out, "Research that cannot be replicated is not science, and cannot be trusted either as part of the profession's accumulated body of knowledge or as a basis for policy." Even when feasible technically, replication is fraught with severe problems. First, replication fails to detect fraud when authors submit dressed up data along with tuned software from which all results neatly follow. Second, replication is rare because empirical researchers have poor incentives to engage in replication exercises ([Ham], p.726). Third, referees of empirical papers usually will not bother to obtain the data and computer code necessary to reestimate the models. Rational researchers weigh the costs of replicating the results of others against the benefits of doing their own research. And fourth, replication that takes place after publication leaves room for false results to influence private behavior and public policy.[8] In contrast to replication, Benford tests are quick, cheap, and capable of routine, automated implementation before publication.

Assume that the proportion Ω ($0 \leq \Omega \leq 1$) in the population of N papers submitted to a journal is fabricated. Let replication be feasible and reveal fraud with probability one.[9] A strategy of "pure replication"[10] randomly selects n ($< N$) papers for replication. Since replication is very costly, sampling is done without replacement. The number of uncovered falsifications follows a hypergeometric distribution $\mathrm{HG}(n, N, N\Omega)$, where $N\Omega$ is the number of fraudulent articles in the population. The expected number of fraudulent papers in the sample is $n\Omega$ with variance $n\Omega(1 - \Omega)(N - n)/(N - 1)$.

The alternative strategy first routinely prefilters all N submitted articles by Benford tests for leading digits anomalies. The expected proportion of rejected articles is the sum of false rejections (false positives) and correct rejections (true positives)

---

soon rejected by the scientific community because the experiment could not be replicated consistently and reliably by other laboratories. In September 2011 a team at CERN announced that neutrinos travel faster than light, contradicting Einstein's Theory of Relativity. Replicating the experiment a few months later, another group at the same laboratory measured the particles traveling at exactly the speed of light. For details see the *Wikipedia* articles on *Cold Fusion* and *Measurements of Neutrino Speed*.

[8] A fraudulent paper in medical research had been cited 227 times before it was eventually withdrawn ([Han], p.25). Even without life-or-death consequences in economics, policy decisions on bank regulation, health issues, or climate change, to name only a few, that are based on faulty research can have huge costs for the society.

[9] Clearly, this assumption is not very realistic. It was mentioned above that in a comprehensive study [McCMH] failed to replicate 90% of the papers investigated. Maybe this means fraud is very widespread.

[10] Here, pure replication means the technical reproduction of numerical results from empirical research using the same data and/or software, in contrast to scientific replication [Ham] where a different data set is used in an attempt to check the robustness of the conclusions reached in a study.

denoted as θ in (12.9). Thus, the probability of a fraudulent paper in the pool of rejected ones is

$$\Phi = \frac{(1 - \beta)\Omega}{\theta}. \tag{12.12}$$

In a second step a sample of n papers is chosen from that pool (assuming $n \leq N\Phi$), again without replacement. Now, the number of uncovered falsifications follows the hypergeometric distribution $HG(n, N, N\Phi)$, with expectation $n\Phi$ and variance $n\Phi(1 - \Phi)(N - n)/(N - 1)$.

Relative efficiency (ρ) of both strategies, defined as

$$\rho = \frac{n\Phi}{n\Omega} = \frac{1 - \beta}{(1 - \beta)\Omega + \alpha(1 - \Omega)}, \tag{12.13}$$

is a function of three parameters: size and power of the tests and the ratio of fabricated papers. Relative efficiency increases with power $(1 - \beta)$ and decreases with size (α) and the degree of falsification (Ω). **Prefiltering** is more effective than "pure replication" if the probability of picking a faulty paper from the pool of rejected ones is greater than the probability of choosing it from the whole population, which means if ρ is greater than 1. This condition simplifies to $(1-\beta-\alpha) > (1-\beta-\alpha)\Omega$. Thus, the strategy of prefiltering the data with Benford tests is more effective than "pure replication" under the weak condition that the Benford tests are unbiased $(1 - \beta > \alpha)$. Prefiltering is most effective if Ω is small. As an example, let $\{\alpha = 0.05, 1 - \beta = 0.3\}$ and $\Omega = 0.1$ (0.2). From (12.13) we get $\rho = 4$ (3). Thus, prefiltering increases the probability of detecting a fraudulent paper by *a factor of 3 to 4* when Ω lies between 10 and 20%, even if the tests have the assumed moderate power.

12.6 CONCLUSIONS

This chapter reviews work on using Benford's Law to detect data anomalies, irregularities, and fraud in government accounting data, macroeconomic forecasts, and published economic research. Widely used macroeconomic *Consensus Forecasts* include a large proportion of excessively rounded real growth and inflation forecasts with potentially severe distortions of their information content. Benford tests for national accounting data of EU-member states detected fraud and manipulation, in particular in the case of Greece. Benford tests applied to regression results published in *Empirica* and *Applied Economics Letters* confirmed Benford's Law for first and second digits of regression coefficients and standard errors in the aggregate. In individual articles a surprisingly large proportion of first digits violate Benford's Law, in contrast with second digits.

Repeatedly surfacing spectacular cases in the press indicate that science is not free of dishonesty and fraud. Evidence is mounting that this is not misbehavior on the part of "a few bad apples" but rather the "tip of an iceberg" [Fan]. Honesty in research is essential for the credibility and support of science in the public. It is illusive to eliminate fabrication, falsification, plagiarism, and other questionable research practices altogether from published research. However, routine applications of Benford tests, preferably prior to publication, could uncover data anomalies and provide first hints of irregularities in empirical economics research. As diagnostic checks to uncover scientific misconduct, Benford tests would increase the efficiency of replication exercises and raise the risk for "scientists behaving badly" [MaAdV].

Chapter Thirteen

Testing for Strategic Manipulation of Economic and Financial Data

Charles C. Moul and John V. C. Nye[1]

After surveying applications of Benford's Law within economics, we consider how a first-digit analysis informs Value-at-Risk data from the U.S. financial sector over the past ten years. We find that Benford's Law fits precrisis data very well but is rejected for postcrisis data. Opportunities and incentives for such misreporting are discussed.

13.1 BENFORD IN ECONOMICS

Although there is a literature on the phenomenon of first digit distributions that goes back to the nineteenth century, the literature on Benford's Law in economics and political economy is mostly of recent vintage. The neglect of Benford's Law in economics is especially strange considering the distinction of its earliest investigators. Apparently the late George Stigler had independently been studying the first digit phenomenon in the 1940s, and work similar to that of the physicist Frank Benford on what became Benford's Law was found among Stigler's papers after his death (Raimi, ([Rai], 1976)). The first published work in a directly economic context was Varian's suggestion (1976) that Benford's Law could and should be used to investigate the validity of macroeconomic data series that are widely used in economics. While violation of Benford's Law does not guarantee the presence of manipulation, the observation that series which might otherwise be expected to be Benford-consistent but which were noticeably inconsistent would serve to alert us of possible anomalies.

However, interest in Benford's Law languished over the next two decades, with economists showing little interest in work devoted to assessing general data validity. For the most part, published work appeared in applied math, accounting, or natural science journals. Nye and Moul ([NyM], 2007) on the unreliability of many of the data series used in international macroeconomic comparisons was probably

[1]Economics Department, Farmer School of Business, Miami University, Ohio; and Nye Mercatus Center and Economics Department, George Mason University and National Research University – Higher School of Economics, Moscow, respectively; We thank Marc Taub for excellent research assistance.

the first work in the field that took the suggestions in Varian ([Va], 1972) to heart. Early work that served as a precursor to what evolved into Nye and Moul ([NyM], 2007), however, could not find a home in an economics journal in the late 1990s. Despite positive referee reports it was commonplace for early drafts submitted as far back as 1997 to be rejected by economics journals on the grounds that a) no referee could be found to judge the paper and b) in any event, assessing data quality through the use of Benford's Law was "not economics." Eventually, Nye and Moul ([NyM], 2007) was accepted by an online economics journal despite the editor's avowed difficulty in finding someone even willing to referee the submission due to lack of familiarity with the issue. Nonetheless, by the mid 2000s, interest in Benford's Law and its potential application to the social sciences, especially economics and political science, had grown, and a number of papers had begun circulating on similar themes.

Nye and Moul ([NyM], 2007) dealt with a seemingly simple issue—the quality of reported national income (GDP/GNP) series at the international level. In addition, it considered the extent to which standard economic transformations successfully preserved the Benfordness of the underlying series. The paper demonstrated that in a number of cases—especially for African nations and those from the communist bloc—strong deviations from Benford's Law occurred even while the data set for all nations and for nations excluding these areas showed strong conformity to Benford's Law. In addition, there were some puzzling and unexplained deviations from Benford's Law in the transformed data typically used by macroeconomists.

Since then Gonzalez-Garcia and Pastor ([GoPa], 2009) have revisited this work and suggested another plausible reason for violation of Benford's Law. Their claim is that macroeconomic structural shifts may account for some violations of Benford's Law that are not due to data quality or fraud. However, this work does not particularly rule out poor data quality and strategic potential for manipulation; it merely gives us an alternative hypothesis to explore. Nor does it deal with the problem of how data that conform to Benford's Law in raw form cease to conform once put through certain macroeconomic transformations such as adjustments for inflation or differing price levels.

In recent years, the biggest payoff to the use of Benford's Law in the social sciences seems to have centered on the evaluation of published statistics by both public and private agencies. Most of the business related papers using Benford's Law were more focused on **accounting issues**—identifying cases in which fraud may have distorted reported statistics. In contrast, the social sciences have found greater use since Nye and Moul ([NyM], 2007) in focusing on the potential inconsistencies with Benford's Law as a test for problems in governmental and other publicly reported data sets. In cases in which the data have good reason to conform to Benford's Law or where it can be demonstrated that most series do conform to Benford's Law, evidence of deviation by a subset of series can be used as a first test for further investigation as to the potential for data tampering or misrepresentation.

For example, Judge and Schechter ([JuSc], 2009) show that widely used survey data from nine commonly used series on agriculture showed very different conformity to Benford's Law when collected in developing nations in comparison to that which came from the United States. They speculate that this might be because

American farmers have better access to their records when answering surveys and use less "guesstimation" than those from developing regions (p. 12). It is of further interest that a later survey from Mexico in 1997 seems to have been of higher quality than earlier surveys. They too argued that Benford's Law could serve as a preliminary check on data series to complement more conventional econometric efforts to correct for unobservable data errors. In general their view is that rounding errors are the most likely causes of developing country failure to conform to Benford's Law.

Rauch et al. ([RauGBE], 2011) studied European Union governmental economic data to see how those series most relevant to compliance with the Stability and Growth Pact criteria might have been affected. Using data submitted by all the euro states, Romania, Latvia, and Belgium do poorly in their ranking of relative conformity to Benford. Their most noticeable finding, however, is that data from Greece showed the greatest deviation from Benford's Law. The fact that Greek data manipulation has been officially confirmed by the European Commission is put forward by the authors as evidence of the value of using Benford's Law as a forensic tool (p. 253). They make clear that these tests are meant to be suggestive, not definitive.

Michalski and Stoltz's [MiSt] (2013) observed that nonconformity of financial data for balance of payments series is not driven just by countries with low institutional quality ratings or by countries from sub-Saharan Africa. Rather, this nonconformity stems at least in part from countries behaving strategically. For economic reasons favorable to their policies, those with fixed exchange rates, high negative net foreign asset positions, negative current account balances, or nations which are generally more vulnerable to capital flow reversals, have more reason to manipulate balance of payments series. They thus provide a strategic rationale for those nations' data being more likely to fail tests of conformity with Benford's Law. We note that this paper's publication at the prestigious Review of Economics and Statistics reverses earlier rejections of precursor work from the 1990s making use of Benford's Law in macroeconomics as being unsuited for economics journals.

These recent papers are notable for their emphasis on issues of data quality and data manipulation with a very different emphasis from forensic work in the accounting literature, but this list is far from exhaustive. Other published work in economics or of interest to scholars in political economy, finance, or related social science research that has appeared in recent years include De Marchi and Hamilton ([DeHa], 2006), Giles ([Gil], 2007), Auffhammer and Carson ([AuCa], 2008), and Diekmann and Jann ([DieJa], 2010).

Our later application touches on the use of Benford in finance, and so we would be remiss not to highlight the notable contributions in that field. Ley ([Ley], 1996) found that first digits of one-day returns from stock indexes mimic the Benford distribution quite well and better than most of Benford's original examples. The large sample sizes (daily observations of Dow Jones Industrial Average 1900–93 and of Standard & Poor's 1926–93) led to formal statistical rejections of the Benford null, but the similarities of the observed and predicted distributions strongly suggest the idea's underlying power.

De Ceuster et al. ([DeDS], 1998) take this insight a step farther with respect to the apparent psychological barriers regarding stock indices. Specifically, the literature had implicitly assumed a uniform distribution of first digits and, based on that foundation, shown that stock indices faced resistance or support at levels ending in zero(s). These levels of course coincide tightly with Benford's insights. By explicitly using a Benford distribution instead of the uniform distribution as the null of typical behavior, De Ceuster et al. find no evidence of such psychological barriers in the Dow Jones Industrial Average, the Financial Times Stock Exchange, or the Nikkei Stock Average.

Corazzo et al. ([CoEZ], 2010) build on this Benford presumption and explore instances when that distribution does not hold. After confirming that S&P 500 daily returns for the 361 continuously present stocks from 1995 to 2007 essentially mimic Benford's Law, they identify the days on which Benford is most decisively rejected. Many, though not all, of these days have clear linkages with news events of the day. The most decisive rejection occurred on February 27, 2007, the day with one of the largest point drops since the markets reopened after September 11, 2001. The authors further point out the curiosity that the vast majority of decisive rejections in their sample occurred after September 11th, suggesting that markets were more anomalous after those events than before.

Given the nascent literature, this field is still ripe for exploitation. Given the strong interest in strategic behavior in economics, it makes sense to use Benford's Law to investigate possible anomalies that suggest manipulation or other interference especially when incentives increase for such tampering. One does not have to believe that all deviations from Benford's Law represent anomalies or manipulation to understand that changes at the margin of the relative deviation of various series from conformity to Benford's Law are a very good preliminary check on the likelihood that changes in conformity correlated to changes in the strategic benefits of manipulation or error prone recording would serve as a "smoking gun" for further investigation. In that light we present a simple application of Benford's Law to banking data related to the recent crisis. Indeed, as this article was being prepared several articles on the use of Benford's Law for fraud detection in banking were completed (Grammatikos and Pappanikolaou [GrPa], Özer and Babacan [ÖzBa], Rauch, Göttsche and F. El Mouaaouy [RauGM], Mir [Mir], and Hoffmarcher and Hornik [HofHo]).

13.2 AN APPLICATION TO VALUE-AT-RISK DATA

Shareholders, executives, and regulators are all interested in the riskiness of portfolios at banks' trading desks. As events continually show, traders tend to push risky portfolios that work well most of the time but occasionally fail in spectacular fashion. Such trading, be it in equities, commodities, or bonds, has historically been associated with investment banks, but a number of large commercial banks also operate their own trading desks. A popular if imperfect measure for assessing the risk of these portfolios is **Value at Risk** (**VAR**). While different banks use somewhat different techniques in calculating VAR, the following example should suffice. Imagine that a trading desk begins the day with an allocation of commodities, debt, equities, and other financial assets. Analysts consider how this portfolio would have performed in each of the trading days of the past year (roughly 250 days). These daily returns (both gains and losses) are then ordered. A 95% (99%) VAR takes the 5th (1st) percentile from the bottom and should then be interpreted as implying that a firm with this portfolio can expect to lose at least this much with 5% (1%) probability.[2] Trading desks therefore provide newly calculated VARs each day as the portfolio and relevant history changes.

The largest investment banks began publishing summary statistics of their VAR measures in the 1990s, and some measure of VAR in financial reporting was quite common by 2002. Measures usually include the total for the trading desk and are often broken down across different asset types (e.g., Fixed Income VAR, Commodities VAR, etc.). The specific statistics varied widely across banks' quarterly and annual financial reports but often included an average (either over the quarter or over the year), extreme values over the period, and the VAR on the last day of the period.

This VAR measure should therefore roughly vary with the size of the bank's investment portfolio. Larger banks will accordingly have larger VARs, and the converse for smaller banks. To the extent that the size and composition of bank investment portfolios change over time, we also expect the VARs to mimic those changes. Given the dispersion among bank sizes and investment portfolios over time that we observe, it is reasonable to believe that Benford should hold in the absence of distortionary tactics. The absence of distortionary tactics, however, should not be taken for granted.

As a number of parties will be supervising the market risk associated with a portfolio, it may be in traders' interests to distort VAR in order to make portfolios seem less risky than they truly are. By doing so, traders put themselves in the position of holding especially risky portfolios. This enables them to reap large gains in profitable states of the world and leave the losses with their banks in the unlucky states. The most straightforward way to minimize VAR would be to smooth daily gains and losses. Suppose that a desk earns $100M one day but fears the next day will bring a sizable loss. By only reporting $50M on the first day, the desk can mitigate its losses the next day. Because VAR looks at infrequent extreme outcomes, a few such actions could notably disguise the true risk of a portfolio.

The apparent risk of portfolios has seen dramatic swings over the past decade. While the U.S. housing bubble and inevitable correction were oft-discussed by 2005 and 2006, relatively few economists at the time publicly made the seemingly obvious connection that much of the collateral (specifically mortgage-backed securities) supporting the U.S. financial sector was vulnerable. The system in which banks insured each other against losses through credit default swaps provided a mistaken level of confidence, but this system had not faced a large-scale panic before. Losses on mortgage-backed securities at Bear Stearns led to its near bankruptcy and then acquisition by JPMorgan (arranged by the New York Fed) in March 2008. A similar story unfolded for Merrill Lynch which was acquired

---

[2]A common criticism is that VAR is often erroneously described as the *most* that a desk's portfolio will lose with some probability.

by Bank of America in September 2008. No buyer could be found for Lehman Brothers at the same time, and its bankruptcy and now worthless commercial paper caused the money markets to lock up. The Federal Reserve then intervened in a myriad ways, and the sole surviving investment banks Goldman Sachs and Morgan Stanley became bank holding companies in the fall of 2008 in order to better access Federal Reserve lines of credit. During this time, U.S. equity markets suffered a correction outside of living memory. From October 2007 to March 2009, the Dow Jones Industrial Average fell 54%, and the more comprehensive Wilshire 5000 and S&P 500 fell 57%. Changes in commodities markets during this time were less uniform, with oil prices down 50%, silver prices flat, and gold prices up 30%.

When compared to the relative calm of the 2002–07 period, this volatility was bound to appear in trading desks' latter-day VAR measures. Such turbulence, however, should not have affected the Benfordness of the measure. Even after the departure of the three investment banks, VAR measures by reporting banks spanned the orders of magnitude necessary for Benford to hold. Furthermore, this increased volatility, as it appeared in the historical record for the VAR constructions, should have provided exactly the sort of variation under which one would expect Benford. If Benfordness is rejected for the VAR sample from 2008–2012, it is not directly because of the aforementioned events, though it may indirectly stem from those events if attitudes or constraints changed.

Given this background, we used the SEC's **Electronic Data-Gathering, Analysis, and Retrieval (EDGAR)** system to collect VAR measures from the financial statements (10-Qs, 10-Ks, and annual reports) of publicly traded commercial and investment banks in the U.S. from 2002.1 to 2012.1. Trading desks are relatively unimportant for smaller commercial banks, and we had limited success in finding VAR measures in those cases. In addition to currently active firms, we also included the three investment banks (Bear Stearns, Merrill Lynch, and Lehman Brothers) that did not survive the financial crisis of 2008 as independent firms. Table 13.1 lists the financial companies (with stock symbol) with quarters of observation and number of VAR observations.

Both VAR averages and daily observations should satisfy Benford's Law, so we limited ourselves to those statistics.[3] Where averages over the past four quarters were provided in lieu of a quarterly average (as was common in annual reports), we inferred the quarterly average. In all, we collected VAR data for 17 banks. As Table 13.1 indicates, investment banks such as Goldman Sachs were much more likely to provide detailed breakdowns of VAR across asset type than commercial banks not known for their trading activity such as Wells Fargo. Because the union of Benford sets is also Benford, we aggregate the VAR measures for the entire trading desk, for the specific assets, for both quarterly average and last day of quarter, and for both 95% and 99% criteria into a single data set.

Our full sample has 3632 observations of VAR. Given our interest in the financial

---

[3] Quarterly minimum and maximum VAR could not be reliably recovered when quarters were aggregated.

Company (Stock Symbol)	Period of Obs.	#VAR Obs.
Bank of America Corp. (BAC)	2002.1–2012.1	243
Bank of New York Mellon Corp. (BK)	2002.1–2012.1	390
BOK Financial Corp (BOKF)	2002.1–2012.1	41
Bear Stearns Companies, Inc. (BSC)	2002.1–2008.1	210
Citigroup Inc. (C)	2002.1–2012.1	418
Goldman Sachs Group, Inc. (GS)	2002.1–2012.1	410
JPMorgan Chase & Co. (JPM)	2002.1–2012.1	582
Key Bank (KEY)	2002.1–2012.1	52
Lehman Brothers Holdings Inc. (LEH)	2002.1–2008.2	220
Merrill Lynch & Co., Inc. (MER)	2002.1–2008.3	262
Morgan Stanley (MS)	2002.1–2012.1	458
PNC Financial Services Group, Inc. (PNC)	2003.2–2012.1	49
Regions Financial Corp. (RF)[a]	2004.3–2011.4	59
Raymond James Financial Inc. (RJF)	2003.2–2012.1	58
SunTrust Banks, Inc. (STI)	2002.4–2012.1	48
State Street Corp. (STT)	2002.1–2012.1	99
Wells Fargo & Co. (WFC)	2004.1–2012.1	33

a) Region's Financial 10-Q in 1Q2012 stated that VAR measures for that quarter were "immaterial."

Table 13.1 List of companies with published VAR, alphabetized by stock symbol.

crisis, we split the sample between before 2008 and after 2008 inclusive (respectively 2288 and 1344 observations). Our distinctions between time periods are intuitive if somewhat arbitrary. Prior to 2007 and the implications of the popped housing bubble becoming obvious, VAR measures were a lower concern both to traders and their watchers. After 2007, the greater market volatility presumably drove up the perception of underlying risk and watchers' concerns but at the same time may have driven down trading desks' appetite for risk. We thus have no ex ante beliefs regarding how the slices of the data will affect the Benfordness of the subsets. To the extent that the χ^2 test does not properly account for sample size, we would expect that the larger, pre-2008 sample would be more likely to reject Benfordness.

Table 13.2 displays the results of simple Benford tests for the various subsamples of the data. In addition to the Benford and observed first-digit likelihoods, we also show the χ^2 statistic against the null of Benfordness and the corresponding p-value. We can decisively reject the null that the entire sample has a Benford distribution; much of this power comes from the mismatch of the prevalence of 2 as the leading digit. A comparison of the Benfordness of the data before 2008 and after 2008 (inclusive), however, is far more insightful. The first digits from the VAR data from 2002.1 to 2007.4 match the Benford distribution of first digits very well. No statistical rejection is possible using the standard χ^2 test. The story changes radically when we examine VAR measures from the time period that included and

	Benford	'02–'12	'02–'07	'08–'12
1	0.3010	0.2902	0.2959	0.2805
2	0.1761	0.2073	0.1823	0.2500
3	0.1249	0.1302	0.1281	0.1339
4	0.0969	0.0903	0.0975	0.0781
5	0.0792	0.0732	0.0795	0.0625
6	0.0669	0.0625	0.0621	0.0632
7	0.0580	0.0490	0.0503	0.0469
8	0.0512	0.0512	0.0511	0.0513
9	0.0458	0.0460	0.0533	0.0335
χ^2		31.7342	6.9173	61.6254
p-value		0.0001	0.5456	0.0000

Table 13.2 Benford test on VAR first digit. The number of observations for the three periods are, respectively, 3632, 2288 and 1344.

followed the financial crisis (2008.1 to 2012.1). Despite the smaller sample size, Benfordness is again rejected. Again, the prevalence of the first digit 2 drives the rejection. While Benford predicts observing 70% more 1s than 2s, the latter sample shows only 12% more 1s than 2s.

As with most Benford results, this finding is more suggestive than conclusive. We have no way to distinguish among the (not mutually exclusive) stories since 2008 that traders have manipulated VAR for private gain or that the government regulators have turned a blind eye to manipulation for the sake of financial confidence or that the crisis served as a structural break. Nevertheless, it is intriguing that Benford tests so cleanly identified the crisis. Other slices of the data (not reported) showed far murkier results using 2007 rather than 2008 as the delineating year. It is our hope that similar analyses may spark new inquiries on comparable topics.

PART V
Applications III: Sciences

Chapter Fourteen

Psychology and Benford's Law

Bruce D. Burns and Jonathan Krygier[1]

There has been very little interaction between psychology and Benford's Law, partly because initially, behavioral data showed that people did not spontaneously generate random numbers conforming to Benford's Law. However in this chapter we outline recent research showing that people can approximate Benford's Law when generating meaningful numbers. This has theoretical implications for decision-making research, practical implications for fraud detection, and may help cast light on Benford's Law as a property of natural data.

14.1 A BEHAVIORAL APPROACH

Researchers both inside and outside the field of psychology have rarely viewed Benford's Law from a behavioral perspective. Even when Dehaene [Deh] pointed to data showing that the words for low digits occur in human language more often than those for high digits, he suggests that this discrepancy may not have anything to do with Benford's Law because "The exact origin of this [Benford's] law is still poorly understood, but one thing is certain: This is a purely formal law, due solely to the grammatical structure of our number system. It has nothing to do with psychology." (Dehaene [Deh], p. 112)

However, we believe that examining how human behavior relates to Benford's Law is of growing importance for both practical and theoretical reasons. For example, because Benford's Law is increasingly utilized in fraud detection, we ideally would like to know whether and when people can spontaneously generate numbers that approximate it. As Bolton and Hand [BolHa] pointed out in their review of statistical fraud detection, "The premise behind fraud detection using tools such as Benford's Law is that fabricating data which conform to Benford's Law is difficult" (p. 238). Thus, the usefulness of Benford's Law as a tool for fraud detection rests on the assumption that people are poor at deliberately generating numbers that conform to it, just as they are generally poor at generating random numbers (Rapoport & Budescu [RapBu]). However, this assumption has not yet been systematically tested.

[1]School of Psychology, The University of Sydney, NSW 2006, Australia. The authors would like to thank Hal Willaby for comments on an earlier draft. The authors were supported by a grant from the University of Sydney, and it is a pleasure to thank them for their generosity.

A more theoretical example of the increasing importance of Benford's Law stems from its potential to cast light on how people generate unknown quantities, a question which has implications for judgment and decision making. Assuming people spontaneously generate numbers that approximate Benford's Law, understanding why and under what conditions they do so could inform models of decision making. Furthermore, if people can act as Benford's Law generators, then they might even be potential models for testing some of the speculations about why Benford's Law appears in nature.

The published empirical literature on whether people generate numbers that approximate Benford's Law is small, perhaps because an early consensus emerged that people do not. In this chapter we review early empirical work along with recent results suggesting that this conclusion may not hold under appropriate conditions. We then consider the implications of these findings and what new questions need to be addressed.

14.2 EARLY BEHAVIORAL RESEARCH

One of the earliest citations of Benford's Law in academic journals was a behavioral test: Hsü [Hsü] (1948) asked participants to write down a "4-digit number that must be original, i.e., created in your own mind." He found no evidence of a fit between the distribution of the **first significant digit** (**FSD**) in the data from his 1044 participants and the frequencies predicted by Benford's Law. However, there was a substantial deviation from a uniform distribution, indicating that people do not generate numbers like dice do. The distribution he found can be seen in Table 14.1. The most common first significant digit was 4, possibly due to the request being for a 4-digit number (i.e., a priming effect from mentioning "4"). Years later, Hill [Hi1] (1988) asked 742 undergraduate students to write down the first 6-digit number that came to mind. The FSD distribution in Table 14.1 shows that digit 1 was generated more often than expected by a uniform distribution, but the data still did not fit Benford's Law. Digit 6 was the most common initial digit, perhaps again due to priming from the request for a 6-digit number.

Kubovy [Kub] further investigated the digit priming effects on number generation by proposing contrasting hypotheses: (1) specifically mentioning a digit as an example could reduce the frequency with which it is produced by making it less representative of a spontaneously generated number; (2) priming due to incidentally mentioning a digit could increase the availability, and hence the production, of a particular digit. For example, he found that a request for a single digit number increased responses of "1" above baseline; however, offering "1" as an example of a potential response dramatically reduced responses of "1." As shown in Table 14.1, a request for a 4-digit number resulted in an initial digit 4 far more often than a (mathematically identical) request for a number between 1000 and 9999. These results support the explanation that the digit 4 peak in Hsu's ([Hsü], 1948) data and digit 6 peak in Hill ([Hi1], 1988) data were due to priming. In none of these experiments did participants produce a distribution of first digits that even approximately fit Benford's Law.

First digit	1	2	3	4	5	6	7	8	9
Hsü [Hsü] 4-digit number "created out of your own mind," n=1044.	13.3	9.2	14.3	15.5	6.6	9.3	12.6	9.1	10.5
Kubovy ([Kub], Exp. 3) "first number between 1000 and 9999 that comes to mind," n=116 (estimates from Burns, 2009)	51.7	5.3	11.7	4.3	10.3	0.8	6.1	5.3	4.4
Hill [Hil] "a 6-digit number out of your own head" n=742	14.7	10.0	10.4	13.3	9.7	15.7	12.0	8.4	5.8
Scott, Barnard & May ([ScoBM], Exp. 1) "number from 1 million–10 million" unelaborated data only, n = 90 (estimates from Krygier [Kry])	5.97	11.9	20.9	8.2	26.1	8.2	8.95	5.2	4.5
Scott, Barnard & May ([ScoBM], Exp. 1) "a number from 1 million–10 million" elaborated data only n = 46 (estimates from Krygier [Kry])	49.0	9.5	16.3	1.4	6.8	2.7	6.8	2.0	5.4

Table 14.1   First-digit percentages (some estimated from graphs) found in studies showing that human-generated numbers fit poorly to Benford's Law. The question asked and the sample size (n) for each study is provided.

Although Scott, Barnard, and May [ScoBM] focused not on Benford's Law but on the effect of digit priming on the distribution of the first digits of responses, they found no evidence supporting Benford's Law when participants were asked to generate random numbers. They did, however, find different distributions when they separated responses containing only one non-zero digit (e.g., 1000), which they called unelaborated responses, from responses with multiple non-zero digits (e.g., 1045), which they called elaborated responses. Table 14.1 shows that for the unelaborated responses, first-digit distributions tended to have a peak at digit 5 and a smaller peak at digit 3, neither of which were evident in the elaborated data. The authors suggest that this may be because the greater executive cognitive processing required to generate elaborated numbers resulted in numbers less influenced by basic number preferences. For the elaborated responses a big peak for digit 1 was observed, but still the distribution did not follow Benford's Law.

The papers listed in Table 14.1 represent all the published behavioral tests of Benford's Law prior to 2007 that we have been able to identify. Although they do appear to show a bias towards digit 1 as a first digit, they strongly suggest that requests for spontaneously generated numbers do not produce a first-digit distribution like that of Benford's Law. This is despite evidence that generated first-digit distributions are sensitive to subtle factors, such as digit priming. Therefore the absence of psychological consideration of Benford's Law appeared to be amply justified by the empirical data.

14.3 RECENT RESEARCH

14.3.1 Diekmann (2007)

The first published study suggesting that people may be able to spontaneously generate first-digit distributions that approximate Benford's Law was by Diekmann [Die] in 2007. He first examined 1457 published regression coefficients from two volumes of the *American Journal of Sociology* and found that their FSD distribution reasonably fit Benford's Law, or at least that their FSD frequencies monotonically declined with digit size. In two subsequent experiments he asked statistics students to fabricate regression coefficients to support a particular hypothesis. Unlike earlier studies, the FSD distribution of the data they generated was a good fit to Benford's Law; the FSD distributions, shown in Table 14.2, failed to reject the null hypothesis that the results were distributed according to Benford's Law. However, Diekmann's sample sizes were small, and he did not address the discrepancy between his and earlier studies.

First digit	1	2	3	4	5	6	7	8	9
Benford ([Ben]) logarithmic distribution	30.1	17.6	12.5	9.7	7.9	6.7	5.8	5.1	4.6
Diekmann ([Die], Exp. 1) "fabricate plausible 4-digit regression coefficients," $n = 10 \times 10$	37.0	21.0	10.0	11.0	9.0	2.0	3.0	6.0	1.0
Diekmann ([Die], Exp. 2) "fabricate plausible 4-digit regression coefficients," $n = 13 \times 10$	26.2	19.2	10.8	5.4	2.3	5.4	5.4	10.8	4.7

Table 14.2   First-digit percentages from Diekmann [Die] compared to Benford [Ben], showing that human-generated numbers can fit reasonably well to Benford's Law. Note that "$n = 10 \times 10$" indicates that 10 participants generated 10 numbers each.

14.3.2 Burns (2009), Study 1

In 2009 Burns [Burns] suggested that the critical difference between earlier studies and Diekmann ([Die], 2007) was that earlier studies did not provide any meaningful context for the numbers participants were asked to generate. Instead, they explicitly asked participants to produce a random number, a choice that avoided a variety of biases that presenting a specific context may have invoked. We know that people are poor at generating truly random numbers and that, when they try, they display their biases about how they expect numbers to be distributed (Rapoport & Budescu [RapBu], 1992). Thus, asking participants for random numbers would appear to be

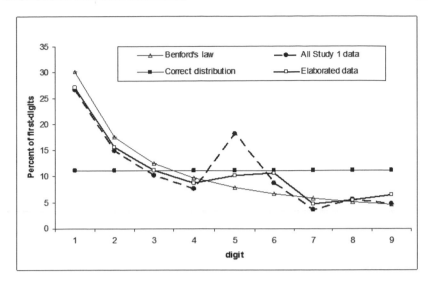

Figure 14.1   Distribution of first digits in Pilot Study 1 data. This shows the distributions
we tested the data against: Benford's Law and the (correct) flat distribution. All
data from Burns [Burns] Study 1 is shown plus the subset of elaborated data.
Note that data is a reasonable, but not perfect fit, to Benford's Law.

an appropriate means of examining whether Benford's Law biases number generation. However, there is no reason to expect truly random numbers to fit to Benford's
Law, and it may be their meaninglessness that obscured evidence of Benford's Law
in earlier studies.

To test whether generating meaningful numbers could produce behavioral evidence for Benford's Law, Burns [Burns] asked students to estimate quantities from
fields similar to those for which Benford [Ben] collected data, such as newspaper
circulations and river lengths. If people are sensitive to statistical relationships in
their wider environment, the best behavioral test of Benford's Law would utilize
fields that in reality fit it, even if participants are not aware of that fit.

In Burns [Burns], Study 1, a computer asked 127 psychology students "to try to
estimate" the answers to a set of nine questions and to "just guess" if all else failed.
Each question referenced a different field, such as the "population of the urban area
of Philadelphia, USA" and the "area drained by the Pearl (Xi Jiang) river." Each of
the nine correct answers had a different leading digit; thus, if a participant answered
all questions correctly, each digit 1 through 9 would be a first digit exactly once.
Therefore, both correct and random answers should yield a flat distribution of first
digits.

The first digits of each participant's answers were extracted, and the percentage
of their nine answers using each digit was calculated. Figure 14.1 shows the mean
percentages for each first digit together with lines representing Benford's Law and
the distribution of the correct answers. The FSD distribution was closer to Ben-

First digit	1	2	3	4	5	6	7	8	9
Benford (1938) logarithmic distribution	30.1	17.6	12.5	9.7	7.9	6.7	5.8	5.1	4.6
Generation (all data)	23.9	15.1	9.7	7.6	19.0	9.1	5.5	5.8	4.4
Generation (elaborated data only)	26.6	16.3	9.1	9.6	11.7	8.4	7.2	5.7	5.2
Selection data	14.3	13.4	9.7	10.1	10.6	9.9	10.5	8.7	12.9

Table 14.3 First-digit percentages for Burns [Burns] Study 2, for the generation task (all data and elaborated responses only) and when selecting from a set of nine potential answers.

ford's Law than to the correct (flat) distribution. As can be seen, the data with the exception of digit 5 was an approximate fit to Benford's Law.

Scott et al. [ScoBM] also found peaks at digit 5 for first digits when they asked participants to generate random numbers. Therefore, Figure 14.1 also includes the distribution for FSD of only elaborated numbers (e.g., excluding the answer "5000"). As can be seen, the peak at digit 5 was reduced, and the fit to Benford's Law was significantly improved.

14.3.3 Burns (2009), Study 2

In Study 2, Burns [Burns] replicated the generation result with a new set of questions drawn from similarly meaningful fields (see Table 14.3) and also gave participants a selection task in which they chose answers from sets provided to them. Given that in reality random numbers should not fit to Benford's Law whereas meaningful numbers like those examined by Benford [Ben] do, it makes sense that attempting to generate random numbers does not yield fits to Benford's Law whereas attempting to generate meaningful numbers does. Conceivably, participants' lifelong exposure to data from real fields which conform to Benford's Law has made them sensitive, perhaps subconsciously, to this distribution. If this is the case, then it would seem likely that Benford's Law should emerge when participants answer questions by *selecting* from numbers with different first digits. Burns's [Burns] Study 2 asked participants at different times to both generate and select numbers as answers to questions about the same fields. Participants selected from amongst nine different answers, each with a different first digit, and the questions were again designed so that correct responses would yield a flat distribution of first digits.

Both generation and selection tasks were presented by computer amongst a set of tasks completed by 335 psychology students. As shown in Table 14.3, the distribution of first digits in the generation task was very similar to that of Burns's [Burns] Study 1, despite utilizing new questions. Again digit 5 deviated the most from Benford's Law, and again the distribution of elaborated responses had a greatly reduced peak for digit 5 and thus more closely approximated Benford's Law. However, Table 14.3 shows that the selection task yielded a much flatter distribution than the generation task, one which was closer to the correct (flat) distribution than to Benford's Law.

14.3.4 Krygier (2009)

Krygier [Kry] in his psychology Honours thesis explored the question of when people will generate FSD distributions that fit to Benford's Law. Expanding on the methodology of Burns [Burns], he manipulated two factors defining the fields for which participants had to generate numbers: familiarity and true fit to Benford. Each participant generated numerical answers to questions in three fields for each of the four conditions familiar/fit (e.g., populations of metropolitan areas), familiar/no fit (e.g., lottery numbers), unfamiliar/fit (e.g., atomic half-lives), and unfamiliar/no fit (e.g., hazard codes). Analysis of his 247 participants' responses could not reject the null hypothesis of no effect of these conditions on degree of fit to Benford's Law. Whereas the distribution of first digits for winning lottery tickets was closer to the flat distribution than to Benford's, the results where complex perhaps because what participants thought was or was not random did not necessarily fit with reality. This was demonstrated by asking a subgroup to rate how random or non-random each of the 12 fields was. The 12 fields' mean randomness ratings correlated negatively with the fields' fit to Benford's Law; the more random a field was perceived to be, the less likely numbers generated for that field were to fit to Benford's Law. More broadly it has been shown experimentally that how people both perceive and generate sequences of events (such as coin tosses) are influenced by whether they perceive them as generated by random processes (Rapoport & Budescu [RapBu]). For example, Burns and Corpus [BurnsCo] showed that the interpretation of streaks of events was influenced by how random people thought the generating process was.

14.4 WHY DO PEOPLE APPROXIMATE BENFORD'S LAW?

The results of the studies described above demonstrated repeatedly that people could generate first-digit distributions that approximate Benford's Law, but they also suggest that this is only the case when they were asked to generate answers to questions from fields for which they thought the answers were non-random. Perhaps this is the case because people encounter actual Benford distributions in their lives.

Recent research in psychology has demonstrated that rapid, subconscious evaluation of environmental cues can proceed without conscious awareness and that these evaluations can influence the interpretation of subsequent, unrelated stimuli (Ferguson, Bargh, & Nayak, [FerBN]). These findings are part of a wider literature on the automaticity affecting many higher mental processes which show effects of environment on both simple and complex behavior that proceed without awareness (Bargh & Ferguson [BarFe]; McCulloch, Ferguson, Kawada, & Bargh [McCFKB]). It is clear that a wide variety of strategies, heuristics, and decision-making biases operate in response to environmental and contextual cues throughout a range of situations. Thus implicit sensitivity to statistical information in the environment provides a possible explanation for why people generate numbers that fit to Benford's Law in meaningful fields. However, attempts to support such an explanation have not yet been successful: Burns [Burns] did not find evidence of Benford's Law when participants had to select answers from nine different numbers, and Krygier [Kry] did not find different FSD distributions when he manipulated familiarity. These studies do not exhaust the possible ways in which this hypothesis could be tested, but for the moment we lack the evidence to support it.

An alternative explanation could draw on Berger and Hill's [BerH4] speculation that Benford's Law exists in nature due to their "mixture of distributions" theorem: Hill [Hi2] noted that, in Benford's [Ben] observations, the more that his data was the result of the combining of data or factors, the better its first digits fit to Benford's Law. Hill ([Hi1], p. 361) stated this as, "If [full-number] distributions are selected at random (in any 'unbiased' way) and random samples are taken from each of these distributions, then the significant-digit frequencies of the combined sample will converge to Benford's distribution, even though the individual [first-digit] distributions selected may not closely follow the law." It is crucial that the sampling be neutral or unbiased as to scale or base. Hill [Hi2] offers a statistical derivation of this proposal, and Berger and Hill [BerH4] further develop it as a mixture of distributions theorem. Therefore it could be hypothesized that people generate numbers with first-digit distributions that approximate Benford's Law because when they draw from their own knowledge to think through possible answers they are effectively combining numbers from different distributions. The finding that elaborated data fit better than unelaborated data to Benford's Law could be seen as consistent with this explanation; Scott et al.'s [ScoBM] analysis suggested that elaborated numbers may be the product of more complex processing, so if complexity leads to accessing multiple sources, it could also lead to mixing from multiple distributions.

14.5 CONCLUSIONS AND FUTURE DIRECTIONS

It appears that people can approximate Benford's Law, but why is not clear. This conclusion has important practical and theoretical implications. In terms of applications of Benford's Law to fraud detection, the evidence suggests that it cannot be assumed that when people make up numbers the resulting distributions will not fit to Benford's Law. However more research is needed into when people are most likely to generate distributions that fit. It appears that people can spontaneously generate numbers approximating to Benford's Law, so fraud tests utilizing Benford's Law could be improved by research into the conditions under which people are likely to generate numbers that fit and the ways in which numbers may deviate from Benford. For example, in our research we have consistently found that the digit 5 tends to be generated with too high a frequency. Therefore perhaps deviations from digit 5 should be given higher weighting in fraud tests, but further study is required to test the generalizability of this finding.

Theoretically Benford's Law has the potential to throw light on how people estimate unknown numbers and thus also on how they make decisions based on such estimates. A common theme in recent research into reasoning and decision making has been that people are influenced by statistical relationships in the environment. This idea is a central part of adaptive approaches to decision making such as that of Gigerenzer and Todd [GiTo], and it underpins apparent automaticities in everyday life (Bargh & Ferguson [BarFe]). Key to these approaches is that people are influenced by statistical relationships even if they have little awareness of them. However, because it is hard to know the precise statistical relationships an individual has experienced over his or her lifetime, rarely is it possible to test whether people are truly acting precisely in accord with an unknown naturally occurring statistical relationship. Therefore Benford's Law may offer an interesting test case, because it is a precise statistical relationship that is both widespread and little known to the public.

Alternatively, if people generate numbers conforming to Benford's Law due to the combining of multiple distributions, then this research also has implications for how people make decisions based on estimates. For example, fit to Benford's Law could be an index of the extent to which multiple sources of information are being combined in a decision. Furthermore, a fit to Benford's Law could even be seen as providing prima facie support for Berger and Hill's [BerH4] "mixture of distributions" theorem. If people are "Benford generators" because they mix distributions, maybe this is a plausible explanation for why other natural processes conform to Benford's Law. It is not possible to test empirically whether natural phenomena fit to Benford's Law due to mixing distributions, but it potentially can be tested for people.

None of our data sets yields an exact fit to Benford's Law. Finding an exact or even very close fit to Benford's Law in either human or natural data would be extremely surprising. Such a fit would imply a lack of any other factors influencing the data; therefore, outside of theoretically generated data, claims of exact fit to Benford's Law tend to be failures to reject the null hypothesis of no difference. With enough data points, data from any field is likely to show some deviation from it. Almost any large data set shows approximate fit rather than exact fit, so human data too should be no better than an approximate fit. What is remarkable is how good the fit of human data is to a function with no free parameters. Whether there are meaningful parameters that substantially improve FSD fit is another issue for future research. For example, fitting the free parameter from Nigrini and Miller's [NiMi1] would improve the fit of human data to Benford's Law. If the same parameter fit multiple data sets then this parameter value might be viewed as a signature for human-generated data.

Our studies have established Benford's Law as a psychological phenomenon, but how and why it is requires more research. We have only just started to scratch the surface of Benford's Law as a behavioral phenomenon.

Chapter Fifteen

Managing Risk in Numbers Games

Mabel C. Chou, Qingxia Kong, Chung-Piaw Teo and Huan Zheng[1]

We apply Benford's law to study how players choose numbers in fixed-odds number lottery games. Empirical data suggests that not all players choose numbers with equal probability in lottery games. Some of them tend to bet on (smaller) numbers that are closely related to events around them (e.g., birthdays, anniversaries, addresses, etc.). In a fixed-odds lottery game, this small-number phenomenon imposes a serious risk on the game operator of a big payout if a very popular number is chosen as the winning number. In this chapter, we quantify this phenomenon and develop a choice model incorporating a modified Benford's law for lottery players to capture the magnitude of the small-number phenomenon observed in the empirical data. In particular, by combining the frequency distribution of digits from two types of players, those who choose all numbers randomly with equal probability and those who go for numbers that are closely related to events around them, we can estimate well the actual frequency distribution of digits in a numbers game. Our study can help lottery operators to customize sales limits for each number and thus control the operational risk of a big payout and the risk of losing rejected bets to underground markets.

15.1 INTRODUCTION

09/09/09 was a happy day. Some 366 couples in Singapore were reported to have gotten married that day as compared to 64 couples on a normal day. The reason? The belief that the auspicious number "999" can bring them eternal love. In Chinese culture, certain numbers are believed by some to be lucky based on the similarity of their pronunciation to that of certain Chinese words. For instance, the number 9 is viewed as lucky for it sounds like the Chinese word for "long-lasting." People from different backgrounds may have different views on lucky numbers. The beliefs in lucky numbers lead to certain numbers being chosen more frequently than others because of their auspiciousness. The betting profiles of the numbers selected in lottery games are thus not uniform, but expected to skew towards the auspicious numbers. What is the resulting impact of this on the game operators?

---

[1] Chou and Teo: National University of Singapore; Kong: Universidad Adolfo Ibáñez; Zheng: Shanghai Jiao Tong University.

In a **fixed-odds lottery game**, the prize is fixed for each ticket, and each winner receives a total payout proportional to the amount of the wagers he/she makes in the game. Hence, the game operator bears the risk of paying out a large sum in prizes if a very popular number is chosen as the winning number. Most game operators handle this issue by imposing a liability limit on the sales of each number—once the accumulated sales on a number hit the liability limit, future bets on that number will be rejected. This raises an associated question: how should a game operator set an appropriate liability limit? This issue is particularly important to legalized game operators as a large chunk of their sales will have to be returned to the government as tax revenues at the end of each year. This prevents the operator from building up a large reserve to absorb the exposure risk if the limit is set too high. On the other hand, if the limit is set too low, the game operator might lose the rejected bets to underground operators.

The choice of the sales limit is intimately related to the ways players select numbers to bet on in the games. There is ample empirical evidence suggesting that players do not choose all numbers with equal probability, but have a tendency to bet on (small) numbers that are closely related to events around them (e.g., birth dates, addresses, etc.). In this chapter, we quantify this phenomenon and examine its relation to Benford's law. By carefully modeling the ways players compose the digits in the numbers game, we refine Benford's law to develop a choice model for lottery players using a handful of parameters only. Surprisingly, this parsimonious choice model is already able to capture some of the most important characteristics of the data in the numbers game.

Exploiting the choice model, we examine the consequences of the small-number phenomenon on the risk profile of the game operator. We show that the relationship between the total sales revenues and the proportion of hot numbers is more stable with the presence of the small-number phenomenon in the choice process. The imposition of a cut-off limit is thus more effective in such an environment. Besides, our analysis suggests that it would be fruitful for the operators to pursue strategies to reduce the effect of the small-number phenomenon, that is, to promote or encourage players to choose numbers randomly.

15.2 PATTERNS IN NUMBER SELECTION: THE SMALL-NUMBER PHENOMENON

There are numerous studies in the gaming literature on the selection of lottery numbers among the players. One group of studies ([Sim, Hen, Hai, ZiBGS]) focuses on the **lotto games** (where players compete to pick, for instance, 6 winning numbers out of 45), and has revealed many interesting behavioral patterns showing how the players select their numbers. The most striking conclusion from these studies is that the players do not select their numbers randomly; that is, not all numbers are chosen with equal likelihood, and there is a tendency to select "auspicious" numbers (for instance, the number 7 is routinely chosen by players in the game in the UK). In particular, small numbers are more popular, as indicated in Table 15.1—the

15 most popular numbers in a 1996 powerball game played in the UK are mostly small numbers([Ti]) .

Rank	Number	Proportion	Rank	Number	Proportion	Rank	Number	Proportion
1	7	0.036	6	12	0.03	11	6	0.028
2	9	0.033	7	8	0.03	12	23	0.027
3	5	0.033	8	4	0.029	13	13	0.026
4	3	0.033	9	10	0.029	14	22	0.026
5	11	0.031	10	2	0.029	15	1	0.026

Table 15.1   The 15 most popular numbers in a 6/45 powerball game.

Another group of studies ([Che, HalDe]) focuses on the **Pick-3** or **Pick-4 numbers game**, where the players compete to pick the 3-digit or 4-digit winning number. [HalDe] also observe that players in Pennsylvania favor small numbers in the 3-digit numbers game. They observe that the bet volumes decrease rapidly from numbers in the 100s to 400s, then slowly to the 900s. A similar phenomenon is also reported by [Che] in his study of the 4-digit game in Massachusetts. A box-plot of the proportion of bets on the first leading digit in a sample 4-digit game is as shown in Figure 15.1. Clearly, there is a strong bias towards the smaller leading digits in this game.

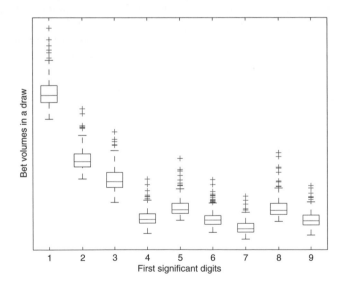

Figure 15.1   Betting volume on the leading digit.

More detailed sales data received on a particular draw in Pennsylvania is clearly presented in [HalDe], which allows us to quantify this phenomenon in the numbers games. Figure 15.2 shows the empirical distribution of the sum-of-three-digits statistic of the numbers chosen by the players in the Pennsylvania game. We compare the empirical distribution against the base case where all the 3-digit numbers are selected with equal probability. Interestingly, the empirical distribution exhibits a leftward shift from the base-case distribution, indicating a general preference for smaller digits in the number selections.

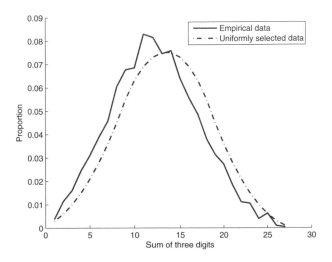

Figure 15.2   Distribution of the sum-of-three-digits statistics in the 3-digit numbers game.

These empirical sets of evidence indeed suggest that players favor small numbers. We call this the **small-number phenomenon** in the numbers game.

There are a few explanations for the small-number phenomenon in lottery games. As stated in many studies ([HalDe, Sim]), a large proportion of players tend to select numbers associated with special dates (e.g., birthdays, anniversaries, etc.), meaningful numbers (e.g., phone numbers, car numbers, address numbers, etc.), and special events (e.g., accidents and murders), and these numbers tend to start with smaller digits. For example, there are only 12 months in a year, so that the numbers 1–12 should be more popular than the numbers 13–45 in many 6/45 lotto games.

[Ben] analyzes the underlying causes of this small-number phenomenon using a geometric method. By simply arranging the set of natural numbers in increasing order, we can count the frequency of each of the first significant digits as it appears in this list. Take the first significant digit 1 as an example. It appears 1112 times if we count from 1 up to 10,000 in this list. Thus, the frequency of numbers with first significant digit 1 is 11.12%, from 1 to 10,000. If we repeat this experiment with numbers from 1 to 10,001, up to 100,000, this frequency count reaches a peak of 55.5% (at 19,999), and then decreases gradually. Figure 15.3 plots the frequency count for numbers from 1 to n, where n ranges from 10,000 to 99,999. The area under the curve accounts for about 30.103% (the probability of having 1 as first significant digit according to Benford's law) of total area.

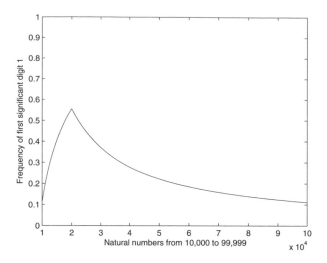

Figure 15.3   Frequency of first significant digit 1.

Another explanation put forth by researchers is the observation that human beings simply cannot choose numbers in a uniform manner. [LoeBr] demonstrate using experimental methods that there is indeed a cognitive bias towards the selection of small numbers by human beings, even when they are told to select numbers "randomly." In one of their studies, a total of 488 subjects were told to "name a sequence of digits with each digit chosen from 1 to 6 as randomly as possible," and they found a surplus of small digits (1, 2, and 3) in all their experiments.

These studies, unfortunately, offer only anecdotal evidence (through surveys and interviews) and rudimentary explanations for the existence of the small-number phenomenon, and do not provide an analytical framework to quantify and model this phenomenon.

15.3 MODELING NUMBER SELECTION WITH BENFORD'S LAW

In a recent paper ([ChKTWZ]), we use Benford's law to develop a choice model for number selection in the Pick-3 games. We assume there are two types of players: **Type I players** pick all numbers randomly with equal probability; **Type II players** pick their "lucky numbers" (arising from events in their daily life, or through superstitious beliefs). We also assume that each player bets $1 on each number chosen. In the following, we investigate the betting profiles generated by putting the two players together.

Definition 15.3.1. *Let β_B and β_N denote the proportions of Type II and Type I players respectively, with*

$$\beta_B + \beta_N = 1. \tag{15.1}$$

Using the sales data published in [HalDe], we assume that the numbers pur-
chased by the Type II players follow Benford's law, and calibrate the proportion
of Type I and Type II players. To ensure that the number selected has exactly 3
digits, we assume that the Type II player may choose to compose a 3-digit number
by padding the number he or she has chosen with leading zeros.[2]

Definition 15.3.2. *Let γ_i denote the proportion of Type II players who are betting
on numbers with i significant digits.*

By definition,

$$\sum_{i=1}^{3} \gamma_i = 1. \tag{15.2}$$

Assumption 15.3.3. *Assume that the Type II player will choose to play the 3-digit
number $d_1 \ldots d_i$ ($d_1 > 0$), with $3 - i$ leading zeros, with probability*

$$\gamma_i \log_{10} \left(1 + \frac{1}{d_1 \times 10^{i-1} + \cdots + d_i} \right). \tag{15.3}$$

Note that this is none other than the classical Benford's law, except that we
weight it with a factor γ_i to account for the proportion of players who bet with
i significant digits.

It is now easy to prove the following proposition.

Proposition 15.3.4. *Under Assumption 15.3.3, the expected proportion of the betting volume on a 3-digit number with first significant digit i, denoted by $\mathbb{E}[S(i)]$,
is*

$$\mathbb{E}[S(i)] = \beta_B \times \log_{10} \left(1 + \frac{1}{i} \right) + \beta_N \times \frac{1}{9}, \quad \text{for all } i = 1, 2, \ldots, 9. \tag{15.4}$$

Note that $\mathbb{E}[S(i)]$ does not depend on γ_j. We can thus use this property to calibrate the values of β_B and β_N by looking at the proportion of bets received for
each significant digit. In the 3D data from Pennsylvania, the proportions of Type
II and Type I players are estimated to be 39.58% ($\beta_B = 0.3958$) and 60.42%
($\beta_N = 0.6042$), according to the least square model.

As shown in Figure 15.4 for the first leading digit comparison, the prediction
from Benford's law captures the general trend in the empirical data, although we
observe a general higher-than-predicted preference for first significant digits 3, 7,
and 8 among the players, whereas the digit 2 has lower than the expected frequency.

To understand the choice preferences beyond the first significant digit, [ChKTWZ]
model an important characteristic in the way players compose the 3-digit numbers

---

[2]Note that this simplifying assumption may not hold in general, as some players may pad the numbers with trailing zeros, and some may simply duplicate the numbers to reach a 3-digit number. [HalDe]
mentioned that triplets like 111 or 888 are very popular in the Pick-3 game in Pennsylvania. Unfortunately, it does not appear possible to incorporate such features into the model, without sacrificing the
simplicity and tractability of the calibration model.

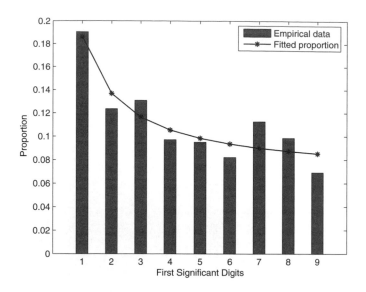

Figure 15.4   Fitted proportion by the model.

in the game. One such common strategy is to combine data from two different series to form a 3-digit number. For example, the number 246 could come from the 24th day of the month of June, or it could come from the address being level 6 of block number 24. The previous model assumes that the 3-digit numbers come from a single data series and hence fails to capture this switching behavior.

We notice that the probability distribution in the first assumption can be written in a different form:

$$\gamma_i \log_{10}\left(1 + \frac{1}{d_1 \times 10^{i-1} + \cdots + d_i}\right)$$
$$= \gamma_i \log_{10}\left(1 + \frac{1}{d_1}\right) \frac{\log_{10}(1 + \frac{1}{d_1 \times 10 + d_2})}{\log_{10}(1 + \frac{1}{d_1})} \cdots \frac{\log_{10}(1 + \frac{1}{d_1 \times 10^{i-1} + \cdots + d_i})}{\log_{10}(1 + \frac{1}{d_1 \times 10^{i-2} + \cdots + d_{i-1}})}.$$

Here, γ_i represents the probability that the Type II player will pick a number with i significant digits, and $\log_{10}(1 + \frac{1}{d_1 \times 10^{i-1} + \cdots + d_i})/\log_{10}(1 + \frac{1}{d_1 \times 10^{i-2} + \cdots + d_{i-1}})$ represents the probability that the ith digit is d_i, given that the first $i - 1$ digits are $d_1 \ldots d_{i-1}$. To model the switching behavior, we refine the recursive approach in the following way.

- As before, $\log_{10}(1 + \frac{1}{d_1})$ represents the probability that the first digit is d_1.

- Let

$$\frac{\log_{10}(1 + \frac{1}{d_1 \times 10^{i-1} + \cdots + d_i})}{\log_{10}(1 + \frac{1}{d_1 \times 10^{i-2} + \cdots + d_{i-1}}) + \lambda}$$

denote the probability that the player continues to generate the ith digit d_i as if it comes from the same data series as the first $i - 1$ digits, with parameter $\lambda > 0$. Note that in this way, the players switch to a different data series with a non-negative probability

$$\frac{\lambda}{\log_{10}(1 + \frac{1}{d_1 \times 10^{i-2} + \cdots + d_{i-1}}) + \lambda}.$$

- If the players switch to a different data series, let p_0 denote the probability that they switch to the digit "0." Otherwise, they switch to digit i, with $i \in \{1, \ldots, 9\}$, with probability $(1 - p_0) \log_{10}(1 + \frac{1}{i})$.

With a slight abuse of notation, we can write

$$\log_{10}\left(1 + \frac{1}{0}\right) := \frac{p_0}{1 - p_0}, \quad \text{and} \quad \lambda := \frac{q}{1 - q}.$$

We can now model the switching behavior in the 3-digit game in the following way.

Assumption 15.3.5. *We assume that the Type II player will choose to play the 3-digit number $d_1 \ldots d_i$ ($d_1 > 0$), with $3 - i$ leading zeros, with probability*

$$\gamma_i \log_{10}\left(1 + \frac{1}{d_1}\right) \frac{(1 - q) \log_{10}(1 + \frac{1}{d_1 \times 10 + d_2}) + q(1 - p_0) \log_{10}(1 + \frac{1}{d_2})}{(1 - q) \log_{10}(1 + \frac{1}{d_1}) + q} \times \cdots$$

$$\times \frac{(1 - q) \log_{10}(1 + \frac{1}{d_1 \times 10^{i-1} + \cdots + d_i}) + q(1 - p_0) \log_{10}(1 + \frac{1}{d_i})}{(1 - q) \log_{10}(1 + \frac{1}{d_1 \times 10^{i-2} + \cdots + d_{i-1}}) + q}.$$

In this way, we can interpret the parameters as follows.

Definition 15.3.6. *Let q denote the switching probability. Let p_0 denote the probability that the digit will be switched to 0.*

Let $\mathbb{E}[S(i, j)]$ denote the expected proportion of bets with first two significant digits i and j respectively.

Proposition 15.3.7. *Under Assumption 15.3.5,*

$$\mathbb{E}[S(i)] = \beta_B \times \log_{10}\left(1 + \frac{1}{i}\right) + \beta_N \times \frac{1}{9}, \quad \text{for all } i = 1, 2, \ldots, 9,$$

$$\mathbb{E}[(S(i, j)] = \beta_B \times \log_{10}\left(1 + \frac{1}{i}\right)$$

$$\times \left(\frac{(1 - q) \log_{10}(1 + \frac{1}{i \times 10 + j}) + q(1 - p_0) \log_{10}(1 + \frac{1}{j})}{(1 - q) \log_{10}(1 + \frac{1}{i}) + q}\right).$$

$$(15.5)$$

Note that the expected proportion of first significant digits remains unchanged under both assumptions. The parameters under Assumption 15.3.5 are calibrated to be $q = 0.9105$, $p_0 = 0.1054$, to best fit the empirical data under the least square model.

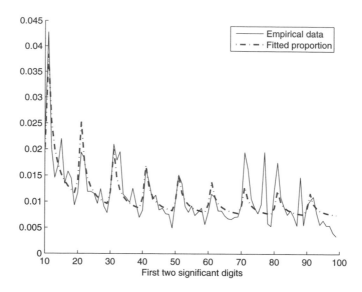

Figure 15.5   Fitted proportion for the first two significant digits.

The expected frequencies of first two significant digits are plotted in Figure 15.5. The frequencies generated from this model closely fit the frequencies of the empirical data. More interestingly, this model is able to capture the small-number phenomenon in the second significant digit of the data series.

The choice model under Assumption 15.3.5 proposed in the earlier section has the ability to track some of the most important characteristics of the betting data in the 3D game. Figure 15.6 depicts the distributions of the sum-of-digits in three data series: the actual data, simulated data from the proposed choice model, and the uniform-choice model.

The estimation of 39.58% Type II and 60.42% Type I players in the population seems right, as it captures the magnitude of the leftward shift in the empirical data reasonably well. Also, note that the choice model does not account for the superstitious beliefs observed in the empirical data (players generally avoid 2 and prefer 7 and 8). This partially explains why the proportions from our model are higher for smaller sum-of-digits (from 3 to 7) and lower for sum-of-digits around 15.

15.4 MANAGERIAL IMPLICATIONS

The small-number phenomenon clearly has important implications for the operational risk management of fixed-odds games. The numbers picked by the Type II players introduce variability and skewness to the distribution of bets on the 3-digit numbers. The winning numbers, on the other hand, are randomly (i.e., uniformly) rolled out by a mechanical device, which implies that the hot numbers are chosen

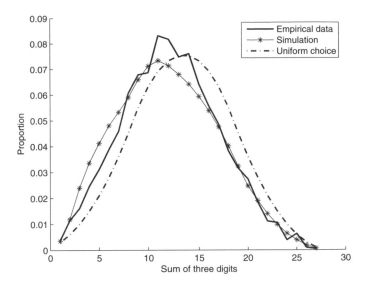

Figure 15.6   Distributions of sum-of-digits in empirical data, simulated data with Assumption 15.3.5, and uniform choice.

with the same probability as other numbers. The mismatch between the winning number distribution and the betting volume distribution leads to a significant operational risk: the operators may face a substantial payout if a popular number happens to be picked as the winning number. This is a phenomenon which often worries the game operators. In Québec, according to [LafSi], "the first drawing caused a prize liability well in excess of the amount received in sales." Fortunately, "over the long run it all evened out and the projected prize percentage was achieved."

15.4.1 Volatility of Prize Liability

Implication: The higher the proportion of players who pick auspicious numbers, the higher the variability in the payout for the game operator.

Consider a game with a prize P and N players, each betting \$1 on a number drawn from a respective distribution. Let $X_{\beta_B}(n)$ denote the amount of bets received on the number n when the proportion of Type II players is equal to β_B. When the winning number for that prize is drawn uniformly among the 999 numbers (from 001 to 999, as we have ruled out the bets on the number 000), the expected payout in our choice model is simply

$$\frac{P}{999} \sum_{n=1}^{999} \mathbb{E}(X_{\beta_B}(n)) \;=\; \frac{P}{999} \times N.$$

The second moment of the payout is

$$P^2\left(\frac{\sum_{n=1}^{999}\mathbb{E}(X_{\beta_B}^2(n))}{999}\right).$$

Hence, the variance of payout is

$$P^2\left(\frac{\sum_{n=1}^{999}\mathbb{E}(X_{\beta_B}^2(n))}{999}-\frac{N^2}{999^2}\right).$$

If all the N players choose their numbers independently, then we have $X_{\beta_B}(n)\sim$ $\mathrm{Bin}(N,p_{\beta_B}(n))$, where $p_{\beta_B}(n)$ denotes the probability that number n is picked in our choice model, given that the proportion of Type II players is β_B. Hence

$$\mathbb{E}(X_{\beta_B}(n)^2)\ =\ N^2 p_{\beta_B}^2(n)+Np_{\beta_B}(n)(1-p_{\beta_B}(n)).$$

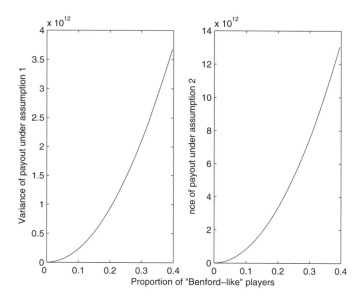

Figure 15.7   Variance of payout as the proportion of Type II players increases.

We can thus analytically compare the variance of the payout, under different values of β_B. As shown in Figure 15.7, under both assumptions, the variability of payout is increasing as the proportion of Type II players increases. When β_B is equal to 0, that is, the demand is evenly distributed, the variance of payout is only 0.003×10^{12}. When β_B increases to 39.58%, the variance of payout under Assumption 15.3.3 is 3.6794×10^{12}, about 1216 times higher than that of the uniform-choice model. Since Assumption 15.3.5 captures more of the volatility of the data, the variance of payout is 13.049×10^{12} in this model, 4313 times bigger than that of the uniform-choice model. For the 3D game in Pennsylvania, we conclude that the standard deviation of the prize payout can be reduced by 65 times if the proportion of the Type II agents (β_B) reduces to 0.

15.4.2 Liability Limit

Implication: Small-number phenomenon allows the use of liability limit as a risk management tool in the numbers games.

One of the key tools used in the risk management of numbers games is to choose a liability limit so that only a moderate proportion of numbers hit the sales limit. Interestingly, the small-number phenomenon plays a crucial role in this issue. Without this, it would be futile for game operators to try figuring out an appropriate liability limit to use in a particular game.

Let D_n denote the (random) demand of a 3-digit number n. The distribution of D_n depends on the proportion of Type I and Type II players in the game. Let C denote the corresponding cut-off limit. Let S_n denote the accepted sales for number n; i.e.,

$$S_n = \min(D_n, C).$$

Note that

$$\mathbb{E}[S_n] = C \cdot P(D_n > C) + \mathbb{E}(D_n | D_n \leq C) \cdot P(D_n \leq C).$$

Let $R(S_1, \ldots, S_N)$ denote the "risk exposure" when sales for the N numbers are given by $(S_1, \ldots, S_N)$. There are several ways to model the risk measure $R(\cdot)$, and it generally depends on the distribution of the winning numbers drawn.

Suppose the expected return given a \$1 bet is r. We use the mean-risk trade-off to model the utility function of the game operator. The expected utility function of the game operator is thus given by

$$r \sum_{n=1}^{N} \mathbb{E}[S_n] - \lambda \mathbb{E}\{R(S_1, \ldots, S_N)\},$$

where λ is an exogenous penalty term for risk exposure.

We can find C by solving the following maximizing problem:

$$\max_{C>0} \ r \sum_{n=1}^{N} [C \cdot P(D_n > C) + \mathbb{E}(D_n | D_n \leq C) \cdot P(D_n \leq C)]$$
$$- \lambda \mathbb{E}\{R(\min(D_1, C), \ldots, \min(D_N, C))\}. \tag{15.6}$$

It can be easily shown that the objective function is convex. Thus, according to the first order condition, the optimal liability limit C satisfies

$$\sum_{n=1}^{N} P(D_n > C) = \frac{\lambda}{r} \mathbb{E}\left[\frac{\partial R(\min(D_1, C), \ldots, \min(D_N, C))}{\partial C}\right]. \tag{15.7}$$

Note that the left-hand side corresponds to the expected number of hot numbers, i.e., the expected number of bet types reaching the cut-off limit in the draw. The cut-off limit can be set by merely choosing a cut-off limit C to control the number of hot numbers.

Suppose the total bets collected are to the value of \$N, and the cut-off limit is \$C for each number. We next estimate the expected number of hot numbers (i.e., the numbers with betting volumes hitting the liability limit).

We define an indicator function $Y_{\beta_B}(n)$ as follows:

$$Y_{\beta_B}(n) = \begin{cases} 1 & \text{if } X_{\beta_B}(n) \geq C; \\ 0 & \text{otherwise.} \end{cases}$$

The expected number of hot numbers with liability limit $\$C$ is

$$\mathbb{E}\left(\sum_{n=1}^{999} Y_{\beta_B}(n)\right) = \sum_{n=1}^{999} P\left(X_{\beta_B}(n) \geq C\right)$$

$$= \sum_{n=1}^{999} \left(1 - \sum_{i=0}^{C-1} \binom{N}{i} (p_{\beta_B}(n))^i (1 - p_{\beta_B}(n))^{N-i}\right).$$

Note we can use a normal distribution $N(Np_{\beta_B}(n), \sqrt{Np_{\beta_B}(n)(1 - p_{\beta_B}(n))})$ to approximate the binomial distribution $Bi(N, p_{\beta_B}(n))$, if N is large enough. Hence, we have

$$\mathbb{E}\left(\sum_{n=1}^{999} Y_{\beta_B}(n)\right) = \sum_{n=1}^{999} \left(1 - \Phi\left(\frac{C - Np_{\beta_B}(n)}{\sqrt{Np_{\beta_B}(n)(1 - p_{\beta_B}(n))}}\right)\right).$$

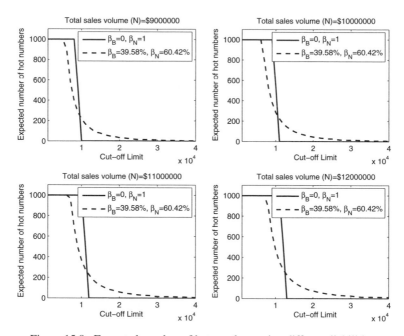

Figure 15.8   Expected number of hot numbers using different liabilities.

We can thus analytically compute the expected number of hot numbers given a liability limit $\$C$, and compare the results using different liability limits. Figure 15.8 shows the expected number of hot numbers under different liability limits in the case that 39.58% of players are Type II and 60.42% of players are Type I, and in the ideal case that all players are Type I agents.

In the ideal case, because all numbers are selected with equal probability, the concentration of measure phenomenon kicks in and the expected number of hot numbers goes through a phase transition—dropping sharply from 999 (all sold out) to 0 (none sold out) for a narrow range of cut-off limit. This is most evident from Figure 15.8: when the total sales are $9M, $10M, $11M, and $12M respectively, the expected number of hot numbers drops sharply to zero when the liability limit is around $10,000, $11,000, $12,000, and $13,000 respectively. In this environment, trying to find the appropriate cut-off limit to control the right level of hot numbers is almost impossible because this number depends critically on the total sales level, a number which normally fluctuates from draw to draw.

In the empirical sales data, we have $\beta_B \approx 0.3958$. In this environment, interestingly, the phase transition phenomenon disappears, and the relationship between the cut-off limit and the expected number of hot numbers is more stable. For a cut-off limit of $1000, the hot numbers fluctuate from 200 to 400 when the total sales level changes from $9M to $12M. The relationship between the total sales and proportion of hot numbers is thus more stable.

15.5 CONCLUSIONS

[ChKTWZ] study the sales data on a particular Pick-3 game and find that players do not uniformly select numbers. Instead, they prefer smaller numbers. For example, numbers starting with 1 or 2 are generally more popular than those starting with 8 or 9. This is called the small-number phenomenon in the numbers game.

One explanation for the small-number phenomenon is that people tend to select numbers associated with special dates (e.g., birthdays, anniversaries), meaningful numbers (e.g., phone, car, or address numbers), and special events (e.g., accidents and murders), and these tend to start with smaller digits. A very interesting natural law, Benford's law, captures this phenomenon. Benford's law describes the distribution of the first non-zero digit in natural data sets. According to this law, about 30% of the data starts with 1, with the proportion decreasing as the digit increases. Only about 4.5% of the data starts with 9. Many natural data sets (for example, accounting data) follow this law.

How does this belief in lucky numbers affect people's cognitive process when they gamble? When people believe that they stand a larger chance to win with certain numbers, instead of perceiving the lottery game as a pure game of chance, they tend to believe that they have some control over the results by choosing these lucky numbers. Consequently, perceived luckiness encourages an illusion of control. This illusion of control in turn makes gambling more attractive and thus, intensifies the gambling behavior, and consequently, lead to a possible addiction.

While the issue of gambling addiction is beyond the scope of this article, it is interesting to note that the behavioral biases towards auspicious and lucky numbers can collectively be modeled using the cold, hard mathematics of Benford's law. Given that roughly 40% of the players in our data set tend to believe/select "lucky numbers," modifying such erroneous beliefs through relevant correction programs may be a solution to prevent problem gambling.

Chapter Sixteen

Benford's Law in the Natural Sciences

David Hoyle[1]

This chapter focuses on the occurrence of Benford's law within the natural sciences, emphasizing that Benford's law is to be expected within many scientific data sets. This is a consequence of the reasonable assumption that a particular scientific process is scale invariant, or nearly scale invariant. We review previous work from many fields showing a number of data sets that conform to Benford's law. In each case the underlying scale invariance, or mechanism that leads to scale invariance, is identified. Having established that Benford's law is to be expected for many data sets in the natural sciences, the second half of the chapter highlights generic potential applications of Benford's law. In addition to detecting potentially fraudulently generated data sets, Benford's law has a role to play in monitoring data consistency and/or data quality. Finally, direct applications of Benford's law are highlighted, whereby the Benford distribution is used in a constructive way rather than simply assessing an already existing data set.

16.1 INTRODUCTION

There are few scientists that do not find Benford's law surprising on first encounter. In part, this is probably due to the fact that many scientists will not have an intuitive feel for what the **first significant digit** (**FSD**) distribution of a real data set should look like. Benford's law tells us that the probability $P(d)$ of the FSD being d is

$$P(d) = \log(1 + d^{-1}). \tag{16.1}$$

Despite its unusual nature, many scientific data sets do indeed appear to conform to Benford's law. The appearance of this unusual law of numbers within so many scientific data sets might be taken by some scientists as suggestive of a new fundamental process of nature. More often it is simply indicative of the natural but important requirement that a process be **scale invariant**. Despite the seemingly more pedestrian explanation for its ubiquitous occurrence, the role of Benford's law within the sciences has many facets which reflect the theoretical studies and applications outlined in earlier chapters. Consequently, potential applications of Benford's law to scientific data include the following.

---

[1] Thorpe Informatics Ltd., Adamson House, Towers Business Park, Wilmslow Rd., Manchester, M20 2YY, UK.

- Benford's law gives the natural FSD distribution for many scientific data sets.

- Benford's law can be used to assess data quality and consistency.

- Benford's law can be used to check for fraudulent scientific data.

- Benford's law can be used to generate appropriate simulated data sets.

We explore each of these issues throughout the chapter. In Sections 16.3.1 and 16.3.2 we review reports of occurrences of Benford's law within the physical sciences, and biosciences respectively. Then in Sections 16.4.1, 16.4.2, and 16.4.3 we discuss examples of applications of Benford's law to scientific data. We start by discussing the mathematical origins of the universality of Benford's law in scientific data. In doing so we potentially gain insight not only into the scientific data but also the processes that generated the data.

16.2 ORIGINS OF BENFORD'S LAW IN SCIENTIFIC DATA

With Benford's law still being a novelty to many scientists, there is often suspicion that conformance with Benford's law should indicate some profound and universal underlying mechanism controlling the science. This belief is reinforced to some degree by the sheer number and diverse range of data sets conforming with Benford's law. The true reason for the appearance of Benford's law in many situations is not a newly discovered fundamental mechanism, but nonetheless a still important one—that of **scale invariance**. The belief that we would not expect to find an intrinsic scale to measurements taken from a number of different sources is hardly a contentious one, and indeed would be considered good scientific principle. Within the natural sciences scale invariance for a quantity is often taken to mean the probability density $p(x)$ follows a power-law, and so takes the form

$$p(x) \sim x^{-\alpha}. \tag{16.2}$$

Probably the most common power-law in scientific data is **Zipf's law** [Zip], for which $\alpha = 1$. Zipf's law is a frequently observed relationship that is as ubiquitous in data sets as Benford's law. Recent explanations have been put forward for the universality of Zipf's law [CorSo, BaBeMi], extending beyond domain- and problem-specific explanations that had already been proposed. Zipf's law and Benford's law are in fact linked at a very simple level, as pointed out by Pietronero et al. [PiTTV]. For the power-law form given in (16.2), the FSD distribution $P(d)$ is easily derived (see for example [PiTTV]). For $\alpha \neq 1$ we have

$$P(d) = \int_d^{d+1} x^{-\alpha} dx = \frac{1}{1-\alpha} \left[(d+1)^{(1-\alpha)} - d^{(1-\alpha)} \right], \tag{16.3}$$

while for $\alpha = 1$ we have instead

$$P(d) = \int_d^{d+1} x^{-1} dx = \log \left(\frac{d+1}{d} \right). \tag{16.4}$$

The derivations of Pietronero et al. [PiTTV] outlined above are appealing but lacking in rigor. As Hill points out there is no proper scale-invariant distribution on the positive reals [Hi4]. However, Hill does rigorously show that Benford's law follows from base invariance of a probability measure over the mantissa of the positive reals, and that base invariance follows from scale invariance of the mantissa [Hi4] (see also Chapter 2). It is a relatively small step to extend the idea that a scientific quantity should have no intrinsic scale to the local notion that the mantissa of the quantity should have no intrinsic scale. Therefore, we should still expect Benford's law to be a natural consequence of our intuitions about collections of scientific measurements being scale invariant.

However, real scientific systems are finite and any finite system cannot truly be scale free. Most real systems have natural constraints on scale, both at the lower and upper ends. Yet, if the effective range is over several orders of magnitude, then effective scale invariance can be observed and we will see close agreement with Benford's law. Clearly, we would expect the precise magnitude of any deviations from Benford's law in the observed FSD of a finite system to be dependent upon the system size.

For finite systems it is also worth asking whether there are other distributions common within the natural sciences that demonstrate close agreement with Benford's law. With Benford's law essentially corresponding to uniform distribution of numbers on a logarithmic scale, one concludes that any random variate whose logarithm displays a broad distribution will result in an FSD distribution close to Benford's law. This is often referred to as the **spread hypothesis**, and while it does imply Benford behavior in some situations, it fails in others; for more on this see Chapter 2 and [BerH3].

An obvious example of a distribution with a "good" spread is the **log-normal distribution**, with density function

$$p(x) \;=\; \frac{1}{x\sqrt{2\pi\sigma^2}} \exp\left(-\frac{1}{2\sigma^2}(\ln x - \mu)^2\right). \tag{16.5}$$

As $\sigma \to \infty$ we expect to see convergence to Benford's law for the FSD distribution. Indeed, it has been commented by Sornette that the tail of a broad log-normal distribution can be mistaken for a power-law, i.e., a scale-invariant distribution [Sor]. We can rewrite the log-normal density function as

$$p(x) \;=\; \frac{e^{-\mu}}{\sqrt{2\pi\sigma^2}}\left(xe^{-\mu}\right)^{-1-\eta(x)}, \tag{16.6}$$

where $\eta(x) = \frac{1}{2\sigma^2}(\ln x - \mu)$. We see that with $\eta(x)$ being a slowly varying function of x, especially as σ becomes large, then the log-normal density function approaches a power law, $p(x) \sim x^{-1}$. Thus we should not be surprised that numbers drawn from a broad log-normal distribution can display close agreement with Benford's law. Within the natural sciences there are many multiplicative mechanisms and processes that naturally lead to a log-normal distribution. Examples range from species abundance—area relationships based upon "broken-stick"-like arguments [Sug], through to multiplicative noise processes in genomic assays [CuKeCh]. In the appropriate asymptotic limit the observation of Sornette outlined above would suggest that such multiplicative processes would lead to exact agreement with Benford's law.

That random multiplicative processes can lead to Benford-like behavior illustrates that there are simple common generative processes that can provide generic mechanisms for explaining the occurrence of Benford's law within scientific data sets. As well as growth via a random multiplicative factor, geometric growth processes (by a fixed factor per unit time) are equally common within the natural sciences. A geometric growth process would lead to an exponential growth curve when measured against elapsed time. For example, growth of bacterial or microbial colonies are often considered to partially follow an exponential curve (often termed the "log-phase") [Monod, ZwJRR]. Uniform random sampling of colony ages (elapsed time) within the log-phase of such colonies will lead to an **exponential distribution** of colony sizes. As has been observed on numerous occasions, the exponential distribution produces an FSD distribution very close to Benford's law [EngLeu].

The preceding discussions explain in part the ubiquity of Benford's law but may give the impression that Benford's law is inevitable. Clearly, there are some scientific data sets which are patently not scale invariant, and we would not expect them to be. Examples include repeated measurements of the same source, i.e., simple technical replication of the measurement process, or a sample of heights of people. In the former example we are measuring only one "thing," while in the latter there are biological and evolutionary constraints that result in human heights being concentrated over a relatively small range.

Finally, it is important to reemphasize that while Benford's law is a fascinating observation, its underlying cause within scientific data does not necessarily represent some newly uncovered profound mechanism. We should be wary of overinterpreting the scientific significance of observing agreement with Benford's law. Certainly, there are plenty of examples where power-law behavior in scientific data has been attributed to novel causes (see for example [MBGHPSS]), when in fact such power-laws can be generated by much more prosaic mechanisms [IsKaCh, Bon.., Vos]. With scale invariance being at the heart of Benford's law, the assigning of significance to observations of Benford's law by some authors [Buc, NiRen, NiWeRe] can cause heated debate and consternation in others [Far]. However, Benford's law is still worthy of study, both in its own right and as the natural FSD distribution for many scientific data sets. Indeed, with Benford's law naturally following from scale invariance it is surprising that Benford's law is not more widely known. By highlighting the ubiquity of Benford's law in scientific data sets and its potential applications we hope to rectify this situation. As Benford's law typically generates considerable interest when it is observed in a data set, it has been tested against an extremely diverse range of scientific data sets. We review some of these data sets in the next section.

16.3 EXAMPLES OF BENFORD'S LAW IN SCIENTIFIC DATA SETS

16.3.1 Physical and Engineering Sciences

Physics on a number of different scales provides a rich source of topics for the study of Benford's law, from atomic and nuclear physics through to fundamental particle physics. Starting at the smallest length scales, Shao and Ma [ShaMa1] found that **hadron** full widths showed agreement with Benford's law. The hadron full width Γ is related to its half-life τ (a measure of particle stability) through

$$\Gamma \times \tau = \hbar, \tag{16.7}$$

where $\hbar = h/2\pi$, with h being **Planck's constant**. Shao and Ma demonstrated the agreement with Benford's law for both **baryons** (particles such as neutrons and protons that are made up of three quarks) and **mesons** (particles made up of a quark and anti-quark pair). As with many data sets, the agreement with Benford's law is improved when the baryon and meson data sets are combined into a single hadron data set.

The agreement between Benford's law and the stability properties of composite particles extends to larger length scales (lower energy scales). Buck et al. found agreement between Benford's law and nuclei α-**decay** half-lives [Buc]. The agreement is evident in both experimentally measured and theoretically calculated half-lives. This would indicate that the occurrence of Benford's law is a result of the stochastic processes underlying α-decay rather than through experimental measurements being potentially dominated by multiplicative experimental noise. The calculated half-lives are not simply copies of their experimental counterparts, since theoretical half-life calculations at the time of the study by Buck et al. were only accurate to within a factor of 2 or 3. This hints at a common underlying mechanism responsible for the emergence of Benford's law, at work within both real α-decay processes and theoretical models. A common mechanism for generating Benford's law within stability properties of nuclei may also extend to β-**decay** processes. Ni and coworkers also found agreement with Benford's law in experimental and theoretical β-decay half-lives [NiRen, NiWeRe]. Although α- and β-decay represent two modes of nuclei transformations, it is not trivially apparent that both should conform to Benford's law since they are controlled by different fundamental forces: α-decay proceeds through emission of an α-particle (equivalently a Helium nucleus) and is governed by the strong and electromagnetic forces, while β-decay proceeds through emission of either an electron or positron and is governed by weak interactions [CotGr].

As well as emerging in the properties of nuclei, Benford's law also arises in atomic physics and hence at still lower energy scales. For example Benford's law can be seen in the properties of the electrons orbiting the nucleus. Specifically, Pain found agreement with Benford's law in the calculated energy spectrum (line strength) of **electronic transitions** (changes between electron energy levels) [Pai]. Pain speculates that the occurrence of Benford's law in this case can be understood in terms of a multiplicative process involving random matrix elements. Random matrices are often used to provide a model Hamiltonian for complex interacting systems, such as electrons orbiting a nucleus. Indeed, Wigner's original work on random matrices was motivated by the need to understand the energy levels of complex nuclei [Wig2]. With the line strength involving products over individual matrix elements a multiplicative process naturally emerges. Pain makes the connection between Benford's law and the **Random matrix theory** (**RMT**) of electronic transitions more explicit by noting that from the RMT approach Porter and Thomas derive an exponential distribution for larger line strengths [PoTh]. As we already commented, an exponential distribution is known to show close, though not exact, agreement with Benford's law. Overall, the RMT explanation of FSD distribution in line strengths is attractive since it also goes some way to a possible explanation of the universality of Benford's law in hadron and nuclear half-lives. It provides a theoretical model that can be shown to lead to Benford's law, yet is used to model physical interactions in composite systems at both atomic and subatomic scales. See Section 3.2.3.1 for more on the connections between random matrix theory and Benford's law, and [FiMil] for a history of its development.

In addition to being prevalent in data sets originating from physical processes operating at very small length scales, we find Benford's law occurring in data sets drawn from much larger length scales. Examples include the density of lightning flashes [MaRRSD], properties of **pulsars** (spinning neutron stars) [ShaMa2], and other astrophysical sources [MorSPZ].

The studies from the physical sciences highlighted above have focused on the agreement of Benford's law with experimentally derived data sets. Within axiomatic subjects such as the physical sciences one can ask the perhaps deeper question, are there aspects of physical laws that imply there will be agreement with Benford's law? To some extent this is addressed in the RMT calculations of Pain [Pai], though if we are to find agreement with Benford's law in a specific physical law it would, by definition, have to be found in one of the fundamental distributions that occur within the physical sciences. The study of stochastic systems governed by probability distributions is the realm of statistical physics [LanLi]. Shao and Ma [ShaMa3] study the agreement with Benford's law of three fundamental distributions from statistical physics, namely the **Boltzmann–Gibbs**, **Fermi–Dirac**, and **Bose–Einstein** distributions. These distributions tell us the probability of finding a system or collection of particles in a state with energy E. The Boltzmann–Gibbs distribution is appropriate for classical systems while the Fermi–Dirac and Bose–Einstein distributions apply to quantum systems, with the caveat that for quantum systems the possible energy levels E are discrete. For **fermionic particles** no two particles may occupy the same energy level, leading to the Fermi–Dirac distribution. Conversely, for **bosonic particles** the occupancy of any energy level is unrestricted and leads to the Bose-Einstein distribution. The three distributions are defined via

$$P_{\text{BG}}(E) \propto \exp(-\beta E) \qquad \text{Boltzmann–Gibbs,} \qquad (16.8)$$

$$P_{\text{FD}}(E) \propto [\exp(\beta E) + 1]^{-1} \qquad \text{Fermi–Dirac,} \qquad (16.9)$$

$$P_{\text{BE}}(E) \propto [\exp(\beta E) - 1]^{-1} \qquad \text{Bose–Einstein,} \qquad (16.10)$$

where $\beta = 1/K_B T$, with K_B being **Boltzmann's constant** and T the absolute temperature. Both the Boltzmann–Gibbs and Fermi–Dirac distributions show close agreement with Benford's law, with the discrepancy showing periodic behavior in β for any choice of FSD. The Bose–Einstein distribution is not normalizable and its FSD distribution is dominated by the singularity at $E = 0$. In this region $P_{\text{BE}}(E) \sim 1/E$, and so Shao and Ma argue that exact agreement with Benford's law is obtained. For the Boltzmann–Gibbs and Fermi–Dirac distributions the close agreement with Benford's law is essentially due to the exponential nature of these distributions, i.e., $P_{\text{FD}} \sim e^{-\beta E}$ for $\beta E \gg 0$ (but see Section 3.5 for a proof that the exponential and related distributions are never exactly Benford). Thus, for energies sampled over a wide range the already known close agreement between the exponential distribution and Benford's law [EngLeu] will be apparent. Shao and Ma have also extended their analysis of the classical Boltzmann–Gibbs distribution to **non-extensive statistical mechanics**, or so-called **Tsallis-statistics** [ShaMa4]. Under Tsallis-statistics the probability of finding a system in a state with energy E is given by one of a family of distributions, $P_q(E)$, parameterized by q,

$$P_q(E) \propto [1 - (1 - q)\beta E]^{1/(1-q)}, \qquad 1 \le q < 2. \tag{16.11}$$

As $q \to 1^+$ we recover the Boltzmann–Gibbs distribution, while as $q \to 2^-$ we recover a power-law distribution. Again Shao and Ma find close agreement between the FSD distribution from $P_q(E)$ and Benford's law, with the agreement fluctuating as β is varied but with the amplitude of such fluctuations decreasing as $q \to 2^-$.

What is often more surprising to newcomers is the agreement between Benford's law and samples of physical constants. Benford's original paper [Ben] and work by others (see [Burke]) illustrate the degree of conformance. However, one would naturally expect the distribution of physical constants, which are themselves measures of scale, to be scale invariant since the opposite implies a fundamental scale for all physical processes. Against this, it has been noted that combinations of some physical constants do appear to be constrained by a scale set by the current age of the universe. The **Large Number Hypothesis (LNH)** of **Dirac** (also called the **Large Number Coincidence**) observes that dimensionless combinations of fundamental physical constants are often in the order of 10^{40} [Dir]. Practically, the large timescale set by the current epoch, and similarly the almost infinitesimal length scale set by the **Planck length** ℓ_p [Pad], do not provide a significant constraint on the list of physical constants considered by Benford [Ben] and others [Burke]. Thus, we still expect the scale-invariance argument to apply to the distribution of physical constants, even in the presence of the LNH.

Finally, we highlight a number of studies involving Benford's law which fall more within the realm of computer science. As well as providing a general review of Benford's law, Torres et al. [TorFGS] demonstrate sizes of files on a computer hard disk follow Benford's law. Again, to the uninitiated the agreement of file sizes with Benford's law appears counterintuitive. There is a natural urge to question what mechanism or physical law could be at play here when surely file sizes are often a result of the human creative processes producing the content. However, the natural explanation is once again that we do not expect there to be an intrinsic scale on which to measure file sizes. Perhaps more intriguing is the observation of Dorogovtsev et al. [DoMeOl] that numbers gleaned from the Internet follow Benford's law. There is no single generating process at work here that we can easily conceptualize. While we may not be able to conceptualize the data generating process it can be argued that we do not expect any intrinsic scale or base for the data, and therefore Benford's law should naturally emerge as a consequence. However, with the numbers obviously coming from a mixture of sources and processes, Hill's derivation in [Hi4] of Benford's law as a collection of samples from a random mixture of distributions would appear as an equally appropriate justification for the pattern of numbers from the World Wide Web.

16.3.2 Biosciences

While the intricate complexities of processes within cells and organisms may appear to make it a harder task to link Benford's law with fundamental concepts within the biosciences, data showing such connections abound within the biosciences. Moreover, much of the focus of biological and biomedical sciences is quite naturally concerned with growth—from individual cells, through to tissues and organs, and finally up to whole populations. It is hardly surprising that many data sets within the biosciences, being measurements of size or scale, show good agreement with Benford's law. In many cases the data reflect an underlying biological multiplicative growth process, but even when the data correspond to measurements of biophysical characteristics we can still find Benford's law at work.

For example, Moret et al. [MorSSZ] studied the properties of folded proteins. A **protein** is produced as a linear sequence, or chain, of amino acids that then subsequently folds into its native three-dimensional structure [Whi]. This 3D shape—its size, pockets, accessibility to other molecules—determines how it interacts with other biomolecules, and consequently determines the function of the protein. Measurements of protein shape are therefore of key interest to structural biologists, and Moret et al. obtained measurements of mass, average radius, and solvent accessible area for a large number of proteins. As might be expected, the standard measurements of size, i.e., mass and average radius, show good agreement with Benford's law, with the mass measurements providing a better fit. However, the measurements of solvent accessible area show a more marked deviation from Benford's law, though 1 is still the most common FSD. (A similar result was found by Nigrini and Miller in studies on stream flow amounts and areas of lakes; see [NiMi1].) Although a measurement of size, the solvent accessible area measures that part of the outer surface of the native fold that can be reached by a water molecule of approximately 1.4Å radius. Therefore, measurements of the solvent accessible area have a natural intrinsic length scale associated with them, and folded proteins that have significant proportions of their outer surface with local radius of curvature below or around 1.4Å may lead to departures from Benford's law.

Similarly, Grandison and Morris [GrMo] found agreement between **kinetic rate constants** (another biophysical quantity) and Benford's law. The rate constants determine how fast **metabolic reactions** that are responsible for production of small essential biomolecules occur within an organism. The agreement with Benford's law is far from perfect, but certainly the frequency of 1 as the first significant digit is much higher than that for any other digit. While this observation may at first sight appear counterintuitive to systems biologists, as Grandison and Morris point out the implication is that the distribution of rate constants is approximately scale invariant. With no a priori reason to assume a preferred timescale for metabolic reactions, scale invariance and agreement with Benford's law are natural consequences. Obviously, physical and biological constraints provide us with intrinsic scales for the rate constants. For example, quantum physics dictates upper limits on how fast electronic transitions can occur within enzyme substrate complexes, while metabolic reactions that occur very slowly may confer such a large disadvantage to the organism that they are effectively selected against and are therefore not observed. Fortunately the range between these limits is still expected to be large, and therefore approximately scale-invariant behavior would be expected.

As with the physical sciences, there is a large diversity of biological data sets which show agreement with Benford's law, spanning a range of physical length scales. The work of Moret et al. [MorSSZ] and Grandison and Morris [GrMo] illustrates the agreement with Benford's law at the level of the properties of an individual biomolecule or complex. We also expect to find agreement with Benford's law in the measurements of the number of biomolecules present. Hoyle et al. [HoRJB] also observed agreement with Benford's law in **messenger RNA (mRNA)** abundance measurements. FSD distributions from **microarray fluorescence intensities** were studied. The intensities provide a proxy for mRNA abundances. It is believed that multiplicative noise dominates the experimental process, leading to a log-normal distribution. Hoyle et al. comment that since a log-normal distribution of sufficiently large variance can often be mistaken for a power-law distribution, the conformance to Benford's law may merely be reflecting the log-normal distribution arising from the multiplicative noise. Equally, though, the actual mRNA abundance of a particular gene is determined by a large number of regulatory factors, commonly acting in a combinatorial fashion. Consequently, the true underlying mRNA levels may themselves be determined essentially by a multiplicative process, again leading to a log-normal-like distribution.

The **post-genomic** biology data sets of the sort considered by Hoyle et al. provide an ideal opportunity against which to test Benford's law. The genome-wide nature of most modern assays provides us with thousands of measurements from a single biological sample, and hence an excellent source of numbers from which to construct reliable estimates of the FSD distribution which can be compared to Benford's law. This is illustrated in Figure 16.1 which shows the FSD distribution for a microarray data set of Mira et al., in this case of a wild-type yeast (*Saccharomyces cervisiae*) strain [MiBC]. The excellent agreement between the microarray FSD distribution and Benford's law is clear. Within the last few years microarray technology has been superseded by **next-generation sequencing** technology, which purports to give direct, digital, and hence more accurate measurement of gene expression [Metz]. Figure 16.1 also shows the FSD distribution for the mapped read counts of a next-generation sequencing data set from the same wild-type yeast strain [NaWWSRGS]. Although the next-generation sequencing data has been obtained by a different laboratory and the yeast culture grown under slightly differing conditions, its conformance to Benford's law is also clearly evident.

As we commented at the beginning this section, multiplicative processes are a key feature in normal growth of biological populations, from individual cells to human populations. Benford's original publication [Ben] examined human population

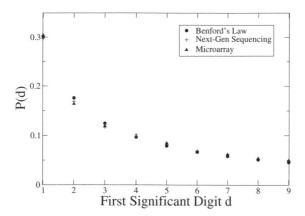

Figure 16.1 First digit distributions for example yeast transcriptome data sets.

sizes at the level of US counties. In a modern global age populations at the country level are just as easily obtainable and the same FSD analysis can be applied. This has been done by Sandon [Sando] who finds good agreement between Benford's law and data from 198 countries, at both the level of population and surface area.

The general mechanisms at work with human populations are on the whole those that apply to populations of smaller organism such as bacteria. Indeed Costas et al. [CosLTF] relate the standard geometric bacterial population growth model to Benford's law. Costas et al. [CosLTF] then confirm Benford's law for cell population sizes of various colonies of the cyanobacterium *Microcystis aeruginos* collected from different locations in southern Spain.

Even when growth of a biological population is uncontrolled, as is the case for developing **tumors**, the presence of Benford's law may still be found. Again, this is due to the underlying processes being multiplicative. Frigyesi et al. [FriGMH] find that the number of chromosomal aberrations present in the cells of various tumors follows **Zipf's law**, i.e., a power-law distribution with exponent −1. Frigyesi et al. [FriGMH] do not explicitly compute the FSD distribution for the number of observed chromosomal aberrations, and so do not make an explicit connection with Benford's law. However, as Pietronero et al. [PiTTV] point out, agreement with Zipf's law implies agreement with Benford's law. It is the chromosomal aberrations that lead to the deviation away from the normal growth dynamics of the cells, and in this case it is suggested that the accumulation of aberrations increases the likelihood of tumor cells becoming more dysfunctional and acquiring new further aberrations. Consequently, the number of chromosomal aberrations is determined through a multiplicative stochastic process with cells from older tumors having larger numbers of chromosomal aberrations.

16.4 APPLICATIONS OF BENFORD'S LAW IN THE NATURAL SCIENCES

16.4.1 Data Quality and Data Consistency

Perhaps the least controversial of the applications of Benford's law within the scientific arena is its use in monitoring data consistency. In this context changes in
the level of conformance with Benford's law are simply used to monitor changes
in the underlying processes producing the data. Hoyle et al. [HoRJB] proposed
taking such an approach with microarray measurements of gene expression. Here
the FSD distribution would be used to monitor the consistency of the experimental
process, at least at the level of an individual researcher in a given laboratory. To
use Benford's law to assess data consistency we do not have to necessarily concern
ourselves with making any statements about whether we believe the data should,
a priori, be Benford or not. Instead we are merely using deviation away from the
Benford distribution as a means of constructing a summary statistic for the patterns
of data observed. In most cases we are using Benford's law in this fashion to identify changes that are outside tolerance levels. Thus, in this manner Benford's law
is being used to monitor not only consistency, but also data quality. We are using
change in conformance with Benford's law as a marker for change in the underlying
data generation process, but not as a means to indicate *what* has changed. However, examples do exist where changes in conformance with Benford's law have
been more directly interpreted. Horn et al. [HoKKS] directly correlate changes in
FSD distributions from **electroencephalography** (**EEG**) data with different states
of anesthesia, or rather concentrations of the Sevoflurane anesthetic given. While a
direct scientific interpretation of the changes in the FSD distribution is being made
here, we must guard against overinterpretation. At the simplest level the degree of
agreement with Benford's law is merely being used to summarize the information
in the EEG trace. Similarly, Hoyle et al. [HoRJB] suggest that large changes in conformance with Benford's law for microarray gene expression data might be indicative of the original mRNA being obtained from a mixed cell-type population. Here
again the degree of agreement with Benford's law is essentially being used as a
practical summary statistic for the original data. For large-scale or high-throughput
data sets this approach to using Benford's law is extremely attractive due to the
relatively low computational complexity of the task. Such an approach has been
used by Brown [Bro1, Bro2] to assess pollutant concentration data sets, as well as
by DeMarchi and Hamilton [DeHa] to monitor the accuracy of self-reported atmo-
spheric releases of toxins from chemical production plants. In both these examples
it is suggested that the data should conform to Benford's law. Therefore, in this
case deviation away from Benford's law indicates poor quality data sets that cannot
be relied upon for their original purpose, and should be ignored or down-weighted
in any subsequent analysis. Orita et al. [OrMNO] give an explicit example of how
conformance to Benford's law can be used to assess the quality or appropriateness
of a data set for subsequent use—in this instance selecting training data for building
statistical models. They first confirm that unbiased selection of data points relating to various pharmaceutical drugs leads to conformance with Benford's law for
physio-chemical properties, such as solubility or toxic dosage levels. Inappropriate

selection of training data can therefore be detected through the use of Benford's law. They also suggest that Benford's law could be used to detect fabricated data. Similarly, for the toxin data of DeMarchi and Hamilton [DeHa], using deviation from Benford's law to assess the quality of self-reported data is suggestive of Benford's law also being used in a surveillance capacity and hence for detection of doctored data. This brings us into a more contentious area, that of scientific fraud.

16.4.2 Data Fraud

As with financial data, Benford's law has been applied to detect fraudulently generated or modified scientific data sets (see also Chapter 17). We distinguish fraudulently generated data sets from those which are merely of poor quality, which were discussed in the previous section. **Fraudulently generated data** sets are those where deviation from Benford's law is as a result of a deliberate attempt to mislead. Poor quality data sets, by contrast, may be misleading in a scientific sense but deviate from Benford's law as a result of poor experimental design or the intrinsically complex and noisy nature of the signal recording process. The scientific domains where fraudulent modification of data may be most likely to occur will be those where the stakes are highest. Such domains include those where significant investment has been made in obtaining the data. A possible example would be late-stage **clinical trials**, where the cost of developing a potential new drug for market can be in excess of one billion dollars [PMDPMLS]. Similarly, those fields where the failure to meet regulatory data standards can incur large and punitive fines will be more prone to fraudulent manipulation of the data. Topical examples of such fields include CO_2 **emissions** and **fisheries catches**, as both of these are examples of fields where countries may be subject to quotas through treaty negotiations that could be influenced by current data measurements. Indeed, inspired by use of Benford's law in survey data, Auffhammer and Carson applied Benford's law to CO_2 emissions [AuCa], with the explicit aim of detecting fraudulent data sets. Similarly, Graham et al. [GrHaPa] applied Benford's law to catch data from several North Atlantic fisheries. In this case catch data from highly regulated fisheries obeyed Benford's law while catch data from less regulated fisheries did not. It is probably fair to assume that Benford's law is unlikely to be uppermost in the mind of anyone producing fraudulent catch returns. This suggests that in the more highly regulated fisheries, where penalties for incorrect data are greater, the returned data properly reflects the actual catches landed. In contrast, for the less regulated fisheries there is less penalty and perhaps a commercial incentive for not accurately reporting the catches landed.

Likewise, commercial, financial, and other indirect pressures or incentives are undoubtedly issues in the reporting of medical and clinical trial data. Fraudulent activity in clinical trial data is a serious issue and one that is perhaps more widespread than many would initially suspect [RaEtAl]. Anomalous patterns in the reporting of medical data analysis is something that has been highlighted by García-Berthou and Alcaraz [GaAl]. Naturally, then, this is an area ripe for analysis of digit distributions. In contrast to the financial case, it is interesting to note that for a number of studies relating to medical data there has been a focus upon **terminal digit** preference [Hay1, Hay2, Hay3, MosDDK]. Unusual patterns in terminal digits may be the result of inappropriate rounding, both inadvertent and malicious, but as commented by Hayes this has the potential to impact upon important patient clinical outcomes [Hay2]. Although Benford's law is often cited when analyzing first digit preferences, a uniform distribution is more commonly used when testing terminal digit preference [GaAl, Hay1], rather than the form of Benford's law appropriate for the particular terminal digit.

Clearly, it is somewhat speculative to say that the data sets discussed above have been produced as a result of attempted fraud. For scientific data sets there are few circumstances where we can categorically say that a data set is "fake." Therefore we might question the validity of using of Benford's law to detect scientific fraud. To establish the utility of Benford's law for fraud detection requires a more rigorous controlled experiment; for more on this see Chapter 14. To this end Diekmann [Die] conducted experiments in which volunteers were asked to fabricate statistical data, in this case **regression** coefficients. The FSD distributions of both the fabricated regression coefficients and a sample of real regression coefficients displayed a monotonic decline, i.e., 1 being the most frequent FSD, with the real coefficients displaying closer agreement with Benford's law [Die]. More interestingly, the distribution of second and third significant digits displayed a greater deviation from Benford's law for the fabricated data, suggesting that application of Benford's law to second and third digits might provide a more sensitive test of scientific fraud. The experiments of Diekmann illustrate the difficulty that humans have in generating random data, a point emphasized by Mosimann et al. [MosWE] and again highlights the potential for Benford's law to detect altered data sets, particularly where scale invariance would be expected.

16.4.3 Direct Application

The use of Benford's law to detect fraudulent scientific data sets rests on the assumption that the scientific data should follow Benford's law. If an assumption of scale invariance is reasonable, then this use of Benford's law is apt. However, in such instances we are essentially only using the expected conformance with Benford's law to detect "outlier" data sets. We are not using our knowledge of Benford's law in any generative or constructive way. This brings us to perhaps the third branch of Benford's law applications in the sciences: directly using the Benford FSD distribution in a constructive way to produce realistic random artificial samples of data, or to improve the efficiency of scientific calculations. Uses of Benford's law in a constructive manner can be found across a wide range of disciplines. For example, Beeli et al. [BeEsJ] use the expected variation in frequency of second significant digits to study correlation between digit frequency and perceived color in **synesthetes** (individuals who associate letters or digits with colors). Similarly, Jolion [Jol] suggests using the Benford distribution to select efficient binary encoding schemes for transmission of data. In this case shorter length binary strings are used to encode the more common FSD 1, 2, and so on; see Chapters 18 and 19 for more on Benford's law in images. This latter example is reminiscent of Knuth's discussion of Benford's law within the context of computer programs

[Knu].

16.5 CONCLUSION

Benford's law rarely fails to surprise and arouse the curiosity of any scientist who encounters it. We have argued that the common occurrence of Benford's law in scientific data sets is simply a consequence of many scientific processes being **scale invariant**, or de facto scale invariant on the typical measurement scales we can access. The inference that a scientific process is scale invariant may not have far reaching consequences, but is still important. If we expect a scientific process to be scale invariant, then significant deviation from Benford's law may indicate the presence of a characteristic length scale or timescale. As scientists we would immediately enquire as to the origin of that length scale/timescale.

Beyond confirming or highlighting an underlying scale-invariant process, it is natural to question what practical use Benford's law is to the working scientist. Within this chapter we have tried to highlight not only the occurrences of Benford's law in scientific data, but also the genuine uses to which Benford's law can be put. Predominantly the most practical applications of Benford's law center on monitoring data quality or data stability. The acceptance of such applications of Benford's law within the sciences or other technological fields will always be difficult. This is in contrast to its acceptance and uptake within forensic accounting (see Chapter 8). The reasons for the limited acceptance within the sciences are manyfold and include the following.

- The lack of a rigorous theory linking changes in FSD frequencies to changes in data quality.

- Inferences from observing agreement with Benford's law can be overhyped.

- Those inferences from observing agreement with Benford's law that can be rigorously established may not be that useful.

- Overtones that Benford's law relates to an assessment of how fraudulent a scientific data set may be, or may not be.

While the theoretical underpinnings of Benford's law may be rigorous (as highlighted by other chapters), the linking of FSD frequencies to data set quality has heretofore been more heuristic. Similarly, connotations of fraud when deviation from Benford's law is observed in scientific data may be due to associations of using Benford's law in other settings. As we have emphasized, the reasons for deviation from Benford's law can be numerous.

These objections should be seen as a challenge. There is an opportunity to put the heuristic applications of Benford's law to scientific data analysis on a more rigorous footing. Equally, there is an opportunity to extend the range of direct scientific applications of Benford's law as new theoretical insights arise.

Chapter Seventeen

Generalizing Benford's Law

Joanne Lee, Wendy K. Tam Cho, and George Judge[1]

We examine and search for evidence of fraud in two clinical data sets from a highly publicized case of scientific misconduct. Departures from Benford's Law sometimes indicate fraud. Our classical Benford analysis along with a presentation of a more general class of Benford-like distributions highlights interesting insights into these cases. In addition, our exposition demonstrates how information-theoretic methods and other data-adaptive methods are promising tools for generating benchmark distributions of first significant digits (FSDs) and examining data sets for departures from expectations.

17.1 INTRODUCTION

Data in many realms have been suspected of being falsified. These instances may be exacerbated in the clinical arena by pressure to obtain certain results, or in the academic arena by pressure to secure grant money. In either case, science is the victim, and the costs can be significant. Accordingly, some method of ensuring data integrity would be a welcome and important advance.

A recent article in the *New York Times Magazine* highlights a well-known case of data falsification from clinical experiments [In]. In this case, data were falsified by Eric Poehlman, a faculty member at the University of Vermont, who pleaded guilty to fabricating more than a decade of data, some connected to federal grants from the National Institutes of Health. Poehlman had authored influential studies on many topics including obesity, menopause, lipids, and aging. In one study, Poehlman had hoped to demonstrate, with patient data over time, that lipid levels deteriorate with age. After a graduate student, DeNino, found that the data did not support this hypothesis, Poehlman tampered with the data, adding fictitious patient data and changing the values for existing patient cases, until evidence of his hypothesis was borne out by the data. DeNino became understandably confused and suspicious when the newly modified data set, ostensibly corrected of mistaken errors, now exhibited a clear trend consistent with Poehlman's original hypothesis.

---

[1]Lee: Researcher, Mathematica Policy Research; Cho: Departments of Political Science and Statistics, and Senior Research Scientist, National Center for Supercomputing Applications at the University of Illinois at Urbana-Champaign; Judge: Department of Agricultural and Resource Economics, University of California at Berkeley.

DeNino proceeded to comb through hundreds of patients' records in the lab and university hospital in an attempt to verify the data, but ultimately found evidence of data tampering.

The implications of this type of data fabrication are severe and have an obvious relationship with how science and health care evolve. Benford's Law has been touted as one way to identify tampered data. While Benford's Law can be powerful, its applicability is also known to be limited to certain classes of data. We examine Benford's Law in the data tampering context and generalize its reach by relating Benford's Law to a family of **first significant digit rules** that can be applied to a wider range of data sets. We also demonstrate how these types of methods would have been helpful in identifying the type of data tampering that occurred in the Poehlman case.

Our primary insight incorporates the mean of the first significant digits in the data set. As a data set's FSD mean changes, we adapt information-theoretic methods that may be used to create alternative null hypotheses for digit proportions. In doing so, we extend the range of Benford's Law to data contexts that may initially seem to not conform to its professed digit distribution.[2] We begin by noting how Benford's Law is connected to Stigler's distribution and extend these connections to the information-theoretic realm. We then develop alternative digit distributions based on maximum entropy principles. Finally, we demonstrate the applicability of these methods on the Poehlman clinical data.

17.2 CONNECTING BENFORD'S LAW TO STIGLER'S DISTRIBUTION

In the mid-1940s, George Stigler, a future economics Nobel Prize winner, claimed that Benford's Law contained a theoretical inconsistency and supplied an alternative distribution of FSDs [Sti]. In Stigler's derivation, the average frequency, F_d, for digit d is

$$F_d = \frac{d \ln(d) - (d+1) \ln(d+1) + (1 + \frac{10}{9} \ln(10))}{9}. \qquad (17.1)$$

Stigler claimed that the difference between his alternative and Benford's Law arose from the hidden assumptions Benford made about the relative frequencies of the largest numbers in statistical tables. Benford assumed that numbers with smaller FSDs occurred more often as bounds for statistical tables. In particular, given a mixture of uniform distributions, $U[0, b)$, the density of the upper bound b is assumed to be proportional to $\frac{1}{b}$. Stigler argued that this assumption was unnecessary in deriving a logarithmic rule, since it neither expanded the scope of the law nor contributed to the theoretical basis for modeling a distribution of first significant digits. In contrast, **Stigler's assumption** is that the largest entries in statistical tables are equally likely to begin with $d = 1, 2, \ldots, 9$, and all other entries are randomly selected from the uniform distribution of numbers smaller than the largest entry.

---

[2]We first examined in [LeCoJu] the theoretical basis of the **Stigler distribution**, and extended his reasoning by incorporating FSD first moment information and information-theoretic methods.

In Stigler's derivation, he defined the rth cycle of numbers as the interval $[10^r, 10^{r+1})$ for some real positive number r. He then found the distribution of FSDs for the highest entry in the cycle and computed the average of the expected frequencies over all highest entries. By the end of the $(r-1)$st cycle, a digit d has been an FSD for $(10^r - 1)/9$ out of $10^r - 1$ numbers. For example, the digit 2, at the end of the cycle $[10,100)$, has been an FSD for $(10^2 - 1)/9 = 11$ numbers out of $10^2 - 1 = 99$ numbers, including those from all previous cycles: 2, 20, 21, ..., 29.

After the $(r-1)$st cycle, d is not a FSD in the next $(d-1)10^r$ numbers. Continuing our example where $r - 1 = 1$, "2" is not an FSD in the interval $[10^2, 10^2 + (2-1)(10^2)) = [100, 200)$. Stigler takes advantage of this alternating pattern based on multiples of d to separate the cycle into smaller intervals. In the first cycle and part of the second cycle, "2" has appeared as an FSD in 11 out of 99 plus 100 numbers. We can write the proportion of FSDs that are d after the $(r-1)$st cycle and the first $(d-1)10^r$ numbers of the rth cycle as

$$\frac{10^r/9}{10^r + (d-1)10^r}. \tag{17.2}$$

The expectation, $F_{d,[10^r,d10^r)}$, of the proportion of FSDs that are d in the interval $[10^r, d10^r)$ is

$$F_{d,[10^r,d10^r)} = \frac{1}{(d-1)10^r} \int_0^{(d-1)10^r} \frac{10^r/9}{10^r + n} dn$$

$$= \frac{1}{9(d-1)} \ln d. \tag{17.3}$$

The FSDs in the next 10^r numbers are all d, so the proportion of d in this cycle is now

$$\frac{10^r/9 + 10^r}{10^r + (d-1)10^r + 10^r}. \tag{17.4}$$

The expectation, $F_{d,[d10^r,(d+1)10^r)}$, of the proportion of FSDs that are d in the interval $[d10^r, (d+1)10^r)$ is

$$F_{d,[d10^r,(d+1)10^r)} = \frac{1}{10^r} \int_0^{10^r} \frac{10^r/9 + n}{10^r + (d-1)10^r + n} dn$$

$$= 1 - \left(d - \frac{1}{9}\right) \ln\left(\frac{d+1}{d}\right). \tag{17.5}$$

Finally, the last $(9-d)10^r$ numbers in the rth cycle contain no numbers with FSDs that are d. Hence, the average proportion, $F_{d,[(d+1)10^r,10^{r+1})}$, of FSDs that are d in the interval $[(d+1)10^r, 10^{r+1})$ is

$$F_{d,[(d+1)10^r,10^{r+1})} = \frac{1}{(9-d)10^r} \int_0^{(9-d)10^r} \frac{10^r/9 + 10^r}{(d+1)10^r + n} dn$$

$$= \frac{10}{9(9-d)} \ln\left(\frac{10}{d+1}\right). \tag{17.6}$$

Following this logic, Stigler found the overall expected proportion of FSDs that are d to be

$$F_d = \frac{d \ln d - (d+1)\ln(d+1) + m}{9}, \tag{17.7}$$

FSD	Stigler's Law	Benford's Law
1	0.241	0.301
2	0.183	0.176
3	0.146	0.125
4	0.117	0.097
5	0.095	0.079
6	0.077	0.067
7	0.061	0.058
8	0.047	0.051
9	0.034	0.046

Table 17.1   Comparison of Benford and Stigler distributions.

where m, the mean of the **Stigler FSD distribution**, is defined as

$$m = \frac{\sum_{d=1}^{9} d^2 \ln(d) - d(d+1) \ln(d+1)}{9 - \sum_{d=1}^{9} d}. \tag{17.8}$$

The resulting frequencies from Stigler's derivation are presented in Table 17.1. The frequencies from Benford's Law are presented for comparison.

While the Stigler and Benford relative frequencies differ, the sets of frequencies are similar in their monotonically decreasing pattern. Because no logarithmic FSD distribution holds generally for all natural data sets, Stigler's Law and Benford's Law might be viewed as members of a family of monotonically decreasing distributions of FSDs, a family that also includes the Power Law, Zipf's Law, and the distributions arising from information-theoretic methods [GrJuSc].

17.3 CONNECTING STIGLER'S LAW TO INFORMATION-THEORETIC METHODS

Thus far, we have discussed Benford's and Stigler's approaches for determining the null-hypothesis distribution of FSDs in tests for fraudulent data. We now discuss how information-theoretic methods produce similar distributions, and highlight their unique ability to easily adapt the specific distribution to moment information from any particular data set. Since phenomena often have unique traits, a distribution that is adaptable to data peculiarities is desirable if such individual idiosyncrasies might affect the particularities of the monotonically decreasing distribution.

To recover the FSD distribution from a sequence of positive real numbers, assume, for the discrete random variable d_i, $i = 1, 2, \ldots, 9$, that at each trial, one of nine digits is observed with probability p_i. Suppose after n trials, we are given first-moment information, $\bar{d}$, the average value of the FSDs,

$$\sum_{j=1}^{9} d_j p_j = \bar{d}. \tag{17.9}$$

Assuming that the only information that exists is this first-moment information, our inverse problem consists of identifying an FSD distribution that reflects the best predictions of the unknown probabilities, $p_1, p_2, \ldots, p_9$. It is readily apparent that since there is one data point and nine unknowns, we have an ill-posed inverse problem where an infinite number of possible discrete probability distributions with $\bar{d} \in [1, 9]$ exist. Based only on the mean, $\sum_{j=1}^{9} d_j p_j = \bar{d}$, and two probability constraints, $\sum_{j=1}^{9} p_j = 1$ and $0 \leq p_j \leq 1$, the problem does not have a unique solution. A function must be inferred from insufficient information when only a feasible set of solutions is specified. In these situations, it is useful to have an approach that allows the investigator to adapt sample-based information recovery methods without committing the FSD function to a particular parametric family of probability densities. The goal is to reduce the infinite-dimensional non-parametric problem to one that is finite-dimensional, ideally without imposing more assumptions than are necessary.

17.3.1 An Information-Theoretic Approach

One way to solve this ill-posed inverse problem for the unknown p_j without making a large number of assumptions or introducing additional information is to formulate it as an extremum problem. This type of extremum problem is, in many ways, analogous to allocating probabilities in a contingency table where p_j and q_j are the observed and expected probabilities respectively of a given event. A solution is achieved by minimizing the divergence between the two sets of probabilities by optimizing a goodness-of-fit (pseudo-distance measure) criterion subject to data-moment constraint(s). One attractive set of divergence measures is the **Cressie–Read power divergence family of statistics** [CrRe, ReACr, Bag],

$$I(\mathbf{p}, \mathbf{q}, \gamma) = \frac{1}{\gamma(1+\gamma)} \sum_{j=1}^{9} \left(p_j \left[\left(\frac{p_j}{q_j} \right)^{\gamma} - 1 \right] \right), \tag{17.10}$$

where γ is an arbitrary unspecified parameter.

Using the Cressie–Read criterion (17.10) to recover the unknown FSD distributions suggests that we seek, given probability expectations $\mathbf{q}$, a solution to the extremum problem

$$\hat{\mathbf{p}} = \arg\min_{\mathbf{p}} \left[I(\mathbf{p}, \mathbf{q}, \gamma) \,\middle|\, \sum_{j=1}^{9} p_j d_j = \bar{d}, \sum_{j=1}^{9} p_j = 1, p_j \geq 0 \right]. \tag{17.11}$$

In the limit, as γ ranges from -1 to 1, two main variations of $I(\mathbf{p}, \mathbf{q}, \gamma)$ have received explicit attention in the literature (see [MitJuMi]). Assuming for expository purposes that the reference distribution is the discrete uniform, i.e., $q_j = 1/9$ for all j, then $I(\mathbf{p}, \mathbf{q}, \gamma)$ converges to an estimation criterion equivalent to **Owen's**

(2001) **empirical likelihood criterion** $\sum_{j=1}^{9} \ln(p_j)$, when $\gamma \to -1$. The empirical likelihood criterion assigns discrete mass across the nine possible FSDs. In the sense of objective function analogies, it is closest to the classical maximum-likelihood approach, and in fact, results in a maximum non-parametric likelihood alternative. The second prominent case for the Cressie–Read statistic corresponds to $\gamma \to 0$ and leads to the maximum entropy or the [Shan] and [Jay] entropy function, $-\sum_{j=1}^{9} p_j \ln(p_j)$.[3]

17.3.2 Maximum Entropy Formulation

Using the Cressie–Read ($\gamma = 0$) criterion for the first digit case, the maximum entropy approach selects probabilities that maximize

$$H(\mathbf{p}) = -\sum_{j=1}^{9} p_j \ln(p_j), \qquad (17.12)$$

subject to the mean $\bar{d}$,

$$\sum_{j=1}^{9} p_j d_j = \bar{d}, \qquad (17.13)$$

and the condition that the probabilities must sum to one,

$$\sum_{j=1}^{9} p_j = 1. \qquad (17.14)$$

The Lagrangian for the extremum problem is

$$L = -\sum_{j=1}^{9} p_j \ln(p_j) + \lambda \left(\bar{d} - \sum_{j=1}^{9} p_j d_j \right) + \eta \left(1 - \sum_{j=1}^{9} p_j \right). \qquad (17.15)$$

Since H is strictly concave, there is a unique interior solution. Solving the first-order conditions yields the maximum entropy exponential result for the jth outcome,

$$\hat{p}_i = \frac{\exp(-d_i \hat{\lambda})}{9 \sum_{j=1}^{9} \exp(-d_j \hat{\lambda})}. \qquad (17.16)$$

In this context, $\hat{p}_i$ are exponentially distributed and the FSD distribution chosen is the one that happens in the most likely way (multiplicity). We note again that $p(\lambda)$ is a member of a canonical exponential family with mean

$$\sum_{j=1}^{9} p_j(\lambda) d_j = \bar{d}. \qquad (17.17)$$

The **Fisher information measure** for λ [GoJM] is

$$I(\lambda) = \sum_{j=1}^{9} p_j(\lambda) d_j^2 - \left(\sum_{j=1}^{9} p_j(\lambda) d_j \right)^2 = \text{Var}(d). \qquad (17.18)$$

---

[3]The maximum entropy criterion distance measure is equivalent to the **Kullback–Leibler (KL) information criterion** and finds the feasible $\hat{p}$ defining the minimum value of all possible expected log-likelihood ratios consistent with, in our case, the FSD mean [Kul]. Solutions for these distance measures cannot be written in a closed form.

17.3.3 Empirical Likelihood and Maximum Entropy Distributions for Various FSD Means

FSD mean	2.00	3.00	3.44	3.55	4.00	4.50
$\hat{p}_1$	0.673	0.395	0.281	0.300	0.208	0.151
$\hat{p}_2$	0.111	0.173	0.175	0.177	0.161	0.137
$\hat{p}_3$	0.061	0.111	0.128	0.125	0.132	0.125
$\hat{p}_4$	0.042	0.082	0.100	0.097	0.111	0.115
$\hat{p}_5$	0.032	0.065	0.082	0.079	0.096	0.107
$\hat{p}_6$	0.026	0.053	0.070	0.067	0.085	0.100
$\hat{p}_7$	0.021	0.046	0.061	0.058	0.076	0.093
$\hat{p}_8$	0.018	0.040	0.054	0.051	0.068	0.088
$\hat{p}_9$	0.016	0.035	0.048	0.046	0.062	0.083

Table 17.2   Estimated empirical likelihood FSD distributions for different FSD means.

Tables 17.2 and 17.3 present the FSD distributions derived from the empirical likelihood and maximum entropy formulations presented above for several different FSD means, including the **Stigler mean** (3.55) and the **Benford mean** (3.44). The maximum entropy solution becomes a uniform distribution when the FSD mean is 5 and becomes monotonically increasing for FSD means above 5. FSD means less than 5 result in distributions that are tilted toward the lower digits and have monotonically decreasing FSD probabilities. The Benford FSD mean of 3.44 yields a maximum entropy distribution similar to Benford's Law, and the Stigler FSD mean of 3.55 yields an maximum entropy distribution similar to Stigler's proposed alternative.

The exponential null hypotheses under maximum entropy have two especially appealing properties. First, the result is achieved while adhering to the principles of Occam's Razor. That is, there are a minimal number of underlying assumptions. Second, the choice is one of maximum multiplicity—in the absence of assumptions, the "best" choice of a distribution among the universe of possible distributions is the one that occurs most frequently.

17.4 CLINICAL DATA

We now turn to an examination of the Poehlman data. We were able to obtain the falsified clinical data with the generous help of John Dahlberg, head of the Data Integrity group at Health and Human Services.[4] For these data, we also have the correct data, i.e., the data before they were changed. The correct data includes 142 observations on changes in insulin levels over a six-year period. The falsified

---

[4]For background on this data set see [In].

FSD mean	2.00	3.00	3.44	3.55	4.00	4.50
$\hat{p}_1$	0.496	0.306	0.250	0.238	0.191	0.148
$\hat{p}_2$	0.251	0.217	0.194	0.188	0.163	0.137
$\hat{p}_3$	0.126	0.153	0.150	0.149	0.140	0.127
$\hat{p}_4$	0.064	0.108	0.117	0.118	0.120	0.118
$\hat{p}_5$	0.032	0.077	0.090	0.093	0.103	0.109
$\hat{p}_6$	0.016	0.054	0.070	0.074	0.088	0.101
$\hat{p}_7$	0.008	0.038	0.054	0.058	0.075	0.094
$\hat{p}_8$	0.004	0.027	0.042	0.046	0.065	0.087
$\hat{p}_9$	0.002	0.019	0.033	0.036	0.055	0.081

Table 17.3   Estimated maximum entropy FSD distributions (with a uniform reference distribution) for different FSD means.

Poehlman data includes 136 observations. The original intent of the study was to analyze the relationship between insulin levels and age.

Distribution	Benford	Insulin$_T$	Insulin$_F$
FSD mean	3.44	3.54	4.03
Correlation	1.00	0.95	0.66
χ^2_8	0.00	10.25	41.80
$\hat{p}_1$	0.301	0.317	0.255
$\hat{p}_2$	0.176	0.148	0.073
$\hat{p}_3$	0.125	0.099	0.102
$\hat{p}_4$	0.097	0.085	0.080
$\hat{p}_5$	0.079	0.134	0.197
$\hat{p}_6$	0.067	0.042	0.139
$\hat{p}_7$	0.058	0.078	0.066
$\hat{p}_8$	0.051	0.063	0.058
$\hat{p}_9$	0.046	0.035	0.029

Table 17.4   The empirical FSD frequencies for Benford and the correct (Insulin$_T$) and falsified (Insulin$_F$) Poehlman data.

17.4.1 The Correct Data

We begin by examining the correct Poehlman data that has 142 observations. The FSD distribution of these data is similar to that proposed by Benford. The distributions of Benford and the correct insulin FSDs are given in Figure 17.1 and Table

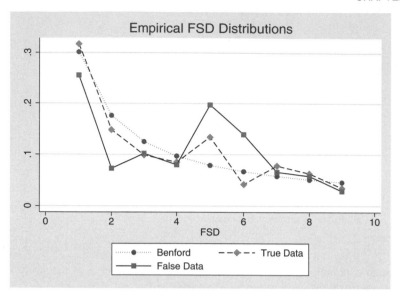

Figure 17.1   Empirical FSD distributions for the Poehlman data.

17.4. The correlation between the Benford and the correct empirical FSD distribution is 0.95. The χ^2 test statistic with 8 degrees of freedom is 10.25. The one-sided 10%, 5%, 1% critical values for χ^2_8 are 13.36, 15.5, and 20.09, respectively. We are not able to reject the null hypothesis of distribution equality between the Benford and empirical FSD distributions.

17.4.2 The Falsified Data

The FSD distribution for the 136 observations of falsified data are also given in Table 17.4 and Figure 17.1. Note the obvious fluctuation of the FSD proportions of the falsified data and their departure from Benford's distribution. The lack of a monotonically decreasing distribution of FSD frequencies in the falsified data contributes to a correlation of only 0.66 between Benford and the falsified data frequencies. The χ^2 test value for the falsified data and Benford's Law is 41.80, which exceeds the 5% level χ^2 value of 15.51, providing a statistical basis for rejecting the null hypothesis of equality. Thus, in addition to the testimony of the graduate student, there is a visual and inferential basis for suspecting manipulation of these research data.

17.4.3 Problem Reformulation for Information-Theoretic Methods

It seems reasonable that the FSD distribution could vary with the measured phenomenon in question. In this section, we use the **empirical likelihood method** and **first moment frequency information** to examine the FSDs.[5] In this context,

---

[5]See [Ow] for an introduction to empirical likelihood methods.

suppose that one of nine digits, $i = 1, \ldots, 9$, is observed with probability p_i, and the information is given in the form of the average value of the FSDs,

$$\sum_{i=1}^{9} d_i p_i = \bar{d}. \tag{17.19}$$

Based on this first moment information and Owen's (2001) empirical likelihood metric, $\frac{1}{9} \sum_{i=1}^{9} \ln p_i$, we can formulate the problem of recovering the unknown and unobservable p_i as the extremum likelihood function,

$$\max_{\mathbf{p}} \left\{ \frac{1}{9} \sum_{i=1}^{9} \ln p_i - \sum_{i=1}^{9} p_i d_i = \bar{d}, \sum_{i=1}^{9} p_i = 1 \right\}. \tag{17.20}$$

The corresponding Lagrange function is

$$L(\mathbf{p}, \eta, \lambda) = \frac{1}{9} \sum_{i=1}^{9} \ln p_i - \eta \left(\sum_{i=1}^{9} p_i - 1 \right) - \lambda \left(\sum_{i=1}^{9} p_i d_i - \bar{d} \right), \tag{17.21}$$

where $\mathbf{p} > 0$ is implicit in the structure of the problem. Solving the corresponding first order condition with respect to p_i leads to the solution

$$p_i^*(\bar{d}, \lambda) = \frac{1}{9} \left(1 + \lambda^*[d_i - d] \right)^{-1}. \tag{17.22}$$

This solution implies that an exponential family of distributions will result as the mean of the FSDs varies over a range implied by actual data sets. In maximizing $\Pi_{i=1}^{9} p_i$ subject to $\sum_{i=1}^{9} p_i = 1$ and $\sum_{i=1}^{9} p_i d_i = \bar{d}$, the p_i are chosen in such a way that the maximum joint probability among all the possible probability assignments is assigned.

As an example, some mean-related empirical likelihood FSD distributions are given for selected FSD means in Table 17.5. Note that the FSD mean for the Benford distribution is 3.44 and, in this case, the empirical likelihood FSD distribution is almost identical to the Benford distribution. If clinical researchers are reluctant to share their data, they might be willing to divulge the mean of their digit frequencies. With this information, empirical likelihood methods offer a viable data evaluation technique. As shown in Table 17.4, the FSD means of the true and false insulin data are 3.54 and 4.03, respectively. The empirical likelihood distributions generated by these means and the Benford FSD mean of 3.44 are also shown in Figure 17.2 and Table 17.5.

The empirical likelihood FSD distribution for the true insulin data with mean 3.54 has a strong visual correlation with the Benford reference distribution. Alternatively, the empirical likelihood FSD distribution for the false insulin data appears visually distinguishable from the empirical likelihood true insulin data and Benford reference distributions. The χ_8^2 values shown in Table 17.5 show that we cannot reject the null hypothesis of distribution equality of the Benford or true insulin data set and the empirical likelihood FSD distributions derived from their means. However, we reject distributional equality for the false insulin data and the empirical likelihood distribution estimated from its mean based on the χ_8^2 value of 34.39.

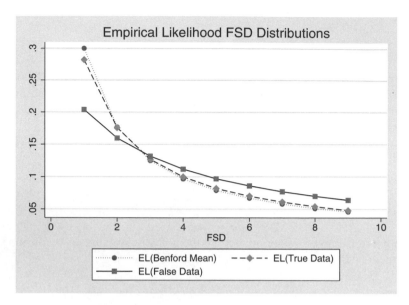

Figure 17.2 Empirical likelihood FSD distributions for Poehlman data.

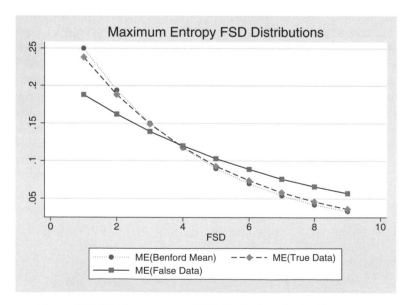

Figure 17.3 Maximum entropy FSD distributions for Poehlman data.

Mean from FSD mean	Benford 3.44	Insulin$_T$ 3.54	Insulin$_F$ 4.03
Correlation	0.95	1.00	0.98
χ_8^2	0.00	9.71	34.39
$\hat{p}_1$	0.30	0.282	0.204
$\hat{p}_2$	0.18	0.176	0.160
$\hat{p}_3$	0.13	0.127	0.132
$\hat{p}_4$	0.10	0.100	0.112
$\hat{p}_5$	0.08	0.082	0.097
$\hat{p}_6$	0.07	0.070	0.086
$\hat{p}_7$	0.06	0.061	0.077
$\hat{p}_8$	0.05	0.054	0.070
$\hat{p}_9$	0.05	0.048	0.064

Table 17.5 Estimated empirical likelihood distributions based on the FSD means of Benford (3.44), true insulin data (3.54), and falsified insulin data (4.03).

We show the comparable maximum entropy FSD distributions in Figure 17.3. The maximum entropy FSD distribution from the true insulin data mean also has a strong visual relationship with the maximum entropy distribution based on the Benford mean, while both of these distributions are easily distinguishable from the maximum entropy distribution based on the mean from false insulin data. Again, we cannot reject the null hypothesis of distribution equality of the Benford or true insulin data set with their maximum entropy FSD distributions, but we can reject equality for the false insulin data and maximum entropy distribution from the false insulin mean.

17.5 SUMMARY AND IMPLICATIONS

In practice, Health and Human Services or any other granting agency would have available only the falsified data from the researcher. However, in this case, the departures of the false insulin FSD distribution from the Benford FSD distribution are significant enough to have likely attracted greater scrutiny of the clinical data under the Benford, Stigler, empirical likelihood, or maximum entropy frameworks. Although scientific fraud is hopefully a rare phenomenon, using an array of FSD distributions including Benford's Law and data-adaptive distributions appears to be a quick and objective way to check data.

Finally, we have found it very difficult to obtain additional clinical data to analyze and regret that we can only report on one set of experimental data. Although this may be a naive request, given the importance of decisions based on clinical data, forming an objective clearing house that analyzes, via Benford and other

Mean from	Benford	Insulin$_T$	Insulin$_F$
FSD Mean	3.44	3.54	4.03
Correlation	0.96	0.96	0.93
χ_8^2	3.82	16.21	34.13
$\hat{p}_1$	0.250	0.238	0.188
$\hat{p}_2$	0.194	0.188	0.162
$\hat{p}_3$	0.150	0.149	0.139
$\hat{p}_4$	0.117	0.118	0.120
$\hat{p}_5$	0.090	0.093	0.103
$\hat{p}_6$	0.070	0.074	0.089
$\hat{p}_7$	0.054	0.058	0.076
$\hat{p}_8$	0.042	0.046	0.066
$\hat{p}_9$	0.033	0.036	0.057

Table 17.6 Estimated maximum entropy distributions based on the FSD means of Benford (3.44), true insulin data (3.54), and falsified insulin data (4.03).

methods, clinical data as they come from researchers would be a useful first step in ensuring data quality.

PART VI

Applications IV: Images

Chapter Eighteen

PV Modeling of Medical Imaging Systems

John Chiverton and Kevin Wells[1]

The Benford distribution is a well-known probability distribution that is applicable to many different naturally occurring sources of data. We have found that it also describes well the distribution of mixtures occurring in medical imaging data due to the Partial Volume (PV) effect. The Benford distribution provides a convenient formulation that is both scale and base invariant, unlike previous formulations based on, for example, the Beta distribution; these required ad hoc manipulation of parameters to obtain the correct distributive shape.

We apply our Bayesian formulation of the PV effect, based on the Benford distribution, to the statistical classification of nuclear medicine imaging data: specifically Positron Emission Tomography (PET) acquired as part of a PET-CT phantom imaging procedure. We describe our PET-CT imaging and post-processing process to derive a gold standard. We use it as a ground truth for the assessment of our Benford classifier formulation. The use of this gold standard shows that the classification of both the simulated and real phantom imaging data are well described by the Benford distribution.

18.1 INTRODUCTION

The Benford distribution is a discrete probability distribution of great interest for many applications including **medical imaging**. This is because it describes the probabilities of occurrence of single digits e.g. 1, 2, 3, ..., 9 in many sources of data. It was first observed in 1881 by Simon Newcomb [New] and later rediscovered again by Frank Benford [Ben] in 1938. The Benford distribution has been found to be applicable to a wide array of different sources of data including books of logarithm tables [Ben], the frequency of digits in tax returns [Nig1], in newspapers [Ben], the atomic weights of molecules [Ben] and even hydrology data [NiMi1].

The purpose of this chapter is to describe another occurrence, which is in the distribution of intensities in medical imaging data, yet another particularly interesting application area. For our model, we use a combination of two Benford distributions for each pair of components in the data, illustrated in Figure 18.1.

---

[1]J. Chiverton is with the School of Engineering, University of Portsmouth, UK and K. Wells is with the Centre for Vision, Speech and Signal Processing, University of Surrey, UK. *john.chiverton@port.ac.uk, k.wells@surrey.ac.uk,* copyright J. Chiverton and K. Wells 2012 (and their

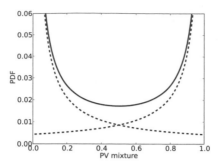

Figure 18.1   Illustration of the Benford **Partial Volume distribution**, composed of two
equal but opposite Benford distributions mapped onto a scale in the range $(0, 1)$.

A **component** is defined here as a region within an image that represents a mass
of tissue with distinctive properties or a physiological region representing a tissue
cluster that exhibits some form of similar biochemical or physiological behavior.
The actual intensity level for an individual component has a probability corresponding to a scaled Kronecker delta function, the intensity of which depends on the
properties of the tissues for **structural imaging** or the amount of **functional activity** at a particular location in the patient. In contrast to this, the Benford distribution
is useful to describe the probabilities for a range of intensities on the boundaries
between the components. These are regions of image data consisting of mixtures of
the intensities of each component due to the inherent blurring effect of the imaging
system. A simplified illustration of this effect, known as the **Partial Volume (PV)**
effect is shown in Figure 18.2.

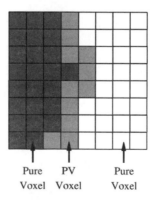

Figure 18.2   Simplified illustration of the PV effect in simulated image data.

The Benford distribution function has a logarithmic form and it is scale and base
invariant [Hi2]. We will show that these properties are particularly relevant for
medical imaging data where points in the image data consisting of a mixture of

components are scaled between the arbitrary intensity levels of the components in the data. We also show, experimentally, that the shape of the Benford distribution closely follows the distribution of the PV intensities.

Medical imaging systems are a particularly interesting application area because they are most often used to acquire diagnostic information about some part of a patient's anatomy or physiology. Some examples of different medical imaging modalities include **Magnetic Resonance Imaging (MRI), planar X-ray imaging** and **X-ray Computed Tomography (CT), Single Photon Emission Tomography (SPECT)** and **Positron Emission Tomography (PET)** image acquisition systems. Some imaging systems result in the emission of some type of signal from within the patient (e.g. MRI, **functional MRI, PET** and **SPECT,** the latter three being used for producing information about physiology rather than anatomy). Other modalities, such as planar X-ray imaging and X-ray CT, are considered as transmission-type modalities where X-rays are transmitted through the patient. The X-rays undergo variable amounts of attenuation dependent on the type of tissues through which they pass and therefore produce images which represent anatomical morphology or shape.

We describe a statistical model based on the use of the Benford distribution as a prior in a Bayesian probabilistic model. This model is applicable to many types of medical imaging data including MRI, SPECT and PET. The described model is applied here to the problem of classifying PET imaging data which is often used for the diagnosis and staging of **cancer** and **brain imaging studies.**

18.1.1 Chapter Overview

The next section, Section 18.2, describes the Partial Volume (PV) effect in more detail which is the name of the imaging artifact that produces the Benford distributed ranges of intensities. Section 18.3 then expands on some theory first proposed in a number of papers [ChWe, WCPBKO] to describe the Benford Partial Volume (PV) distribution. Our Benford PV distribution is composed of two equal but symmetrically opposing Benford distributions, illustrated in Figure 18.1.

Our statistical model of the PV effect, using the Benford distribution is applied to PET imaging data using a Bayes classifier formulation also described in Section 18.3. We use the Benford PV distribution as a prior on the intensity distribution of PV **voxels,** which are three-dimensional (3-D) volumetric data points or volume elements, the 3-D equivalent of the 2-D pixel. An experimental methodology is then described in Section 18.4 together with simulation-based experimental techniques. The experimental methodology includes a summary of a PET-CT phantom imaging study originally described in [WCPBKO] that was used to generate and simulate some imaging data that we can use to rigorously test our classifier. The generated imaging data are then classified with the Benford PV distribution Bayes classifier and results and discussion are presented in Section 18.5. Conclusions then follow in Section 18.6 to close this chapter.

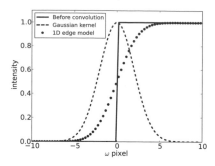

Figure 18.3 Illustration of 1-D edge model before and after blurring effects due to the image
acquisition process inherent in any imaging system. In this case a Gaussian
Point Spread Function (PSF) acts on the idealized edge to produce a blurred
representation of the boundary between two regions in an image.

18.2 THE PARTIAL VOLUME EFFECT

The Partial Volume (PV) effect can be understood as the degree of blurring or
mixing of information that occurs at boundaries between objects because of the
finite spatial resolution inherent to an **image acquisition process**. A finite number
of three-dimensional (3-D) volumetric data points or **voxels** are therefore utilized to
represent the continuous information in the true object. The signal representing this
information will have been subjected to a number of processes that predominantly
only allow lower frequencies to pass, thus limiting the amount of higher frequency
information.

The **Point Spread Function (PSF)** is often used to describe imaging and signal processing systems as it characterizes a system's response to an infinitesimally
small point source signal. Once the PSF is known, responses to any other signal can
be determined via mathematical operations such as convolution; see e.g. [GoWo].
A **Gaussian** function is often used as a common imaging system PSF, representing
the natural blurring produced by the process of **image acquisition** in a particular
imaging system. An example PSF together with the effect on an idealized 1-D **edge
model** can be seen in Figure 18.3.

This spreading in the digital domain typically carries across multiple voxels. The
broader the PSF the more the imaging information about an object will be spread
or blurred. It is thus logical to bin the discretized image into voxel dimensions
based on the width of a PSF. This helps to reduce redundancy in representation of
the imaging data and to quantify the resolution of the imaging process. Indeed,
Haacke et al. [HaBTV] (for MRI) state that the optimal size of a voxel can be
approximated by the **Full-Width at Half Maximum (FWHM)** of the PSF. A Sinc
function is commonly used for MRI but the FWHM for the common Gaussian PSF
is approximately 2.35σ, where σ is the standard deviation of the Gaussian function.

The PV effect is therefore directly related to the amount of spreading induced by
the action of the PSF of the image acquisition process. The PV effect describes an

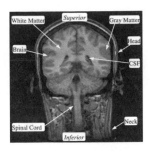 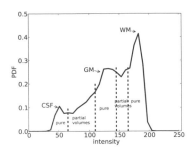

Figure 18.4   An example of a coronal T1 MR image slice of the human head where the different brain tissues can be seen and a histogram of the intensities for the brain tissues: **CerebroSpinal Fluid (CSF)**, **Gray Matter (GM)** and **White Matter (WM)**. The histogram is annotated with intensity ranges that are likely to correspond to pure tissues and intensities of voxels composed of a mixture of different tissues (Partial Volumes). Data from [CenMA].

artifact that is common to all systems that attempt to represent a continuous signal with a finite number of samples, although it is described by various different terms for different fields of application. A simplified illustration of the PV effect in an image is given in Figure 18.2.

There are several factors that govern the PV effect, which is inextricably linked with the blurring process associated with discrete signal quantization. This can be in 1, 2 or 3 dimensions, as well as other physical and technology-limiting effects associated with capturing the continuous signal in a band-limited acquisition system. The effect that PV voxels have on the statistical properties of imaging data is to reduce the possible independence of statistical classes of interest and merge them into (conceptually) a single entity. For example, a statistical class in a brain MR imaging application may include a particular range of intensities that might be typical for **White Matter (WM)**; see e.g. Figure 18.4.

Similarly, a different range of intensities could be used to describe the **Gray Matter (GM)** voxels. If the PV effect was not present and ignoring the effect of other artifacts, i.e., an idealistic imaging device, then the WM and GM classes would possess distinct intensity bands (ignoring noise and GM regions such as the Basal Ganglia that are lighter in appearance in T1 MRI). But due to factors such as noise and the PV effect, the boundary between the GM and WM intensities is usually quite ambiguous. A particular intensity may then originate from GM, WM or a mixture of the two classes. This is also true for **functional imaging** data where biological activity rather than morphological information is being imaged.

The PV effect is therefore a significant factor in medical imaging data and quantitative estimates of particular tissues or tissue volumes (e.g. tumor volume) or physiological activity (e.g. image maps of glucose utilization or rates of protein synthesis) often benefit by incorporating the PV effect in quantitative analysis.

18.3 MODELING OF THE PV EFFECT

A **noiseless single channel medical image** $I : \mathbb{R}^d \to \mathbb{R}$ can be considered to be
the result of a blurring convolution process of an idealized image space f with a
system PSF h:

$$I(\omega) = (h * f)(\omega), \tag{18.1}$$

where $\omega \in \Omega \subset \mathbb{R}^d$ represents spatial location so that Ω defines the limits of image
space. The idealized image space function $f : \mathbb{R}^d \to T$ has a range that consists
of a finite number (n) of tissues or activity values $T = \{T_1, T_2, \ldots, T_n : T_i \in \mathbb{R}\}$;
thus $f : \mathbb{R}^d \to (T \subset \mathbb{R}^n)$.

The real-valued PSF, h, is characterized by a shape (such as a multidimensional
Gaussian function) and the Full-Width at Half Maximum (FWHM) of that shape.
The PSF action is to blur the distinct class membership of T depending on their
spatial configuration in f. The PSF has the following properties: $h : \mathbb{R}^d \to \mathbb{R}$ and
$\int_{\omega \in \Omega'} h(\omega)d\omega = 1$ where $\Omega' = \{\omega | h(\omega) \neq 0, \omega \in \mathbb{R}^d\}$. In practice h is taken to
have compact support because $h(\omega) \approx 0$ for $\| \omega \| \gg 0$.

The blurring convolution of the discrete ideal image data f with the PSF h can
be defined either of two ways:

$$(h * f)(\omega) = \int_{\omega' \in \Omega'} h(\omega')f(\omega - \omega')d\omega'$$

$$= \int_{\omega' \in \Omega'} h(\omega - \omega')f(\omega')d\omega'. \tag{18.2}$$

This convolution blends together tissue values $T_i \in T$ so that each point may
contain a mixture of each tissue in a manner dependent on their spatial arrangement,
such that these values will be in the range $[\min(T), \max(T)]$. This can be seen by
the following simple 1-D **edge model**.

Theorem 18.3.1 (Simple 1-D edge model). *A noiseless 1-D edge model ($d = 1$)
consisting of two contiguous tissues or activity levels T_1 and T_2 can be described
by (including the PV affected interface or edge)*

$$I(\omega) = T_1 + \frac{(T_2 - T_1)}{2} \left(\mathrm{erf}\left(\frac{\omega - \omega_{\mathrm{b}}}{\sqrt{2\sigma^2}} \right) + 1 \right), \tag{18.3}$$

where ω_{b} is the location of the interface.

Proof. The convolution in (18.1) can be performed, as defined here with $d = 1$ and
$n = 2$, so that $I : \mathbb{R} \to \mathbb{R}$ and $T = \{T_1, \; T_2\}$ via

$$I(\omega) = (f * h)(\omega) = \int_{-\infty}^{\infty} f(\omega')h(\omega - \omega')d\omega'. \tag{18.4}$$

This integral can be simplified if we consider the PV distribution generated at a
single interface, e.g. $\omega = 0$. If $T_1 = 0$ and $T_2 = 1$ so that $f(\omega) = H(\omega)$ (the

Heaviside step function), the convolution then becomes

$$I(\omega) = \int_{-\infty}^{\omega} h(\omega')d\omega'.$$ (18.5)

A Gaussian function is a commonly assumed shape for an imaging system PSF, which takes the form

$$h(\omega) = \exp\left(-\frac{\|\omega\|^2}{2\sigma^2}\right).$$ (18.6)

The result of the convolution in (18.5) with (18.6) as the PSF is

$$I(\omega) = \frac{1}{2}\left(\operatorname{erf}\left(\frac{\omega}{\sqrt{2\sigma^2}}\right) + 1\right),$$ (18.7)

where $\operatorname{erf}(x) = (2/\sqrt{\pi})\int_{-\infty}^{t}\exp(-u^2)du$ is the **Gaussian error function**. Equation (18.7) is now a 1-D edge model where the edge is located across $\omega = 0$ and with signal intensities of 0 and 1; see Figure 18.3 for an illustration.

For arbitrary values of T_1 and T_2, $I(\omega)$ scales linearly and the location of the interface can be shifted by ω_b, thus resulting in (18.3). $\square$

Theorem 18.3.1 succinctly describes a 1-D edge model that can be generalized to higher dimensions via cuts that are orthogonal to an edge through the data. We have previously shown in [ChWe] that the density of intensities described by (18.7) are well approximated by the Benford PV distribution, discussed shortly in Section 18.3.1. However, it is difficult to describe any further details without a slightly different approach and resulting formulation, which we now give.

Definition 18.3.2. *A discrete class membership vector* $\mathbf{c}(\omega) = (c_1\ c_2\ \ldots\ c_n)^{\mathrm{T}}$ *that does not model the PV effect indicates* which *class is present at a particular location ω prior to the action of a PSF. It can therefore form a standard basis in* $\mathbb{R}^n$ *with the properties that for $c_i \leq n$ we have $c_i \in \{0, 1\}$ and $\sum_i c_i = 1$.*

We may also define a class membership vector that does describe the PV effect.

Definition 18.3.3. *The partial volume vector* $\alpha(\omega) = (\alpha_1\ \alpha_2\ \ldots\ \alpha_n)^{\mathrm{T}}$ *consists of $\alpha_i \leq n$ such that $\alpha_i \in [0, 1]$ and $\sum_i \alpha_i = 1$.*

Corollary 18.3.4. *For $n = 2$ the resulting 2-D PV vector* $\alpha = (\alpha_1\ \alpha_2)^{\mathrm{T}}$ *can be fully specified by a single scalar variable $\alpha = \alpha_1 = 1 - \alpha_2$ where $\alpha \in [0, 1]$.*

An alternative, but more involved description for f, the idealized image space, is then

$$f(\omega) = \mathbf{c}(\omega)^{\mathrm{T}}\mathbf{T}$$ (18.8)

where $\mathbf{T} = (T_1\ T_2\ \ldots\ T_n)^{\mathrm{T}}$. This can then be used in (18.2):

$$(h * f)(\omega) = \int_{\omega' \in \Omega'} h(\omega')\mathbf{c}(\omega - \omega')^{\mathrm{T}}\mathbf{T}d\omega'$$

$$= \underbrace{\left(\int_{\omega' \in \Omega'} h(\omega')\mathbf{c}(\omega - \omega')^{\mathrm{T}}d\omega'\right)}_{\alpha(\omega)^{\mathrm{T}}}\mathbf{T} = \alpha(\omega)^{\mathrm{T}}\mathbf{T}.$$ (18.9)

Substituting (18.9) into (18.1) yields

$$I(\boldsymbol{\omega}) = \boldsymbol{\alpha}(\boldsymbol{\omega})^{\mathrm{T}}\mathbf{T}. \tag{18.10}$$

This formulation can be converted into matrix form:

$$S = G\mathbf{T}, \tag{18.11}$$

where S is a column vector with elements $I(\boldsymbol{\omega})$ and G is a matrix where each row is a PV vector $\boldsymbol{\alpha}(\boldsymbol{\omega})^{\mathrm{T}}$. Both S and G can contain elements corresponding to all the image data, i.e., every $\boldsymbol{\omega} \in \Omega$, or, commonly, a subset of Ω as a **Region Of Interest (ROI)**.

This formulation was used by Rousset et al. [RouMA] and others who referred to G as a **Geometric Transfer Matrix (GTM)**. Matrix inversion techniques can be used to find G directly, but do not explicitly take into account the noise which is also present in the image data. Different combinations of mixtures may also be more likely, which is not modeled by just inverting the GTM (unlike a prior in a Bayesian formulation, which we will see shortly).

A **noise model** should also be considered, which may be statistically represented as a Gaussian, Poisson or Rician noise source depending on the particular imaging modality (e.g. CT, PET or MRI in the above examples, respectively). A Poisson noise model has a mean parameter λ that governs the shape of the resulting noise. Similarly the Rician distribution's shape depends on its mean value. Image sources with a noise distribution process η would usually need to be modeled with

$$I_\eta(\boldsymbol{\omega}) = \eta((h * f)(\boldsymbol{\omega})). \tag{18.12}$$

A Gaussian noise source η_G can often be treated independently of the underlying signal if the imaging modality produces noise with constant variance independent of the mean, hence

$$I_{\eta_G}(\boldsymbol{\omega}) = (h * f)(\boldsymbol{\omega}) + \eta_G. \tag{18.13}$$

The noise present in the image data can thus otherwise contaminate the estimated G, often resulting in magnified levels of noise in the PV estimates. Therefore [RouMA] went on to describe a technique to take into account the noise in the signal based on upper bounds that used the standard deviations of the noise.

We use a different technique based on a prior distribution $p(\boldsymbol{\alpha})$ that takes the form of the Benford distribution for the PV mixtures. We embed the prior in a Bayesian statistical estimation framework where the noise distribution can be explicitly modeled as in (18.12) instead of formulating our work solely around (18.11).

18.3.1 Benford PV Distribution

A possible way of defining a probabilistic model of the PV distribution is one based on a linear combination of functions κ_i corresponding to the mixing of two (or more) components.

Consider the PV distribution being composed of two equal but opposite distributions, reflected about the point $\alpha = 0.5$:

$$p(\alpha) = C\left(\kappa_1(\alpha) + \kappa_2(\alpha)\right), \tag{18.14}$$

where $\kappa_1(\alpha) = \kappa_2(1 - \alpha)$.

We have found (see [ChWe, WCPBKO]) through mathematical analysis, simulation and physical experimentation that the shape of the Benford distribution closely matches the shape of the distribution of PV intensities. Thus, the forms considered here for κ_1 and κ_2 are distributions based on the Benford distribution. The Benford distribution was proposed by Frank Benford who observed that the leading digit β of many natural sources of numerical data can be found to follow a probability mass function (PMF) of the form

$$P(\beta) = \log_b \left(1 + \frac{1}{\beta}\right), \tag{18.15}$$

where the leading digits, $\beta \in \{1, 2, \ldots, b-1\}$, are constrained to the base of the logarithm. The following definition, repeated here for convenience is originally defined in (2.3).

Definition 18.3.5. *The Benford distribution can also be extended to any number of significant digits, $\boldsymbol{\beta} = (\beta_1 \ \beta_2 \ \ldots \ \beta_m)^{\mathrm{T}}$:*

$$P(\boldsymbol{\beta}) = \log_{10}\left(1 + \left(\sum_{q=1}^{m} \beta_q 10^{m-q}\right)^{-1}\right). \tag{18.16}$$

We shall refer to this as the Benford distribution, which can be applied to numbers of arbitrary precision m.

For example, (18.16) gives the same probability for the numbers 1.34 and 134. Thus, the distribution can be applied to any decimal range quantified by the precision parameter m. We can use this, in modified form, to calculate probabilities for *discrete* PV quantities in the range $(0, 1)$.

Definition 18.3.6. *A discrete PV random variable is specified with a fixed number of decimal places, i.e., $\xi \times 10^\lambda \in \mathbb{N}$, where $\xi \in [0, 1]$ and λ describes the **precision** or the number of decimal places of ξ.*

This definition for a discrete PV random variable is predicated on a computer-based PV variable of finite precision, usually based on the precision used in the software storing the digital samples. This now provides us with sufficient information to describe the following.

Conjecture 18.3.7. *The discrete PV distribution $P(\xi)$ can be realized as a combination of Benford distributions, reflected about the axis $\xi = 0.5$, each representing the spread of one class being distributed across the other, so that*

$$P(\zeta) \propto \log_{10}\left(1 + \frac{1}{\xi 10^\lambda}\right) + \log_{10}\left(1 + \frac{1}{(1-\xi)10^\lambda}\right). \tag{18.17}$$

Proof. We give the construction, but do not determine the actual shape of the distribution.

Let us consider a discrete form for the two class PV distribution given by (18.14):

$$P(\xi) \propto (\kappa_1(\xi) + \kappa_2(\xi)). \tag{18.18}$$

The κ_1 and κ_2 can be replaced by the Benford distribution of Definition 18.3.5:

$$P(\xi) \propto \log_{10}\left(1 + \frac{1}{v_1(\xi)}\right) + \log_{10}\left(1 + \frac{1}{v_2(\xi)}\right). \qquad (18.19)$$

The functions $v_1(\xi)$ and $v_2(\xi)$ now have to take a form equivalent to $\sum_{q=1}^{m} \beta_q 10^{m-q}$ in Definition 18.3.5. It is easy to see, because a discrete PV random variable has the properties given in Definition 18.3.6, that we can say

$$v_1(\xi) = \xi 10^{\lambda},$$
$$v_2(\xi) = (1 - \xi)10^{\lambda}. \qquad (18.20)$$

$\square$

Equation (18.17) represents a discrete PV distribution that can be quoted to any degree of precision λ. This is potentially very useful for the representation of the PV distribution, consisting of two equal but opposite Benford distributions, as illustrated in Figure 18.1.

We may also consider a number of interesting properties of this Benford PV distribution.

Lemma 18.3.8. *The shape of the Benford PV distribution is invariant to the base of the logarithm b and the precision λ, where*

$$P(\xi) \propto \log_b\left(1 + \frac{1}{\xi 10^{\lambda}}\right) + \log_b\left(1 + \frac{1}{(1 - \xi)10^{\lambda}}\right). \qquad (18.21)$$

Proof. This is easy to see because we know that $\log_b(x) = \frac{\log_c(x)}{\log_c(b)}$ where $\log_c(b) = C$ is a constant. According to (18.15), $\beta \in \{1, 2, \ldots, b-1\}$ and $b = 10^{\lambda}$, hence if b can take any value without changing the shape of the distribution then the precision λ can also take any value. $\square$

Conjecture 18.3.9. *The Benford PV distribution can be recast as a continuous density function*

$$p(\alpha) \propto \log_b\left(1 + \frac{1}{\alpha 10^{\lambda}}\right) + \log_b\left(1 + \frac{1}{(1 - \alpha)10^{\lambda}}\right), \qquad (18.22)$$

where the value of the normalizing constant is a function of the precision λ. We also believe that any integer precision in α or $(1 - \alpha)$, after multiplication by 10^{λ}, will have negligible effect on the shape of the distribution.

18.3.2 Classification of Medical Image Data

The PV distributions described above, $p(\alpha)$ and $P(\xi)$, can be used as prior distributions for the mixing of bicomponent regions of a medical image such as the boundaries between two tissues or two contiguous regions of activity (see e.g. [WCPBKO]).

Bayes' theorem can then provide a convenient approach to calculating the **posterior** PDF or PMF of PV mixing variables ξ for a voxel with a particular intensity I (omitting ω for brevity). For the discrete case we have

$$P(\xi|I) = \frac{p(I|\xi)\pi(\xi)}{p(I)}, \qquad (18.23)$$

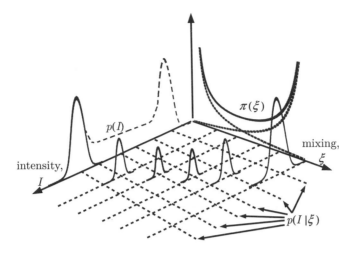

Figure 18.5 Illustration of a classification space for a two-class problem, consisting of the marginal distribution $p(I)$ composed of individual intensity distributions for particular PV mixing values $p(I|\xi)$ each weighted by the Benford PV distribution $\pi(\xi)$.

where $p(I|\xi)$ is the intensity-based **likelihood** PDF for a particular mixture value ξ, and $p(I)$ is the marginal intensity PDF. Here $\pi(\xi)$ is a **prior** that combines both the (Benford) PV distribution and Kronecker delta components for the non-PV components, defined shortly. Figure 18.5 illustrates the PV classification space for a two-class PV classification problem. The intensity likelihood $p(I|\xi)$ is dependent on the noise distribution assumed for the imaging data.

For a Gaussian **noise distribution**, and assuming linear mixing, the mean can be calculated with

$$\mu_\xi = \xi_j \mu_j + \xi_k \mu_k, \tag{18.24}$$

and the variance is

$$\sigma_\xi^2 = (\xi_j \sigma_j)^2 + (\xi_k \sigma_k)^2, \tag{18.25}$$

where μ_j, μ_k and σ_j, σ_k are the pure (unmixed or "non-PV affected") component means and standard deviations for voxel classes i and j respectively. The marginal PDF $p(I)$ is calculated here via the numerical method of Riemann sums (see e.g. [RobCa]).

The expectation over ξ w.r.t. the posterior PDF for ξ corresponds to a value of ξ that minimizes the **mean square error**. Therefore, in order to obtain an estimate of the PV content for a voxel with an intensity I we calculate (for the discrete case)

$$\xi_E = \mathrm{E}\left[\xi|P(\xi|I)\right] = \sum_\xi \frac{\xi}{p(I)} p(I|\xi)\pi(\xi). \tag{18.26}$$

Here $\pi(\xi)$ is a combination of $P(\xi)$ for a pair of classes i, j and scaled Kronecker

 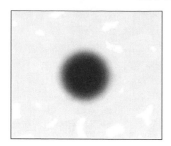

Figure 18.6 Illustration of exemplar classifier output for a hot insert, where the insert contains a greater concentration of the radio-pharmaceutical to simulate increased biological uptake often associated with pathological conditions such as a tumor. The left image corresponds to the insert mixing value ($\xi = \xi_1$) and the right corresponds to the complementary mixing component (($1 - \xi$) = ξ_2). Brightness is proportional to the magnitude of the mixing present for hot insert (left) and background (right). © 2007 IEEE. Reprinted, with permission, from K. Wells et al., *Quantifying the Partial Volume Effect in PET Using Benford's Law*, IEEE TNS **54** (5:1), 2007 pp. 1616–1625.

delta functions $\delta(1 - \xi_i)$ so that

$$\pi(\boldsymbol{\xi}) = \delta(1 - \xi_j)P_j(\mu_j) + \delta(1 - \xi_i)P_i(\mu_i) + P(\xi_i, \xi_j)P_{i,j}(\mu_i, \mu_j)U(\xi_i, \xi_j), \tag{18.27}$$

where $U(\xi_i, \xi_j) = 1$ for $0 < \xi_i, \xi_j < 1$ and 0 otherwise to *switch off* the PV distribution at the extreme values of $\boldsymbol{\xi}$, $P(\mu_i)$, $P(\mu_j)$ are the pure prior probabilities and $P(\mu_i, \mu_j)$ is the PV prior probability.

Equation (18.26), together with (18.27), provides a realizable approach to estimating PV content of an image $I(\boldsymbol{\omega})$ for $\boldsymbol{\omega} \in \Omega$ or subregion $\boldsymbol{\omega} \in \mathcal{R} \subset \Omega$, known as a **Region Of Interest (ROI)**. The intensity values $I(\boldsymbol{\omega})$ associated with each voxel $\boldsymbol{\omega}$ are used to calculate expected PV content $\boldsymbol{\xi}_E(\boldsymbol{\omega})$ for each voxel via (18.26) together with estimated prior probabilities (pure and PV), means and standard deviations.

These estimated PV vectors $\boldsymbol{\xi} = (\xi_1 \ \ldots \ \xi_n)^{\mathrm{T}}$ can be used to generate image maps for each PV scalar variable ξ_i. An example with $n = 2$ can be seen in Figure 18.6 for the classification of a *hot* insert which has a greater concentration of radio-pharmaceutical in comparison to the surrounding medium.

This approach can be utilized for the classification of 3-D medical imaging data into regions with similar anatomical or physiological properties, wherein some voxels contain mixtures of more than one region's characteristics known as Partial Volume (PV) voxels. However, this formulation is only valid for the classification of data consisting of target objects (and background) with dimensions greater than at least twice the Full-Width at Half Maximum of the PSF. This means that the approach will only work for data which contains a proportion of "pure" voxels (i.e., unmixed or unaffected by the PV effect). The present formulation considers only $n = 2$ but can be simply extended to $n > 2$ by modifying (18.27) and extending the sums in (18.24) and (18.25) to include additional components.

18.3.3 Classification Error

The **performance** of a classifier on PV imaging data can be quantified via measures such as the **Root Mean Square (RMS) error** between the estimated PV values and the ground truth PV errors. However classification performance is dependent on the amount of noise present in the imaging data and the separation of the pure unmixed classes in the intensity space (in the absence of noise). Therefore an RMS error based on the differences between estimated signal levels and ground truth signal levels could also be used for a region consisting of a majority of classes i and j:

$$L = \frac{\sqrt{\frac{1}{N} \sum_{\boldsymbol{\omega}} (\mu_T(\boldsymbol{\omega}) - \mu_E(\boldsymbol{\omega}))^2}}{|\mu_i - \mu_j|}, \tag{18.28}$$

where μ_i and μ_j are the signal levels for voxels composed purely of classes i and j. The PV signal levels μ_T and μ_E are defined here by the following.

Definition 18.3.10. *Signal level is the intensity value of a voxel in the absence of a noise component. Hence for a voxel $\boldsymbol{\omega}$,*

$$\mu(\boldsymbol{\omega}) = \sum_{i=0}^{n} \xi_i(\boldsymbol{\omega})\mu_i, \tag{18.29}$$

where μ_i is the signal level for class i.

An informative measure of the differences between two components in imaging data that takes into account these classification-dependent properties is the **Contrast to Noise Ratio (CNR)**, defined as

$$\mathrm{CNR} = \frac{|\mu_i - \mu_j|}{\sigma} \qquad \text{where} \qquad \sigma = \sqrt{\frac{\sigma_i^2 + \sigma_j^2}{2}}. \tag{18.30}$$

18.4 MATERIALS AND METHODS

We use the results of a PET-CT phantom imaging study,[2] originally described in [WCPBKO], to test the validity of the Benford model on simulated and experimental (real) imaging data.

A **PET-CT scanner** is a particularly useful diagnostic instrument because it can generate a dual representation of the subject: one based on a **functional representation** (using **Positron Emission Tomography (PET)**) and the other based on a **structural or morphological representation** (CT) corresponding to anatomy. **Functional imaging** data alone is often difficult to interpret because of the lack of corresponding structural information, but nonetheless is useful because subtle changes in physiology, such as higher glucose utilization rate, often precede subsequent gross changes in morphology anatomy. For example, the start of a highly

---

[2]**PET-CT scanners** are a type of dual modality medical imaging scanner, wherein patients first undergo an X-ray CT scan to produce an image of their anatomy, and are then injected with a radioactive compound which is then imaged to show the distribution of some aspect of their physiology such as glucose utilization.

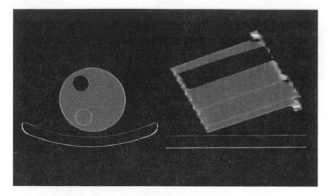

Figure 18.7   CT cross-sectional and sagittal slices through the CT phantom data used in
this work. The (upper) air-filled (cold) insert and the (lower) "hot" water-filled
insert are clearly defined. The obvious gap between the phantom and the couch
is due to the angle at which the phantom was propped in the axial direction, in
order to experimentally better sample the partial volume voxel distribution. ©
2007 IEEE. Reprinted, with permission, from K. Wells et al., *Quantifying the
Partial Volume Effect in PET Using Benford's Law*, IEEE TNS **54** (5:1), 2007
pp. 1616–1625.

visible tumor is preceded by the accumulation of a cluster of cancer cells and their
ability to form into an invasive structure. Integrating both imaging modalities in a
single scanner allows registration of the two separate data sets combining structural
and functional information in a single image. A potential disadvantage of using the
dual PET-CT scanner is due to the physics of the **image acquisition process**: PET
imaging is an inherently lower spatial resolution imaging process than CT. However, this property can be used quite advantageously for PV analysis.

Real imaging data is affected by noise and the objective of classifying the data is
to determine the quantity of a tissue or activity concentration for an ROI of voxels.
However it is difficult to directly determine the validity of the Benford PV distribution model because of the hidden nature of the PV content. We therefore study
the validity of the Benford model with the use of a test object, or a phantom, to
control the amount of PV mixing present. First we undertake a simulation based
on part of the phantom's geometry, and then compare this with experimental PET
measurements made on the phantom.

The simulated geometry used here represents a cylindrical phantom containing
a radioactive solution at a particular concentration in an aqueous solution, with a
water-filled insert at higher activity concentration and a cold air-filled insert. In this
case the hot insert represents the case of an area of high radioactive concentration
in the body, such as the brain or a lesion, and the air filled cavity represents a
lung-like region. A Computed-Tomography (CT) image of the phantom with the
same geometry was used to derive the ground truth data set. Elementary image
processing techniques were used to define three binary templates, corresponding to
the two inserts and another for the main cylinder; see Figure 18.7.

The PET imaging process produces data with relatively larger voxel dimensions

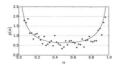

Figure 18.8 Normalized histogram produced from a simulated noise-free PV data set. The result of an idealized noiseless situation composed of two peaks (off scale) corresponding to high frequency occurrences of the means of the unmixed voxel regions in the ground truth data and a function linking the two peaks together, corresponding to the simulated PV voxels. The continuous line demonstrates excellent agreement between the Benford PV distribution and the synthetic data. © 2007 IEEE. Reprinted, with permission, from K. Wells et al., *Quantifying the Partial Volume Effect in PET Using Benford's Law*, IEEE TNS **54** (5:1), 2007 pp. 1616–1625.

in comparison to a CT imaging process where the PET data was generated with voxel dimensions of $4.00 \times 4.00 \times 4.00$ mm^3. A Gaussian function with FWHM of 9.89 mm was found to provide a good characterization of the PET PSF validated by experiment (see [WCPBKO]) to represent the actual PSF of the imaging system used here (Phillips Gemini PET-CT system) when imaging Ga-68 as a particular radioisotope. This was assumed to be stationary across the entire 3-D volume. Spatial variations in the PSF do exist for some PET image acquisition setups but for the system used here, i.e., one based on a 3-D camera, these variations have been found to be relatively insignificant; see [GrPaFl].

The CT data was reconstructed with voxel sizes of $1.17 \times 1.17 \times 5.00$ mm^3. It was subsequently processed with a number of image processing steps including region growing and thresholding to produce the image maps that could be used as the ground truth. The image maps were then filtered with the measured PET PSF and reregistered so that the individual compartments in the CT data resulted in a data volume with a one-to-one correspondence of voxels in the two types of data, i.e., CT derived ground truth versus PET. This ground truth was then used to assess the performance of the PV classifier.

A PV histogram for the noiseless PSF filtered data set was calculated and compared with the Benford PV distribution and found to exhibit excellent agreement between the two as shown in Figure 18.8. A similar process was used to generate a simulated PET data set where Gaussian noise in the noise simulated data set was generated using parameters estimated from the real PET imaging data.

18.4.1 Classification and Assessment

The PV classifier described in Section 18.3.2, was then applied to the problem of classifying the PV content for individual voxels in the PET imaging data. ROIs were defined for the hot insert and the cold insert, as illustrated in Figure 18.9.

The quality of the classification process was quantified via the **Root Mean Square (RMS) error** L between the signal levels in the ground truth $\mu_T(\boldsymbol{\omega}) = \sum \xi_{T,i}(\boldsymbol{\omega})\mu_i$ and the estimated signal levels $\mu_E(\boldsymbol{\omega}) = \sum \xi_{E,i}(\boldsymbol{\omega})\mu_i$; see (18.28) for more in-

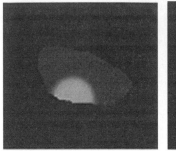 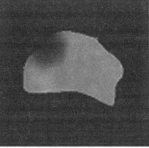

Figure 18.9   Exemplar ROI templates used to define the experimental PV analysis area
              in each slice: hot insert on left and cold air insert on right. The underlying
              grayscale image data has been independently scaled to better show phantom
              insert structures. These ROIs ensure air/outer wall mixing effects are avoided.
              Combining the ROI template with the CT-derived insert templates allowed the
              data simulator to reproduce the same imaging geometry used for experimental
              analysis. The apparent change in background gray level is due to grayscale au-
              toscaling within the display software, used to create these ROI templates. ©
              2007 IEEE. Reprinted, with permission, from K. Wells et al., *Quantifying the
              Partial Volume Effect in PET Using Benford's Law*, IEEE TNS **54** (5:1), 2007
              pp. 1616–1625.

formation. Next L was calculated for the ROI volumes $\mathcal{R}$ shown in Figure 18.9.
These were then subdivided into RMS errors for PV voxels alone and RMS errors
for pure voxels alone. This subdivision provides an idea of the effectiveness of the
classifier on these respective classes of voxels.

18.5 RESULTS AND DISCUSSION

18.5.1 Partial Volume Simulation

As described earlier, a number of data volumes were simulated: (i) a noiseless data
set convolved with the PET measured PSF, and (ii) a noise-affected data set also
convolved with the PET measured PSF with noise parameters measured from the
real PET data set. The noise-affected data set was classified and the classification
accuracy using the RMS measure described above for each class can be seen in
Table 18.1.

The results from classifying the simulated data represent the lower limit possible
on these types of errors. We also plot the histogram of the simulated data and
compare it with the Benford PV model in Figure 18.10, demonstrating an excellent
fit. These results are for the simulated data only which do not include the effects
of other imaging artifacts such as from attenuation or scatter. The effects of these
were more fully investigated in [WCPBKO].

Parameter source	CNR	Overall RMS error	PV voxel RMS error	Non-PV voxel RMS error
Hot insert	72.48	0.01	0.01	0.00
Cold insert	33.49	0.02	0.03	0.02

Table 18.1 RMS differences (quoted to 2 decimal places) in classification performance between simulated data sets using voxel class parameters taken from corresponding PET data set.

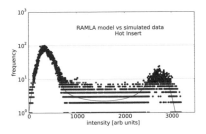

Figure 18.10 Histogram from ROI around simulated hot insert. Simulated data using noise parameters derived from real data (symbols) provide excellent agreement with the scaled **marginal** distribution utilizing the Benford model (continuous line). Background peak appears leftmost, insert is represented by rightmost peak. © 2007 IEEE. Reprinted, with permission, from K. Wells et al., *Quantifying the Partial Volume Effect in PET Using Benford's Law*, IEEE TNS **54** (5:1), 2007 pp. 1616–1625.

Parameter source	CNR	Overall RMS error	PV voxel RMS error	Non-PV voxel RMS error
Hot insert	72.48	0.03	0.04	0.00
Cold insert	33.49	0.08	0.17	0.01

Table 18.2   The RMS classification errors between ground truth (processed CT data) and classified experimental PET data set.

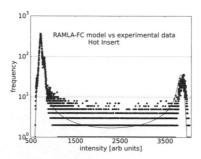

Figure 18.11   Histogram of the experimental data obtained from the ROI applied to the hot insert (symbols) capturing both insert and background. Note excellent fit of data points compared to the assumed marginal distribution (continuous line). Background peak appears leftmost, insert is represented by rightmost peak. © 2007 IEEE. Reprinted, with permission, from K. Wells et al., *Quantifying the Partial Volume Effect in PET Using Benford's Law*, IEEE TNS **54** (5:1), 2007 pp. 1616–1625.

18.5.2 PET-CT Phantom Imaging Study

A similar classification experiment and performance characterization was undertaken using the real PET imaging data. RMS error measurements were then calculated by using the CT derived ground truth data. The RMS PV errors for these experiments can be seen in Table 18.2.

As might be expected, the errors obtained from the experimental data are higher than the errors obtained for the simulated data. The registration process, although mostly intrinsic to the PET-CT phantom study, may have caused a systematic bias on the errors. A similar comment can be made about the wall thickness for the inserts in the phantom. Other potential sources of experimental error include effects from imaging artifacts such as attenuation. The fit accuracy of the Benford PV model for the real data set can be seen in Figure 18.11.

The classification results for the cold insert are somewhat worse for the PV voxels in comparison to the simulated results. The fit of the model to the cold insert histogram data, shown in Figure 18.12, confirms that the model is not as applicable to the cold insert part of the data. A potential problem with the classification of the cold insert is the assumption of a spatially invariant PSF. However, in PET the width of the PSF is dependent on the local environment that the radiation is immersed in, and clearly in this case the density changes significantly between the

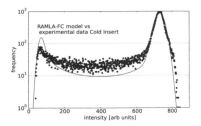

Figure 18.12 Histogram of the experimental data obtained from the ROI applied to the cold insert (symbols) compared to the assumed marginal distribution (continuous line). Background peak appears rightmost, insert is represented by left peak. © 2007 IEEE. Reprinted, with permission, from K. Wells et al., *Quantifying the Partial Volume Effect in PET Using Benford's Law*, IEEE TNS **54** (5:1), 2007 pp. 1616–1625.

water and the air-filled insert. Nevertheless, the PV classification on the experimental data, even for the cold insert, is still less than 8% overall (PV and pure voxel errors combined).

18.6 CONCLUSIONS

This chapter has introduced some core concepts in medical imaging, particularly those related to an imaging artifact known as the PV effect. We have described the scene for a statistical model based on Benford's Law, referred to here as the Benford Partial Volume (PV) distribution. We have gone on to describe a simple classifier based on Bayes theory which uses the Benford PV distribution as a prior. We have presented the experimental methodology and results of applying the proposed classifier to the classification of PV voxels in a PET-CT phantom imaging study originally published in [WCPBKO].

The results show that we have experimentally found the Benford PV distribution to be a reasonable modeling tool for the classification of imaging data affected by the PV artifact. Furthermore the Benford PV distribution provides a convenient formulation to describe the PV artifact. We have considered the case of when the number of classification classes is $n = 2$, but it is a trivial exercise to extend this to any number of classes, where for each additional classification class another individual Benford distribution can be introduced.

Chapter Nineteen

Application of Benford's Law to Images

Fernando Pérez-González, Tu-Thach Quach, Chaouki T. Abdallah, Gregory L. Heileman and Steven J. Miller[1]

This chapter analyzes the application of Benford's law to pictures taken from nature with a digital camera. While the values output by the image capture device embedded in the camera, i.e., the pixels, do not follow Benford's law, we show that if they are transformed into a domain that better approximates the human visual system then the resulting values satisfy a generalized form of Benford's law. This can be used for image forensic applications, such as detecting whether an image has been modified to carry a hidden message (steganography) or has been compressed with some loss of quality.

19.1 INTRODUCTION

"An image is worth a thousand words," and even more so since the advent of digital technologies that have made it much easier not only to capture images but also to store and deliver them. Images and videos constitute the dominant type of traffic in communication networks, and several recent analyses predict that by 2016 more than 50% of the total Internet traffic will correspond to images/video.

Considering that many natural phenomena seem to follow Benford's law and that images are often nothing but "snapshots of nature," it is pertinent to wonder whether images (at least those taken from nature) obey Benford's law. Intriguingly, very few works have addressed this question. The first is a paper by J. M. Jolion [Jol], who showed that Benford's law holds reasonably well for the gradient magnitude of images and in pyramidal decompositions based on the Laplace transform. Even though [Jol] has great experimental value, we show (cf. Section 19.5) that

---

[1]Pérez-González: Department of Signal Theory and Communications, University of Vigo, EE Telecomunicacion, Campus Universitario, 36310 Vigo, Spain; Quach: Sandia National Laboratories, Albuquerque, NM; Miller: Department of Mathematics and Statistics, Williams College, Williamstown, MA; Abdallah and Heileman: Electrical & Computer Engineering Department, University of New Mexico, Albuquerque, NM. Research supported by the European Union under project REWIND (Grant Agreement Number 268478), the European Regional Development Fund (ERDF) and the Spanish Government under projects DYNACS (TEC2010-21245-C02-02/TCM) and COMONSENS (CONSOLIDER-INGENIO 2010 CSD2008-00010), the Galician Regional Government under projects "Consolidation of Research Units" 2009/62, 2010/85 and SCALLOPS (10PXIB322231PR), and the Iberdrola Foundation through the Prince of Asturias Endowed Chair in Information Science and Related Technologies.

his analysis for the gradient magnitude of images hinges on some assumptions that may need to be revised. E. Acebo and M. Sbert [AceSbe] proposed a Benford's law test to determine whether synthetic images were generated using physically realistic methods, although the fact that many real images do not follow Benford's law (see Section 19.3) raises some doubts of the suitability of this approach. Acebo and Sbert's work also lacks a supporting theory to explain the connections (if any) of images and Benford's law. A major step forward, expanded in this chapter, was taken by Pérez-González et al. in [PéAH1], [PéAH2], where a rigorous link between well-established statistical models for images and Benford's law was uncovered.

In this chapter we show that while images in the "pixel" domain do not conform to Benford's law, the situation changes quite dramatically when they are transformed using the **Discrete Cosine Transform** (DCT). Some key observations regarding the logarithm of DCT coefficients lead us to present a generalization of Benford's law, based on Fourier analysis, that yields a much closer fit to the observed digits frequencies. We also give a theoretical justification for why images in the DCT domain satisfy the generalized law; such explanation relies on the fact that DCT coefficients may be very well modeled by Generalized Gaussian Distributions.

We also revisit Jolion's derivation for the gradient of images, showing that some of the approximations made there do not hold. We give an alternative explanation which is again based on Generalized Gaussian Distributions for both the horizontal and vertical components of the gradient, and the so-called β-*Rayleigh* distribution which is used to model the gradient magnitude. This in turn serves to justify why closeness to Benford's law is not as large for the gradient magnitude as it is for the DCT coefficients. Moreover, with regard to Benford's law, the DCT is "purer" than the gradient in the sense that the former is, unlike the latter, an invertible transform.

Finally, we discuss some potential applications in image forensics, by showing how our Fourier-based formulation may be used to detect whether an image has been compressed or if it carries a hidden message.

19.2 BACKGROUND

In this section we recall some of the known properties that affect random variables in the context of Benford's law. We refer the reader to Chapter 2 for rigorous proofs of those properties. We use the tilde to denote the transformation of a random variable that reduces modulo 1 its base-10 logarithm, i.e., given V, we construct $\tilde{V} := \log_{10} |V| \bmod 1$. Given a real positive number x, we will say that $\log_{10} |x| \bmod 1$ is defined in the **Benford domain**.

Theorem 19.2.1. *A random variable X is Benford if and only if the random variable $\tilde{X} = \log_{10} |X| \bmod 1$ is uniform in $[0, 1)$.*

See Theorem 2.4.2 and its proof.

Theorem 19.2.2 (Scale invariance). *Suppose that X is Benford; then the random variable $Z = \alpha X$, for an arbitrary $\alpha \in \mathbb{R}^+$, is Benford if only if X is Benford.*

Proof. See Section 2.4.2 for a proof. Alternatively, consider the random variable $\log_{10}|Z| \bmod 1 = \log_{10}\alpha + \log_{10}|X| \bmod 1$. This corresponds to a cyclic shift of $\log_{10}\alpha \bmod 1$ on the probability density function (pdf) of $\tilde{X}$. Now, for this shift to have the same probabilities regardless of the value of α one must have that $\tilde{X}$ is uniform in $[0, 1)$.   □

Theorem 19.2.3 (Product of independent random variables). *Let X be Benford, and let Y be another random variable independent of X. Then the random variable $Z = X \cdot Y$ is Benford.*

Proof. This corresponds to Theorem 2.6.3. For the sketch of an alternative proof, notice that $\log_{10}|Z| = \log_{10}|X| + \log_{10}|Y|$. Therefore, the pdf of $\log_{10}|Z|$ will be the convolution of those of $\log_{10}|X|$ and $\log_{10}|Y|$. However, we are interested in the modulo 1 reduction of $\log_{10}|Z|$; its pdf can be obtained by performing the convolution between the pdf of $\tilde{X} = \log_{10}|X| \bmod 1$ and that of $\log_{10}|Y|$, and reducing the result modulo 1. Alternatively, one can do the circular convolution over $[0, 1)$ of the pdf's of $\tilde{X}$ and $|\tilde{Y}| = \log_{10}|Y| \bmod 1$. But when $\tilde{X}$ is uniform in $[0, 1)$ the circular convolution is always uniform in $[0, 1)$, regardless of the distribution of $\tilde{Y}$.   □

The product interpretation connects Benford's law to *mixtures of random variables*. Mixtures of random variables are relevant in image processing following the work of Hjorungnes et al. [HjLR], where a Laplacian distribution (often used to model the coefficients of a blockwise discrete cosine transform; cf. Section 19.3) is written as a mixture of Gaussians whose variance is controlled by an exponential distribution. Thus, if $f_X(x)$ denotes a zero-mean unit-variance Gaussian pdf, the mixture pdf takes the form

$$f_Z(z) = \int_0^\infty f_X(z|\sigma^2) f_{\Sigma^2}(\sigma^2) d\sigma^2 \tag{19.1}$$

where $f_{\Sigma^2}(\sigma^2)$ is an exponential, i.e., $f_{\Sigma^2}(\sigma^2) = \lambda \exp(-\lambda\sigma^2)$. It can be shown that the variance of Z coincides with the mean of the mixing density, i.e., $1/\lambda$. Interestingly, mixtures of the general form given in (19.1) can be written in such a way that Theorem 19.2.3 is straightforwardly applied. Indeed, the random variable Z whose pdf is $f_Z(z)$ is obtained through (19.1) can be written as $Z = X \cdot \Sigma$, with Σ^2 the random variable that controls the variance. It is therefore easy to conclude that if either X or Σ conform to Benford's law, then Z will also do so.

19.3 APPLICATION OF BENFORD'S LAW TO IMAGES

Given the seemingly good match of natural phenomena distributions to Benford's law, it is reasonable to ask whether this will be so for **images**. Unfortunately, it is well known that image luminances possess a histogram that does not admit a closed form, as there is a strong variation from picture to picture. Hence, it is highly unlikely that Benford's law, or any generalization, will be applicable here. Our experiments confirm that gray-level images do not satisfy Benford's law. To illustrate, consider the image "Man" (of size 1024×1024 pixels) shown in Figure 19.1 for

Figure 19.1 Figure "Man" used in the experiments. Image taken from http://sipi.usc.edu/database/copyright.php (number 5.3.01).

which the histogram of the variable $\log_{10} |X| \bmod 1$ is shown in Figure 19.2(a). Clearly this histogram falls short of being constant, which would guarantee compliance to Benford's law. Consequently, the Leading Digit distribution is quite different from that proposed by Benford, as plotted in Figure 19.2(b).

One however obtains quite different results when one considers the coefficients of the blockwise DCT transform, as it is found that they match a Benford distribution reasonably well.

Given a gray-level image $x(u,v)$, $u, v \in \{0, 1, \ldots, N-1\}$, of size $N \times N$ pixels,[2] where x is the luminance, u and v denote respectively the horizontal and vertical coordinates, its DCT is defined by the pair

$$b(i,k) = C(i)C(k) \sum_{u=0}^{N-1} \sum_{v=0}^{N-1} x(u,v) \cos\left[\frac{\pi}{N}i\left(u+\frac{1}{2}\right)\right] \cos\left[\frac{\pi}{N}k\left(v+\frac{1}{2}\right)\right],$$

$$(19.2)$$

$$x(u,v) = \sum_{i=0}^{N-1} \sum_{k=0}^{N-1} C(i)C(k)b(i,k) \cos\left[\frac{\pi}{N}i\left(u+\frac{1}{2}\right)\right] \cos\left[\frac{\pi}{N}k\left(v+\frac{1}{2}\right)\right],$$

$$(19.3)$$

where $C(0) = 1/\sqrt{N}$ and $C(m) = \sqrt{2}/\sqrt{N}$, for $m = 1, 2, \ldots, N-1$.

---

[2] For simplicity, we limit our exposition to square images.

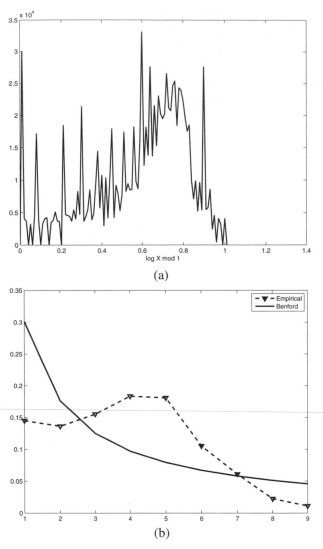

Figure 19.2 Histogram of the luminance values of "Man" in Benford ($\log_{10} |\cdot|$ mod 1) domain (a). Distribution of the Leading Digit corresponding to "Man" (b).

The transform pair given above is referred to as the **full-frame DCT**. In many instances, e.g. in the **JPEG compression** standard, a blockwise DCT is preferred. In such cases, the equations above still apply with $x(u, v)$ and $b(i, k)$ respectively the input and output blocks of size $N \times N$. The $b(i, k)$ are called the **DCT coefficients**, and due to their different properties, a distinction is often made between the zero-frequency (**DC coefficient**) (i.e., that corresponding to $i = k = 0$), and the remainder, which are known as the **AC coefficients**.

The popularity of the DCT arises from its ability to remove redundancy between neighboring pixels, which in compression applications leads to uncorrelated coefficients which can be quantized independently. The other desirable feature of the DCT is its *energy compaction*, which leads to packing the input pixels into as few coefficients as possible, allowing the remaining ones to be discarded, with little perceptual impact.

Figure 19.3(a) shows the histogram of the variable $\log_{10}|X| \bmod 1$, with X given in the block-DCT domain, for the image "Man," while Figure 19.3(b) represents the distribution of the Leading Digit, which now lies closer to Benford's distribution. The DCT block size used in these figures is 8×8; however, similar results are obtained by considering other block sizes as well as other images. A crucial observation from Figure 19.3(a) is that the histogram is not quite flat, but instead can be modeled with a constant plus a sinusoidal term. This somehow surprising phenomenon was observed in the hundreds of images we analyzed, thus suggesting a generalization of Benford's law to accommodate the extra term.

The crucial question is therefore why DCT coefficients follow this generalized form of Benford's law. The following fact is known about images in the DCT domain: the AC coefficients of a block-based DCT can be accurately modeled by a **Generalized Gaussian Distribution** (GGD). For instance, for the 8×8-block DCT, let $b^{(m)}(i,k)$, $i, k \in \{0, \ldots, 7\}$, $i + k \neq 0$, denote the (i, k)th AC coefficient of the DCT of the mth block. Then $b^{(m)}(i, k)$ for all m can be thought of as being drawn from a GGD. A GGD has the form

$$f_X(x) = Ae^{-|\beta x|^c}, \tag{19.4}$$

where A and β are expressed in terms of c and the standard deviation σ as follows:

$$\beta = \frac{1}{\sigma}\left(\frac{\Gamma(3/c)}{\Gamma(1/c)}\right)^{1/2}; \quad A = \frac{\beta c}{2\Gamma(1/c)}. \tag{19.5}$$

The parameter c is the *shaping factor*. The particular cases of $c = 2$ and $c = 1$ correspond to the well-known Gaussian and Laplacian distributions, respectively. Unfortunately, for real images the parameters σ and c vary with the frequency indices (i, k). This implies that the coefficients we are modeling should be considered as being generated by a mixture of GGDs, in which the parameters are governed by a certain rule. This might complicate the derivation of a statistical model for the variables in the Benford domain; fortunately, in Section 19.5 we will see how this difficulty can be overcome by averaging with respect to the parameters.

19.4 A FOURIER-SERIES-BASED MODEL

The sinusoidal character of the histogram in Figure 19.3(a) suggests that a Fourier representation for the pdf of the variable $\tilde{X} = \log_{10}|X| \bmod 1$ is plausible (cf.

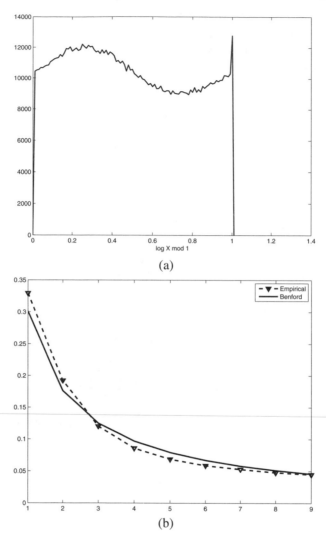

Figure 19.3 (a) Histogram of the DCT values of "Man" in Benford ($\log_{10} |\cdot|$ mod 1) domain. (b) Distribution of the Leading Digit corresponding to "Man." Block size is 8×8.

Chapter 3).[3] For a further justification, let $X' = \log_{10} |X|$, so that

$$f_{\tilde{X}}(x) = \sum_{n=-\infty}^{\infty} f_{X'}(x-n) = f_{X'}(x) * \sum_{n=-\infty}^{\infty} \delta(x-n), \quad x \in [0,1), \quad (19.6)$$

where $*$ denotes convolution and $\delta(x)$ is Dirac's delta. Taking the Fourier transform on both sides of the equation above (extending it in a periodic way to the whole real

---

[3]We follow the "engineering" notation throughout this chapter, with $j := \sqrt{-1}$, a_n denoting the nth Fourier series coefficient and $\Phi(\omega)$ the Fourier transform.

line), we can write

$$\Phi_{\tilde{X}}(\omega) = \int_{-\infty}^{\infty} f_{\tilde{X}}(x)e^{-j\omega x}dx$$

$$= \Phi_{X'}(\omega) \sum_{k=-\infty}^{\infty} \delta(\omega - 2\pi k)$$

$$= \sum_{k=-\infty}^{\infty} \Phi_{X'}(2\pi k)\delta(\omega - 2\pi k), \tag{19.7}$$

where $\Phi_{X'}(\omega)$ denotes the Fourier transform of $f_{X'}(x)$ and $\Phi_{\tilde{X}}(\omega)$ represents the Fourier transform of the periodically extended $f_{\tilde{X}}(x)$. Given the fact that the Fourier transform of $\cos(\omega_0 x)$ is $(\delta(\omega - \omega_0) + \delta(\omega + \omega_0))/2$, it follows that the pdf of the variable in the modular logarithmic domain is

$$f_{\tilde{X}}(x) = 1 + 2\sum_{n=1}^{\infty} |a_n|\cos(2\pi n x + \phi_n), \quad x \in [0,1) \tag{19.8}$$

where $a_n = |a_n|e^{j\phi_n} := \Phi_{X'}(2\pi n)$.

Note that this representation is valid as long as the conditions for convergence of the Fourier series are met. However, the cases of specific interest to us are those for which the magnitude of the Fourier coefficients $|a_n|$ is small for moderate and large n. Note that the case of a pure Benford random variable corresponds to $|a_n| = 0$ for all $n \geq 1$.

We want to show that a GGD random variable can be accurately modeled in the Benford domain by a distribution composed of a constant and one sinusoidal term. To this end, we compute the Fourier series coefficients of the distribution of $\tilde{X}$ as

$$a_n = \int_{-\infty}^{\infty} f_{X'}(x)e^{-j2\pi n x}dx$$

$$= 2A \int_{0}^{\infty} \exp(-(\beta z)^c)\exp(\ j2\pi n \log z/\log 10)dz$$

$$= \frac{2A}{\beta}e^{j(2\pi n \log \beta)/\log 10} \int_{0}^{\infty} \exp(-(z)^c)z^{-j2\pi n/\log 10}dz$$

$$= \frac{2A}{\beta c}e^{j(2\pi n \log \beta)/\log 10}\Gamma\left(\frac{-j2\pi n + \log 10}{c\log 10}\right). \tag{19.9}$$

We refer to the a_n as the **Benford–Fourier coefficients**. The following two properties will prove to be useful later, and they are both a consequence of $2A/(\beta c) = 1/\Gamma(1/c)$.

- The magnitudes of the coefficients depend only on c and n but are independent of the variance.

- The only effect of the variance is a phase change. Suppose that a certain phase is achieved for σ'; then it is easy to see that the same shift is achieved for any $\sigma = 10^k\sigma'$, with k integer.

The most important observation however is that the magnitudes of the coefficients in (19.9) decrease very rapidly with n. In fact, using equation (8.326) of [GrRy] it is immediately shown that

$$|a_n|^2 = \prod_{k=0}^{\infty} \left(1 + \frac{(2\pi n)^2}{\log^2(10)(ck+1)^2} \right)^{-1},$$ (19.10)

which quickly converges to the true value, and if truncated provides a good approximation. Moreover, from (19.10) it is easy to see that the Fourier series coefficients monotonically increase with the shaping factor c.

To get an idea of the magnitude of a_n, we have evaluated (19.10) for different values of c. The results for a Gaussian (i.e., $c = 2$) and a Laplacian (i.e., $c = 1$) are represented in Table 19.1. For $c = 0.5$ we observe that the magnitudes are so small that even $|a_1| = 0.00614761$.

n	1	2	3	4
$c = 2$	0.165849	0.0194532	0.00228155	0.00026759
$c = 1$	0.0569	0.00110	$1.866 \cdot 10^{-5}$	$2.964 \cdot 10^{-7}$

Table 19.1 Magnitude of a_n for different values of the shaping factor c.

The main consequence of all these evaluations is that for all values of the shaping gain smaller than 2 (which are typical in images), the approximation

$$f_{\tilde{X}}(x) \approx \check{f}(x) = 1 + 2|a_1|\cos(2\pi x + \phi_1), \quad x \in [0, 1)$$ (19.11)

is reasonable. However, unless c is small the Benford property does not hold (even in an approximate way), so one must keep the term corresponding to the first coefficient of the Fourier series expansion in order to ensure a small approximation error. In Figure 19.4 we plot the theoretical Leading Digit distribution corresponding to the "Man" image after using the approximation in (19.11), where a_1 is obtained by computing the histogram (with 100 bins) of the samples $\{\log_{10}|b^{(m)}(i, k)| \bmod 1; m = 1, \ldots, M; i = 0, \ldots, 7, k = 0, \ldots, 7\}$, where $b^{(m)}(i, k)$ is the (i, k)th coefficient of the mth 8×8-block of the DCT and M is the number of blocks. Now the excellent agreement with the empirical distribution is remarkable.

We are interested in bounding the error made by retaining only one coefficient of the Fourier series expansion, i.e.,

$$\varepsilon = \int_0^1 |f_{\tilde{X}}(x) - \check{f}(x)|^2 dx = \sum_{n=2}^{\infty} |a_n|^2,$$

where the second line is a direct consequence of Parseval's relation.

In Appendix 19.9 we show that $\varepsilon < g(c)$, where

$$g(c) = 0.0421603 \cdot \exp\left(\frac{c+1}{c}\right) \cdot \exp\left(-\frac{2\pi^2}{c\log 10}\right) \cdot \left[1 - \exp\left(-\frac{2\pi^2}{c\log 10}\right)\right]^{-1/2}.$$ (19.12)

The bound $g(c)$ can be numerically evaluated to yield Table 19.2, and the plot in Figure 19.4.

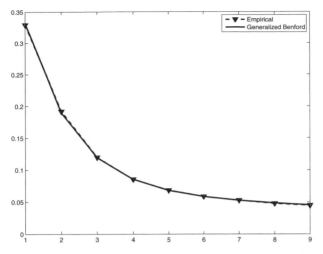

Figure 19.4 Empirical digit distribution and generalized Benford law. $|a_1| = 0.067$; $\phi_1 = -1.221$ radians.

c	$g(c)$
0.50	$2.70 \cdot 10^{-8}$
0.75	$4.20 \cdot 10^{-6}$
1.00	$5.24 \cdot 10^{-5}$
1.25	$2.38 \cdot 10^{-4}$
1.50	$6.55 \cdot 10^{-4}$
1.75	$1.35 \cdot 10^{-3}$
2.00	$2.32 \cdot 10^{-3}$

Table 19.2 Error bound $g(c)$ vs. shape parameter c.

Note that although the Fourier coefficients a_n may be obtained by projecting the histogram of $\log_{10}|X|$ onto the complex exponential $\exp(-j2\pi nx)$, a simpler procedure which does not require the explicit calculation of the histogram can be devised by realizing that the first equality in (19.9) corresponds to $\mathbb{E}\left[e^{-j2\pi n \log_{10}|X|}\right]$. Therefore a_n can be estimated through either of the following sample averages:

$$
a_n \approx \frac{1}{MN^2} \sum_{m=1}^{M} \sum_{i=0}^{N-1} \sum_{k=0}^{N-1} \exp\left(-j2\pi n \log_{10}|b^{(m)}(i,k)|\right)
$$

$$
= \frac{1}{MN^2} \sum_{m=1}^{M} \sum_{i=0}^{N-1} \sum_{k=0}^{N-1} |b^{(m)}(i,k)|^{-j2\pi n/\log 10}, \tag{19.13}
$$

where M is the number of $N \times N$ DCT blocks.

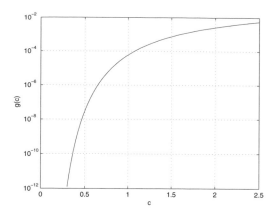

Figure 19.5 Bound in (19.12) vs. shape parameter c.

19.4.1 Experimental Results with a Large Database

We have validated our approach with the **UCID (Uncompressed Color Image Database)** which consists of 1338 uncompressed images representing a variety of natural scenes [SchSti]. To measure the goodness of fit between the actual Leading Digit distributions and the model under test, we have selected two measures. The **Kullback–Leibler divergence (KLD)** is defined, for two discrete probability mass functions $p(i), q(i)$, as

$$D(p\|q) = \sum_i p(i) \log_2 \frac{p(i)}{q(i)}, \tag{19.14}$$

and is measured in bits. Notice that the KLD is not symmetric. On the other hand, the χ^2-divergence is defined as

$$\chi^2(p, q) = \sum_i \frac{(p(i) - q(i))^2}{p(i)}. \tag{19.15}$$

For both measures, $q(i)$ and $p(i)$, $i = 1, \ldots, 9$, denote respectively the empirical and theoretical distributions of the Leading Digit.

The results are summarized in Table 19.3. The measures (KLD and χ^2) are computed for every image in the database and then averaged. As discussed in Section 19.3, images in the pixel domain do not obey Benford's law or our proposed generalization. In fact, seven images of the database had to be left out for this computation because at least one of the digits in $\{1, \ldots, 9\}$ was missing as a Leading Digit when all pixels were swept. Turning our attention to the generalization, we see that it gives a more accurate prediction of the Leading Digit distribution: with respect to Benford's law, we afford a reduction in both KLD and χ^2 of two orders of magnitude. Finally, we compare the results obtained with the two proposed methods for calculating a_1: one based on projecting the histogram and the other using (19.13). We conclude that (19.13) is not only less computationally demanding but also more accurate.

	KLD (mean)	KLD (std)	χ^2 (mean)	χ^2 (std)
Benford (pixel domain)	0.3627	0.4085	0.4920	0.5106
Benford (8×8-DCT)	0.0023	0.0028	0.0032	0.0038
Benford (16×16-DCT)	0.0018	0.0024	0.0025	0.0032
Generalized Benford (16×16-DCT). Hist.	$3.2 \cdot 10^{-5}$	$2.2 \cdot 10^{-5}$	$4.5 \cdot 10^{-5}$	$3.0 \cdot 10^{-5}$
Generalized Benford (16×16-DCT). (19.13)	$3.1 \cdot 10^{-5}$	$2.1 \cdot 10^{-5}$	$4.3 \cdot 10^{-5}$	$2.9 \cdot 10^{-5}$

Table 19.3 Goodness of fit for Benford's law and the proposed generalization for images in the pixel and the DCT domains (UCID database).

The divergence measures presented above can be used to conceive an image-dependent hypothesis test in which the null hypothesis corresponds to the probabilities of the Leading Digits following Benford's law (or our generalization) according to the observed data. To this end, it is common to use Pearson's test statistic, which amounts to multiplying the χ^2-divergence in (19.15) by the number of summands in (19.15). In addition, it is worth mentioning that when $p(i) \approx q(i)$, for all $i = 1, \ldots, 9$, a Taylor series expansion of (19.14) yields

$$D(p\|q) \approx \frac{\chi^2(p, q)}{2 \log 2}, \tag{19.16}$$

with $\log$ denoting the natural (base e) logarithm. Noticing this relation, for our hypothesis-testing purposes it suffices to focus on Pearson's statistic. In Pearson's hypothesis test, the statistic is compared to a chi-square distribution with as many degrees of freedom as the cardinality of the sample space minus 1, which in our case becomes 8. From a table of the cumulative distribution function of a chi-square distribution with 8 degrees of freedom, it is immediate to see that for a significance level of 5%, the null hypothesis is rejected if the statistic is larger than 15.51 [PapPU].

We have carried out the hypothesis test for each of the images in the UCID database. For the 16×16-block DCT against Benford's law, the mean Pearson's statistic is 486.35 with a standard deviation of 631.92. For a 5% significance level, the null hypothesis is rejected in 1283 images of the 1338 (95.89% of the cases). In contrast, for our proposed generalization where a_1 is computed using (19.13), the mean test statistic is 8.33 with a standard deviation of 5.70. For the same significance, now we only reject 112 out of the 1338 images (8.37% of the cases). We can conclude that the generalized law performs much better in terms of explaining the distribution of the Leading Digit. This in turn paves the way for some applications which will be presented in Section 19.7.

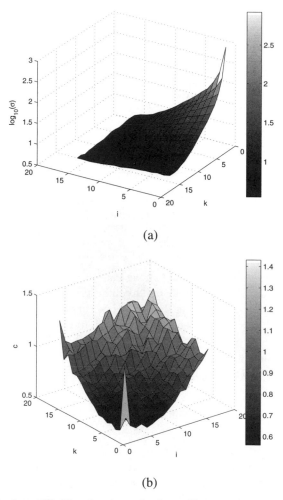

Figure 19.6 Maximum-likelihood estimated values of $\log_{10} \sigma$ (in (a)) and of c (in (b)) for the 16×16-block DCT of "Man."

19.5 RESULTS CONCERNING ENSEMBLES OF DCT COEFFICIENTS

So far, we have shown that GGD's are closely approximated as in (19.11). However, as we have remarked, different DCT coefficients will have different parameters σ and c. Figure 19.6 shows the values of σ and c estimated through **Maximum Likelihood** for each DCT coefficient of the "Man" image when blocks of size 16×16 are considered.

Suppose that these two parameters are modeled as being drawn from a joint distribution $f_{C, \Sigma}(c, \sigma)$. Then the pdf of the variable in the modular logarithmic domain can be written as

$$f_{\tilde{X}}(x) = 1 + \int_0^\infty \int_0^\infty f_{C,\Sigma}(c,\sigma) \sum_{\substack{n=-\infty \\ n \neq 0}}^{\infty} a_n(c,\sigma) e^{j2\pi nx} dc \cdot d\sigma$$

$$= 1 + 2\text{Re} \left\{ \sum_{n=1}^{\infty} \overline{a}_n e^{j2\pi nx} \right\}, \tag{19.17}$$

where $\overline{a}_n$ is the mean value of a_n averaged over the joint distribution of C and Σ. Then as long as the averaged Benford–Fourier coefficients $\overline{a}_n$ are such that their magnitude is small for $n > 1$, the approximation of the form (19.11) is valid.

We can now state the following.

Lemma 19.5.1. *Let X be a GGD with shaping parameter c drawn from a distribution $f_C(c)$. Then*

$$|\overline{a}_n| \leq \int_0^\infty f_C(c) |a_n(c)| dc. \tag{19.18}$$

Proof. We have

$$\overline{a}_n = \int_0^\infty \int_0^\infty a_n(c,\sigma) f_{\Sigma|C}(\sigma|c) f_C(c) d\sigma \, dc. \tag{19.19}$$

Since for any complex random variable X, $|\mathbb{E}\{X\}| \leq \mathbb{E}\{|X|\}$, we can write

$$|\overline{a}_n| \leq \int_0^\infty f_C(c) \int_0^\infty |a_n(c,\sigma)| f_{\Sigma|C}(\sigma|c) dc \, d\sigma$$

$$= \int_0^\infty |a_n(c)| f_C(c) dc, \tag{19.20}$$

where the last equality follows from the fact that for a GGD $|a_n(c,\sigma)|$ is invariant with σ. □

As a corollary, suppose that c is such that for all i, k, $c(i,k) \leq c^+$ for some real number c^+. Then for all (i,k) and all n,

$$|a_n(c,\sigma)| \leq |a_n(c^+)|, \tag{19.21}$$

which follows from the previous lemma and the fact that for a GGD, $|a_n(c)|$ monotonically increases with c. Inequality (19.21) suggests that for values of c^+ less than 2, as is customary in practice,[4] the approximation given in (19.11) is valid.

The previous discussion has important implications in video applications: if all frames of a video sequence can be modeled as in (19.11), then the whole sequence also satisfies this property. Therefore, our generalized form of Benford's law applies to video sequences as well, provided that one works with the block-DCT coefficients of each frame. This occurs, for instance, in the prevalent MPEG-2 video compression standard.

---

[4]For instance, in the case of the 16 × 16-block DCT of "Man" we have found that the maximum-likelihood estimated values for c are in the range $[0.56, 1.43]$.

The problem with the bound in (19.21) is that the value of c^+ clearly leads to an overestimate of $|a_n(c)|$. One alternative is to use the expansion given in [PapPU] that allows us to write

$$\mathbb{E}\{|a_n(c)|\} \approx |a_n(\mu_c)| + \left.\frac{d|a_n(c)|^2}{dc^2}\right|_{c=\mu_c} \cdot \sigma_c^2/2. \tag{19.22}$$

It is possible to calculate the second derivative of $|a_n(c)|$ as a combination of gamma, digamma and trigamma functions and then evaluate the function above. For instance, in the "Man" image (with 8×8 blocks) we have $\mu_c = 0.866406$ and $\sigma_c^2 = 0.047179$ and hence, doing the calculations, $|a_1(c)| \approx 0.04381$, $|a_2(c)| \approx 0.0013457$ and $|a_3(c)| \approx 3.11 \cdot 10^{-5}$.

The previous results have focused on the influence of the distribution shaping parameter c upon the Benford–Fourier coefficients. We discuss next the effect of the GGD standard deviation being a random variable and show that in general it also produces a reduction in the coefficients' magnitude.

Lemma 19.5.2. *Let X be a zero-mean generalized Gaussian r.v. such that its standard deviation Σ is itself a Benford r.v., i.e., $\tilde{\Sigma} \sim U[0,1)$. Then $\tilde{X} \sim U[0,1)$.*

Proof. First, notice that this result is a particular case of the mixture of random variables discussed in Section 19.2. However, the proof that we give here illustrates some aspects that will be interesting later. For a fixed standard deviation σ we can write the Benford–Fourier coefficients in (19.9) as

$$a_n = K \exp(-j2\pi n \log_{10} \sigma), \ n \geq 1, \tag{19.23}$$

where K is a complex number independent of σ. Let $V = \log_{10} \Sigma$; then marginalizing with respect to V we have

$$\begin{aligned}
a_n &= K \int_{-\infty}^{\infty} \exp(-j2\pi nv) f_V(v) dv \\
&= K \sum_{k=-\infty}^{\infty} \int_0^1 \exp(-j2\pi n(v+k)) f_V(v+k) dv \\
&= K \int_0^1 \exp(-j2\pi nv) \sum_{k=-\infty}^{\infty} f_V(v+k) dv \\
&= 0
\end{aligned} \tag{19.24}$$

for all $n \geq 1$. The last equality is because $\sum_{k=-\infty}^{\infty} f_V(v+k)$, for $v \in [0,1)$, is the pdf of $V \bmod 1 = \tilde{\Sigma}$, and that $\mathbb{E}\{e^{-j2\pi nX}\} = 0$ if $X \sim U[0,1)$. $\square$

The former result can be relaxed in a number of ways if we focus on a_1 which, as we have seen, is for practical values of the GGD parameters the only significant coefficient. Thus, to guarantee that $a_1 = 0$ it is enough that

$$\mathbb{E}\{\exp(-j2\pi \log_{10} \Sigma)\} = 0, \tag{19.25}$$

where the expectation is taken with respect to the standard deviation Σ. The following example illustrates such a situation.

Example 19.5.3. *Consider four zero-mean generalized Gaussian r.v.'s with shaping parameter $c = 1$ and standard deviations $1, 10^{0.25}, 10^{0.5}$ and $10^{0.75}$. Then the average is such that $a_1 = 0$.*

Lemma 19.5.2 and Example 19.5.3 show that as long as the individual variances spread over different orders of magnitude, the average of the Leading Digits is closer to Benford's distribution than the individual random variances.

Another result along the same line of research is the following.

Lemma 19.5.4. *Let X be a zero-mean generalized Gaussian r.v. such that its standard deviation Σ is such that $\log_{10} \Sigma$ is uniformly distributed in $[t_0, t_1)$. Let $|a_{n,*}|$ denote the magnitude of the nth coefficient obtained for any $\sigma \in [10^{t_0}, 10^{t_1})$. Then*

$$|\mathbb{E}[a_n]| \leq \frac{|a_{n,*}|}{\pi n(t_1 - t_0)}. \tag{19.26}$$

Proof. From (19.23) we have

$$
\begin{aligned}
\mathbb{E}[a_n] &= \frac{K}{(t_1 - t_0)} \int_{t_0}^{t_1} \exp(-j2\pi nv)dv, n > 1 \\
&= \frac{jK}{2\pi n(t_1 - t_0)}[\exp(-j2\pi nt_0) - \exp(-j2\pi nt_1)], \tag{19.27}
\end{aligned}
$$

and then (19.26) follows from the fact that the term within brackets in (19.27) has an upper bound of 2.   □

Note that the decrease in (19.26) depends on the interval spanned by the standard deviation Σ in a base-10 logarithmic scale. Of course, the result in (19.26) is an upper bound, so it is possible that in some cases the magnitude is even smaller, as when $(t_1 - t_0)$ is an integer, in which case we recover Lemma 19.5.2.

Another interesting question is what happens when one averages the Leading Digits of different images. Notice that this is the approach followed by Fu et al. in [FuSS], who empirically showed that the average of the Leading Digits from the images in the UCID database followed Benford's distribution when working in the DCT domain. This result might seem surprising at first since we have already shown that they do not follow Benford's distribution, but rather a generalized distribution. We have repeated the experiment and found that the KLD is $6.2909 \cdot 10^{-4}$ and the χ^2 is $8.6758 \cdot 10^{-4}$ (compare these results with the second row of Table 19.3). To understand the apparent contradiction, notice that while we are averaging the goodness of fit from *individual* images, Fu et al. were measuring the goodness of fit for the *average Leading Digit* in the database. This is why we do not report the standard deviation of the measurements, as there is only one.

We discuss next the rationale behind Fu et al.'s result and then comment on its possible pitfalls. Given two continuous random variables X, Y, $D(X||Y)$ denotes the KLD, which is measured similarly to (19.14), by replacing the sum by an integral.

Lemma 19.5.5. *Let X_1 and X_2 be two zero-mean generalized Gaussian random variables with the same shaping parameter c and different variances. Let Y be a*

Benford r.v., i.e., $\tilde{Y} \sim U[0, 1)$. Finally, let $\tilde{X}$ be a random variable with pdf $f_{\tilde{X}}(x)$
such that

$$f_{\tilde{X}}(x) = \frac{f_{\tilde{X}_1}(x) + f_{\tilde{X}_2}(x)}{2}. \tag{19.28}$$

Then $D(\tilde{X}||\tilde{Y}) < D(\tilde{X}_1||\tilde{Y})$ and $D(\tilde{X}||\tilde{Y}) < D(\tilde{X}_2||\tilde{Y})$.

Proof. From (19.9) we know that the Benford–Fourier coefficients of X_1 and X_2 in the Benford domain, which we denote by respectively $a_{n,1}$ and $a_{n,2}$, have the same magnitude but different phase. Let a_n denote the nth Fourier coefficient of X in the Benford domain. Then by the triangle inequality, we have that $|a_n| < |a_{n,1}|$ and $|a_n| < |a_{n,2}|$, for all $n > 1$. The rest of the proof is straightforward. $\square$

To fully interpret the implications of this result, notice that the distribution of the *average of the Leading Digits* of the random variables X_1 and X_2 is obtained by integrating the distribution of $\tilde{X}$ in intervals of the form $[\log_{10} k, \log_{10}(k + 1))$, $k = 1, \ldots, 9$. The lemma states that $\tilde{X}$ gets closer to uniform (i.e., a Benford r.v.) than the individual variables. The main consequence of this result is that if we average the Leading Digits of two generalized Gaussian random variables having the same shaping parameter, the resulting random variable will *likely*[5] get closer to the Benford distribution.

We now understand why it is plausible that the average of the Leading Digits of many images in the DCT domain is very close to Benford's law, and that this does not contradict our findings regarding the Leading Digit distribution for *individual* images. This fact should not be overlooked in forensic applications: it is not reasonable to expect that the Leading Digits of a given image in the DCT domain conform to Benford's law, so it is not possible to devise a useful test that simply measures the closeness of the Leading Digit distribution to Benford's.

19.6 JOLION'S RESULTS REVISITED

Jolion [Jol] experimentally showed that the **gradient magnitude of images** follows Benford's law approximately. As we discussed in the Introduction, this was the first time a connection between images and Benford's law was made. He attempted an analytical justification for the observed behavior, but his explanation is not entirely satisfactory, as we discuss in this section.

Jolion focuses his attention on the response of the bidimensional gradient to image edges, as he argues that the gradient of images is dominated by such response. A detailed analysis is provided for the one-dimensional case: a derivative-Gaussian operator is applied to a step function. Since the convolution with a step can be seen as an integral which in turn cancels the derivative out, the result is a one-dimensional Gaussian pulse, whose amplitude distribution ultimately determines the pdf of the gradient of the image. The probability of a quantized (using truncation) Gaussian pulse is analytically obtained as follows.

---

[5]Notice that the fact that the KLD to a Benford random variable decreases does not necessarily imply that the KLD of the Leading Digits is reduced.

Let $h(x) = \alpha \exp(-x^2/(2\sigma_K^2))$ be the Gaussian pulse.[6] The probability that $h(x)$ is truncated to integer $m > 0$ (assuming that values in x are taken uniformly) is proportional to u_m such that

$$m + 1 = \alpha \exp\left(-\frac{(x_m + u_m)^2}{2\sigma_K^2}\right), \quad 0 < m < \alpha - 1, \tag{19.29}$$

where x_m is given by the negative[7] solution to

$$m = \alpha \exp\left(-\frac{x_m^2}{2\sigma_K^2}\right), \quad 0 < m < \alpha - 1. \tag{19.30}$$

It is easy to invert the equations above to find

$$u_m = \sigma_K \sqrt{2} \left(\sqrt{\log\left(\frac{\alpha}{m}\right)} - \sqrt{\log\left(\frac{\alpha}{m+1}\right)}\right). \tag{19.31}$$

After some approximations, Jolion argues that

$$u_m \approx \frac{\sigma_K}{\sqrt{2}\log\alpha} \cdot \frac{1}{m}, \tag{19.32}$$

so the probability distribution of the gradient follows a $1/x$-law and hence the Leading Digits obey Benford's law (cf. Section 4.3).

The problem with Jolion's derivation is that some approximations are valid for small m while others make sense only if m is close to α. In Figure 19.7 we plot the true distribution in (19.31) and Jolion's approximation (19.32). We remark that while the true distribution is U-shaped, the proposed approximation is monotonic. In fact, it can be shown that the true distribution achieves the minimum for some $m \in [\alpha/\sqrt{e} - 2, \alpha/\sqrt{e} + 1]$.

Moreover, the approximate distribution at which Jolion arrives is discrete (i.e., a probability mass function) and not continuous, so even if it were of the form $1/m$, the upper limit α would need to be quite large for the Leading Digit to approximately conform to Benford's law.

To avoid discretizing the gradient amplitude and later matching the resulting distribution to a continuous pdf, we have derived a closed-form expression for the pdf of the continuous amplitude starting with the same assumptions as Jolion. We stress here the fact that the process of obtaining the Leading Digit distribution intrinsically includes a truncation, so quantizing the amplitude prior to computing the Leading Digit is redundant.

Consider then a uniform random variable X with support $[-a, a)$. This random variable is mapped onto Y using a Gaussian transformation:

$$Y = h(X) = \alpha e^{-X^2/2\sigma_K^2}, \quad -a \leq X < a. \tag{19.33}$$

---

[6]Notice that $h(x)$ is not a pdf, so there is no constraint on the value of α.
[7]Without loss of generality, we assume that the solution is negative, as the other case follows similarly.

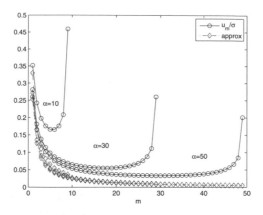

Figure 19.7   Distribution of m and Jolion's approximation for different values of α.

The pdf of Y is[8]

$$f_Y(y) = \frac{\sigma_K^2/a}{\alpha h^{-1}(y)e^{-(h^{-1}(y))^2/2\sigma_K^2}}$$

$$= \frac{\sigma_K/a}{y(-2\log(y/\alpha))^{1/2}}, \quad h(a) \le y < \alpha. \qquad (19.34)$$

So we see that in addition to the factor $1/y$, there is another which will be significant for values of y close to α, and which Jolion's approximations neglect.

Now consider the variable $Z = \log Y$ (we use natural logarithms here for convenience, without affecting the general conclusions). Recall that when Z is reduced modulo 1 we should expect a uniform-like behavior for Y to follow Benford's law. The pdf of this new variable is

$$f_Z(z) = \frac{\sigma_K/a}{(-2\log(e^z/\alpha))^{1/2}} \qquad (19.35)$$

$$= \frac{\sigma_K/a}{\sqrt{2(-z+\log\alpha)}}, \quad \log(h(a)) \le z < \log\alpha. \qquad (19.36)$$

The function $f_Z(z)$ is monotonically increasing within its support; this has the important implication that the modulo 1 reduced variable can *never* be uniform and, furthermore, that Benford's law does not hold in a strict sense for a random variable with the distribution in (19.34). Moreover, it can be shown that the magnitudes of the Benford–Fourier coefficients $|a_n|$ only decrease as $1/\sqrt{n}$, making less plausible an approximation like (19.11).

For the bidimensional case, the gradient has horizontal and vertical components, and the response to an edge with orientation θ_P radians produces two projected components having the form $\alpha\cos(\theta_P)\exp(-m^2/(2\sigma_K^2))$ and $\alpha\sin(\theta_P)$

---

[8]Although the equation $x = h^{-1}(y)$ has two solutions, we consider only the positive one, as both give identical results; hence, we multiply the numerator of (19.34) by 2 to account for this fact.

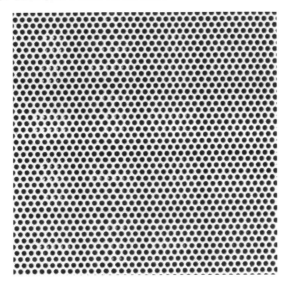

Figure 19.8 The "grid" image.

$\exp(-m^2/(2\sigma_K^2))$ so that the magnitude of the vector with these two components is again $\alpha \exp(-m^2/(2\sigma_K^2))$, thus recovering the response (and the distribution it induces) for a one-dimensional edge.

To test the validity of the claims made regarding the magnitude of the gradient, we have computed its histogram for images containing periodic patterns and simple edges. The "grid" image shown in Figure 19.8 is one such example.

The resulting histogram corresponding to the magnitude of the gradient for a Gaussian kernel with $\sigma_K = 1$ is represented in Figure 19.9, where the U shape from (19.31) is apparent. Notice that since the histogram computes the total number of occurrences in each interval of the gradient magnitude, it is reasonable to compare it with the probabilities for the uniform intervals u_m given by (19.31).

Obviously, a real-world image is not only composed of edges, but also noise and fine structure (e.g., textures). The gradient response to these features is modeled by a Rayleigh distribution; in any event, Jolion claims that since the contribution of noise/fine structure is low (except for noise/highly textured images), then the pdf is dominated by the pdf corresponding to edges. Unfortunately, were this the case, our analysis above would show that gradient magnitude of images *does not* follow Benford's law or our generalization. This is not consistent with Jolion's and our own observations (see results for the UCID database later in this section). We must then disregard a justification based solely on the gradient response to simple edges, all having the same height.

A more plausible explanation follows from our observation that both the horizontal and vertical components of the gradient can be reasonably modeled by a GGD. In Figure 19.10 we represent the transformed histogram of the gradient components of the image "Man" after performing the mapping $u \mapsto -(\log|\beta u|)^{1/c}$ using the

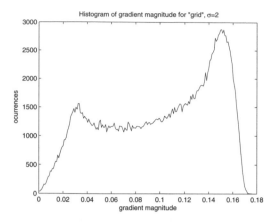

Figure 19.9 Histogram of the gradient magnitude corresponding to the "grid" obtained for a Gaussian gradient kernel with $\sigma_K = 2$.

values of σ and c obtained by means of maximum-likelihood estimation and then substituting in (19.5) to compute β. The estimated values are $c_x = 0.55$, $c_y = 0.56$, $\sigma_x = 0.0301$ and $\sigma_y = 0.0262$, where the subindices x, y stand for the horizontal and vertical directions, respectively. The transformed histograms in Figure 19.10 have been normalized so that their peak value is zero. In interpreting the plots, notice that a true GGD with the given parameters would appear as two straight lines with slopes $\pm\beta$. Those are also plotted in the figure for reference. The closeness of the transformed histogram to such lines, except for those bins which correspond to unlikely values, allows us to conclude that the GGD is a reasonable model for the gradient components.

We can recover now a result from [GoAPA], where it is shown in the context of Gabor coefficients that if their real and imaginary parts are GGD with the same c and σ parameters, its magnitude can be modeled by a β-Rayleigh distribution, which is a generalization of a Rayleigh using a shaping factor c so that the latter corresponds to $c = 2$. This distribution has the form

$$f_X(x) = A x^{c/2} e^{-|\beta x|^c}, \quad x > 0, \tag{19.37}$$

where β is defined as in (19.5) and now

$$A = \frac{c\beta^{(c+2)/2}}{\Gamma\left(\frac{2+c}{2c}\right)}. \tag{19.38}$$

It is interesting now to compute the Benford–Fourier coefficients for the distribution in (19.37). To this end, we can repeat the derivation in Section 19.4 to show that

$$a_n = \frac{A}{\beta c} e^{j(2\pi n \log \beta)/\log 10} \Gamma\left(\frac{-j4\pi n + \log 10(c+2)}{2c \log 10}\right). \tag{19.39}$$

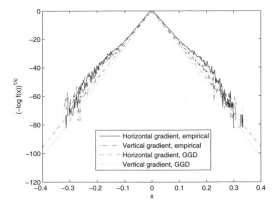

Figure 19.10 Transformed histogram (400 bins) of the gradient components (Gaussian kernel with $\sigma_K = 0.8$) for the image "Man" and transformed analytical pdf's.

The squared magnitude of the coefficients is in this case

$$|a_n|^2 \;=\; \beta^c \prod_{k=0}^{\infty} \left(1 + \frac{(2\pi n)^2}{\log^2(10)(ck + \frac{c+2}{2})^2} \right)^{-1}, \tag{19.40}$$

where we can readily see that now the magnitude is no longer independent of the standard deviation σ. However, the rate of decay for the magnitude of the coefficients is identical to those corresponding to the DCT, so the same reasoning for keeping only the first coefficient of the series also holds.

Experimentally, we have applied a Gaussian gradient kernel with $\sigma_K = 0.8$ on the images from the UCID database, and we have found the results summarized in Table 19.4.

	KLD (mean)	KLD (std)	χ^2 (mean)	χ^2 (std)
Benford	0.0166	0.032	0.0308	0.0979
Generalized Benford	0.0056	0.0273	0.0108	0.0562

Table 19.4 Goodness of fit for Benford's law and the proposed generalization for the gradient magnitude of images (UCID database).

We can see that even though the proposed generalization produces a considerable improvement in predicting the Leading Digit distribution with respect to Benford's law, neither matches the excellent fit achieved with the DCT coefficients. We conjecture that the reason for this behavior is that while in the DCT case we are averaging the Benford–Fourier coefficients coming from a number of GGDs with different parameters (cf. Figure 19.6), and thus having different phases, in the case of the gradient such averaging does not take place.

These conclusions are confirmed by the corresponding Pearson's hypothesis test (cf. Section 19.4.1) which gives the following values when the observed Leading Digit are compared with Benford's law: the mean statistic is $6.06 \cdot 10^3$ with standard deviation of $1.93 \cdot 10^4$. For a 5% significance level, we reject 1335 out of the 1338 images in the UCID database. If the generalized law is considered instead, the mean statistic becomes $2.12 \cdot 10^3$ with standard deviation $1.10 \cdot 10^4$, and still 1175 images are rejected.

Interestingly enough, the same arguments employed here can also be applied to the magnitude of **Gabor coefficients** [GoAPA], which find widespread use in facial recognition applications. Preliminary experiments performed on face image databases show that the magnitude of those coefficients also follows our generalized law.

19.7 IMAGE FORENSICS

Benford's law has been successfully applied to detect fraud in tax data [Cars, Nig5]. The test is based on the assumption that real data follow Benford's law on the basis that they come from many independent sources with different scales (much the same as data in a newspaper, which also approximately satisfies the law). Another application in **forensics** is to detect data manipulation in scientific data. Diekmann [Die] showed that the first digits of regression coefficients from sociological analyses closely approximate the Benford distribution. Diekmann went on to conduct an experiment in which he asked students in a statistics course to fabricate regression coefficients; while the hypothesis of the Leading Digit from the fabricated data not following Benford's law could not be rejected, this was not the case when the second most significant digit was considered, with a statistically significant difference. Schaefer et al. [ScSMG] investigated the use of Benford's law to analyze data from surveys, where there is always the risk that the interviewers fabricate the data. They focused on data from the German Socio-Economic Panel which contained proven fakes, detected after a second wave of the survey. Once again, Benford's law served to spot those interviewers who had cheated: by analyzing the full set of answers to specific questions known to conform to Benford's law (such as net-income or tax data), it was possible to easily detect fabrications.

In view of the above applications, it is natural to ask whether Benford's law may find any use in image forensics. Here we focus on *image compression detection* and *image steganography*. In the first application, we are interested in knowing whether a given image has been lossy-compressed in the DCT domain and then converted back to the pixel domain. Detecting this compression is thus important for knowing whether an image someone is buying has the maximum possible quality or instead has lost some features along the compression process. On the other hand, the purpose of steganography is to detect the existence of a message hidden in an image. Next, we show how the Benford–Fourier coefficients can be used in both forensic applications.

19.7.1 Image Compression

Detecting whether an image has been compressed is important to both forensic analysis and commercial applications. Our goal is to determine whether a given bitmap image has been JPEG compressed. **JPEG compression** is a popular and widely used **image compression** standard [Wall]. The major steps in **JPEG compression** are as follows. An image is first transformed using a block-based DCT. The block size is typically 8×8. The DCT coefficients for each block are then quantized using a quantization table based on the quality of the compression. Higher quality uses less quantization and preserves more details of the original image. Lower quality drives many of the coefficients toward zero. The last step uses entropy encoding to efficiently store the quantized coefficients. The quantized coefficients are referred to as JPEG coefficients. Fu et al. [FuSS] proposed to detect JPEG compressed images using a generalized Benford law describing the probability of the Leading Digit d as

$$p(d) \;=\; N \log_{10} \left(1 + \frac{1}{s + d^q} \right), \tag{19.41}$$

where N is a normalization factor which makes $p(d)$ a probability distribution, s and q are model parameters for different images and JPEG compression factors. If $s = 0$ and $q = 1$, the distribution becomes the Benford distribution.

It was observed that the JPEG coefficients of compressed images with quality factor of 100 follow this generalized law. However, for images that have been double compressed, i.e., a JPEG image is JPEG compressed again, the first digit distribution of the JPEG coefficients deviates from the generalized law. Therefore, to detect whether an image has been compressed, the candidate image is JPEG compressed with quality factor 100. For compressed images, this process results in double compression and Fu et al.'s generalized law is violated. On the other hand, for uncompressed images, the distribution of the JPEG coefficients should follow the generalized law.

In a similar spirit, we use our generalized Benford law to detect compressed images. Our image set consists of 1000 uncompressed gray-level images from the UCID database. For each uncompressed image, we generate two JPEG compressed images using quality factors 75 and 95. For each image (both compressed and uncompressed), we compute $|a_n|$. We randomly choose half of the uncompressed images and the corresponding compressed images for training a Gaussian kernel **Support Vector Machine (SVM)** classifier [Joa] and the other half are used for testing. SVM is a pattern recognition algorithm which learns to classify patterns from trained data. In general, SVM relies on kernels to separate data into linearly separable regions for classification. Typical kernels include linear, polynomial and Gaussian. Some kernels require user-supplied parameters. In the case of the Gaussian kernel, σ_K is required. The choice of these parameters can significantly affect the performance of classification. As such, it is typical that these parameters are optimized by searching the parameter space for values that result in the best classification performance. The danger here is that the resulting classifier is highly dependent on the data used in training and testing. The optimized parameters may not hold in a more general case and the achieved accuracy may be overly optimistic.

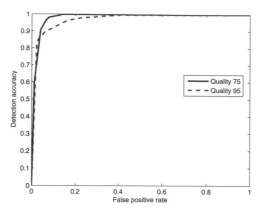

Figure 19.11   The ROC curve of the Benford-based JPEG compression detector using only $|a_1|$ and $|a_2|$ as features for two different JPEG quality factors 75 and 95.

To avoid this pitfall, we use the default kernel parameter $\sigma_K = 1$.

The performance of a compression detector is measured using its **Receiver Operating Characteristic (ROC)** curve. This is a plot of the correct detection probability versus the false positive probability. The correct detection probability is the number of detected compressed images over the total number of compressed images, and the false positive probability is the number of uncompressed images classified as compressed over the total number of uncompressed images.

In Figure 19.11, we plot the ROC curve of our compression detector using two features: $|a_1|$ and $|a_2|$. For concreteness, we calculate the average probability of error:

$$p_e = \frac{1}{2} \left(p_{\text{FP}} + p_{\text{FN}} \right), \tag{19.42}$$

where p_{FP} is the false positive probability and p_{FN} is the false negative probability. The best average probabilities of error are 0.049 and 0.083 for JPEG quality factors 75 and 95, respectively. The result matches our intuition that heavily compressed images are more detectable. The performance of our compression detector can be improved by using more Benford–Fourier coefficients $|a_n|$ as features. In fact, using five coefficients, the average probabilities of error become zero for both JPEG quality factors.

For comparison, we repeated the same experiment using the method proposed by Fu et al. [FuSS]. Ideally, we would like to use the same set of images and SVM parameters. Since we do not know the SVM parameters used in [FuSS], we use the default parameters to match our previous experiment. The ROC curve is shown in Figure 19.12. The best average probabilities of errors are 0.005 and 0.097 for JPEG quality factors 75 and 95, respectively. We conclude that our compression detector performs significantly better using fewer features (five in our case vs. the nine Leading Digits in Fu et al.'s detector).

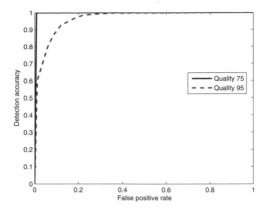

Figure 19.12   The ROC curve of the compression detector proposed by Fu et al. for two different JPEG quality factors 75 and 95.

19.7.2 Image Steganography

Steganography is the art of communicating covertly using innocuous looking *cover* objects. Steganography has a rich and long history. In 499 BC, Histiaeus shaved the head of his slave and tattooed a message on his head. Once his hair grew back, he was sent to deliver the message to Aristagoras, who shaved the slave's head again to read the message, which instructed Aristagoras to revolt against the Persians. Other examples of steganography include wax tablet, invisible ink and microdots.

It must be emphasized that steganography differs from cryptography as the latter is only concerned with protecting the content and not its presence. In cryptography, an observer is fully aware of the fact that secret messages are being exchanged between the involved parties. This may be sufficient for the observer to take action. In steganography, the messages are concealed in such a way as not to arouse the suspicion of an unintended observer.

Steganography is related to another data-hiding technique called watermarking (see Chapter 10 of [CozMil] for more on these two subjects). However, their goals are different. Watermarking aims to be robust against distortions and is subject to removal attacks. The presence of a watermark is generally easy to detect. In most cases, the use of watermarking is known.

In modern steganography, multimedia cover objects are often used to hide messages. Digital images, in particular, are widely used due to their popularity and accessibility. In digital image steganography, hidden messages are embedded into cover images to produce *stego* images. Perhaps due to a false sense of security, the most popular embedding method is least-significant bit (LSB) replacement, where some the LSBs of the cover image are replaced by the message bits. By using only the LSBs, the stego image is visually imperceptible.

Detecting the presence of steganographic content is called *steganalysis*. The premise of steganalysis is based on the fact that embedding changes the cover im-

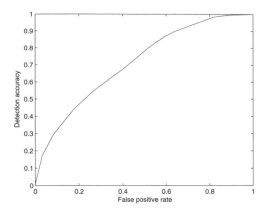

Figure 19.13   The ROC curve of the Benford-based steganalysis detector. The payload is 1
            bpp.

age leaving traces of evidence that can be used to detect the presence of a hidden
message. This is easily seen when a forensic analyst has access to the cover image.
If an analyst receives another image that looks like the cover image, she can simply
tell whether the image has been modified by comparing the two images. In practice, the cover images are typically not available to an analyst. This complicates
the situation and turns steganalysis into a difficult problem.

Virtually all modern steganalysis detectors rely on some features derived from
known cover images and stego images and pattern recognition algorithms to detect
steganographic contents. These features can be viewed as a compressed representation of all cover and stego images. Pattern recognition algorithms can then use
this compressed representation to learn to classify cover and stego images. As a
consequence, the most critical task in steganalysis is to identify the right features
to use for classification. This process is largely an art. In some cases, knowledge
of the embedding algorithm can help guide the process.

The performance of a steganalysis detector, or rather, of the features, is measured using the **Receiver Operating Characteristic (ROC)**. The ROC shows the
performance of the detector by comparing its false positive probability versus its
detection probability. The detection probability is the number of correctly identified
stego images over the total number of stego images. The false positive probability
is the number of cover images classified as stego over the total number of cover
images.

Our goal is to demonstrate that Benford's law can be used for steganalysis. We do
not aim to create a state-of-the-art steganalyzer as that would not be possible with
the current scheme. Rather, we only attempt to add to a growing list of forensic
applications based on Benford's law.

Our image set consists of 1000 uncompressed gray-level images from the UCID
database. For each image, we generate a stego image using LSB replacement and a

payload of 1 bit per pixel (bpp). The payload quantifies the size of the message. For an image of size 512-by-512, with a payload of 1 bpp, the total number of message bits is 262,144. Half of the cover images and their corresponding stego images are used for training and the other half are used for testing. We use a Gaussian kernel SVM with $\sigma_K = 1$, a widely used algorithm in state-of-the-art steganalysis detectors, as our pattern recognition algorithm. The features we use are $|a_1|$ through $|a_{50}|$.

In Figure 19.13, we plot the ROC curve of our steganalysis detector. It is clear that the Benford-based detector can detect steganographic content better than pure guessing, which would result in a straight 45 degree line. The best average probability of error using (19.41) for the Benford-based steganalysis detector is $p_e = 0.359$. This is an interesting result and it suggests that some of the statistical differences between cover and stego images are indeed captured by the generalized Benford law.

19.8 SUMMARY

The gradual and inevitable advance toward an all-digital world has brought about the undesirable feature of expediting the manipulation or even the fabrication of digital assets. There is then an increasing need for simple tools that allow us to identify those misuses as a first step to a more detailed and costly analysis. Benford's law is an excellent candidate which, in fact, is already being used in some commercial software packages for the analysis of financial fraud. Here we have shown how a generalization of Benford's law can be employed for forensic purposes in images, that is, for detecting whether a certain natural image contains a hidden message, and for determining when a given image has been previously compressed. We have done so by proving that GGDs follow a generalized form of Benford's law and, furthermore, that this extends to combinations of GGDs, opening the gate to video forensic applications.

Our generalization of Benford's law heavily relies on a Fourier series expansion of the data pdf in the Benford domain. This expansion had been previously used to justify convergence of an infinite product of random variables to the Benford distribution [Boy], but to the best of our knowledge, it has been applied, for the first time here, to improve the predictive value of Benford's law. Given the fact that an exponential distribution can be seen as a particular instance of a one-sided Generalized Gaussian, our results immediately extend to this case. In fact, as half-life decays of radioactive particles are known to follow an exponential distribution, it is clear that Benford's law alone is not sufficient to predict the distribution of the Leading Digit. Other phenomena following exponential distributions, such as the arrival time between packets on the Internet, can be accurately predicted by our proposed generalization.

There are several extensions to our work that are worth mentioning. On the one hand, our theoretical analysis can be adapted to include the so-called Generalized Gamma distributions, which offer slightly better image modeling capabilities. The approximation in (19.11) is still valid for this case. Another extension concerns other domains: for instance, it has been observed that the output of the Discrete Wavelet Transform (DWT) can also be modeled by Generalized Gaussian Distributions; hence it is reasonable to expect that our generalization of the Benford distribution is also applicable in this domain. Preliminary experiments carried out by the authors clearly indicate that this is the case (with a third-order symlet filter, the chi-square-based test rejects Benford's law for 1199 images of the UCID database as opposed to only 223 for the generalized law). Other experimental results, not reported here, lead to the conclusion that our generalization also applies to speech and music signals, thus paving the way for forensic applications dealing with these kinds of sources.

19.9 APPENDIX

We wish to bound

$$b_n = \prod_{k=0}^{\infty} \left(1 + \frac{\alpha_n^2}{(ck+1)^2}\right)^{-1}, \quad \alpha_n = \frac{2\pi n}{\log 10}.$$

If $\frac{c+1}{\alpha_n} < 1$ then below we show that

$$b_n \leq \frac{\log^2 10}{\log^2 10 + (2\pi n)^2} \cdot \exp\left(\frac{c+1}{c}\right) \cdot \exp\left(-\frac{\pi^2 n}{c \log 10}\right).$$

We have

$$b_n = \prod_{k=0}^{\infty} \left(\frac{(ck+1)^2 + \alpha_n^2}{(ck+1)^2}\right)^{-1}$$

$$= \prod_{k=0}^{\infty} \frac{(ck+1)^2}{(ck+1)^2 + \alpha_n^2}$$

$$= \frac{1}{1+\alpha_n^2} \prod_{k=1}^{\infty} \left(1 - \frac{\alpha_n^2}{(ck+1)^2 + \alpha_n^2}\right),$$

$$\log b_n = -\log(1+\alpha_n^2) + \sum_{k=1}^{\infty} \log\left(1 - \frac{\alpha_n^2}{(ck+1)^2 + \alpha_n^2}\right). \quad (19.43)$$

Using

$$\log(1-u) = -\sum_{\ell=1}^{\infty} \frac{u^\ell}{\ell}, \quad (19.44)$$

we find

$$\log b_n = -\log(1+\alpha_n^2) - \sum_{k=1}^{\infty} \sum_{\ell=1}^{\infty} \frac{1}{\ell} \left(\frac{\alpha_n^2}{(ck+1)^2 + \alpha_n^2}\right)^\ell. \quad (19.45)$$

Thus we obtain an upper bound for $\log b_n$ (and hence an upper bound for b_n) by keeping only the $\ell = 1$ term above (each summand is positive, but is hit by a negative sign). Thus

$$\log b_n \leq -\log(1 + \alpha_n^2) - \sum_{k=1}^{\infty} \frac{\alpha_n^2}{(ck+1)^2 + \alpha_n^2}. \qquad (19.46)$$

A better estimate is obtainable by using the Euler–MacLaurin formula. We have

$$\sum_{k=1}^{\infty} \frac{\alpha_n^2}{(ck+1)^2 + \alpha_n^2} \geq \int_{x=1}^{\infty} \frac{\alpha_n^2\, dx}{(cx+1)^2 + \alpha_n^2}; \qquad (19.47)$$

the reason for this is that the integrand is monotonically decreasing, and the sum is basically the upper sum approximation. Because of the minus sign, we thus increase the bound on $\log b_n$ if we replace the sum by this integral. We find

$$\log b_n \leq -\log(1 + \alpha_n^2) - \int_{x=1}^{\infty} \frac{\alpha_n^2\, dx}{(cx+1)^2 + \alpha_n^2}$$
$$= -\log(1 + \alpha_n^2) - \int_{x=1}^{\infty} \frac{dx}{1 + \left(\frac{cx+1}{\alpha_n}\right)^2}. \qquad (19.48)$$

We change variables. Let $u = (cx+1)/\alpha_n$ (so $dx = \frac{\alpha_n}{c}\, du$). Thus

$$\log b_n \leq -\log(1 + \alpha_n^2) - \frac{\alpha_n}{c} \int_{\frac{c+1}{\alpha_n}}^{\infty} \frac{du}{1 + u^2}. \qquad (19.49)$$

As the antiderivative of $(1 + u^2)^{-1}$ is $\arctan(u)$ and $\arctan(\infty) = \pi/2$, we find

$$\log b_n \leq -\log(1 + \alpha_n^2) - \frac{\alpha_n}{c}\left(\arctan(\infty) - \arctan\left(\frac{c+1}{\alpha_n}\right)\right)$$
$$= -\log(1 + \alpha_n^2) - \frac{\alpha_n \pi}{2c} + \frac{\alpha_n}{c}\arctan\left(\frac{c+1}{\alpha_n}\right). \qquad (19.50)$$

Thus we are left with estimating the remaining arc-tangent. The Taylor series of arc-tangent is

$$\arctan(x) = \sum_{\ell=0}^{\infty} \frac{(-1)^\ell x^{2\ell+1}}{2\ell + 1}. \qquad (19.51)$$

We assume from now on that $\frac{c+1}{\alpha_n} < 1$. As we are only concerned with $n \geq 2$, this is a weak condition, and holds whenever $c < 4.45$. Using $0 < x < 1$ implies $0 < \arctan(x) < x$ (this follows because we have an alternating sum of terms which decrease in absolute value), so we have

$$\log b_n \leq -\log(1 + \alpha_n^2) - \frac{\alpha_n \pi}{2c} + \frac{c+1}{c}. \qquad (19.52)$$

Exponentiating yields

$$b_n \leq \frac{1}{1 + \alpha_n^2} \cdot \exp\left(-\frac{\alpha_n \pi}{2c}\right) \cdot \exp\left(\frac{c+1}{c}\right). \qquad (19.53)$$

Plugging in the value for α_n yields

$$b_n \leq \frac{\log^2 10}{\log^2 10 + (2\pi n)^2} \cdot \exp\left(\frac{c+1}{c}\right) \cdot \exp\left(-\frac{\pi^2 n}{c\log 10}\right). \qquad (19.54)$$

The Cauchy–Schwarz inequality states that

$$\left| \sum_{n=2}^{\infty} c_n \gamma_n \right| \leq \sqrt{\sum_{n=2}^{\infty} c_n^2} \sqrt{\sum_{n=2}^{\infty} \gamma_n^2}. \tag{19.55}$$

For us,

$$c_n = \frac{\log^2 10}{\log^2 10 + (2\pi n)^2}, \quad \gamma_n = \exp\left(-\frac{\pi^2 n}{c \log 10}\right). \tag{19.56}$$

Lemma 19.9.1. *We have*

$$\sum_{n=2}^{\infty} c_n^2 \leq \frac{\log 10}{8\pi} \left(\pi - 2\arctan\left(\frac{2\pi}{\log 10}\right) - \sin\left(2\arctan\left(\frac{2\pi}{\log 10}\right)\right) \right)$$
$$\approx 0.00177749,$$
$$\sum_{n=2}^{\infty} \gamma_n^2 = \exp\left(-\frac{4\pi^2}{c \log 10}\right) \cdot \left[1 - \exp\left(-\frac{2\pi^2}{c \log 10}\right)\right]^{-1}. \tag{19.57}$$

Proof. We have

$$\sum_{n=2}^{\infty} c_n^2 = \sum_{n=2}^{\infty} \left(\frac{1}{1 + \left(\frac{2\pi n}{\log 10}\right)^2} \right)^2$$
$$= \left(1 + \left(\frac{2 \cdot 2\pi}{\log 10}\right)^2\right)^{-2} + \sum_{n=3}^{\infty} \left(\frac{1}{1 + \left(\frac{2\pi n}{\log 10}\right)^2} \right)^2$$
$$\leq \left(1 + \left(\frac{2 \cdot 2\pi}{\log 10}\right)^2\right)^{-2} + \int_{2}^{\infty} \left(1 + \left(\frac{2\pi x}{\log 10}\right)^2\right)^{-2} dx \tag{19.58}$$

(as the integrand is monotonically decreasing, thus we only increase the integral by starting at 2 instead of 3). We first change variables by letting $y = 2\pi x / \log 10$, and find

$$\sum_{n=3}^{\infty} c_n^2 \leq \frac{\log 10}{2\pi} \int_{4\pi/\log 10}^{\infty} \frac{dy}{(1 + y^2)^2}. \tag{19.59}$$

We now change variables by letting $y = \tan\theta$, so $dy = \sec^2\theta\, d\theta$:

$$\sum_{n=3}^{\infty} c_n^2 \leq \frac{\log 10}{2\pi} \int_{\arctan(4\pi/\log 10)}^{\pi/2} \frac{\sec^2\theta\, d\theta}{(1 + \tan^2\theta)^2}$$

$$= \frac{\log 10}{2\pi} \int_{\arctan(4\pi/\log 10)}^{\pi/2} \frac{\sec^2\theta\, d\theta}{\sec^4\theta}$$

$$= \frac{\log 10}{2\pi} \int_{\arctan(4\pi/\log 10)}^{\pi/2} \cos^2\theta\, d\theta$$

$$= \frac{\log 10}{2\pi} \left[\frac{\theta}{2} + \frac{\sin(2\theta)}{4} \right]_{\theta=\arctan(4\pi/\log 10)}^{\pi/2}$$

$$= \frac{\log 10}{8\pi} \left(\pi - 2\arctan\left(\frac{4\pi}{\log 10}\right) - \sin\left(2\arctan\left(\frac{4\pi}{\log 10}\right)\right) \right)$$

$$\approx 0.00072228. \tag{19.60}$$

Thus

$$\sum_{n=2}^{\infty} c_n^2 \leq \left(1 + \left(\frac{4\pi}{\log 10}\right)^2\right)^{-2}$$

$$+ \frac{\log 10}{8\pi} \left(\pi - 2\arctan\left(\frac{4\pi}{\log 10}\right) - \sin\left(2\arctan\left(\frac{4\pi}{\log 10}\right)\right) \right)$$

$$\approx 0.00177749. \tag{19.61}$$

If instead we numerically evaluate the sum, we get approximately 0.00140459 (we could easily get closer to this result by keeping more terms). In particular, we see our estimation is quite good (we are off by about 26%).

We now turn to the sum of γ_n^2:

$$\sum_{n=2}^{\infty} \gamma_n^2 = \sum_{n=2}^{\infty} \exp\left(-\frac{2\pi^2 n}{c \log 10}\right)$$

$$= \exp\left(-\frac{4\pi^2}{c \log 10}\right) \sum_{\ell=0}^{\infty} \left(\exp\left(-\frac{2\pi^2}{c \log 10}\right)\right)^{\ell}$$

$$= \exp\left(-\frac{4\pi^2}{c \log 10}\right) \cdot \left[1 - \exp\left(-\frac{2\pi^2}{c \log 10}\right)\right]^{-1}, \tag{19.62}$$

where the last line follows by the geometric series formula. □

Lemma 19.9.2. *We have*

$$\sum_{n=2}^{\infty} b_n \leq 0.0422 \cdot \exp\left(\frac{c+1}{c}\right) \cdot \exp\left(-\frac{2\pi^2}{c \log 10}\right) \cdot \left[1 - \exp\left(-\frac{2\pi^2}{c \log 10}\right)\right]^{-1/2}. \tag{19.63}$$

Proof. This follows immediately from the previous lemma and the Cauchy–Schwarz inequality. □

We can now give some good estimates on $\sum_{n\geq 2} b_n$ for various values of c. Instead of using our exact bound of about 0.001777 for the sum of c_n^2 we instead use the numerical bound of about 0.001405. We have

$$\sum_{n=2}^{\infty} b_n \leq g(c), \tag{19.64}$$

where

$$g(c) = 0.0422 \cdot \exp\left(\frac{c+1}{c}\right) \cdot \exp\left(-\frac{2\pi^2}{c\log 10}\right) \cdot \left[1 - \exp\left(-\frac{2\pi^2}{c\log 10}\right)\right]^{-1/2}.$$
$$\tag{19.65}$$

PART VII

Exercises

Chapter Twenty

Exercises

20.1 A QUICK INTRODUCTION TO BENFORD'S LAW

A couple of important points.

- There are many problems that would fit in multiple chapters. To help both the instructors and the readers, we have decided to collect them here. Thus, some of the exercises in this chapter will be far more accessible after reading later parts of the book.

- In Mathematica, if you define the following function you can then use it to find the first digit:

  ```
  firstdigit[x_] := Floor[10^Mod[Log[10,x],1]]
  ```

  (a similar function is definable in other languages, but the syntax will differ slightly).

Exercise 20.1.1. *If X is Benford base 10, find the probability that its significand starts 2.789.*

Exercise 20.1.2. *If X is Benford base 10, find the probability that its significand starts with 7.5 (in other words, its significand is in $[7.5, 7.6)$).*

Exercise 20.1.3. *If X is Benford base 10, find the probability that its significand has no 7s in the first k digits (thus a significand of 1.701 would have no 7 in its first digit, but it would have a 7 in its first two digits.*

Exercise 20.1.4. *Consider α^n for various α and various ranges of n; for example, take $\alpha \in \{2, 3, 5, 10, \sqrt{2}, \sqrt{5}, \sqrt{10}, \pi, e, \gamma\}$ (here γ is the Euler–Mascheroni constant; see*
http://en.wikipedia.org/wiki/Euler-Mascheroni_constant
for a description and properties), and let n go from 1 to N, where $N \in \{10^3, 10^5, 10^7\}$. Which of these data sets do you expect to be Benford? Why or why not? Read up about chi-square goodness of fit tests (see for example
http://en.wikipedia.org/wiki/Pearson_chi_square) and
compare the observed frequencies with the Benford probabilities.

Exercise 20.1.5. *Revisit the previous problem with more values of N. The problem is that there we looked at three snapshots of the behavior; it is far more interesting to plot the chi-square values as a function of N, for N ranging from say 100 to 10^7 or more. You will see especially interesting behavior if you look at the first digits of π^n.*

Exercise 20.1.6. *We have seen that the Benford behavior of a sequence is related to equidistribution of its logarithm. Thus, in the previous problem it may be useful to look at a log-log plot. Thus instead of plotting the chi-square value against the upper bound N, plot the logarithm of the chi-square value against $\log N$.*

Exercise 20.1.7. *Frequently taking logarithms helps illuminate relationships. For example, Kepler's third law (see* `http://www.physicsclassroom.com/class/circles/Lesson-4/Kepler-s-Three-Laws`*) says that the square of the time it takes a planet to orbit a sun is proportional to the cube of the semimajor axis. Find data for these quantities for the eight planets in our system (or nine if you count Pluto!) and plot them, and then do a log-log plot. A huge advantage of log-log plots is that linear relations are easy to observe and estimate; try to find the best fit line here, and note that the slope of the line should be close to 1.5 (if T is the period and L is the length of the semimajor axis, Kepler's third law is that there is a constant C such that $T^2 = CL^3$, or equivalently $T = CL^{3/2}$, or $\log T = \frac{3}{2}\log L + \log C$). Revisit the original plot, and try to see that it supports T^2 is proportional to L^3!*

Exercise 20.1.8. *Prove the log-laws: if $\log_b x_i = y_i$ and $r > 0$ then*

- $\log_b b = 1$ *and* $\log_b 1 = 0$ *(note* $\log_b x = y$ *means* $x = b^y$*);*

- $\log_b(x^r) = r\log_b x;$

- $\log_b(x_1 x_2) = \log_b x_1 + \log_b x_2$ *(the logarithm of a product is the sum of the logarithms);*

- $\log_b(x_1/x_2) = \log_b x_1 - \log_b x_2$ *(the logarithm of a quotient is the difference of the logarithms; this follows directly from the previous two log-laws);*

- $\log_c x = \log_b x / \log_b c$ *(this is the change of base formula).*

Exercise 20.1.9. *The last log-law (the change of base formula) is often forgotten, but is especially important. It tells us that if we can compute logarithms in one base then we can compute them in any base. In other words, it suffices to create just* one *table of logarithms, so we only need to find one base where we can easily compute logarithms. What base do you think that is, and how would you compute logarithms of arbitrary positive real numbers?*

Exercise 20.1.10. *The previous problem is similar to issues that arise in probability textbooks. These books only provide tables of probabilities of random variables drawn from a normal distribution,[1] as one can convert from such a table to probabilities for any other random variable. One such table is online here:*

---

[1] The random variable X is normally distributed with mean μ and variance σ^2 if its probability density function is $f(x; \mu, \sigma) = \exp\left(-(x-\mu)^2/(2\sigma^2)\right)/\sqrt{2\pi\sigma^2}$.

http://www.
mathsisfun.com/data/standard-normal-distribution-table.html. Use a stan-
dard table to determine the probability that a normal random variable with mean
$\mu = 5$ *and variance* $\sigma^2 = 16$ *(so the standard deviation is* $\sigma = 4$*) takes on a*
value between -3 *and* 7*. Thus, similarly to the change of base formula, there is an*
enormous computational saving as we only need to compute probabilities for one
normal distribution.

Exercise 20.1.11. *Prove* $\frac{d}{dx} \log_b x = \frac{1}{x \log b}$*. Hint: First do this when* $b = e$*, the*
base of the natural logarithms; use $e^{\log x} = x$ *and the chain rule.*

Exercise 20.1.12. *Revisit the first two problems, but now consider some other se-*
quences, such as $n!$*,* $\cos(n)$ *(in radians of course as otherwise the sequence is*
periodic), n^2*,* n^3*,* $n^{\log n}$*,* $n^{\log \log n}$*,* $n^{\log \log \log n}$*,* n^n*. In some situations* $\log_4$ *does*
not mean the logarithm base 4, but rather four iterations of the logarithm function.
It might be interesting to investigating $n^{\log_{f(n)} n}$ *under this definition for various*
integer-valued functions f*.*

Exercise 20.1.13. *Revisit the previous problem but for some recurrence relations.*
For example, try the Fibonacci numbers ($F_{n+2} = F_{n+1} + F_n$ *with* $F_0 = 0$ *and*
$F_1 = 1$*) and some other relations, such as the following.*

- *Catalan numbers:* $C_n = \frac{1}{n+1} \binom{2n}{n}$*; these satisfy a more involved recurrence*
  (see
  http://en.wikipedia.org/wiki/Catalan_number).

- *Squaring Fibonaccis:* $G_{n+2} = G_{n+1}^2 + G_n^2$ *with* $G_0 = 0$ *and* $G_1 = 1$*.*

- F_p *where* p *is a prime (i.e., only look at the Fibonaccis at a prime index).*

- *The logistic map:* $x_{n+1} = rx_n(1 - x_n)$ *for various choices of* r *and starting*
  values x_0 *(see http://en.wikipedia.org/wiki/Recurrence_relation).*

- *Newton's method for the difference between the nth prediction and the true*
  value. For example, to find the square root of α *we use* $x_{n+1} = \frac{1}{2}\left(x_n + \frac{\alpha}{x_n}\right)$*,*
  and thus we would study the distribution of leading digits of $|\sqrt{\alpha} - x_n|$*. One*
  could also look at other roots, other numbers, or more complicated functions.
  For more on Newton's method, see http://mathworld.wolfram.com
  /NewtonsMethod.html.

- *The* $3x + 1$ *Map:* $x_{n+1} = 3x_n + 1$ *if* x_n *is odd and* $x_n/2$ *if* x_n *is even*
  (though some authors use a slightly different definition, where for x_n *even,*
  one instead lets $x_{n+1} = x_n/2^d$*, where* d *is the highest power of 2 dividing*
  x_n*). It is conjectured that no matter what positive starting seed* x_0 *you take,*
  eventually x_n *cycles among 4, 2, and 1 for n sufficiently large (or is identi-*
  cally 1 from some point onward if we use the second definition). We return to
  this problem in Chapter 3.

For the remaining problems, whenever a data set satisfies Benford's Law we mean the *strong* version of the law. This means the cumulative distribution function of the significand is $F_X(s) = \log_{10}(s)$ for $s \in [1, 10)$, which implies that the probability of a first digit of d is $\log_{10}(1 + 1/d)$.

Exercise 20.1.14. *If a data set satisfies (the strong version of) Benford's Law base 10, what are the probabilities of all pairs of leading digits? In other words, what is the probability the first two digits are $d_1 d_2$ (in that order)? What if instead our set were Benford base b?*

Exercise 20.1.15. *Let X be a random variable that satisfies (the strong version of) Benford's Law. What is the probability that the second digit is d? Note here that the possible values of d range from 0 to 9.*

Exercise 20.1.16. *Building on the previous problem, compute the probability that a random variable satisfying the strong version of Benford's Law has its kth digit equal to d. If we denote these probabilities by $p_k(d)$, what is $\lim_{k \to \infty} p_k(d)$? Prove your claim.*

Exercise 20.1.17. *Find a data set that is spread over several orders of magnitude, and investigate its Benfordness (for example, stock prices or volume traded on a company that has been around for decades).*

Exercise 20.1.18. *Look at some of the data sets from the previous exercises that were* not *Benford, and see what happens if you multiply them together. For example, consider $n^2 \cdot \cos(n)$ (in radians), or $n^2 \sqrt{10}^n \cos(n)$, or even larger products. Does this support the claim in the chapter that products of random variables tend to converge to Benford behavior?*

Exercise 20.1.19. *Let $\mu_{k;b}$ denote the mean of significands of k digits of random variables perfectly satisfying Benford's Law, and let μ_b denote the mean of the significands of random variables perfectly following Benford's Law. What is $\mu_{k;b}$ for $k \in 1, 2, 3$? Does $\mu_{k;b}$ converge to μ_b? If yes, bound $|\mu_{k;b} - \mu_b|$ as a function of k.*

Exercise 20.1.20. *Benford's Law can be viewed as the distribution on significands arising from the density $p(x) = \frac{1}{x \log(10)}$ on $[1, 10)$ (and 0 otherwise). More generally, consider densities $p_r(x) = C_r / x^r$ for $x \in [1, 10)$ and 0 otherwise with $r \in (-\infty, \infty)$, where C_r is a normalization constant so that the density integrates to 1. For each r, calculate the probability of observing a first digit of d, and calculate the expected value of the first digit.*

20.2 A SHORT INTRODUCTION TO THE MATHEMATICAL THEORY OF BENFORD'S LAW

For a more detailed development of this material, see *An Introduction to Benford's Law* [BerH5] by Berger and Hill.

20.3 FOURIER ANALYSIS AND BENFORD'S LAW

20.3.1 Problems from Introduction to Fourier Analysis

The following exercises are from the chapter "An Introduction to Fourier Analysis," from the book *An Invitation to Modern Number Theory* (Princeton University Press, Steven J. Miller and Ramin Takloo-Bighash). This chapter is available online on the web page for this book (go to the links for Chapter 3).

Exercise 20.3.1. *Prove e^x converges for all $x \in \mathbb{R}$ (even better, for all $x \in \mathbb{C}$). Show the series for e^x also equals*

$$\lim_{n \to \infty} \left(1 + \frac{x}{n} \right)^n, \tag{20.1}$$

which you may remember from compound interest problems.

Exercise 20.3.2. *Prove, using the series definition, that $e^{x+y} = e^x e^y$ and calculate the derivative of e^x.*

Exercise 20.3.3. *Let f, g, and h be continuous functions on $[0, 1]$, and $a, b \in \mathbb{C}$. Prove*

1. $\langle f, f \rangle \geq 0$, and equals 0 if and only if f is identically zero;

2. $\langle f, g \rangle = \overline{\langle g, f \rangle}$;

3. $\langle af + bg, h \rangle = a\langle f, h \rangle + b\langle g, h \rangle$.

Exercise 20.3.4. *Find a vector $\vec{v} = \begin{pmatrix} v_1 \\ v_2 \end{pmatrix} \in \mathbb{C}^2$ such that $v_1^2 + v_2^2 = 0$, but $\langle \vec{v}, \vec{v} \rangle \neq 0$.*

Exercise 20.3.5. *Prove x^n and x^m are not perpendicular on $[0, 1]$. Find a $c \in \mathbb{R}$ such that $x^n - cx^m$ is perpendicular to x^m; c is related to the projection of x^n in the direction of x^m.*

Exercise 20.3.6 (Important). *Show for $m, n \in \mathbb{Z}$ that*

$$\langle e_m(x), e_n(x) \rangle = \begin{cases} 1 & \text{if } m = n, \\ 0 & \text{otherwise.} \end{cases} \tag{20.2}$$

Exercise 20.3.7. *Let f and g be periodic functions with period a. Prove $\alpha f(x) + \beta g(x)$ is periodic with period a.*

Exercise 20.3.8. *Prove any function can be written as the sum of an even and an odd function.*

Exercise 20.3.9. *Show*

$$\langle f(x) - \hat{f}(n)e_n(x), e_n(x) \rangle = 0. \tag{20.3}$$

This agrees with our intuition: after removing the projection in a certain direction, what is left is perpendicular to that direction.

Exercise 20.3.10. *Prove*

1. $\langle f(x) - S_N(x), e_n(x) \rangle = 0$ *if* $|n| \leq N$;

2. $|\widehat{f}(n)| \leq \int_0^1 |f(x)| dx$;

3. *Bessel's Inequality: if* $\langle f, f \rangle < \infty$ *then* $\sum_{n=-\infty}^{\infty} |\widehat{f}(n)|^2 \leq \langle f, f \rangle$;

4. *Riemann–Lebesgue Lemma: if* $\langle f, f \rangle < \infty$ *then* $\lim_{|n| \to \infty} \widehat{f}(n) = 0$ *(this holds for more general f; it suffices that $\int_0^1 |f(x)| dx < \infty$);*

5. *Assume f is differentiable k times; integrating by parts, show $|\widehat{f}(n)| \ll \frac{1}{n^k}$ and the constant depends only on f and its first k derivatives.*

Exercise 20.3.11. *Let $h(x) = f(x) + g(x)$. Does $\widehat{h}(n) = \widehat{f}(n) + \widehat{g}(n)$? Let $k(x) = f(x)g(x)$. Does $\widehat{k}(n) = \widehat{f}(n)\widehat{g}(n)$?*

Exercise 20.3.12. *If $\langle f, f \rangle, \langle g, g \rangle < \infty$ then the dot product of f and g exists: $\langle f, g \rangle < \infty$ (see Remark 11.2.4 of [MiT-B]). Do there exist $f, g : [0, 1] \to \mathbb{C}$ such that $\int_0^1 |f(x)| dx, \int_0^1 |g(x)| dx < \infty$ but $\int_0^1 f(x)\overline{g}(x) dx = \infty$? Is $f \in L^2([0, 1])$ a stronger or an equivalent assumption to $f \in L^1([0, 1])$?*

Exercise 20.3.13. *Define*

$$A_N(x) = \begin{cases} N & \text{for } |x| \leq \frac{1}{N}, \\ 0 & \text{otherwise.} \end{cases} \tag{20.4}$$

Prove A_N is an approximation to the identity on $[-\frac{1}{2}, \frac{1}{2}]$. If f is continuously differentiable and periodic with period 1, calculate

$$\lim_{N \to \infty} \int_{-\frac{1}{2}}^{\frac{1}{2}} f(x) A_N(x) dx. \tag{20.5}$$

Exercise 20.3.14. *Let $A(x)$ be a non-negative function with $\int_{\mathbb{R}} A(x) dx = 1$. Prove $A_N(x) = N \cdot A(Nx)$ is an approximation to the identity on $\mathbb{R}$.*

Exercise 20.3.15 (Important). *Let $A_N(x)$ be an approximation to the identity on $[-\frac{1}{2}, \frac{1}{2}]$. Let $f(x)$ be a continuous function on $[-\frac{1}{2}, \frac{1}{2}]$. Prove*

$$\lim_{N \to \infty} \int_{-\frac{1}{2}}^{\frac{1}{2}} f(x) A_N(x) dx = f(0). \tag{20.6}$$

Exercise 20.3.16. *Prove the two formulas above. The geometric series formula will be helpful:*

$$\sum_{n=N}^{M} r^n = \frac{r^N - r^{M+1}}{1 - r}. \tag{20.7}$$

Exercise 20.3.17. *Show that the Dirichlet kernels are not an approximation to the identity. How large are $\int_0^1 |D_N(x)| dx$ and $\int_0^1 D_N(x)^2 dx$?*

Exercise 20.3.18. *Prove the Weierstrass Approximation Theorem implies the original version of the Weierstrass Theorem.*

Exercise 20.3.19. *Let $f(x)$ be periodic function with period 1. Show*

$$S_N(x_0) = \int_{-\frac{1}{2}}^{\frac{1}{2}} f(x)D_N(x-x_0)dx = \int_{-\frac{1}{2}}^{\frac{1}{2}} f(x_0-x)D_N(x)dx. \quad (20.8)$$

Exercise 20.3.20. *Let $\widehat{f}(n) = \frac{1}{2^{|n|}}$. Does $\sum_{-\infty}^{\infty} \widehat{f}(n)e_n(x)$ converge to a continuous, differentiable function? If so, is there a simple expression for that function?*

Exercise 20.3.21. *Fill in the details for the above proof. Prove the result for all f satisfying $\int_0^1 |f(x)|^2 dx < \infty$.*

Exercise 20.3.22. *If $\int_0^1 |f(x)|^2 dx < \infty$, show Bessel's Inequality implies there exists a B such that $|\widehat{f}(n)| \leq B$ for all n.*

Exercise 20.3.23. *Though we used $|a+b|^2 \leq 4|a|^2 + 4|b|^2$, any bound of the form $c|a|^2 + c|b|^2$ would suffice. What is the smallest c that works for all $a, b \in \mathbb{C}$?*

Exercise 20.3.24. *Let $f(x) = \frac{1}{2} - |x|$ on $[-\frac{1}{2}, \frac{1}{2}]$. Calculate $\sum_{n=0}^{\infty} \frac{1}{(2n+1)^2}$. Use this to deduce the value of $\sum_{n=1}^{\infty} \frac{1}{n^2}$. This is often denoted $\zeta(2)$ (see Exercise 3.1.7 of [MiT-B]). See [BoPo][2] for connections with continued fractions, and [Kar][3] for connections with quadratic reciprocity.*

Exercise 20.3.25. *Let $f(x) = x$ on $[0, 1]$. Evaluate $\sum_{n=1}^{\infty} \frac{1}{n^2}$.*

Exercise 20.3.26. *Let $f(x) = x$ on $[-\frac{1}{2}, \frac{1}{2}]$. Prove $\frac{\pi}{4} = \sum_{n=1}^{\infty} \frac{(-1)^{n+1}}{(2n-1)^2}$. See also Exercise 3.3.29, and see Chapter 11 of [BorB][4] or [Schum][5] for a history of calculations of π.*

Exercise 20.3.27. *Find a function to determine $\sum_{n=1}^{\infty} \frac{1}{n^4}$.*

Exercise 20.3.28. *Show the Gaussian $f(x) = \frac{1}{\sqrt{2\pi\sigma^2}} e^{-(x-\mu)^2/2\sigma^2}$ is in $\mathcal{S}(\mathbb{R})$ for any $\mu, \sigma \in \mathbb{R}$.*

Exercise 20.3.29. *Let $f(x)$ be a Schwartz function with compact support contained in $[-\sigma, \sigma]$ and denote its Fourier transform by $\widehat{f}(y)$. Prove for any integer $A > 0$ that $|\widehat{f}(y)| \leq c_f y^{-A}$, where the constant c_f depends only on f, its derivatives and σ. As such a bound is useless at $y = 0$, one often derives bounds of the form $|\widehat{f}(y)| \leq \frac{\widetilde{c_f}}{(1+|y|)^A}$.*

---

[2]E. Bombieri and A. van der Poorten, *Continued fractions of algebraic numbers.* Pages 137–152 in *Computational Algebra and Number Theory (Sydney, 1992)*, Mathematical Applications, Vol. 325, Kluwer Academic, Dordrecht, 1995.

[3]A. Karlsson, *Applications of heat kernels on Abelian groups: $\zeta(2n)$, quadratic reciprocity, Bessel integral*, preprint.

[4]J. Borwein and P. Borwein, *Pi and the AGM: A Study in Analytic Number Theory and Computational Complexity*, John Wiley and Sons, New York, 1987.

[5]P. Schumer, *Mathematical Journeys*, Wiley-Interscience, John Wiley & Sons, New York, 2004.

Exercise 20.3.30. *Consider*

$$f(x) = \begin{cases} n^6 \left(\frac{1}{n^4} - |n - x| \right) & \text{if } |x - n| \leq \frac{1}{n^4} \text{ for some } n \in \mathbb{Z}, \\ 0 & \text{otherwise.} \end{cases} \qquad (20.9)$$

Show $f(x)$ is continuous but $F(0)$ is undefined. Show $F(x)$ converges and is well defined for any $x \notin \mathbb{Z}$.

Exercise 20.3.31. *If $g(x)$ decays like $x^{-(1+\eta)}$ for some $\eta > 0$, then $G(x) = \sum_{n \in \mathbb{Z}} g(x + n)$ converges for all x, and is continuous.*

Exercise 20.3.32. *For what weaker assumptions on f, f', f'' does $\sum_{n \in \mathbb{Z}} f(n) = \sum_{n \in \mathbb{Z}} \widehat{f}(n)$?*

Exercise 20.3.33. *One cannot always interchange orders of integration. For simplicity, we give a sequence a_{mn} such that $\sum_m (\sum_n a_{m,n}) \neq \sum_n (\sum_m a_{m,n})$. For $m, n \geq 0$ let*

$$a_{m,n} = \begin{cases} 1 & \text{if } n = m, \\ -1 & \text{if } n = m + 1, \\ 0 & \text{otherwise.} \end{cases} \qquad (20.10)$$

Show that the two different orders of summation yield different answers (the reason for this is that the sum of the absolute value of the terms diverges).

Exercise 20.3.34. *Find a family of functions $f_n(x)$ such that*

$$\lim_{n \to \infty} \int_{-\infty}^{\infty} f_n(x) dx \neq \int_{-\infty}^{\infty} \lim_{n \to \infty} f_n(x) dx \qquad (20.11)$$

and each $f_n(x)$ and $f(x)$ is continuous and $|f_n(x)|, |f(x)| \leq M$ for some M and all x.

Exercise 20.3.35. *Let f, g be continuous functions on $I = [0, 1]$ or $I = \mathbb{R}$. Show if $\langle f, f \rangle, \langle g, g \rangle < \infty$ then $h = f * g$ exists.* Hint: *Use the Cauchy–Schwarz inequality. Show further that $\widehat{h}(n) = \widehat{f}(n)\widehat{g}(n)$ if $I = [0, 1]$ or if $I = \mathbb{R}$. Thus the Fourier transform converts convolution to multiplication.*

Exercise 20.3.36. *Let X_1, X_2 be independent random variables with density p. Prove*

$$\text{Prob}(X_1 + X_2 \in [a, b]) = \int_a^b (p * p)(z) dz. \qquad (20.12)$$

Exercise 20.3.37 (Important). *If for all $i = 1, 2, \ldots$ we have $\langle f_i, f_i \rangle < \infty$, prove for all i and j that $\langle f_i * f_j, f_i * f_j \rangle < \infty$. What about $f_1 * (f_2 * f_3)$ (and so on)? Prove $f_1 * (f_2 * f_3) = (f_1 * f_2) * f_3$. Therefore convolution is associative, and we may write $f_1 * \cdots * f_N$ for the convolution of N functions.*

Exercise 20.3.38. *Suppose $X_1, \ldots, X_N$ are i.i.d.r.v. from a probability distribution p on $\mathbb{R}$. Determine the probability that $X_1 + \cdots + X_N \in [a, b]$. What must be assumed about p for the integrals to converge?*

Exercise 20.3.39. *One useful property of the Fourier transform is that the derivative of $\widehat{g}$ is the Fourier transform of $2\pi i x g(x)$; thus, differentiation (hard) is converted to multiplication (easy). Explicitly, show*

$$\widehat{g}'(y) = \int_{-\infty}^{\infty} 2\pi i x \cdot g(x) e^{-2\pi i x y} dx. \tag{20.13}$$

If g is a probability density, note $\widehat{g}'(0) = -2\pi i \mathbb{E}[x]$ and $\widehat{g}''(0) = -4\pi^2 \mathbb{E}[x^2]$.

Exercise 20.3.40. *If $B(x) = A(cx)$ for some fixed $c \neq 0$, show $\widehat{B}(y) = \frac{1}{c}\widehat{A}\left(\frac{y}{c}\right)$.*

Exercise 20.3.41. *Show that if the probability density of $X_1 + \cdots + X_N = x$ is $(p * \cdots * p)(x)$ (i.e., the distribution of the sum is given by $p * \cdots * p$), then the probability density of $\frac{X_1 + \cdots + X_N}{\sqrt{N}} = x$ is $(\sqrt{N}p * \cdots * \sqrt{N}p)(x\sqrt{N})$. By Exercise 20.3.40, show*

$$FT\left[(\sqrt{N}p * \cdots * \sqrt{N}p)(x\sqrt{N})\right](y) = \left[\widehat{p}\left(\frac{y}{\sqrt{N}}\right)\right]^N. \tag{20.14}$$

Exercise 20.3.42. *Show for any fixed y that*

$$\lim_{N\to\infty}\left[1 - \frac{2\pi^2 y^2}{N} + O\left(\frac{y^3}{N^{3/2}}\right)\right]^N = e^{-2\pi^2 y^2}. \tag{20.15}$$

Exercise 20.3.43. *Show that the Fourier transform of $e^{-2\pi^2 y^2}$ at x is $\frac{1}{\sqrt{2\pi}} e^{-x^2/2}$. Hint: This problem requires contour integration from complex analysis.*

Exercise 20.3.44. *Modify the proof to deal with the case of p having mean μ and variance σ^2.*

Exercise 20.3.45. *For reasonable assumptions on p, estimate the rate of convergence to the Gaussian.*

Exercise 20.3.46. *Let p_1, p_2 be two probability densities satisfying*

$$\int_{-\infty}^{\infty} x p_i(x) dx = 0, \quad \int_{-\infty}^{\infty} x^2 p_i(x) dx = 1, \quad \int_{-\infty}^{\infty} |x|^3 p_i(x) dx < \infty. \tag{20.16}$$

Consider $S_N = X_1 + \cdots + X_N$, where for each i, X_1 is equally likely to be drawn randomly from p_1 or p_2. Show the Central Limit Theorem is still true in this case. What if we instead had a fixed, finite number of such distributions $p_1, \ldots, p_k$, and for each i we draw X_i from p_j with probability q_j (of course, $q_1 + \cdots + q_k = 1$)?

Exercise 20.3.47 (Gibbs Phenomenon). *Define a periodic with period 1 function by*

$$f(x) = \begin{cases} -1 & \text{if } -\frac{1}{2} \leq x < 0, \\ 1 & \text{if } 0 \leq x < \frac{1}{2}. \end{cases} \tag{20.17}$$

Prove that the Fourier coefficients are

$$\widehat{f}(n) = \begin{cases} 0 & \text{if } n \text{ is even}, \\ \frac{4}{n\pi i} & \text{if } n \text{ is odd}. \end{cases} \tag{20.18}$$

Show that the Nth partial Fourier series $S_N(x)$ converges pointwise to $f(x)$ wherever f is continuous, but overshoots and undershoots for x near 0. Hint: *Express the series expansion for $S_N(x)$ as a sum of sines. Note $\frac{\sin(2m\pi x)}{2m\pi} = \int_0^x \cos(2m\pi t)dt$. Express this as the real part of a geometric series of complex exponentials, and use the geometric series formula. This will lead to*

$$S_{2N-1}(x) = 8\int_0^x \Re\left(\frac{1}{2i}\frac{e^{4n\pi it}-1}{\sin(2\pi t)}\right)dt = 4\int_0^x \frac{\sin(4n\pi t)}{\sin(2\pi t)}dt, \quad (20.19)$$

which is about 1.179 (or an overshoot of about 18%) when $x = \frac{1}{4n\pi}$. What can you say about the Fejér series $T_N(x)$ for x near 0?

Exercise 20.3.48 (Nowhere Differentiable Function). *Weierstrass constructed a continuous but nowhere differentiable function! We give a modified example and sketch the proof. Consider*

$$f(x) = \sum_{n=0}^{\infty} a^n\cos(2^n \cdot 2\pi x), \quad \frac{1}{2} < a < 1. \quad (20.20)$$

Show f is continuous but nowhere differentiable. Hint: *First show $|a| < 1$ implies f is continuous. Our claim on f follows from noting that if a periodic continuous function g is differentiable at x_0 and $\widehat{g}(n) = 0$ unless $n = \pm 2^m$, then there exists C such that for all n, $|\widehat{g}(n)| \leq Cn2^{-n}$. To see this, it suffices to consider $x_0 = 0$ and $g(0) = 0$. Our assumptions imply that $(g, e_m) = 0$ if $2^{n-1} < m < 2^{n+1}$ and $m \neq 2^n$. We have $\widehat{g}(2^n) = (g, e_{2^n}F_{2^n-1}(x))$ where F_N is the Fejér kernel. The claim follows from bounding the integral $(g, e_{2^n}F_{2^n-1}(x))$. In fact, more is true: Baire showed that, in a certain sense, "most" continuous functions are nowhere differentiable! See, for example, [Fol].*[6]

Exercise 20.3.49 (Isoperimetric Inequality). *Let $\gamma(t) = (x(t), y(t))$ be a smooth closed curve in the plane; we may assume it is parametrized by arc length and has length 1. Prove the enclosed area A is largest when $\gamma(t)$ is a circle.* Hint: *By Green's Theorem*

$$\oint_\gamma xdy - ydx = 2\text{Area}(A). \quad (20.21)$$

The assumptions on $\gamma(t)$ imply $x(t), y(t)$ are periodic functions with Fourier series expansions and $\left(\frac{dx}{dt}\right)^2 + \left(\frac{dy}{dt}\right)^2 = 1$. Integrate this equality from $t = 0$ to $t = 1$ to obtain a relation among the Fourier coefficients of $\frac{dx}{dt}$ and $\frac{dx}{dt}$ (which are related to those of $x(t)$ and $y(t)$); (20.21) gives another relation among the Fourier coefficients. These relations imply $4\pi\text{Area}(A) \leq 1$ with strict inequality unless the Fourier coefficients vanish for $|n| > 1$. After some algebra, one finds this implies we have a strict inequality unless γ is a circle.

Exercise 20.3.50 (Applications to Differential Equations). *One reason for the introduction of Fourier series was to solve differential equations. Consider the vibrating string problem: a unit string with endpoints fixed is stretched into some*

---

[6]G. Folland, *Real Analysis: Modern Techniques and Their Applications*, 2nd edition, Pure and Applied Mathematics, Wiley-Interscience, New York, 1999.

*initial position and then released; describe its motion as time passes. Let $u(x,t)$
denote the vertical displacement from the rest position x units from the left endpoint at time t. For all t we have $u(0,t) = u(1,t) = 0$ as the endpoints are fixed.
Ignoring gravity and friction, for small displacements Newton's laws imply*

$$\frac{\partial^2 u(x,t)}{\partial x^2} = c^2 \frac{\partial^2 u(x,t)}{\partial t^2}, \tag{20.22}$$

*where c depends on the tension and density of the string. Guessing a solution of the
form*

$$u(x,t) = \sum_{n=1}^{\infty} a_n(t) \sin(n\pi x), \tag{20.23}$$

solve for $a_n(t)$.

*One can also study problems on $\mathbb{R}$ by using the Fourier transform. Its use stems
from the fact that it converts multiplication to differentiation, and vice versa: if
$g(x) = f'(x)$ and $h(x) = xf(x)$, prove that $\widehat{g}(y) = 2\pi i y \widehat{f}(y)$ and $\frac{d\widehat{f}(y)}{dy} = -2\pi i \widehat{h}(y)$. This and Fourier inversion allow us to solve problems such as the heat
equation*

$$\frac{\partial u(x,t)}{\partial t} = \frac{\partial^2 u(x,t)}{\partial x^2}, \quad x \in \mathbb{R}, \; t > 0 \tag{20.24}$$

with initial conditions $u(x,0) = f(x)$.

20.3.2 Problems from Chapter 1: Revisited

Many of the problems from Chapter 1 are appropriate here as well. In addition to
reexamining those problems, consider the following.

Exercise 20.3.51. *Is the sequence $a_n = n^{\log n}$ Benford?*

Exercise 20.3.52. *In some situations $\log_4$ does not mean the logarithm base 4, but
rather four iterations of the logarithm function. Investigate $n^{\log_{f(n)} n}$ under this
definition for various integer-valued functions f.*

20.3.3 Problems from Chapter 3

Exercise 20.3.53. *Assume an infinite sequence of real numbers $\{x_n\}$ has its logarithms modulo 1, $\{y_n = \log_{10} x_n \bmod 1\}$, satisfying the following property: as
$n \to \infty$ the proportion of y_n in any interval $[a,b] \subset [0,1]$ converges to $b - a$ **if**
$b - a > 1/2$. Prove or disprove that $\{x_n\}$ is Benford.*

Exercise 20.3.54. *As $\sqrt{2}$ is irrational, the sequence $\{x_n = n\sqrt{2}\}$ is uniformly
distributed modulo 1. Is the sequence $\{x_n^2\}$ uniformly distributed modulo 1?*

Exercise 20.3.55. *Does there exist an irrational α such that α is a root of a
quadratic polynomial with integer coefficients and the sequence $\{\alpha^n\}_{n=1}^{\infty}$ is Benford base 10?*

Exercise 20.3.56. *We showed a geometric Brownian motion is a Benford-good process; is the sum of two independent geometric Brownian motions Benford-good?*

The next few questions are related to a map we now describe. We showed that, suitably viewed, the $3x + 1$ map leads to Benford behavior (or is close to Benford for almost all large starting seeds). Consider the following map. Let $R(x)$ be the number formed by writing the digits of x in reverse order. If $R(x) = x$ we say x is palindromic. If x is not a palindromic number set $P(x) = x + R(x)$, and if x is palindromic let $P(x) = x$. For a given starting seed x_0 consider the sequence where $x_{n+1} = P(x)$. It is not known whether there are any x_0 such that the resulting sequence diverges to infinity, though it is believed that almost all such numbers do. The first candidate to escape is 196; for more see http://en.wikipedia.org/wiki/Lychrel_number (this process is also called "reverse-and-add," and the candidates are called Lychrel numbers).

Exercise 20.3.57. *Consider the reverse-and-add map described above applied to a large starting seed. Find as good of a lower bound as you can for the number of seeds between 10^n and 10^{n+1} such that the resulting sequence stabilizes (i.e., we eventually hit a palindrome).*

Exercise 20.3.58. *Come up with a model to estimate the probability a given starting seed in 10^n and 10^{n+1} has its iterates under the reverse-and-add map diverge to infinity. Hint: x plus $R(x)$ is a palindrome if and only if there are no carries when we add; thus you must estimate the probability of having no carries.*

Exercise 20.3.59. *Investigate the Benfordness of sequences arising from the reverse-and-add map for various starting seeds. Of course the calculation is complicated by our lack of knowledge about this map, specifically we don't know even one starting seed that diverges! Look at what happens with various Lychrel numbers. For each N can you find a starting seed x_0 such that it iterates to a palindrome after N or more steps?*

Exercise 20.3.60. *Redo the previous three problems in different bases. Your answer will depend now on the base; for example, much more is known base 2 (there we can give specific starting seeds that iterate to infinity).*

Exercise 20.3.61. *Use the Erdös–Turan Inequality to calculate upper bounds for the discrepancy for various sequences, and use those results to prove Benford behavior. Note you need to find a sequence where you can do the resulting computation. For example, earlier we investigated $a_n = n^{\log n}$; are you able to do the summation for this case?*

Exercise 20.3.62. *Consider the analysis of products of random variables. Fix a probability p (maybe $p = 1/2$), and independent identically distributed random variables $X_1, \ldots, X_n$. Assume as $n \to \infty$ the product of the X_i's becomes Benford. What if now we let $\widetilde{X}_n$ be the random variable where we toss n independent coins, each with probability p, and if the ith toss is a head then X_i is in the product (if the product is empty we use the standard convention that it is then 1). Is this process Benford?*

Exercise 20.3.63. *Redo the previous problem, but drop the assumption that the random variables are identically distributed.*

Exercise 20.3.64. *Redo the previous two problems, but now allow the probability that the ith toss is a head to depend on i.*

Exercise 20.3.65. *Consider*

$$\phi_m \;=\; \begin{cases} m & \text{if } |x - \tfrac{1}{8}| \le \tfrac{1}{2m}, \\ 0 & \text{otherwise}; \end{cases} \tag{20.25}$$

this is the function from Example 3.3.5 and led to non-Benford behavior for the product. Can you write down the density for the product?

Exercise 20.3.66. *In the spirit of the previous problem, find other random variables where the product is not Benford.*

Exercise 20.3.67. *Consider a Weibull random variable with a scale parameter α of 1 and translation parameter β of 0; so $f(x;\gamma) = x^{\gamma-1}\exp(x^\gamma)$ for $x \ge 0$ and is zero otherwise. Investigate the Benfordness of chaining random variables here, where the shape parameter γ is the output of the previous step.*

Exercise 20.3.68. *The methods of [JaKKKM] led to good bounds for chaining exponential and uniform random variables. Can you obtain good, explicit bounds in other cases? For example, consider a binomial process with fixed parameter p.*

Exercise 20.3.69. *Apply the methods of Cuff, Lewis, and Miller (for the Weibull distribution) to other random variables. Consider the generalized Gamma distribution (see*

http://en.wikipedia.org/wiki/Generalized_gamma_distribution

for more information), where the density is

$$f(x;a,d,p) \;=\; \frac{p/d^a}{\Gamma(d/p)} x^{d-1} \exp\left(-(x/a)^p\right)$$

for $x > 0$ and 0 otherwise, where a, d, p are positive parameters.

For the next few problems, let $f_r(x) - 1/(1 + |x|^r)$ with $r > 1$.

Exercise 20.3.70. *Show that for $r > 1$, $\int_{-\infty}^{\infty} f_r(x)\,dx$ is finite, and $\int_{-\infty}^{\infty} f_r(x)\,dx = \frac{2\pi}{r}\csc\left(\frac{\pi}{r}\right)$.*

Exercise 20.3.71. *Verify the Fourier transform identity used in our analysis:*

$$p_r\left(e^{b+y}\right) e^{b+y} \;=\; \frac{1}{2}\sin\left(\frac{\pi}{r}\right) e^{2\pi i b y}\csc\left(\frac{\pi}{r}(1 - 2\pi i y)\right),$$

where $b \in [0, 1]$.

20.4 BENFORD'S LAW GEOMETRY

Exercise 20.4.1. *Perform a chi-square goodness-of-fit test on the data values in Table 4.1.*

Exercise 20.4.2. *Let the random variable X have the Benford distribution as defined in this chapter. Find $\mathbb{E}[X]$. Next, generate one million Benford random variates and compute their sample mean. Perform this Monte Carlo experiment several times to ensure that the sample means are near $\mathbb{E}[X]$.*

Exercise 20.4.3. *Let $T \sim \text{exponential}(1)$. Find the probability mass function of the leading digit to three-digit accuracy. Compare your results to those in Table 4.2.*

Exercise 20.4.4. *Redo the previous exercise, but instead of finding the probability mass function of the leading digit, find the cumulative distribution function of the significand (i.e., find the probability of observing a significand of at most s).*

Exercise 20.4.5. *Determine the set of conditions on a, b, and c associated with $W \sim \text{triangular}(a, b, c)$ which result in $T = 10^W$ following Benford's Law.*

Exercise 20.4.6. *Use R to confirm that the cumulative distribution function $F_x(x) = \text{Prob}(X \le x) = \log_{10}(x + 1)$ results in a probability mass function that gives the distribution specified in Benford's Law. What is the range of x?*

Exercise 20.4.7. *Use R to determine whether the cumulative distribution function $F_x(x) = \text{Prob}(X \le x) = x^2$ (for some range for x) results in a probability mass function that gives the distribution specified in Benford's Law. If yes, what is the range for x?*

Exercise 20.4.8. *Which of the following distributions of W follow Benford's Law?*

- $f_W(w) \sim U(0, 3.5)$.

- $f_W(w) \sim U(17, 117)$.

- $f_W(w) = w^3 - w^2 + w$ *for* $0 \le w \le 1$, *and* $1 - w^3 + w^2 - w$ *for* $1 \le w \le 2$.

- $f_W(w) = \sqrt{w}$ *for* $0 \le w \le 1$, *and* $1 - \sqrt{w - 1}$ *for* $1 \le w \le 2$.

Exercise 20.4.9. *Let b_1 and b_2 be two different integers exceeding 1. Is there a probability density p on an interval I such that if a random variable X has p for its probability density function then X is Benford in both base b_1 and b_2? What if the two bases are allowed to be real numbers exceeding 1? Prove your claims.*

20.5 EXPLICIT ERROR BOUNDS VIA TOTAL VARIATION

Exercise 20.5.1. *Find $\text{TV}(\sin(x), [-\pi, \pi])$.*

Exercise 20.5.2. *Confirm that $\text{TV}(h, \mathbb{J}) = \text{TV}^+(h, \mathbb{J}) + \text{TV}^-(h, \mathbb{J})$.*

Exercise 20.5.3. *Let Y_o and Z be independent random variables such that Y_o has a density f_o with $\mathrm{TV}(f_o) < \infty$ and Z has distribution π. Verify that $Y := Y_o + Z$ has density $f(y) = \int f_o(y - z)\,\pi(dz)$ with $\mathrm{TV}(f) \leq \mathrm{TV}(f_o)$.*

Exercise 20.5.4. *Show that an absolutely continuous probability density f on $\mathbb{R}$ satisfies*

$$\mathrm{TV}(f)^2 \leq \int \frac{f'(x)^2}{f(x)}\,dx.$$

Exercise 20.5.5. *Let $\gamma_{a,\sigma}$ be the density of the Gamma distribution $\mathrm{Gamma}(a,\sigma)$ with shape parameter $a > 0$ and scale parameter $\sigma > 0$, i.e.,*

$$\gamma_{a,\sigma}(x) = \sigma^{-a} x^{a-1} \exp(-x/\sigma)/\Gamma(a)$$

for $x > 0$, and $\gamma_{a,\sigma} = 0$ on $(-\infty, 0]$.

1. Show that for $a \geq 1$,

$$\mathrm{TV}(\gamma_{a,\sigma}) = \sigma^{-1}\,\mathrm{TV}(\gamma_{a,1}) \quad and \quad \mathrm{TV}(\gamma_{a,1}) = 2((a-1)/e)^{a-1}/\Gamma(a).$$

2. It is well known that $\Gamma(t+1) = (t/e)^t \sqrt{2\pi t}(1+o(1))$ as $t \to \infty$ (this is Stirling's formula). What does this imply for $\mathrm{TV}(\gamma_{a,1})$? Show that $\mathrm{TV}(\gamma_{a,\sigma}) \to 0$ as $\sqrt{a}\,\sigma \to \infty$ and $a \geq 1$.

Exercise 20.5.6. *Let X be a strictly positive random variable with density h on $(0,\infty)$. Verify that $Y := \log_B(X)$ has density f given by $f(y) = \log(B)B^y h(B^y)$ for $y \in \mathbb{R}$.*

Exercise 20.5.7. *Let X be a random variable with distribution $\mathrm{Gamma}(a,\sigma)$ for some $a, \sigma > 0$; see Exercise 20.5.5.*

1. Determine the density $f_{a,\sigma}$ of $Y := \log_B(X)$. Here you should realize that $f_{a,\sigma}(y) = f_{a,1}(y - \log_B(\sigma))$. Show then that

$$\mathrm{TV}(f_{a,\sigma}) = 2\log(B)(a/e)^a/\Gamma(a).$$

What happens as $a \to \infty$?

2. To understand why the leading digits of X are far from Benford's Law for large a, verify that $X = \sigma(a + \sqrt{a}Z_a)$ for a random variable Z_a with mean zero and variance one. (Indeed, the density of Z_a converges uniformly to the standard Gaussian density as $a \to \infty$.) Now investigate the distribution of $Y = \log_B(X)$ as $a \to \infty$.

20.6 LÉVY PROCESSES AND BENFORD'S LAW

Exercise 20.6.1. *Provide an example of a non-continuous cadlag function.*

Exercise 20.6.2. *Prove that a Weiner process is also a Lévy process.*

Exercise 20.6.3. *Prove that a Poisson process is also a Lévy process.*

Exercise 20.6.4. *Prove that the exponential Lévy process* $\{\exp(X_t)\}$ $(t \in \mathbb{R})$ *is a martingale with respect to* $(\mathcal{F}_t) := \sigma\{X_s : s \leq t\}$ *if and only if* $\mathbb{E}[\exp(X_t)] = 1$.

Exercise 20.6.5. *Let* $f(t) = \mathbb{E}[\exp(it\xi)], g(t) = \mathbb{E}[\exp(it\eta)]$ $(t \in \mathbb{R})$ *be the characteristic functions of (real-) valued random variables* ξ, η $(i = \sqrt{-1})$. *Recall that* $\exp(it) = \cos t + i \sin t$ $(t \in \mathbb{R})$ *and* $\mathbb{E}[\exp(it\xi)] := \mathbb{E}[\cos(t\xi)] + i\mathbb{E}[\sin(t\xi)]$ $(t \in \mathbb{R})$. *Finally,* $\overline{a + ib} := a - ib$ $(a, b \in \mathbb{R})$ *denotes the complex conjugate of* $a + ib$. *Note that* $|f|^2(t) = f(t) \cdot \bar{f}(t)$. *Show the following.*

1. *f is continuous, $f(0) = 1$, and $|f(t)| \leq 1$, $t \in \mathbb{R}$.*

2. *$\bar{f}$ is a characteristic function.*

3. *$f \cdot g$ is a characteristic function. Hence, $|f|^2$ is a characteristic function.*

4. *Let $h_1, h_2, \ldots$ be characteristic functions. If $a_1 \geq 0, a_2 \geq 0, \ldots$ are real numbers such that $a_1 + a_2 + \cdots = 1$, then $a_1 h_1 + a_2 h_2 + \cdots$ is a characteristic function.*

5. *Show that every characteristic function h is non-negative definite, i.e., for all $n \geq 2$, real $t_1, \ldots, t_n$, and complex $a_1, \ldots, a_n$ we have that*

$$\sum_{j=1}^{n} \sum_{k=1}^{n} h(t_j - t_k) a_j \bar{a}_k \geq 0.$$

Exercise 20.6.6. *Show that, for each real number $p > 0$, $f(z) := \cos(2\pi pz)$ $(z \in \mathbb{R})$ is a characteristic function. Deduce that $g(z) := (\cos(2\pi pz))^2$ $(z \in \mathbb{R})$ is a characteristic function.*

Exercise 20.6.7. (This exercise gives an example of a characteristic function which "wildly fluctuates.") *It follows from Exercises 20.6.6 and 20.6.5(4) that*

$$h(z) := \sum_{k=1}^{\infty} 2^{-k} (\cos(2\pi 7^k z))^2, \quad z \in \mathbb{R}$$

is a characteristic function. Show that h is of infinite total variation over each non-degenerate interval $[a, b]$, i.e.,

$$\sup \left\{ \sum_{k=1}^{n} |h(z_{k+1}) - h(z_k)| \right\} = \infty,$$

the supremum taken over all $n \geq 1$ and real numbers $a \leq z_1 < z_2 < \cdots < z_{n+1} \leq b$.

Hint: *It suffices to prove the claim for intervals $[r + 7^{-N}, r + 2 \cdot 7^{-N}]$ (being convenient for calculations!) where $N \geq 1$ is an integer and $r \geq 0$ a real number. Let $k \geq N + 1$ and denote by $I(k)$ the set of integers j such that $1 + (r + 7^{-N})7^k < j \leq ((r + 2 \cdot 7^{-N})7^k)$. For $j \in I(k)$ put $t_{2j-1}(k) = (j - 1/4)7^{-k}, t_{2j}(k) = j \cdot 7^{-k}$. Show, by using the inequalities $|a + b| \geq |a| - |b|$ and $|(\cos b)^2 - (\cos a)^2| \leq 2|b - a|$ $(a, b \in \mathbb{R})$ that*

$$\sum_{j \in I(k)} |h(t_{2j}(k)) - h(t_{2j-1}(k))| \geq 2(1 - \pi/5)7^{-N}(7/2)^k + \text{const.}$$

Exercise 20.6.8.   *1. Try to guess how the integral $\int_a^b f(z) \exp(itz) dz$ behaves as $t \to \infty$ if $f : [a, b] \to \mathbb{R}$ is a step function of the form $f(t) = \sum_{j=1}^m c_j \mathbb{I}_{[b_{j-1}, b_j)}(t)$ where $a \le b_0 < b_1 < \cdots < b_m \le b$.*

2. Verify your guess when f is an indicator function of an interval.

3. How does the above integral behave when f is continuous on $[a, b]$?

Exercise 20.6.9. *Show that a Lévy measure Q satisfies $Q(\mathbb{R} \mathbf{r} (-\alpha, \alpha)) < \infty$ for all $\alpha > 0$.*

Exercise 20.6.10. *Let X be a Lévy process having Lévy measure Q. Show that, for fixed $c > 0$ and $s \ge 0$, the process X^* given by $X_t^* = X_{ct+s} - X_s$ $(t \ge 0)$ is a Lévy process having Lévy measure $Q^* = cQ$.*

Exercise 20.6.11. *Let $N = (N_t)$ $(t \ge 0)$ be a Poisson process with parameter $\lambda > 0$.*

1. Verify that the generating triple of N is given by $(\lambda, 0, Q^*)$ where Q^* has total mass λ concentrated on $\{1\}$.

2. Verify (6.15) directly for $X = N$, i.e.,

$$Q^*(A) = c^{-1} \mathbb{E}[\#\{s < t \le s + c : \Delta N_t \in A \mathbf{r} \{0\}\}]$$

holds for all $c > 0, s \ge 0$, and every Borel set $A \subset \mathbb{R}$.

Exercise 20.6.12. *Let $T_t = \sum_{j=1}^{N_t} \zeta_j$ $(t \ge 0)$ denote the compound Poisson process of Example 6.1.21. (Here, (N_t) is a Poisson process with parameter $\lambda > 0; \zeta_1, \zeta_2, \ldots$ are independent random variables with a common distribution Q_1 such that $Q_1(\{0\}) = 0$. Furthermore, the processes (ζ_n) and (N_t) are independent of each other.)*

1. Show that the characteristic function g_t of T_t $(t \ge 0)$ is given by

$$g_t(z) = \exp\left[\lambda t \int_{\mathbb{R}} (e^{izx} - 1) Q_1(dx)\right]$$

for all $z \in \mathbb{R}$ and $t \ge 0$.

2. It can be shown (see the reference in Example 6.1.21) that (T_t) is a Lévy process. Determine its generating triple (β, σ^2, Q).

Exercise 20.6.13. *Let W be a (standard) Brownian motion (BM). Show that, for each $c > 0, W^* = (cW_{t/c^2})$ is a BM (scaling property).*

Exercise 20.6.14. *Let $\xi \sim N(\mu, \sigma^2)$ where $\mu \in \mathbb{R}$ and $\sigma > 0$.*

1. Deduce from (6.26) that the characteristic function of ξ is given by

$$\mathbb{E}[\exp(iz\xi)] = \exp(i\mu z - \sigma^2 z^2 / 2), \quad z \in \mathbb{R}.$$

2. *Deduce from the result in (1) that, for all $\mu, z \in \mathbb{R}$ and $\sigma > 0$,*

$$\int_{-\infty}^{\infty} \cos(zx) \exp(-(x-\mu)^2/(2\sigma^2))dx = \sqrt{2\pi\sigma^2} \cos(\mu z) \exp(-\sigma^2 z^2/2)$$

and

$$\int_{-\infty}^{\infty} \sin(zx) \exp(-(x-\mu)^2/(2\sigma^2))dx = \sqrt{2\pi\sigma^2} \sin(\mu z) \exp(-\sigma^2 z^2/2).$$

Exercise 20.6.15. *Let $W = (W_t)$ be a BM. Put*

$$S_{t,u} := \sup_{0 \le s \le u} |W_{t+s} - W_t|, \ t \ge 0, u > 0.$$

1. *Show that $S_{t,u}$ is a random variable. (This requires a little argument since the definition of $S_{t,u}$ involves uncountably many random variables!)*
   Hint: *Recall that all sample paths of W are continuous.*

2. *Show that $W_n/n \to 0$ $(n \to \infty)$ a.s.*

3. *Since, for each fixed $t \ge 0, (W_{u+t} - W_t) \ (u \ge 0)$ is a BM, it follows that*

   for each $t > 0, S_{t,1}$ has the same distribution as $S_{0,1}$. $(*)$

   Furthermore, we have that

   $$P(S_{0,1} \ge a) \le 2\exp(-a^2/2), \ a \ge 0 \qquad\qquad (**)$$

   (see, e.g., [KaSh]). Use (2) as well as $(*)$ and $(**)$ to show that

   $$W_t/t \to 0 \ (t \to \infty) \ a.s.$$

   Hint: *Use the Borel–Cantelli Lemma.*

Exercise 20.6.16. *Let $\xi_1, \xi_2, \ldots$ be independent random variables defined on some probability space $(\Omega, \mathcal{F}, P)$, which have a common distribution given by $P(\xi_n = +1) = p, P(\xi_n = -1) = 1 - p =: q \ (n \ge 1)$, where $0 < p < 1$. Put $S_n := \xi_1 + \cdots + \xi_n, \ n \ge 0 \ (S_0 = 0)$, and let $(\mathcal{F}_n) \ (n \ge 0)$ be the filtration generated by (ξ_n). (Note that $\mathcal{F}_0 = \{\emptyset, \Omega\}$.)*

1. *Show that $Y_n := (q/p)^{S_n} \ (n \ge 0)$ is an $(\mathcal{F}_n)$-martingale.*

2. *Put $c(\alpha) := \mathbb{E}[\exp(\alpha\xi_1)] = p\exp(\alpha) + q\exp(-\alpha) \ (\alpha \in \mathbb{R})$. Show that, for every fixed $\alpha \in \mathbb{R}$,*

   $$Z_n := \exp(\alpha S_n)/(c(\alpha))^n \ (n \ge 0)$$

   is an $(\mathcal{F}_n)$-martingale.

Exercise 20.6.17. *Let $\xi_1, \xi_2, \ldots$ be independent random variables defined on the same probability space, which have a common distribution given by $P(\xi_n = +1) = P(\xi_n = -1) = 1/2$. Put $S_0 = 0$ and $S_n = \xi_1 + \cdots + \xi_n \ (n \ge 1)$ which means that (S_n) is a simple symmetric random walk on $\mathbb{Z}$, starting at 0. Let $(\mathcal{F}_n)$ be the filtration generated by (ξ_n). Show that following two sequences are $(\mathcal{F}_n)$-martingales:*

1. $(S_n^3 - 3nS_n)$.

2. $(S_n^4 - 6nS_n^2 + 3n^2 + 2n)$.

Hint: *Note that* $\mathbb{E}[\xi_n|\mathcal{F}_{n-1}] = \mathbb{E}[\xi_n] = 0$ *a.s. (since* ξ_n *is independent of* $\mathcal{F}_{n-1}$*), and that* $\mathbb{E}[S_{n-1}^2\xi_n|\mathcal{F}_{n-1}] = S_{n-1}^2\mathbb{E}[\xi_n] = 0$ *a.s. (since* S_{n-1} *is* $\mathcal{F}_{n-1}$*-measurable). Note that* $S_n = S_{n-1} + \xi_n$.

Exercise 20.6.18. *Let* $(\Omega, \mathcal{F}, P)$ *be a probability space and let* $(\mathcal{F}_n)$ $(n \geq 0)$ *be any filtration on* $(\Omega, \mathcal{F})$. *In the sequel let* $Z = (Z_n)$ $(n \geq 0)$ *and* $H = (H_n)$ $(n \geq 1)$ *be sequences of random variables defined on* $(\Omega, \mathcal{F})$ *such that* Z *is adapted and* H *is predictable which means that, for all* $n \geq 1$, H_n *is* $\mathcal{F}_{n-1}$*-measurable. The sequence* $H \bullet Z$ *given by*

$$(H \bullet Z)_n := \sum_{j=1}^n H_j(Z_j - Z_{j-1}), \ n \geq 0 \ ((H \bullet Z)_0 = 0)$$

is called the H*-transform of* Z *or the (discrete) stochastic integral of* H *with respect to* Z. *Now let* Z *be an* $(\mathcal{F}_n)$*-martingale and assume that* $H_j(Z_j - Z_{j-1}) \in L^1$, $j = 1, 2, \ldots$. *Show that* $H \bullet Z$ *is an* $(\mathcal{F}_n)$*-martingale.*
Hint: *Use the iteration property of conditional expectations (see Example 6.1.29).*

Exercise 20.6.19. *Let* $W = (W_t)$ *be a BM and let* $(\mathcal{F}_t)$ *be the filtration generated by* W. *Show that the following processes are* $(\mathcal{F}_t)$*-martingales:*

1. (W_t).

2. $(W_t^2 - t)$.

3. $(W_t^4 - 6tW_t^2 + 3t^2)$.

Hint: *Note that* $W_t - W_s$ *is independent of* $\mathcal{F}_s$ $(0 \leq s \leq t)$.

Exercise 20.6.20. *Let* (N_t) *be a Poisson process with parameter* $\lambda > 0$, *and put* $M_t = N_t - \lambda t$ $(t \geq 0)$. *Let* $(\mathcal{F}_t)$ *be the filtration generated by* (N_t).

1. *Show that* (M_t) *is an* $(\mathcal{F}_t)$*-martingale.*
   Hint: $N_t - N_s$ *is independent of* $\mathcal{F}_s$ $(0 \leq s < t)$.

2. *Show that* $(M_t^2 - \lambda t)$ *is an* $(\mathcal{F}_t)$*-martingale.*
   Hint: *Write* $M_t^2 - M_s^2 = (M_t - M_s)^2 + 2M_s(M_t - M_s)$ $(0 \leq s < t)$.

Exercise 20.6.21. *Let* (N_t) *be a Poisson process with parameter* $\lambda > 0$, *and let* $c > 0$ *be any constant.*

1. *Determine the constant* $\mu(c)$ *such that the process* $(\exp(cN_t + \mu(c)t))$ $(t \geq 0)$ *is a martingale with respect to the filtration* $(\mathcal{F}_t)$ *generated by* (N_t).
   Hint: *Use Theorem 6.1.30 and Exercise 20.6.11.*

2. *Verify directly that the process obtained in (1) is an* $(\mathcal{F}_t)$*-martingale.*
   Hint: *Use that* $\mathbb{E}[\exp(c(N_t - N_s))|\mathcal{F}_s] = \mathbb{E}[\exp(c(N_t - N_s))]$ *a.s.* $(0 \leq s < t)$ *since* $N_t - N_s$ *is independent of* $\mathcal{F}_s$.

Exercise 20.6.22. *Let ξ have a binomial distribution with parameters $n \geq 1$ and $0 \leq p \leq 1$, i.e.,*

$$P(\xi = k) = \binom{n}{k} p^k (1-p)^{n-k}, \ k = 0, 1, \ldots, n.$$

1. *Use Azuma's inequality (Theorem 6.3.1) to prove the following inequality which is due to H. Chernoff (Ann. Math. Statist.* **23** *(1952), 493–507):*

$$P(|\xi - np| \geq t) \leq 2 \exp(-2t^2/n), \ t \geq 0, \ n \geq 1. \tag{$*$}$$

   Hint: *ξ has the same distribution as a sum of suitable 0–1 random variables $\xi_1, \ldots, \xi_n$.*

2. *Verify $(*)$ directly for $n = 1$.*

Exercise 20.6.23. *Prove (6.147).*
Hint: *First note that $|g(z)| =: \exp(I(z))$, where*

$$I(z) := \int_0^z \frac{\cos x - 1}{x} \left(\log\left(\frac{z}{x}\right) \right)^r dx, \ z \geq 0, \ r > 0.$$

Then (6.147) says that

$$I(z) \leq \frac{1}{2(r+1)} \left(1 - (\log(2z/(3\pi)))^{r+1} \right), \ z \geq 4\pi, \ r > 0. \tag{$*$}$$

In order to prove $(*)$ note that the cosine is ≤ 0 on the intervals $J(k) := [(2k-1)\pi - \pi/2, (2k-1)\pi + \pi/2]$, and that

$$J(k) \subset [0, z] \quad iff \quad 1 \leq k \leq k(z) := \lfloor z/(2\pi) + 1/4 \rfloor. \tag{$**$}$$

Hence

$$I(z) \leq -\sum_{k=1}^{k(z)-1} \int_{J(k)} \frac{1}{x} \left(\log\left(\frac{z}{x}\right) \right)^r dx.$$

Using $(**)$ and comparing with a certain Riemann integral finally yields $(*)$.

Exercise 20.6.24. *A process $Z_t = Z_0 \exp(X_t)$, $t \geq 0$ ($Z_0 > 0$) is observed at time points $t = 0, 1, 2, \ldots, T$, where (X_t) is a Lévy process of jump-diffusion type as in Example 6.5.2. Let $H_0(2)$ denote the null hypothesis which says that there exist $\alpha \in \mathbb{R}, c \geq 2, \lambda \geq 0$ and a distribution Q_1 on $\mathbb{R}$ satisfying $Q_1(\{0\}) = 0$ such that (X_t) is associated with α, c, λ, and Q_1. (Note that $H_0(2)$ has a meaning different from that at the beginning of Section 6.5!) Let $H_0(2)$ be rejected if $|\tilde{L}_T/T - p_{10}(1)| \geq 0.1$ (see (6.100) and (6.150)). Let the level of significance be 0.1. (Note that the rejection of $H_0(2)$ entails the rejection of the null hypothesis that (Z_t) is a Black–Scholes process having volatility ≥ 2; see (6.27).) How large has T to be? (Answer: $T \geq 1715$.)*

Exercise 20.6.25. *A process $Z_t = Z_0 \exp(X_t), t \geq 0$ ($Z_0 > 0$) is observed at the time points $t = 0, 1, 2, \ldots, T$, where $(X_t) = \alpha t + T_t, t \geq 0$. Here, $\alpha \in \mathbb{R}$; (T_t) is a compound Poisson (or CP-)process associated with $\lambda > 0$ and $Q_1 = N(\mu, \sigma^2)$ (see*

Example 6.1.21). Suppose that the null hypothesis $H_0(\lambda^*, \sigma^*)$ $(\lambda^* > 0, \sigma^* > 0)$ is to be tested, which says that there exist $\alpha \in \mathbb{R}, \mu \in \mathbb{R}, \lambda \geq \lambda^*$, and $\sigma \geq \sigma^*$ such that $X_t = \alpha t + T_t$ $(t \geq 0)$, and (T_t) is a CP-process associated with λ and Q_1. Verify that the test outlined in Exercise 20.6.24, which rejects $H_0(\lambda^*, \sigma^*)$ if $|\tilde{L}_T/T - p_{10}(1)| \geq 0.1$, is not applicable no matter how the level of significance $0 < p_0 < 1$ is chosen.
Hint: *Show that there does not exist any (finite) constant Σ^* satisfying (6.153) (g being the characteristic function of X_1, (X_t) being an arbitrary Lévy process satisfying $H_0(\lambda^*, \sigma^*)$). Use Exercise 20.6.14(2).*

Exercise 20.6.26. *Suppose we observe a process $Z_t = Z_0 \exp(\mu t + c X_t), t \geq 0$ $(Z_0 > 0)$ at time points $t = 0, 1, \ldots, T$. Let (X_t) be a gamma process with parameters α and Δ, and consider (as in Example 6.5.5) the null hypothesis $H_0(c^*, \alpha^*, \Delta^*)$ where $B = 10, c_* = \alpha^* = 1, \Delta^* = 2, p_0 = v = 0.1, m = 1, d_1 = 1$, and $\lambda(10) = (2\pi/\log 10)^2$ (recall that $\log$ is the natural logarithm).*

1. *Show that in this special case we can choose $\Sigma^* = (\log 10)^2/24$.*

2. *How large has the time horizon T to be? (Answer: $T \geq 2129$ (instead of $T \geq 2582$ as in Example 6.5.5!).)*

Exercise 20.6.27. *Prove the following elementary result (Lemma 6.6.7): Let $a_1, a_2, \cdots$ be real numbers such that $0 \leq a_n < 1$ $(n \geq 1)$ and $\sum_{n=1}^{\infty} a_n < \infty$. Then*

$$\sum_{n=1}^{\infty} a_n^t \to 0 \ (t \to \infty).$$

Exercise 20.6.28. *Prove the claim in Example 6.1.28.*

Exercise 20.6.29. *Prove the iteration property of conditional expectations (see Example 6.1.29).*

Exercise 20.6.30. *Prove Lemma 6.2.1.*

20.7 BENFORD'S LAW AS A BRIDGE BETWEEN STATISTICS AND AC-COUNTING

An auditor decides to run a Benford's Law test on a data set that consists of 1000 legitimate expense records from a business, plus a number of fraudulent transactions that an employee is making to a front for a business set up in a relative's name. Because the employees of the business have to obtain special approval for expenditures over $10,000, the fraudulent transactions are all for amounts between $9000 and $9999. For the 1000 legitimate expenditures, we have this data:

First Digit	Observed
1	314
2	178
3	111
4	92
5	88
6	59
7	56
8	56
9	46

Exercise 20.7.1. *Using the Benford Law test at*

> *http://web.williams.edu/Mathematics/sjmiller/public_html/benford*
> */chapter01/MillerNigrini_ExcelBenfordTester_Ver401.xlsx*

(or any other suitable software), verify that the data conforms reasonably well to Benford's Law.

Exercise 20.7.2. *Use trial and error (or some more clever approach) to determine how many fraudulent transactions with first digit 9 would need to be added to the 1000 legitimate observations above in order for the hypothesis that the data follows Benford's Law to be rejected at a five percent significance level. Does this seem plausible?*

Exercise 20.7.3. *What is the role of sample size in the sensitivity of Benford's Law? Suppose there are 10,000 legitimate observations instead of 1000, but the ratios for legitimate observations remains the same, i.e., the number of observations for each digit is multiplied by 10. Try the problem again. What changes?*

Exercise 20.7.4. *In which of the following situations is an auditor most likely to use Benford's Law?*

- *An analysis of a fast food franchise's inventory of hamburgers.*

- *An audit of a Fortune 500 company's monthly total revenue over the fiscal year.*

- *An analysis of a multibillion dollar technology company's significant assets.*

Exercise 20.7.5. *Give an additional example of a way that including Benford's Law in an introductory-level statistics class will meet the four goals of the GAISE report of 2005.*

Exercise 20.7.6. *Determine whether the following situations are Type I errors, Type II errors, or neither.*

- *An auditor uses Benford's Law to analyze the values of canceled checks by a business in the past fiscal year. The auditor finds that there are significant spikes in the data set, with 23 and 37 appearing as the first two digits more often than expected. After further investigation, it was found that there were valid non-fraudulent explanations for the variations in the first digits.*

- *An auditor finds that a company's reported revenue does not follow Benford's Law. Further investigation is taken, and it is found that a manager has been rounding up her weekly sales to the nearest thousand to earn an incentive based on a weekly sales benchmark. The manager claims that the inflated sales were an accounting error.*

- *An owner of a business has falsely claimed to give his employees bonuses on each paycheck based on their monthly sales in order to lower his income taxes. An auditor examines the data, but is unable to confidently claim that the data does not follow Benford's Law. Rather than waste funds on a costly investigation, the auditor chooses not to investigate the owner.*

Exercise 20.7.7. *What are the negative effects of a Type I error in an audit? A Type II error? In what situations might one be more dangerous than the other?*

Exercise 20.7.8. *What are some of the reasons listed in the chapter that might explain why a data set should not be expected to follow Benford's Law?*

Exercise 20.7.9. *Give an example of a reason other than fraud that explains why a data set that is expected to conform to Benford's Law does not.*

20.8 DETECTING FRAUD AND ERRORS USING BENFORD'S LAW

Exercise 20.8.1. *Do the following data sets meet the requirements described by Nigrini in order to be expected to follow Benford's Law? Explain why or why not.*

- *The 4-digit PIN numbers chosen by clients of a local bank.*

- *The annual salaries of graduates from a public university.*

- *Numeric student ID numbers assigned by a school.*

- *The distances in miles between Washington, DC and the 500 most populated cities in the United States (excluding Washington, DC).*

- *Results to a survey of 1000 students asked to provide a number in between 1 and 1,000,000.*

- *The number of tickets bought for all events held in a particular stadium over the past five years.*

Exercise 20.8.2. *Take a company which has been at the heart of a scandal (for example, Enron) and investigate some of its publicly available data.*

Exercise 20.8.3. *An audit of a small company reveals a large number of transactions starting with a 5. Come up with some explanations other than fraud. Hint: There are two cases: it is the same amount to the same source each time, and it isn't.*

20.9 CAN VOTE COUNTS' DIGITS AND BENFORD'S LAW DIAGNOSE ELECTIONS?

Exercise 20.9.1. *If X satisfies Benford's Law, then the mean of its second digit is 4.187. What is the mean of the kth digit?*

Exercise 20.9.2. *If X satisfies Benford's Law, multiply by an appropriate power of 10 so that it has k integer digits. What is the probability the last digit is d? What is the probability the last two digits are equal? What is the probability the last two digits differ by 1?*

Exercise 20.9.3. *Find some recent voting data (say city or precinct totals) and investigate the distribution of the first and second digits.*

20.10 COMPLEMENTING BENFORD'S LAW FOR SMALL N: A LOCAL BOOTSTRAP

Exercise 20.10.1. *Do you agree with the assessment that Nigrini's conditions for applying Benford's Law are mostly satisfied? Why or why not?*

Exercise 20.10.2. *Why does having a large $\sigma(\log_{10} x_i)$ and a large $\sigma(\log_{10} w_{i,j})$ ensure that the $v_{i,j}$ first-digit distribution approaches Benford's Law?*

Exercise 20.10.3. *What does it mean for bootstrap methods to be considered "conservative"? Identify some of the ways in which bootstrap methods are conservative.*

Exercise 20.10.4. *There are many conservative statistics. Look up the Bonferroni adjustment for multiple comparisons, as well as alternatives to that.*

Exercise 20.10.5. *How would a local bootstrap realization change if the value of Δ were changed?*

Exercise 20.10.6. *Confirm that if $c_{bK7} > 99.924\%$, then $c_{eK7} > 99.99960\%$.*

20.11 MEASURING THE QUALITY OF EUROPEAN STATISTICS

Exercise 20.11.1. *In which of the following two scenarios would χ^2 be larger?*

- *The first-digit frequencies are mostly identical to the expected Benford distribution, but the digit 1 appears 31.1% of the time and the digit 2 appears 16.6% of the time (compared with the expected values of approximately 30.1% and 17.6%, respectively).*

- *The first-digit frequencies are mostly identical to the expected Benford distribution, but the digit 8 appears 6.12% of the time and the digit 2 appears 3.58% of the time (compared with the expected values of approximately 5.12% and 4.58%, respectively).*

Exercise 20.11.2. *What is μ_b, the value of the mean of the Benford distribution of first digits base b?*

Exercise 20.11.3. *What is the value of a^* if $\mu_e = 3.5$?*

Exercise 20.11.4. *Using Figure 11.1, confirm the values of χ^2, χ^2/n, and d^* for the distribution of first digits for Greek social statistics in the year 2004.*

Exercise 20.11.5. *Using Figure 11.1 and the formula for distance measure a^* used by Judge and Schechter, calculate the value of the mean of the data set (μ_e) in the year 2004. Confirm this value by using the formula $\mu_e = \frac{\sum_{i=1}^{9} n\mathrm{Prob}(D_1=i)}{n}$.*

The final problem uses data on two fictitious countries, which is available online

```
http://web.williams.edu/Mathematics/sjmiller/public_html
              /benford/chapter11/
```

(some of the additional readings on that web page may be useful as well).

Exercise 20.11.6. *Calculate the values χ^2, χ^2/n, d^*, and a^* and compare the results for both countries. Which one of these two countries should be examined closer? Are the outcomes consistent?*

20.12 BENFORD'S LAW AND FRAUD IN ECONOMIC RESEARCH

Exercise 20.12.1. *Use (12.1) to find $f(6)$ and $F(6)$ for Benford's Law.*

Exercise 20.12.2. *If X is a Benford variable defined on $[1, 10)$, then what is the probability that the second digit is 5 given that the first digit is also 5?*

Exercise 20.12.3. *Use (12.4) to confirm that when using Benford's Law for Rounded Figures, $\mathrm{Prob}(D_1 = 8) = 0.054$.*

Exercise 20.12.4. *If X is a Benford variable defined on $[1, 10)$, given that the first digit is 8, what is the probability that the second digit is 0 when rounding to two significant digits? What is the probability that the second digit is 2?*

Exercise 20.12.5. *Using Benford's Law for Rounded Figures as the frequencies of first digits for a data set of 300 observed values, calculate Q_1, Q_2, M_1, and M_2 using (12.6) and (12.7).*

Exercise 20.12.6. *Should the Q_1 test or the M_1 test be used for attempting to detect variations in Benford's Law?*

- *What if the data set in question has a mean of 3.44?*

- *Which test should be used for detecting variations in the Generalized Benford Law?*

Exercise 20.12.7. *The Federal Tax Office (FTO) knows that $\Omega = 10\%$ of tax declarations of small and medium enterprises are falsified. The FTO checks the first digits using Benford's Law. Random samples of tax declarations are drawn and the null hypothesis (H_o) "Conformity to Benford's Law" is tested at the $\alpha = 5\%$ level of significance.*

- *Using (12.9), what rejection rate of $H_o(\theta)$ would you expect if the probability of a Type II error β lies in the interval [0.05, 0.75]?*

- *The FTO obtained the rejection rate $\theta = 0.12$. Use (12.9) to calculate the probability β of a Type II error.*

- *The FTO arranges for an audit at the taxable enterprise if the Benford test rejects H_o for a certain tax declaration at the $\alpha = 5\%$ level. What is the probability that such an audit will be provoked erroneously? And what is the probability to forbear an audit erroneously?*

Exercise 20.12.8. *A sample of scientific articles is taken, and 17% are found to have regression coefficients with a doubtful distribution of first digits. Use (12.10) to calculate $\hat{\Omega}$.*

20.13 TESTING FOR STRATEGIC MANIPULATION OF ECONOMIC AND FINANCIAL DATA

Exercise 20.13.1. *What are some of the potential reasons given in Section 13.1 for why data sets that are expected to follow Benford's Law fail to do so?*

Exercise 20.13.2. *Did Benford's Law prove financial misreporting during the financial crisis? Justify your assertion.*

Exercise 20.13.3. *What are some of the potential motives that banks have for manipulating VAR data?*

20.14 PSYCHOLOGY AND BENFORD'S LAW

Exercise 20.14.1. *Using (11.1) in Section 11.3, find χ^2 for the elaborated and unelaborated data from Scott, Barnard, and May's study found in Table 14.1.*

Exercise 20.14.2. *What distribution of leading digits would you expect if people were asked to randomly give an integer from 1 to N? How does your answer depend on N? Try an experiment with some of your friends and family.*

20.15 MANAGING RISK IN NUMBERS GAMES: BENFORD'S LAW AND THE SMALL-NUMBER PHENOMENON

Exercise 20.15.1. *What are the risks associated with a high liability limit in a fixed-odds lottery game? What if the limit is too small?*

Exercise 20.15.2. *From the data obtained in Table 15.1, determine the probability that a given number on a ticket for the UK powerball game is a single digit.*

Exercise 20.15.3. *Figure 15.1 shows the proportion of tickets in a Pennsylvania Pick-3 game with a given first digit. Explain why there are several outliers larger than the mean proportion and no outliers smaller than the mean proportion.*

Exercise 20.15.4. *What is the probability that a Type I player chooses the number 345 in a Pick-3 game?*

Exercise 20.15.5. *Let Alice be a Type II player in a Pick-3 game who bets on a number with three significant digits 80% of the time, a number with two significant digits 15% of the time, and a number with one significant digit 5% of the time. What is the probability that Alice bets on the number 345? The number 45? The number 5?*

Exercise 20.15.6. *In the Pennsylvania Pick-3 game, the least square model indicates that 60.42% of the players are Type I players and 39.58% of the players are Type II players. Based on this model, use (15.4) to calculate the expected proportion of the betting volume on a three-digit number with first significant digit 4.*

Exercise 20.15.7. *Let Bob be a Type II player in a Pick-3 game who bets on a number with three significant digits 80% of the time, but also has a tendency to exhibit switching behavior; that is, he will switch later digits with probability 0.9105, and switch the digit to 0 with probability 0.1054. What is the probability that Bob bets on the number 345?*

Exercise 20.15.8. *Use (15.5) to calculate the probability that Bob chooses a three-digit number in between 520 and 529 inclusive.*

Exercise 20.15.9. *Calculate the variance using the equation in Section 15.4.1 under the scenario that all players randomly select a three-digit number.*

20.16 BENFORD'S LAW IN THE NATURAL SCIENCES

Exercise 20.16.1. *Demonstrate that (16.3) holds for $\alpha = 2$.*

Exercise 20.16.2. *Rewrite the log-normal distribution density function (16.5) as the log-normal density function (16.6).*

Exercise 20.16.3. *Show that as σ grows larger, the log-normal density function approaches the power law $p(x) = C_\sigma x^{-1}$, where C_σ is a constant depending on σ.*

Exercise 20.16.4. *Provide examples not mentioned in the chapter of scientific data sets that are not effectively scale invariant.*

Exercise 20.16.5. *Explain the intuition behind why the following distributions are approximately Benford:*

- *The Boltzman–Gibbs distribution (16.8).*

- *The Fermi–Dirac distribution (16.9).*

- *The Bose–Einstein distribution (16.10).*

Exercise 20.16.6. *Obtain a physics textbook (or a CRC handbook, or...) and find a list of physical constants. Perform a chi-square test to determine whether the list of constants follows Benford's Law as expected.*

Exercise 20.16.7. *Sandon found agreement between Benford's Law and population and surface area data for the countries of the world. Find a source that provides the population density of each country. Then determine if population density follows Benford's Law. This can be done using a chi-square test. In general, should the ratio of two Benford random variables be Benford?*

20.17 GENERALIZING BENFORD'S LAW: A REEXAMINATION OF FALSIFIED CLINICAL DATA

Exercise 20.17.1. *Use (17.1) to calculate the average frequency of first digits in Stigler's distribution of first significant digits. Check to see that the distribution matches the values displayed in Table 17.1.*

Exercise 20.17.2. *Verify (17.3), (17.5), and (17.6). Then verify that the sum of the three subsets matches (17.7).*

Exercise 20.17.3. *Calculate the mean of the Stigler FSD distribution and Benford FSD distribution to confirm that they are equivalent to 3.55 and 3.44, respectively.*

Exercise 20.17.4. *For the Estimated Maximum Entropy FSD distribution for data with an FSD mean of 3.44 shown in Table 17.3, find $H(p)$ and ensure that the criteria from (17.13) and (17.14) are reached.*

- *If the Estimated Maximum Entropy FSD distribution is accurate, then the listed probabilities will maximize $H(p)$. First, determine whether replacing $\hat{p}_1$ with 0.231 and $\hat{p}_2$ with 0.2 still allows (17.13) and (17.14) to hold. Now find $H(p)$. Is $H(p)$ larger or smaller than before?*

Exercise 20.17.5. *If the FSD mean is 5, what will be the estimated maximum entropy FSD distribution? What is $\mathrm{Var}(d)$ according to (17.18)?*

Exercise 20.17.6. *Examining the Poehlman data in Table 17.4, calculate the difference for each digit FSD distribution given by Benford's Law.*

Exercise 20.17.7. *The estimated empirical likelihood distributions given an FSD mean will maximize $\sum_{i=1}^{9} p_i$. To test this, ensure that the product of the p_i's from Table 17.5 are greater than the empirical data found in Table 17.4.*

Exercise 20.17.8. *A researcher is trying to decide whether a data set follows Benford's Law or Stigler's Law. What values of the mean of the leading digit suggest Benford over Stigler? What values suggest Stigler over Benford?*

20.18 PARTIAL VOLUME MODELING OF MEDICAL IMAGING SYSTEMS USING THE BENFORD DISTRIBUTION

Exercise 20.18.1. *What is the PV effect? What implications does the PV effect have for medical imaging?*

Exercise 20.18.2. *Prove Corollary 18.3.4.*

Exercise 20.18.3. *What advantages are there to describing the PV effect using matrices as in (18.11)?*

Exercise 20.18.4. *What are the differences between a Rician noise model described by (18.12) and a Gaussian noise model described in (18.13)?*

Exercise 20.18.5. *Use (18.22) to calculate $p(\alpha)$ for $\alpha = 0.50$, where α has two digits of precision.*

Exercise 20.18.6. *How is the contrast to noise ratio (CNR) affected if both the distance between the signal levels of two components and the standard deviation of each class is doubled?*

20.19 APPLICATION OF BENFORD'S LAW TO IMAGES

Exercise 20.19.1. *In (19.9) one of the factors is $\Gamma\left(\frac{-j2\pi n + \log 10}{c \log 10}\right)$, where $j = \sqrt{-1}$. Estimate how rapidly this tends to zero as $|n| \to \infty$ as a function of c (if you wish, choose some values of c to get a feel of the behavior).*

Exercise 20.19.2. *In (19.19) we find that $|a_n(c, \sigma)| \le |a_n(c^+)|$ for all n; investigate how close these can be for various choices of c and σ.*

Exercise 20.19.3. *In Example 19.5.3 we found four zero-mean Gaussians with shaping parameter $c = 1$ with four different standard deviations and $a_1 = 0$. Can you find six zero-mean Gaussians with shaping parameter $c = 1$ and six different standard deviations with $a_1 = 0$? What about eight? More generally, can you find $2m$ such Gaussians for m a positive integer?*

Bibliography

[Abd]   H. Abdi, *Bonferroni and Sidak Corrections for Multiple Comparisons*, (2007), in (N. J. Salkind ed.), Encyclopedia of Measurement and Statistics. Thousand Oaks, USA: Sage.

[AbdTh]   M. Abdolmohammadi and J. Thibodeau, *Auditing*, in Encyclopedia of Business and Finance, 2nd edition, Macmillan, New York, 2007.

[AbrSteg] M. Abramowitz and I. A. Stegun, *Handbook of Mathematical Functions with Formulas, Graphs, and Mathematical Tables*, Dover, New York, 1964.

[AceSbe] E. D. Acebo and M. Sbert, *Benford's Law for Natural and Synthetic Images*, Workshop on Computational Aesthetics in Graphics, Visualization and Imaging, (2007), 169–176.

[Adh]   A. K. Adhikari, *Some Results on the Distribution of the Most Significant Digit*, Sankhyā: Indian Journal of Statistics, Series B **31** (1969), 413–420.

[AdhSa]   A. K. Adhikari and B. P. Sarkar, *Distribution of Most Significant Digit in Certain Functions Whose Arguments Are Random Variables*, Sankhyā: Indian Journal of Statistics, Series B **30** (1968), 47–58.

[Al]   P. C. Allaart, *An Invariant-Sum Characterization of Benford's Law*, **34** (1997), 288–291.

[AnRoSt] T. Anderson, L. Rolen and R. Stoehr, *Benford's Law for Coefficients of Modular Forms and Partition Functions*, Proc. Am. Math. Soc. **139** (2011), no. 5, 1533–1541.

[An]   Anonymous, *Weibull Survival Probability Distribution*, last modified May 11, 2009.
          http://www.wepapers.com/Papers/30999/Weibull_
          Survival_Probability_Distribution.

[ArJa]   L. Arshadi and A. H. Jahangir, *Benford's Law Behavior of Internet Traffic*, Journal of Network and Computer Applications **40** (2014), 194–205.

[AuCa]   M. Auffhammer and R. Carson, *Forecasting the Path of China's CO2 Emissions Using Province Level Information*, Journal of Environmental Economics and Management **47** (2008), no. 1, 47–62.

[AuHi]   K. Auspurg, T. Hinz, *What Fuels Publication Bias?*, Jahrbücher für Nationalökonomie und Statistik (Journal of Economics and Statistics) **231**, no. 5 and 6 (2011), 636–660.

[BaBeMi] S. Baek, S. Bernhardsson and P. Minnhagen, *Zipf's Law Unzipped*, New Journal of Physics **13** (2011), 043004, doi:10.1088/1367-2630/13/4/043004

[Bag]   K. A. Baggerly, *Empirical Likelihood as a Goodness-of-Fit Measure*, Biometrika, 1998.

[BaBoDS] J. Baik, A. Borodin, P. Deift and T. Suidan, *A Model for the Bus System in Cuernevaca (Mexico)*, Math. Phys. 2005, 1–9. Online at http://arxiv.org/abs/math/0510414.

[BaPiYo] P. Baine, J. Pitman and M. Yor, *Probability Laws Related to the Jacobi Theta and Riemann Zeta Functions, and Brownian Excursions*, Bulletin of the AMS **38**, No. 4, June 2001, 435–465.

[Bak]   A. Baker, *The Theory of Linear Forms in Logarithms*, Transcendence Theory: Advances and Applications (A. Baker and D. W. Masser, eds), Academic Press, 1977.

[BarFe]   J. A. Bargh and M. J. Ferguson, *Beyond Behaviorism: On the Automaticity of Higher Mental Processes*, Psychological Bulletin **126** (2000), 925–945.

[BaHeSl] T. Barrale, R. Hendel and M. Sluys, *Sequences of Initial Digits of Fibonacci Numbers*, Aportaciones Matematicas, Investigacion **20** (2011), 25–42.

[BaLi]   P. Baxandall and H. Liebeck, *Vector Calculus*, Clarendon Press, Oxford, 1986.

[Be]   R. Beals, *Notes on Fourier Series*, Lecture Notes, Yale University, 1994.

[BebSca] B. Beber and A. Scacco, *The Devil Is in the Digits*, (2009), Washington Post, 20 June 2009.

[BeEsJ]   G. Beeli, M. Esslen and L. Jäncke, *Frequency Correlates in Grapheme-Color Synaesthesia*, Psychological Science **18** (2007), no. 9, 788–792.

[Ben]   F. Benford, *The Law of Anomalous Numbers*, Proceedings of the American Philosophical Society **78** (1938), 551–572.

[Ber1]   A. Berger, *Chaos and Chance*, deGruyter, Berlin, 2001.

[Ber2]   A. Berger, *Multi-dimensional Dynamical Systems and Benford's Law*, Discrete Contin. Dyn. Syst. **13** (2005), 219–237.

[Ber3]   A. Berger, *Benford's Law in Power-like Dynamical Systems*, Stoch. Dyn. **5** (2005), 587–607.

[Ber4]   A. Berger, *Some Dynamical Properties of Benford Sequences*, J. Difference Equ. Appl. **17** (2011), no. 2, 137–159.

[BerBH]   A. Berger, L. A. Bunimovich and T. Hill, *One-Dimensional Dynamical Systems and Benford's Law*, Trans. Amer. Math. Soc. **357** (2005), no. 1, 197–219.

[BerH1]   A. Berger and T. P. Hill, *Newton's Method Obeys Benford's Law*, American Mathematical Monthly **114** (2007), 588–601.

[BerH2]   A. Berger and T. P. Hill, *Benford Online Bibliography*, http://www.benfordonline.net.

[BerH3]   A. Berger and T. P. Hill, *Benford's Law Strikes Back: No Simple Explanation in Sight for Mathematical Gem*, Mathematical Intelligencer **33** (2011), 85–91.

[BerH4]   A. Berger and T. P. Hill, *A Basic Theory of Benford's Law*, Probab. Surv. **8** (2011), 1–126.

[BerH5]   A. Berger and T. P. Hill, *An Introduction to Benford's Law*, Princeton University Press, Princeton, 2015.

[BerHKR]   A. Berger, T. P. Hill, B. Kaynar and A. Ridder, *Finite-State Markov Chains Obey Benford's Law*, to appear in SIAM J. Matrix Analysis.

[Berg]   J. O. Berger, *Statistical Decision Theory and Bayesian Analysis*, 2nd edition, Springer Series in Statistics, 1985.

[BerRin]   D. Berman and T. Rintoul, *Preliminary Analysis of the Voting Figures in Iran's 2009 Presidential Election* (Ali Ansari ed.), London: Chatham House, St Andrews, 2009.

[Ber]   J. Bertoin, *Lévy Processes*, Cambridge University Press, Cambridge, 1996.

[Bh]   R. N. Bhattacharya, *Speed of Convergence of the n-Fold Convolution of a Probability Measure on a Compact Group*, Z. Wahrscheinlichkeitstheorie verw. Geb. **25** (1972), 1–10.

[Bi1]   P. Billingsley, *Convergence of Probability Measures*, John Wiley & Sons, New York, 1968.

[Bi2]   P. Billingsley, *Prime Numbers and Brownian Motion*, Amer. Math. Monthly **80** (1973), 1099–1115.

[Bi3]   P. Billingsley, *Probability and Measure*, 3rd edition, John Wiley & Sons, New York, 1995.

[BlS]   F. Black and M. Scholes, *The Prices of Options and Corporate Liabilities*, J. Political Economy **81** (1973), 637–659.

[BlGs] A. Blais and T. Gschwend, *Strategic Defection Across Elections, Parties and Voters*, in Citizens, Context, and Choice: How Context Shapes Citizens' Electoral Choices (R. Dalton and C. Anderson, eds), Oxford University Press, New York, 2011, 176–195.

[BlNa] A. Blais and R. Nadeau, *Measuring Strategic Voting: A Two-Step Procedure*, Electoral Studies **15** (1996), no. 1, 39–52.

[BlNaGN] A. Blais, R. Nadeau, E. Gidengil and N. Nevitte, *Measuring Strategic Voting in Multiparty Plurality Elections*, Electoral Studies **20** (2001), no. 2, 343–352.

[BohJe] H. Bohr and B. Jessen, *On the Distribution of the Values of the Riemann Zeta-Function*, Amer. J. Math. **58** (1936), 35–44.

[BolHa] R. Bolton and D. Hand, *Statistical Fraud Detection: A Review*, Statistical Science **17** (2002), no. 3, 235–255.

[BomPo] E. Bombieri and A. van der Poorten, *Continued Fractions of Algebraic Numbers*, in Computational Algebra and Number Theory (Sydney, 1992), Mathematical Applications, **325**, Kluwer Academic, Dordrecht, 1995, 137–152.

[Bon..] S. Bonhoeffer, A. Herz, M. Boerlijst, S. Nee, M. Nowak and R. May, *No Signs of Hidden Language in Noncoding DNA*, Physical Review Letters **76** (1996), no. 11, 1977.

[BorB] J. Borwein and P. Borwein, *Pi and the AGM: A Study in Analytic Number Theory and Computational Complexity*, John Wiley and Sons, New York, 1987.

[BowAz] A. Bowman and A. Azzalini, *Applied Smoothing Techniques for Data Analysis: The Kernel Approach with S-Plus Illustrations*, Clarendon Press, Oxford, 1997.

[BoyDi] W. Boyce and R. DiPrima, *Elementary Differential Equations and Boundary Value Problems*, 7th edition, John Wiley & Sons, New York, 2000.

[Boy]   J. Boyle, *An Application of Fourier Series to the Most Significant Digit Problem*, Amer. Math. Monthly **101** (1994), 879–886.

[Br]   L. Breiman, *Probability*, Addison–Wesley, Reading, MA, 1968.

[BrDu] J. Brown and R. Duncan, *Modulo One Uniform Distribution of the Sequence of Logarithms of Certain Recursive Sequences*, Fibonacci Quarterly **8** (1970), 482–486.

[Bro1]   R. Brown, *Benford's Law and the Screening of Analytical Data: The Case of Pollutant Concentrations in Ambient Air*, Analyst **130** (2005), 1280–1285.

[Bro2]   R. Brown, *The Use of Zipf's Law in the Screening of Analytical Data: A Step Beyond Benford*, Analyst **132** (2007), 344–349.

[Buc]   B. Buck, A. Merchant and S. Perez, *An Illustration of Benford's First Digit Law Using Alpha Decay Half-Lives*, European Journal of Physics **14** (1993), 59–63.

[Bun1]   Der Bundeswahlleiter, *Ergebnisse der Wahlbezirksstatistik: Die Wahlleiter des Bundes und der Länder, Auszugsweise Vervielfältigung und Verbreitung mit Quellenangaben gestattet*, Wahl zum 16, Deutschen Bundestag am 18, September 2005, Wiesbaden: im Auftrag der Herausgebergemeinschaft.

[Bun2]   Der Bundeswahlleiter, *Ergebnisse der Wahlbezirksstatistik: Die Wahlleiter des Bundes und der Länder, Auszugsweise Vervielfältigung und Verbreitung mit Quellenangaben gestattet*, Wahl zum 17, Deutschen Bundestag am 27, September 2009, Wiesbaden: im Auftrag der Herausgebergemeinschaft.

[Bun3]   Der Bundeswahlleiter, *Federal Elections Act*, accessed March 21, 2012. URL
           `http://www.bundeswahlleiter.de/en/bundestagswah`
           `len/downloads/rechtsgrundlage%n/bundeswahlgeset`
           `z_engl.pdf`.

[Bun4]   Der Bundeswahlleiter, *Ergebnisse der Wahlbezirksstatistik: Die Wahlleiter des Bundes und der Länder, Auszugsweise Vervielfältigung und Verbreitung mit Quellenangaben gestattet*, Wahl zum 15, Deutschen Bundestag am 22, September 2002, Wiesbaden: im Auftrag der Herausgebergemeinschaft.

[Burke]   J. Burke and E. Kincanon, *Benford's Law and Physical Constants: The Distribution of Initial Digits*, American Journal of Physics **59** (1991), no. 10, 952.

[BurRA]   L. E. Burman, W. R. Reed, J. Alm, *A Call for Replication Studies*, Public Finance Review **38**, no. 6 (2010), 787–793.

[Burns]   B. D. Burns, *Sensitivity to Statistical Regularities: People (Largely) Follow Benford's Law*, in (N. A. Taatgen & H. van Rijn, eds.), Proceedings of the 31st Annual Conference of the Cognitive Science Society (2000), 2872–2877. Austin, TX: Cognitive Science Society.

[BurnsCo] B. D. Burns and B. Corpus, *Randomness and Inductions from Streaks: "Gambler's Fallacy" versus "Hot Hand."*, Psychonomic Bulletin & Review **11** (2004), 179–184.

[CanSe] F. Cantu and S. Saiegh, *Fraudulent Democracy? An Analysis of Argentina's Infamous Decade Using Supervised Machine Learning*, Political Analysis **19** (2011), no. 4, 409–433.

[Car]   L. Carleson, *On the Convergence and Growth of Partial Sums of Fourier Series*, Acta Math. **116** (1966), 135–157.

[Carr]   K. J. Carroll, *On the Use and Utility of the Weibull Model in the Analysis of Survival Data*, Controlled Clinical Trials **24** (2003), no. 6, 682–701.

[Cars]   C. A. Carslaw, *Anomalies in Income Numbers: Evidence of Goal-Oriented Behavior*, Accounting Review **LXIII** (1988), no. 2, 321–327.

[Cart]   Carter Center, *Observing the Venezuela Presidential Recall Referendum: Comprehensive Report*, 2005.

[Cas]   J. W. S. Cassels, *An Introduction to Diophantine Approximation*, Cambridge University Press, London 1957.

[Cast]   J. Castaneda, *Perpetuating Power: How Mexican Presidents Were Chosen*, New Press, New York, 2000.

[CenMA] Center for Morphometric Analysis at Massachusetts General Hospital, "20 Normal MR Brain Data Sets and Their Manual Segmentations", `http://www.cma.mgh.harvard.edu/ibsr/`, Accessed 2006.

[Che]   H. Chernoff, *How to Beat the Massachusetts Number Game: An Application of Some Basic Ideas in Probability and Statistics*, Mathematical Intelligencer **3** (1999), 166–175.

[ChKo]   P. Chhibber and K. Kollman, *The Formation of National Party Systems: Federalism and Party Competition in Canada, Great Britain, India, and the United States*, Princeton University Press, Princeton, NJ, 2004.

[ChWe]   J. P. Chiverton and K. Wells, *Mixture Effects in FIR Low-Pass Filtered Signals*, IEEE Signal Processing Letters, 13(6), June 2006, 369–372.

[ChGa]   W. K. T. Cho and B. J. Gaines, *Breaking the (Benford) Law: Statistical Fraud Detection in Campaign Finance*, American Statistician **61** (2007), 218–223.

[ChKTWZ] M. C. Chou, Q. X. Kong, C. P. Teo, Z. Z. Wang and H. Zheng, *Benford's Law and Number Selection in Fixed-Odds Numbers Game*, Journal of Gambling Studies **25** (2009), no. 4, 503–521.

[ChT]   Y. S. Chow and H. Teicher, *Probability Theory. Independence, Interchangeability, Martingales*, 3rd edition, Springer, 1997.

[ClTh]   R. J. Cleary and J. C. Thibodeau, *Applying Digital Analysis Using Benford's Law to Detect Fraud: The Dangers of Type I Errors*, Auditing: A Journal of Practice and Theory **24** (2005), no. 1, 77–81.

[Conr]   J. B. Conrey, *The Riemann Hypothesis*, Notices of the AMS, March 2003, 341–353.

[ConT]   R. Cont and P. Tankov, *Financial Modelling with Jump Processes*, Chapman & Hall, London, 2004.

[ConG]   J. H. Conway and R. Guy, *The Book of Numbers*, Springer, Berlin, 1996.

[CorSo]   B. Corominas-Murtra and R. Solé, *Universality of Zipf's Law*, Physical Review E **82** (2010), 011102, doi:10.1103/PhysRevE.82.011102.

[CoEZ]   M. Corazza, A. Ellero and A. Zorzi, *Checking Financial Markets via Benford's Law: The S&P 500 Case*. In Mathematical and Statistical Methods for Actuarial Sciences and Finance (Corazza and Pizzi, eds.), Springer, 2010, 93–101.

[CoBr]   O. Corzoa and N. Brachob, *Application of Weibull Distribution Model to Describe the Vacuum Pulse Osmotic Dehydration of Sardine Sheets*, LWT - Food Science and Technology **41** (2008), no. 6, 1108–1115.

[CosLTF]   E. Costas, V. López-Rodas, F. J. Toro and A. Flores-Moya, *The Number of Cells in Colonies of the Cyanobacterium Microcystis aeruginosa Satisfies Benford's Law*, Aquatic Botany **89** (2008), DOI 10.1016/j.aquabot.2008.03.011.

[CotGr]   W. N. Cottingham and D. A. Greenwood, *An Introduction to Nuclear Physics*, Cambridge University Press, Cambridge, 2nd edition, 2001.

[Cox1]   G. Cox, *Strategic Voting Equilibria Under the Single Nontransferable Vote*, American Political Science Review **88** (1994), no. 3, 608–621.

[Cox2]   G. Cox, *Making Votes Count: Strategic Coordination in the World's Electoral Systems*, Cambridge University Press, New York, 1996.

[CozMil]   M. Cozzens and S. J. Miller, *The Mathematics of Encryption: An Elementary Introduction*, AMS Mathematical World series **29**, Providence, RI, 2013, 332 pages.

[CrRe]   N. A. C. Cressie and T. R. C. Read, *Multinomial Goodness-of-Fit Tests*, Journal of the Royal Statistics Society Series B, 1984.

[CuLeMi]   V. Cuff, A. Lewis and S. J. Miller, *The Weibull Distribution and Benford's Law*, preprint.

[CuKeCh]   X. Cui, M. Kerr and G. Churchill, *Transformations for cDNA Microarray Data*, Statistical Applications in Genetics and Molecular Biology **2** (2003), Article 4.

[DM]   D. Danilov and J. R. Magnus, *On the Harm That Ignoring Pretesting Can Cause*, Journal of Econometrics **122** (2004), 27–46.

[Da]   H. Davenport, *Multiplicative Number Theory*, 2nd edition, Graduate Texts in Mathematics **74**, Springer, New York, 1980, revised by H. Montgomery.

[DaNa]   H. A. David and H. N. Nagaraja, *Order Statistics*, 3rd edition, Wiley Interscience, Hoboken, NJ, 2003.

[DeDS]   M. J. K. De Ceuster, G. Dhaene and T. Schatteman, *On the Hypothesis of Psychological Barriers in Stock Markets and Benford's Law*, Journal of Empirical Finance **5** (1998), no. 3, 263–279.

[DeMyOr] J. Deckert, M. Myagkov and P. Ordeshook, *Benford's Law and the Detection of Election Fraud*, Political Analysis **19** (2011), no. 3, 245–268.

[Deh]   S. Dehaene, *The Number Sense: How the Mind Creates Mathematics* (1997), Oxford University Press, Oxford, UK.

[DeHa]   S. De Marchi and J. Hamilton, *Trust, but Verify: Assessing the Accuracy of Self-Reported Pollution Data*, Journal of Risk and Uncertainty **32** (2006), no. 1, 57–76.

[Dia]   P. Diaconis, *The Distribution of Leading Digits and Uniform Distribution mod 1*, Ann. Probab. **5** (1979), 72–81.

[DiaS-C] P. Diaconis and L. Saloff-Coste, *Comparison Theorems for Reversible Markov Chains*, Ann. Appl. Probab. **3** (1993), no. 3, 696–730.

[Die]   A. Diekmann, *Not the First Digit! Using Benford's Law to Detect Fraudulent Scientific Data*, Journal of Applied Statistics **34** (2007), no. 3, 321–329.

[DieJa]   A. Diekmann and B. Jann, *Benford's Law and Fraud Detection: Facts and Legends*, German Economic Review **11** (2010), no. 3, 397–401.

[Dir]   P. Dirac, *Cosmological Models and the Large Numbers Hypothesis*, Proceedings of the Royal Society of London A **338** (1974), no. 1615, 439–446.

[DoMeOl] S. Dorogovtsev, J. Mendes and J. Oliveira, *Frequency of Occurrence of Numbers in the World Wide Web*, Physica A **360** (2006), 548–556.

[DrNi]   P. D. Drake and M. J. Nigrini, *Computer Assisted Analytical Procedures Using Benford's Law*, Journal of Accounting Education **18** (2000), 127–146.

[DuFr]   L. Dubins and D. Freedman, *Random Distribution Functions*, Proc. Fifth Berkeley Sympos. Math. Statist. and Probability (Berkeley, Calif., 1965/66), Vol. II: Contributions to Probability Theory, Part 1 (1967), 183–214, Univ. California Press, Berkeley, CA.

[DueLeu]　L. Dümbgen and C. Leuenberger, *Explicit Bounds for the Approxima-tion Error in Benford's Law*, Electronic Communications in Probability **13** (2008), 99–112.

[DurHP]　C. Durtschi, W. Hillison and C. Pacini, *The Effective Use of Benford's Law to Assist in Detecting Fraud in Accounting Data*, Journal of Foren-sic Accounting, 2004.

[EdH..]　X. Edelsbrunner, K. Huan, B. Mackall, S. J. Miller, J. Powell, C. Turnage-Butterbaugh and M. Weinstein, *Benford's Law, the Cauchy Distribution and Financial Data*, preprint 2014.

[Ele1]　Elections Canada, *38th General Election – Poll-by-Poll Results – Raw Data*, 2006.
http://www.elections.ca/scripts/resval/ovr_2004.asp?prov=&lang=e.

[Ele2]　Elections Canada, *39th General Election – Poll-by-Poll Results – Raw Data*, 2006.
http://www.elections.ca/scripts/resval/ovr_39ge.asp?prov=&lang=e.

[Ele3]　Elections Canada, *40th General Election – Poll-by-Poll Results – Raw Data*, 2010.
http://www.elections.ca/scripts/resval/ovr_40ge.asp?prov=&lang=e.

[Ele4]　Elections Canada, *41st General Election – Poll-by-Poll Results – Raw Data*, 2012.
http://www.elections.ca/scripts/resval/ovr_41ge.asp?prov=&lang=e.

[EngLeu]　H. A. Engel and C. Leuenberger, *Benford's Law for Exponential Ran-dom Variables*, Statistics & Probability Letters **63** (2003), 361–365.

[EUC1]　European Commission, *Eurostat Newsrelease Euroindicators*, Brussels, February 6, 2012.

[EUC2]　European Commission, *Communication from the Commission to the European Parliament and the Council, Towards Robust Quality Man-agement for European Statistics*, Brussels, April 15, 2011.

[EUC3]　European Commission, *Report on Greek Government Deficit and Debt Statistics*, Brussels, 2010.

[Fan]　D. Fanelli, How Many Scientists Fabricate and Falsify Research? A Systematic Review and Meta-Analysis of Survey Data, PLoS ONE 4, no. 5 (2009), 1–11.

[Far]   F. Farkas and G. Gyürky, *The Significance of Using the Newcomb-Benford Law as a Test of Nuclear Half-Life Calculations*, Acta Physica Polonica B **41** (2010), no.6, 1213–1221.

[Fef]   C. Fefferman, *Pointwise Convergence of Fourier Series*, Ann. of Math. Ser. 2 **98** (1973), 551–571.

[Fel]   W. Feller, *An Introduction to Probability Theory and Its Applications*, 2nd edition, Vol. II, John Wiley & Sons, New York, 1971.

[FerBN]   M. J. Ferguson, J. A. Bargh and D. A. Nayak, *After-Affects: How Automatic Evaluations Influence the Interpretation of Subsequent, Unrelated Stimuli*, Journal of Experimental Social Psychology **41** (2005), 182–191.

[FiMil]   F. W. K. Firk and S. J. Miller, *Nuclei, Primes and the Random Matrix Connection*, Symmetry **1** (2009), 64–105; doi:10.3390/sym1010064.

[FisVa]   M. Fisz and V. S. Varadarajan, *A Condition of Absolute Continuity of Infinitely Divisible Distribution Functions*, Z. Wahrscheinlichkeitstheorie verw. Gebiete **1** (1963), 335–339.

[Fol]   G. Folland, *Real Analysis: Modern Techniques and Their Applications*, 2nd edition, Pure and Applied Mathematics, Wiley-Interscience, New York, 1999.

[Fre]   B. S. Frey, *Economists in the PITS?* International Review of Economics **56**, no. 4 (2009), 335–346.

[FriGMH]   A. Frigyesi, D. Gisselsson, F. Mitelman and M. Höglund, *Power Law Distribution of Chromosome Aberrations in Cancer*, Cancer Research **63** (2003), 7094–7097.

[Fry]   S. Fry, *How Political Rhetoric Contributes to the Stability of Coercive Rule: A Weibull Model of Post-Abuse Government Survival.* Paper presented at the annual meeting of the International Studies Association, Le Centre Sheraton Hotel, Montreal, Quebec, Canada, Mar 17, 2004.

[FuSS]   D. Fu, Y. Q. Shi and W. Su, *A Generalized Benford's Law for JPEG Coefficients and its Applications in Image Forensics*, Security, Steganography, and Watermarking of Multimedia Contents IX **6505**, 2007.

[GAISE]   GAISE College Report, *Guidelines for Assessment and Instruction in Statistics Education*, American Statistical Association, 2005. http://www.amstat.org/education/gaise/.

[GaAl]   E. García-Berthou and C. Alcaraz, *Incongruence between Test Statistics and P Values in Medical Papers*, BMC Medical Research Methodology **4** (2004), 13.

[GeWi]   C. L. Geyer and P. P. Williamson, *Detecting Fraud in Datasets Using Benford's Law*, Computation in Statistics: Simulation and Computation **33** (2004), no. 1, 229–246.

[GiTo]   G. Gigerenzer and P. M. Todd, *Simple Heuristics That Make Us Smart*, ABC Research Group (1999), Oxford University Press, London.

[Gil]   D. E. Giles, *Benford's Law and Naturally Occurring Prices in Certain eBay Auctions*, Applied Economics Letters **14** (2007), no. 3, 157–161.

[GoJM]   A. Golan, G. Judge and D. Miller, *Maximum Entropy Econometrics: Robust Estimation with Limited Data*, John Wiley & Sons, 1996.

[GonPa]   J. Gonzales-Garcia and G. Pastor, *Benford's Law and Macroeconomic Data Quality*, International Monetary Fund, Working Paper, available at `http://papers.ssrn.com/sol3/papers.cfm?abstract_id=1356437`, 2009.

[GoPa]   J. Gonzalez-Garcia and P. Gonzalo, *Benford's Law and Macroeconomic Data Quality*, IMF Working Paper WP/09/10, 2009.

[GoAPA]   D. González-Jiménez, E. Argones-Rua, F. Pérez-González and J. L. Alba-Castro, *Benford's Law for Natural and Synthetic Images*, IEEE International Conference on Image Processing (2010), 1245–1248.

[GoWo]   R. C. Gonzalez and R. E. Woods, *Digital Image Processing*, Addison–Wesley, 1992.

[GrLaW]   M. Graber, A. Launov, K. Wälde, *Publish or Perish? The Increasing Importance of Publications for Prospective Economics Professors in Austria, Germany and Switzerland*, German Economic Review **9**, no. 4 (2008), 457–472.

[GrRy]   I. S. Gradshteyn and I. M. Ryshik, *Tables of Integrals, Series, and Products*, Academic Press, New York, 1980.

[GrKnP]   R. L. Graham, D. E. Knuth and O. Patashnik, *Concrete Mathematics. A Foundation for Computer Science*, 2nd edition, Addison–Wesley, Reading MA, 1994.

[GrHaPa]   S. Graham, J. Hasseldine and D. Paton, *Statistical Fraud Detection in a Commercial Lobster Fishery*, New Zealand Journal of Marine and Freshwater Research **43** (2009), no. 1, 457–463.

[GrPa]   T. Grammatikos and N. I. Papanikolaou, *Using Benford's Law to Detect Fraudulent Practices in Banking*, SSRN Scholarly Paper No. ID 2352775, 2013, Rochester, NY: Social Science Research Network.

[GrMo]   S. Grandison and R. Morris, *Biological Pathway Kinetic Rate Constants Are Scale-Invariant*, Bioinformatics **24** (2008), no. 6, 741–743.

[GrPaFl] R. Gregory, M. Partridge and M. A. Flower, *Performance Evaluation of the Philips Gemini PET/CT System*, IEEE Trans. Nucl. Sci., 53(1), Feb. 2006., 93–101,

[GrJuSc] M. Grendar, G. Judge and L. Schechter, *An Empirical Non-parametric Likelihood Family of Data-Based Benford-Like Distributions*, Physica A: Statistical Mechanics and its Applications, 2006.

[Gsch] T. Gschwend, *Ticket-Splitting and Strategic Voting under Mixed Electoral Rules: Evidence from Germany*, European Journal of Political Research **46** (2007), no. 1, 1–23.

[GüTd] S. Günnel and K.-H. Tödter, *Does Benford's Law Hold in Economic Research and Forecasting?*, Empirica **36** (2009), 273–292.

[HaBTV] E. M. Haacke, R. W. Brown, M. R. Thomas and R. Venkateson, *Magnetic Resonance Imaging Physical Principles and Sequence Design*, Wiley-Liss, New York, USA, 1999.

[Hai] J. Haigh, *The Statistics of National Lottery*, Journal of the Royal Statistical Society, Series A **160** (1997), no. 2, 187–206.

[HalDe] A. R. Halpern and S. D. Devereaux, *Lucky Numbers: Choice Strategies in the Pennsylvania Daily Number Game*, Bulletin of the Psychonomic Society **27** (1989), no. 2, 167–170.

[Ham] D. S. Hamermesh, *Viewpoint: Replication in Economics*, Canadian Journal of Economics **40**, no. 3 (2007), 715–733.

[Ha] R. W. Hamming, *On the distribution of numbers*, Bell Syst. Tech. J. **49** (1970), 1609–1625.

[Han] D. Hand, *Deception and Dishonesty with Data: Fraud in Science*, Significance **4** (2007), 22–25.

[HaWr] G. H. Hardy and E. Wright, *An Introduction to the Theory of Numbers*, 5th edition, Oxford Science Publications, Clarendon Press, Oxford, 1995.

[Has] W. K. Hastings. *Monte Carlo Sampling Methods Using Markov Chains and Their Applications*, Biometrika **57** (1970), no. 1, 97–109.

[Hay1] S. Hayes, *Does Terminal Digit Preference Occur in Pathology?*, Journal of Clinical Pathology **61** (2008), no. 8, 975–976.

[Hay2] S. Hayes, *Terminal Digit Preference Occurs in Pathology Reporting Irrespective of Patient Management Implication*, Journal of Clinical Pathology **61** (2008), no. 9, 1071–1072.

[Hay3] S. Hayes, *Benford's Law in Relation to Terminal Digit Preference*, Journal of Clinical Pathology **62** (2009), no. 6, 574–575.

[Hej]   D. Hejhal, *On a Result of Selberg Concerning Zeros of Linear Combinations of L-Functions*, International Math. Res. Notices **11** (2000), 551–557.

[Hen]   N. Henze, *A Statistical and Probabilistic Analysis of Popular Lottery Tickets*, Statistica Neerlandica **51** (1997), no. 2, 155–163.

[HePa]   M. Herrmann and F. Pappi, *Stategic Voting in German Constituencies*, Electoral Studies **27** (2008), no. 2, 228–244.

[HeSt]   E. Hewitt and K. Stromberg, *Real and Abstract Analysis*, Springer, Berlin, 1969.

[Hi1]   T. P. Hill, *Random-Number Guessing and the First Digit Phenomenon*, Psychological Reports **6** (1988), 967–971.

[Hi2]   T. P. Hill, *Base-Invariance Implies Benford's Law*, Proceedings of the American Mathematical Society **123** (1995), 887–895.

[Hi3]   T. Hill, *The First-Digit Phenomenon*, American Scientist **86** (1996), 358–363.

[Hi4]   T. Hill, *A Statistical Derivation of the Significant-Digit Law*, Statistical Science **10** (1996), 354–363.

[HiSi]   M. Hindry and J. Silverman, *Diophantine Geometry: An Introduction*, Graduate Texts in Mathematics **201**, Springer, New York, 2000.

[HjLR]   A. Hjorungnes, J. M. Lervik and T. A. Ramstad, *Entropy Coding of Composite Sources Modeled by Infinite Gaussian Mixture Distributions*, IEEE Workshop on Digital Signal Processing (1996), 235–238.

[HofHo]   P. Hofmarcher and K. Hornik, *First Significant Digits and the Credit Derivative Market During the Financial Crisis*, Contemporary Economics **7** (2013), no. 2, 21–29.

[HogMC]   R. V. Hogg, J. W. McKean and A. T. Craig, *Introduction to Mathematical Statistics*, 6th edition, Prentice–Hall, Englewood Cliffs, NJ, 2005.

[Hol]   P. J. Holewijn, *On the Uniform Distribution of Sequences of Random Variables*, Z. Wahrscheinlichkeitstheorie verw. Geb. **14** (1969), 89–92.

[HoKKS]   B. Horn, M. Kreuzer, E. Kochs and G. Schneider, *Different States of Anesthesia Can Be Detected by Benford's Law*, Journal of Neurosurgical Anesthesiology **18** (2006), no. 4, 328–329.

[HoRJB]   D. Hoyle, M. Rattray, R. Jupp and A. Brass, *Making Sense of Microarray Data Distributions*, Bioinformatics **18** (2002), no. 4, 576–584.

[Hsü]   E. H. Hsü, *An Experimental Study on "Mental Numbers" and a New Application*, Journal of General Psychology **38** (1948), 57–67.

[Hu]   W. Hurlimann, *Benford's Law from 1881 to 2006*,
         http://arxiv.org/pdf/math/0607168.

[Ins1]   Instituto Federal Electoral, *Estadísticas y Resultados Electorales*,
         http://www.ife.org.mx/computos2006/distritales/
         ReporteDiputadoMR%5B*%5D%5B*%5D.html, where * stands
         for various numerals (accessed July 13, 2006).

[Ins2]   Instituto Federal Electoral, *CG391/2011 Resolución del Consejo
         General del Instituto Federal Electoral sobre la Solicitud de Reg-
         istro del Convenio de Coalición Total para Postular Candidato
         a Presidente de los Estados Unidos Mexicanos, así como Can-
         didatos a Senadores y Diputados por el Principio de May-
         oría Relativa, Presentado por los Partidos Políticos Nacionales
         de la Revolución Democrática, del Trabajo y Movimiento Ciu-
         dadano, para Contender en el Proceso Electoral Federal 2011-2012*,
         http://www.ife.org.mx/docs/IFE-v2/DS/DS-CG/DS-
         SesionesCG/CG-resoluciones/201%1/noviembre/CGex
         201111-28/CGe281111rp3.pdf.

[Ins3]   Instituto Federal Electoral, *CG73/2012 Resolución del Consejo Gen-
         eral del Instituto Federal Electoral sobre la Solicitud de Modifi-
         cación del Convenio de Coalición Parcial Denominada "Compro-
         miso por México" para Postular Candidatos a Presidente de los Es-
         tados Unidos Mexicanos, Senadores y Diputados por el Principio de
         Mayoría Relativa, Presentada por los Partidos Políticos Nacionales
         Revolucionario Institucional, Verde Ecologista de México y Nueva
         Alianza para Contender en el Proceso Electoral Federal 2011-2012*,
         http://www.ife.org.mx/docs/IFE-v2/DS/DS-
         CG/DS-SesionesCG/CG-resoluciones/201%2/Febrero/
         CEex201202-08/CGe80212rp6.pdf.

[Ins4]   Instituto Federal Electoral, *Clausulado Integral del Con-
         venio de la Coalición Parcial Denominada "Compromiso
         por México" Celebrado por el Partido Revolucionario
         Institucional y el Partido Verde Ecologista de México*,
         http://www.ife.org.mx/docs/IFE-v2/DEPPP/DEPPP-
         ConveniosCoalicion/alianzas_D%EPPP/alianzas_
         DEPPP-pdf/2011-2012/clausula-
         convenio-compromexico.pdf

[Ins5]   Instituto Federal Electoral, *Estadísticas y Resultados Electorales.
         Descarga de la base de datos de los Cómputos Distritales*, URL:
         http://www.ife.org.mx/portal/site/ifev2/Estadis
         ticas_y_Resultados_Electorales (accessed October 19,
         2012).

[In]   J. Interlandi, *An Unwelcome Discovery*, New York Times Magazine,
         October 2006.

[IRS]   Internal Revenue Service, *General Description Booklet for the 1988 Individual Public Use Tax File*, U.S. Department of the Treasury, Statistics of Income Division, 1989. Available at http://www.nber.org/~taxsim/gdb/.

[IsKaCh]   N. Israeloff, M. Kagalenko and K. Chan, *Can Zipf Distinguish Language From Noise in Noncoding DNA?*, Physical Review Letters **76** (1996), no. 11, 1976.

[Iv]   A. Ivić, *On Small Values of the Riemann Zeta-Function on the Critical Line and Gaps between Zeros*, Lietuvos Matematikos Rinkinys **42** (2002), 31–45.

[IwKo]   H. Iwaniec and E. Kowalski, *Analytic Number Theory*, AMS, Providence, RI, 2004.

[Iw]   H. Iwasawa, *Gaussian Integral Puzzle*, Math. Intelligencer **31** (3) (2009), 38–41.

[JaThYe]   M. Jameson, J. Thorner and L. Ye, *Benford's Law for Coefficients of Newforms*, preprint 2014. http://arxiv.org/abs/1407.1577.

[JaKKKM] D. Jang, J. U. Kang, A. Kruckman, J. Kudo and S. J. Miller, *Chains of Distributions, Hierarchical Bayesian Models and Benford's Law*, Journal of Algebra, Number Theory: Advances and Applications **1**, no. 1 (March 2009), 37–60.

[JanRu]   E. Janvresse and T. de la Rue, *From Uniform Distribution to Benford's Law*, Journal of Applied Probability **41** (2004), no. 4, 1203–1210.

[Jay]   E. T. Jaynes, *Information Theory and Statistical Mechanics*, Phys. Rev., 1957.

[JesWi]   B. Jessen and A. Wintner, *Distribution Functions and the Riemann Zeta Function*, Transactions of the AMS **38** (1935), 48–88.

[Joa]   T. Joachims, *Making Large-Scale SVM Learning Practical*, Advances in Kernel Methods - Support Vector Learning, MIT Press, 1999.

[Jol]   J. M. Jolion, *Images and Benford's Law*, Journal of Mathematical Imaging and Vision **14** (2001), 73–81.

[JuSc]   G. Judge and L. Schechter, *Detecting Problems in Survey Data Using Benford's Law*, Journal of Human Resources **44** (2009), no. 1, 1–24.

[Jud]   H. F. Judson, *The Great Betrayal: Fraud in Science*, Harcourt, Orlando, FL, 2004.

[KaMeb]   K. Kalinin and W. Mebane, *Understanding Electoral Frauds through Evolution of Russian Federalism: From "Bargaining Loyalty" to "Signaling Loyalty"*. Paper presented at the 2011 Annual Meeting of the Midwest Political Science Association, Chicago, IL, March 31–April 2, 2011.

[Ka]   O. Kallenberg, *Foundations of Modern Probability*, 2nd edition, Springer, Berlin, 2002.

[KaSh]   I. Karatzas and S. E. Shreve, *Brownian Motion and Stochastic Calculus*, Graduate Texts in Mathematics **113**, Springer, 1988.

[Kar]   A. Karlsson, *Applications of Heat Kernels on Abelian Groups: $\zeta(2n)$, Quadratic Reciprocity, Bessel Integral*, preprint.

[KaSa]   N. Katz and P. Sarnak, *Zeros of Zeta Functions and Symmetries*, Bull. AMS **36** (1999), 1–26.

[Kat]   Y. Katznelson, *An Introduction to Harmonic Analysis*, 3rd edition, Cambridge University Press, Cambridge, 2004.

[KeSn]   J. P. Keating and N. C. Snaith, *Random Matrix Theory and $\zeta(1/2 + it)$*, Comm. Math. Phys. **214** (2000), no. 1, 57–89.

[Kemp]   J. Kemperman, *Bounds on the Discrepancy Modulo 1 of a Real Random Variable*, Institute of Mathematical Statistics Bulletin **4**, no. 1 (1975), 138.

[Kh]   A. Khintchin, *Zur Theorie der unbeschränkt teilbaren Verteilungsgesetze*, Mat. Sbornik **44** (1937), 79–119.

[Ki..]   G. King, B. Palmquist, G. Adams, M. Altman, K. Benoit, C. Gay, J. Lewis, R. Mayer and E. Reinhardt, *The Record of American Democracy, 1984–1990*, Harvard University, Cambridge, MA, 1997.

[Kles]   J. Klesner, *The July 2006 Presidential and Congressional Elections in Mexico*, Electoral Studies **26** (2007), no. 4, 803–808.

[Knu]   D. Knuth, *The Art of Computer Programming, Volume 2: Seminumerical Algorithms*, Addison–Wesley, 3rd edition, 1997.

[Kol]   A. N. Kolmogorov, *Une série de Fourier-Lebesgue divergent partout*, Comptes Rendus **183** (1926), 1327–1328.

[KonMi]   A. Kontorovich and S. J. Miller, *Benford's Law, Values of L-Functions and the $3x + 1$ Problem*, Acta Arith. **120** (2005), 269–297.

[KonSi]   A. Kontorovich and Ya. G. Sinai, *Structure Theorem for (d, g, h)-Maps*, Bull. Braz. Math. Soc. (N.S.) **33** (2002), no. 2, 213–224.

[Kosh]   T. Koshy, *Fibonacci and Lucas Numbers with Applications*, Wiley-Interscience, New York, 2001.

[Koss1]   A. E. Kossovsky, *Towards a Better Understanding of the Leading Digits Phenomena*, preprint. http://arxiv.org/abs/math/0612627.

[Koss2]   A. E. Kossovsky, *Benford's Law*, World Scientific, 2014.

[Kou]   S. Kou, *A Jump-Diffusion Model for Option Pricing*, Management Science **48** (2002), 1086–1101.

[KouWa]   S. Kou and H. Wang, *Option Pricing under a Double Exponential Jump-Diffusion Model*, Management Science **50** (2004), 1178–1192.

[Kr]   V. M. Kruglov, *A Note on Infinitely Divisible Distributions*, Theory Probab. Appl. **15** (1970), 319–324.

[Kry]   J. Krygier, *Psychological Relevance of Benford's Law: Sensitivity to Statistical Regularities and Implications for Theories of Number Representation*, Unpublished Honours thesis (2009), University of Sydney, Australia.

[Kub]   M. Kubovy, *Response Availability and the Apparent Spontaneity of Numerical Choices*, Journal of Experimental Psychology: Human Performance and Performance **2** (1977), 359–364.

[KuiNi]   L. Kuipers and H. Niederreiter, *Uniform Distribution of Sequences*, John Wiley & Sons, 1974.

[Kul]   S. Kullback, *Information Theory and Statistics*, Wiley Publication in Mathematical Statistics, 1959.

[Lad]   R. Ladouceur, D. Dube, I. Giroux, N. Legendre and C. Gaudet, *Cognitive Biases and Playing Behavior on American Roulette and the 6/49 Lottery* (1996), unpublished manuscript, Université Laval.

[LafSi]   J. M. Lafaille and G. Simonis, *Dissected Re-assembled: An Analysis of Gaming*, 2005.

[Lag1]   J. Lagarias, *The $3x+1$ Problem and its Generalizations*, Organic Mathematics (Burnaby, BC, 1995), 305–334, CMS Conf. Proc. **20**, Amer. Math. Soc., Providence, RI, 1997.

[Lag2]   J. Lagarias, *The $3x + 1$ Problem: An Annotated Bibliography*, http://arxiv.org/abs/math/0309224.

[Lag3]   J. Lagarias, *The Ultimate Challenge: The $3x + 1$ Problem*, American Mathematical Society, Providence, RI, 2010.

[LagSo]   J. Lagarias and K. Soundararajan, *Benford's Law for the* $3x + 1$ *Function*, J. London Math. Soc. **74** (2006), ser. 2, no. 2, 289–303.

[LanLi]   L. D. Landau and E. M. Lifshitz, *Statistical Physics*, 3rd edition, Butterworth-Heinemann, Oxford, 1996.

[Lan]   D. Lando, *Credit Risk Modeling*, Princeton University Press, Princeton, NJ, 2004.

[Lang1]   S. Lang, *Undergraduate Analysis*, 2nd edition, Springer, New York, 1997.

[Lang2]   S. Lang, *Complex Analysis*, Graduate Texts in Mathematics, vol. 103, Springer, New York, 1999.

[Lau]   A. Laurinčikas, *Limit Theorems for the Riemann Zeta-Function on the Critical Line II*, Lietuvos Mat. Rinkinys **27** (1987), 489–500.

[LedTa]   M. Ledoux and M. Talagrand, *Probability in Banach Spaces*, Springer, Berlin, 1991.

[LeDuc1]   L. LeDuc, *The Federal Election in Canada, June 2004*, Electoral Studies **24** (2005), no. 2, 338–344.

[LeDuc2]   L. LeDuc, *The Federal Election in Canada, January 2006*, Electoral Studies **26** (2007), no. 3, 716–720.

[LeDuc3]   L. LeDuc, *The Federal Election in Canada, October 2008*, Electoral Studies **28** (2009), no. 2, 326–329.

[LeDuc4]   L. LeDuc, *The Federal Election in Canada, May 2011*, Electoral Studies **31** (2012), no. 1, 239–242.

[LeCoJu]   J. Lee, W. Cho and G. Judge, *Stigler's Approach to Recovering the Distribution of First Significant Digits in Natural Data Sets*, Statistics and Probability Letters **80** (2010), no. 2, 82–88.

[LeScEv]   L. M. Leemis, B. W. Schmeiser and D. L. Evans, *Survival Distributions Satisfying Benford's Law*, American Statistician **54** (2000), no. 4, 236–241.

[Lév1]   P. Lévy, *Théorie de l'addition des variables aléatoires*, Gauthier-Villars, Paris, 1937.

[Lév2]   P. Lévy, *L'addition des variables aléatoires définies sur une circonférence*, Bull. de la S.M.F. **67** (1939), 1–41.

[Ley]   E. Ley, *On the Peculiar Distribution of the U.S. Stock Indices Digits*, American Statistician **50** (1996), no. 4, 311–313.

[LiBEM]   J. A. List, C. D. Bailey, P. J. Euzent and T. L. Martin, *Academic Economists Behaving Badly? A Survey on Three Areas of Unethical Behavior*, Economic Inquiry **39**, no. 1 (2001), 162–170.

[LoeBr]   T. Loetscher and P. Brugger, *Exploring Number Space by Random Digit Generation*, Exp. Brain Res. **180** (2007), 655–665.

[Lop]   J. A. G. Lopez, *Fraude Electoral?*, Doble Hélice Ediciones, Chihuahua, Mexico, 2009.

[LoRSST] T. Louwers, B. Ramsey, D. Sinason, J. Strawser and J. Thibodeau, *Auditing & Assurance Services*, 4th edition, Burr Ridge, IL, Irwin–McGraw–Hill, 2011.

[Loy]   R. M. Loynes, *Some Results in the Probabilistic Theory of Asympototic Uniform Distributions Modulo* 1, Z. Wahrscheinlichkeitstheorie verw. Geb. **26** (1973), 33–41.

[LePeWi] D. A. Levin, Y. Peres and E. L. Wilmer, *Markov Chains and Mixing Times*, AMS, Providence, RI, 2009.

[Luo]   W. Luo, *Zeros of Hecke L-Functions Associated with Cusp Forms*, Acta Arith. **71** (1995), no. 2, 139–158.

[MaRa]   N. Madras and D. Randall, *Markov Chain Decomposition for Convergence Rate Analysis*, Ann. Appl. Probab. **12** (2002), no. 2, 581–606.

[Mag]   B. Magaloni, *Voting for Autocracy: Hegemonic Party Survival and Its Demise in Mexico*, Cambridge University Press, New York, 2006.

[ManHu]   B. Mandelbrot and R. L. Hudson, *The (Mis)behavior of Markets. A Fractal View of Risk, Ruin, and Reward*, Basic Books, New York, 2004.

[MaRRSD] P. Manoochehrnia, F. Rachidi, M. Rubinstein, W. Schulz and G. Diendorfer, *Benford's Law and Its Application to Lightning Data*, IEEE Transactions on Electromagnetic Compatibility **52** (2010), no. 4, 956–961.

[MBGHPSS] R. Mantegna, S. Buldyrev, A. Goldberger, S. Havlin, C. Peng, M. Simons and H. Stanley, *Linguistic Features of Noncoding DNA Sequences*, Physical Review Letters **73** (1994), no. 23, 3169–3172.

[MaAdV]   B. C. Martinson, M. S. Anderson and R. de Vries, *Scientists Behaving Badly*, Nature **435** (2005), 737–738.

[McCDo]   J. McCann and J. Domínguez, *Mexicans React to Electoral Fraud and Political Corruption: An Assessment of Public Opinion and Voting Behavior*, Electoral Studies **17** (1998), no. 4, 483–503.

[McCFKB] K. C. McCulloch, M. J. Ferguson, C. C. K. Kawada, J. A. Bargh, *Taking a Closer Look: On the Operation of Nonconscious Impression Formation*, Journal of Experimental Social Psychology **44** (2008), 614–623.

[McCMH] B. D. McCullough, K. A. McGeary and T. D. Harrison, *Lessons from the JMCB Archive*, Journal of Money, Credit, and Banking **38**, no. 4, (2006), 1093–1107.

[McCVi] B. D. McCullough and H. D. Vinod, *Verifying the Solution from a Nonlinear Solver: A Case Study*, American Economic Review **93** (2003), 873–892.

[McD] C. McDiarmid, *On the Method of Bounded Differences*, in Surveys in Combinatorics (J. Siemons ed.), 148–188, London Math. Soc. Lecture Notes Series **141**, Cambridge University Press, Cambridge, 1989.

[McLPe] G. McLachlan and D. Peel, *Finite Mixture Models*, John Wiley & Sons, New York, 2000.

[McSABF] B. McShane, M. Adrian, E. T. Bradlow and P. S. Fader, *Count Models Based on Weibull Interarrival Times*, Journal of Business and Economic Statistics **26** (2006), no. 3, 369–378.

[Meb1] W. Mebane, *Election Forensics: Vote Counts and Benford's Law*, Paper prepared for the 2006 Summer Meeting of the Political Methodology Society, UC-Davis, July 20–22, 2006.

[Meb2] W. R. Mebane, Jr., *Note on the Presidential Election in Iran, June 2009*, (2009), http://www.umich.edu/~wmebane/note29jun2009.pdf.

[Meb3] W. Mebane, *Election Fraud or Strategic Voting?*, Paper prepared for the 2010 Annual Meeting of the Midwest Political Science Association, Chicago, IL, April 22–25, 2010.

[Meb4] W. Mebane, *Fraud in the 2009 Presidential Election in Iran?*, Chance **23** (2010), no. 1, 6–15.

[Meb5] W. Mebane, *Comment on 'Benford's Law and the Detection of Election Fraud'*, Political Analysis **19** (2011), no. 3, 269–272.

[Meb6] W. Mebane, *Second-Digit Tests for Voters' Election Strategies and Election Fraud*, Paper prepared for the 2012 Annual Meeting of the Midwest Political Science Association, Chicago, IL, April 11–14, 2012.

[Meb7] W. Mebane, *Election Forensics*, Book MS, 2013.

[MebKa1] W. Mebane and K. Kalinin, *Electoral Falsification in Russia: Complex Diagnostics Selections 2003-2004, 2007-2008* (in Russian), Russian Electoral Review **2** (2009), 57–70.

[MebKa2] W. Mebane and K. Kalinin, *Electoral Fraud in Russia: Vote Counts Analysis Using Second-Digit Mean Tests*, Paper prepared for the 2010 Annual Meeting of the Midwest Political Science Association, Chicago, IL, April 22–25, 2010.

[MebKe] W. Mebane and T. Kent, *Second Digit Implications of Voters' Strategies and Mobilizations in the United States during the 2000s*, Paper prepared for the 2013 Annual Meeting of the Midwest Political Science Association, Chicago, IL, April 11–14, 2013.

[Mer] R. Merton, *Theory of Rational Option Pricing*, Bell J. Economics and Management Science **4** (1973), 141–183.

[Mer2] R. Merton, *Option Pricing When Underlying Stock Returns Are Discontinuous*, J. Financial Economics **3** (1976), 125–144.

[MeRRTT] N. Metropolis, A. W. Rosenbluth, M. N. Rosenbluth, A. H. Teller and E. Teller, *Equation of State Calculations by Fast Computing Machines*, Journal of Chemical Physics **21** (1953), no. 6, 1087–1092.

[Metz] M. Metzker, *Sequencing Technologies - The Next Generation*, Nature Review Genetics **11** (2010), 31–46.

[MiSt] T. Michalski and G. Stoltz, *Do Countries Falsify Economic Data Strategically? Some Evidence That They Might*, Review of Economics and Statistics **95** (2013), no. 2, 591–616.

[Miko] P. G. Mikolaj, *Environmental Applications of the Weibull Distribution Function: Oil Pollution*, Science 2 **176** (1972), no. 4038, 1019–1021.

[Mik] T. Mikosch, *Non-Life Insurance Mathematics*, Springer, 2004.

[Mi1] S. J. Miller, *When the Cramér-Rao Inequality Provides No Information*, Communications in Information and Systems **7** (2007), no. 3, 265–272.

[Mi2] S. J. Miller, *A Derivation of the Pythagorean Won-Loss Formula in Baseball*, Chance Magazine **20** (2007), no. 1, 40–48.

[MiNi1] S. J. Miller and M. Nigrini, *The Modulo 1 Central Limit Theorem and Benford's Law for Products*, International Journal of Algebra **2** (2008), no. 3, 119–130.

[MiNi2] S. J. Miller and M. Nigrini, *Order Statistics and Benford's Law*, International Journal of Mathematics and Mathematical Sciences **2008** (2008), Article ID 382948, 19 pages. doi:10.1155/2008/382948.

[MiT-B] S. J. Miller and R. Takloo-Bighash, *An Invitation to Modern Number Theory*, Princeton University Press, Princeton, NJ, 2006.

[Mir]   T. A. Mir, *The Leading Digit Distribution of the Worldwide Illicit Financial Flows*, preprint 2012, http://arxiv.org/abs/1201.3432.

[MiBC]   N. Mira, J. Becker and I. Sá-Correia, *Genomic Expression Program Involving the Haa1p-Regulon in Saccharomyces cerevisiae Response to Acetic Acid*, Omics **14** (2010), 587–601.

[MitJuMi]   R. C. Mittelhammer, G. Judge and D. J. Miller, *Econometric Foundations*, Cambridge University Press, 2000.

[Monod]   J. Monod, *The Growth of Bacterial Cultures*, Annual Review of Microbiology **3** (1949), 371–394.

[Mon]   H. Montgomery, *The Pair Correlation of Zeros of the Zeta Function*, Analytic Number Theory, Proc. Sympos. Pure Math. **24**, Amer. Math. Soc., Providence, 1973, 181–193.

[MorSPZ]   M. Moret, V. de Senna, M. Pereira and G. Zebende, *Newcomb-Benford Law in Astrophysical Sources*, International Journal of Modern Physics C **17** (2006), no. 11, 1597–1604.

[MorSSZ]   M. Moret, V. de Senna, M. Santana and G. Zebende, *Geometric Structural Aspects of Proteins and Newcomb-Benford Law*, International Journal of Modern Physics C **20** (2009), no. 12, 1981–1988.

[Morr]   K. E. Morrison,*The Multiplication Game*, Mathematics Magazine **83** (2010), 100–110.

[MosDDK]   J. Mosimann, J, Dahlberg, N. Davidian and J. Krueger, *Terminal Digits and the Examination of Questioned Data*, Accountability in Research **9** (2002), no. 2, 75–92.

[MosWE]   J. Mosimann, C. Wiseman and R. Edelman, *Data Fabrication: Can People Generate Random Digits?*, Accountability in Research **4** (1995), no. 4, 31–55.

[MyOrSh]   M. Myagkov, P. Ordeshook and D. Shaikin, *The Forensics of Election Fraud: With Applications to Russia and Ukraine*, Cambridge University Press, New York, 2009.

[NaWWSRGS] U. Nagalakshmi, Z. Wang, K. Waern, C. Shou, D. Raha, M. Gerstein and M. Snyder, *The Transcriptional Landscape of the Yeast Genome Defined by RNA Sequencing*, Science **320** (2008), 1344–1349.

[NaSh]   K. Nagasaka and J. S. Shiue, *Benford's Law for Linear Recurrence Sequences*, Tsukuba J. Math. **11** (1987), 341–351.

[New]   S. Newcomb, *Note on the Frequency of Use of the Different Digits in Natural Numbers*, Amer. J. Math. **4** (1881), 39–40.

[NiRen]   D. Ni and Z. Ren, *Benford's Law and Half-Lives of Unstable Nuclei*, European Physical Journal A **38** (2008), 251–255.

[NiWeRe]   D. Ni, L. Wei and Z. Ren, *Benford's Law and β-Decay Half-Lives*, Communications in Theoretical Physics **51** (2009), 713–716.

[Nig1]   M. Nigrini, *The Detection of Income Tax Evasion through an Analysis of Digital Frequencies*, PhD thesis, University of Cincinnati, OH, USA, 1992.

[Nig2]   M. Nigrini, *Using Digital Frequencies to Detect Fraud*, The White Paper **8** (1994), no. 2, 3–6.

[Nig3]   M. Nigrini, *Digital Analysis and the Reduction of Auditor Litigation Risk*, in Proceedings of the 1996 Deloitte & Touche / University of Kansas Symposium on Auditing Problems (M. Ettredge ed.), University of Kansas, Lawrence, KS, 1996, 69–81.

[Nig4]   M. Nigrini, *A Taxpayer Compliance Application of Benford's Law*, Journal of the American Taxation Association, 1996.

[Nig5]   M. Nigrini, *The Use of Benford's Law as an Aid in Analytical Procedures*, Auditing: A Journal of Practice & Theory **16** (1997), no. 2, 52–67.

[Nig6]   M. Nigrini, *Adding Value with Digital Analysis*, Internal Auditor, February, (1999), 21–23.

[Nig7]   M. Nigrini, *I've Got Your Number: How a Mathematical Phenomenon Can Help CPAs Uncover Fraud and Other Irregularities*, Journal of Accountancy (May 1999), 79–83.

[Nig8]   M. Nigrini, *An Assessment of the Change in the Incidence of Earnings Management Around the Enron-Andersen Episode*, Review of Accounting and Finance **4** (2005), no. 1, 92–110.

[Nig9]   M. Nigrini, *Data-Driven Forensic Investigation: Using Microsoft Access and Excel to Detect Fraud and Data Irregularities*, John Wiley & Sons, Hoboken, NJ, 2011.

[NiMi1]   M. Nigrini and S. J. Miller, *Benford's Law Applied to Hydrology Data – Results and Relevance to Other Geophysical Data*, Mathematical Geology **39** (2007), no. 5, 469–490.

[NiMi2]   M. Nigrini and S. J. Miller, *Data Diagnostics Using Second Order Tests of Benford's Law*, Auditing: A Journal of Practice and Theory **28** (2009), no. 2, 305–324. doi:10.2308/aud.2009.28.2.305.

[NiMit]   M. Nigrini and L. Mittermaier, *The Use of Benford's Law as an Aid in Analytical Procedures*, Auditing: A Journal of Practice & Theory **16** (1997), no. 2, 52–67.

[NyM]   J. Nye and C. Moul, *The Political Economy of Numbers: On the Application of Benford's Law to International Macroeconomic Statistics*, B.E. Journal of Macroeconomics, 2007.

[OrMNO]   M. Orita, A. Moritomo, T. Niimi and K. Ohno, *Use of Benford's Law in Drug Discovery Data*, Drug Discovery Today **15** (2010), 328–331.

[Ost]   S. Osterloh, *Accuracy and Properties of Germany Business Cycle Forecasts*, Applied Economics Quarterly **54**, no. 1 (2008), 27–57.

[Ow]   A. B. Owen, *Empirical Likelihood*, Chapman and Hall, 2001.

[ÖzBa]   G. Özer and B. Babacan, *Benford's Law and Digital Analysis: Application on Turkish Banking Sector*, Business and Economics Research Journal **4** (2013), no. 1, 29–41.

[Pad]   T. Padmanabhan, *Physical Significance of Planck Length*, Annals of Physics **165** (1985), no. 1, 38–58.

[Pai]   C. Pain, *Benford's Law and Complex Atomic Spectra*, Physical Review E **77** (2008), 012102, doi:10.1103/PhysRevE.77.012102.

[PapPU]   A. Papoulis, S.U. Pillai and S. Unnikrishna, *Probability, Random Variables, and Stochastic Processes*, McGraw–Hill, New York, 2002.

[Pat]   S. J. Patterson, *An Introduction to the Theory of the Riemann Zeta-Function*, Cambridge Studies in Advanced Mathematics 14, Cambridge University Press, 1995.

[PMDPMLS] S. Paul, D. Mytelka, C. Dunwiddie, C. Persinger, B. Munos, S. Lindborg and A. Schact, *How to Improve R&D Productivity: The Pharmaceutical Industry's Grand Challenge*, Nature Reviews Drug Discovery **9** (2010), 203–214.

[PéAH1]   F. Pérez-González, C. T. Abdallah and G. L. Heileman, *A Generalization of Benford's Law and Its Application to Images*, European Control Conference, (2007).

[PéAH2]   F. Pérez-González, C. T. Abdallah and G. L. Heileman, *Benford's Law in Image Processing*, IEEE International Conference on Image Processing (2007), 405–408.

[PerTo1] L. Pericchi and D. Torres, *La Ley de Newcomb-Benford y sus aplicaciones al Referendum Revocatorio en Venezuela*, Reporte Técnico no-definitivo 2a. versión: Octubre 01, 2004.

[PerTo2] L. Pericchi and D. Torres, *Quick Anomaly Detection by the Newcomb-Benford Law, with Applications to Electoral Processes Data from the USA, Puerto Rico and Venezuela*, Statistical Science **26** (2011), no. 4, 502–516.

[PiTTV] L. Pietronero, E. Tosatti, V. Tosatti and A. Vespignani, *Explaining the Uneven Distribution of Numbers in Nature: The Laws of Benford and Zipf*, Physica A: Statistical Mechanics and its Applications, 2001.

[Pin] R. Pinkham, *On the Distribution of First Significant Digits*, Annals of Mathematical Statistics **32**, no. 4 (1961), 1223–1230.

[PoTh] C. Porter and R. Thomas, *Fluctuations of Nuclear Reaction Widths*, Physical Review **104** (1956), 483–491.

[Q–] M. Quarantelli, K. Berkouk, A. Prinster, B. Landeau, C. Svarer, L. Balkay, B. Alfano, A. Brunetti, J.-C. Baron and M. Salvatore, *Integrated Software for the Analysis of Brain PET/SPECT Studies with Partial-Volume Effect Correction*, J. Nucl. Med. **45**, (2004), 192–201.

[R] R Development Core Team, *R: A Language and Environment for Statistical Computing*, R Foundation for Statistical Computing, Vienna, Austria, 2011. http://www.R-project.org

[Rai] R. A. Raimi, *The First Digit Problem*, Amer. Math. Monthly **83** (1976), no. 7, 521–538.

[Ran] D. Randall, *Rapidly Mixing Markov Chains with Applications in Computer Science and Physics*, Computing in Science and Engineering **8** (March 2006), no. 2, 30–41.

[RaEtAl] J. Ranstam, M. Buyse, S. George, S. Evans, N. Geller, B. Scherrer, E. Lesaffre, G. Murray, L. Edler, J. Hutton, T. Colton and P. Lachenbruch, *Fraud in Medical Research: An International Survey of Biostatisticians*, Controlled Clinical Trials **21** (2000), 415–427.

[RapBu] A. Rapoport and D. V. Budescu, *Generation of Random Series in Two-Person Strictly Competitive Games*, Journal of Experimental Psychology: General **121** (1992), 352–363.

[RauGBE] B. Rauch, M. Göttsche, G. Brähler and S. Engel, *Fact and Fiction in EU-Governmental Economic Data*, German Economic Review **12** (2011), no. 2, 243–255.

[RauGBK] B. Rauch, M. Göttsche, G. Brähler and T. Kronfeld, *Deficit versus Social Statistics: Empirical Evidence for the Effectiveness of Benford's Law*, Applied Economics Letters **21** (2014), no. 3, 147–151.

[RauGM] B. Rauch, M. Göttsche and F. El Mouaaouy, *LIBOR Manipulation-Empirical Analysis of Financial Market Benchmarks Using Benford's Law*, 2013. Available at SSRN 2363895.

[ReaCr] T. R. C. Read and N. A. C. Cressie, *Goodness-of-Fit Statistics for Discrete Multivariate Data*, Springer, 1988.

[Rei]   R. D. Reiss, *Approximate Distributions of Order Statistics*, Springer, New York, 1989.

[RevYo]   D. Revuz and M. Yor, *Continuous Martingales and Brownian Motion*, 3rd edition, Springer, Berlin, 1999.

[Rob]   H. Robbins, *On the Equidistribution of Sums of Independent Random Variables*, Proc. Amer. Math. Soc. **4** (1953), 786–799.

[RobCa]   C. P. Robert and G. Casella, *Monte Carlo Statistical Methods*, Springer, New York, 2004.

[RobSt]   C. J. Roberts and T. D. Stanley, *Meta-Regression Analysis: Issues of Publication Bias in Economics*, Blackwell Publishing, Oxford, UK, 2005.

[Rod]   R. Rodriguez, *First Significant Digit Patterns from Mixtures of Uniform Digits*, American Statistician, 2004.

[RorBr]   C. Rorden and M. Brett, *Stereotaxic Display of Brain Lesions*, Behav. Neurol. **12** (2000), 191–200.

[Rot]   K. Roth, *Rational Approximations to Algebraic Numbers*, Mathematika **2** (1955), 1–20.

[Rou]   B. F. Roukema, *A First-Digit Anomaly in the 2009 Iranian Presidential Election*, Journal of Applied Statistics **41** (2014), 164–199. DOI:10.1080/02664763.2013.838664. http://arXiv.org/abs/0906.2789.

[RouMA]   O. G. Rousset, Y. Ma and A. C. Evans, *Correction for Partial Volume Effects in PET: Principle and Validation*, Journal of Nuclear Medicine **39(5)**, May 1998, 904–911.

[Roy]   H. L. Royden, *Real Analysis* 3rd edition, Macmillan, New York, 1988.

[Rud]   W. Rudin, *Principles of Mathematical Analysis*, 3rd edition, International Series in Pure and Applied Mathematics, McGraw–Hill, New York, 1976.

[Sak]   H. Sakamoto, *On the Distributions of the Product and the Quotient of the Independent and Uniformly Distributed Random Variables*, Tôhoku Math. J. **49** (1943), 243–260.

[Sala1]   Sala Superior, *Cómputo Final de la Elección Presidencial, Declaración de Validez del Proceso Electoral, y de Presidente Electo de los Estados Unidos Mexicanos*, Tribunal Electoral del Poder Judicial de la Federación, Comisión Calificadora: Magistrados Constancio Carrasco Daza, Flavio Galván Rivera y Salvador Olimpo Nava Gomar, August 31, 2012.

http://www.ife.org.mx/documentos/proceso_2011-
2012/documentos/computo_final_calificacion_
jurisdiccional.pdf.

[Sala2]   Sala   Superior,   *Juicio   de   Inconformidad*,   Tribunal
          Electoral   del   Poder   Judicial   de   la   Federación,   Ex-
          pediente: SUP-JIN-359/2012,   August   30,   2012,
          http://www.ife.org.mx/documentos/proceso_2011-
          2012/documentos/SUP-JIN-359-2012.pdf.

[S-i-M]   X. X. Sala-i-Martin, *I Just Ran Two Million Regressions*, American
          Economic Review **87**, no.2 (1997), 178–183.

[SamTa]   G. Samorodnitsky and M. Taqqu, *Stable Non-Gaussian Random Pro-
          cesses*, Chapman & Hall, London, 1994.

[Sande]   R.   Sandels,   *Massive   Fraud   in   Mexico's   Presidential   Elec-
          tions*,   Centre   for   Research   on   Globalization,   October   15,   2012.
          http://www.globalresearch.ca/massive-fraud-in-
          mexicos-presidential-elections/5308401.

[Sando]   F. Sandon, *Do Populations Conform to the Law of Anomalous Num-
          bers?*, Population **57** (2002), 753–761.

[Sat1]   K.-I. Sato, *Absolute Continuity of Multivariate Distributions of Class
          L*, J. Multivariate Analysis **12** (1982), 89–94.

[Sat2]   K.-I. Sato, *Lévy Processes and Infinite Divisible Distributions*, Cam-
          bridge University Press, Cambridge, 2005.

[ScSMG]   Ch. Schaefer, J. P. Schrapler, J. P. Muller and G. G. Wagner, *Automatic
          Identification of Faked and Fraudulent Interviews in the German SOEP*,
          Journal of Applied Social Science Studies **125** (2005), 183–193.

[SchSti] G. Schaefer and M. Stich, *UCID: An Uncompressed Colour Image
          Database*, Storage and Retrieval Methods and Applications for Mul-
          timedia **5307** (2004), 472–480.

[Sc1]   P. Schatte, *On Sums Modulo 2π of Independent Random Variables*,
          Math. Nachr. **110** (1983), 243–261.

[Sc2]   P. Schatte, *On the Asymptotic Uniform Distribution of Sums Reduced
          mod 1*, Math. Nachr. **115** (1984), 275–281.

[Sc3]   P. Schatte, *On the Asymptotic Logarithmic Distribution of the Floating-
          Point Mantissas of Sums*, Math. Nachr. **127** (1986), 7–20.

[SchWil] H. T. Schreuder and M. S. Williams, *Reliability of confidence intervals
          calculated by bootstrap and classical methods using the IA 1-ha plot de-
          sign*, in Gen. Tech Rep. RMRS-GTR-57 (2000), Fort Collins, CO: U.S.
          Department of Agriculture, Forest Service, Rocky Mountain Research
          Station.

[Schum]   P. Schumer, *Mathematical Journeys*, Wiley-Interscience, John Wiley & Sons, New York, 2004.

[Schür1]   K. Schürger, *Wahrscheinlichkeitstheorie*, Oldenbourg, München, 1998.

[Schür2]   K. Schürger, *Extensions of Black-Scholes Processes and Benford's Law*, Stochastic Processes and Their Applications **118** (2008), 1219–1243.

[ScoBM]   S. K. Scott, P. J. Barnard and J. May, *Specifying Executive Representations and Processes in Number Generation Tasks*, Quarterly Journal of Experimental Psychology: Section A **54** (2001), 641–664.

[Sel1]   A. Selberg, *Contributions to the Theory of the Riemann Zeta-Function*, Arch. Math. Naturvid. **48** (1946), no. 5, 89–155.

[SelBaBe]   T. Sellke, M. Bayarri and J. Berger, *Calibration of p Values for Testing Precise Null Hypotheses*, American Statistician **55** (2001), no. 1, 62–71.

[Ser]   J. P. Serre, *A Course in Arithmetic*, Springer, 1996.

[Shan]   C. E. Shannon, *A Mathematical Theory of Communication*, Bell Technical Journal, 1948.

[ShaMa1]   L. Shao and B. Ma, *First Digit Distribution of Hadron Full Width*, Modern Physics Letters A **24** (2009), no. 40, 3275–3282.

[ShaMa2]   L. Shao and B. Ma, *Empirical Mantissa Distributions of Pulsars*, Astroparticle Physics **33** (2010), 255–262.

[ShaMa3]   L. Shao and B. Ma, *The Significant Digit Law in Statistical Physics*, Physica A **389** (2010), 3109–3116.

[ShaMa4]   L. Shao and B. Ma, *First Digit Law in Nonextensive Statistics*, Physical Review E **82** (2010), 041110, doi:10.1103/PhysRevE.82.041110

[ShiHeTh]   S. Shikano, M. Herrmann and P. Thurner, *Strategic Voting under Proportional Representation: Threshold Insurance in German Elections*, West European Politics **32** (2009), no. 3, 634–656.

[ShiMa]   S. Shikano and V. Mack, *When Does the Second-Digit Benford's Law-Test Signal an Election Fraud? Facts or Misleading Test Results*, Jahrbücher fur Nationalökonomie und Statistik **231** (2009), no. 5–6, 719–732.

[Shi]   A. N. Shiryayev, *Probability*, Springer, Berlin, 1984.

[Sim]   J. Simon, *An Analysis of the Distribution of Combinations Chosen by UK National Lottery Players*, Journal of Risk and Uncertainty **17** (1999), no. 3, 243–276.

[Sin]   Ya. G. Sinai, *Statistical $(3x + 1)$ Problem*, Comm. Pure Appl. Math. **56** (2003), no. 7, 1016–1028.

[Sinc1]   A. J. Sinclair, *Algorithms for Random Generation and Counting. A Markov Chain Approach*, in Progress in Theoretical Computer Science, Birkhäuser, Boston, MA, 1993.

[Sinc2]   A. J. Sinclair, *Convergence Rates for Monte Carlo Experiments*, in Numerical Methods for Polymeric Systems (Minneapolis, MN, 1996), 1–17 (S. G. Whittington ed.), IMA Vol. Math. Appl. **102**, Springer, New York, 1998.

[Sm]   S. W. Smith, *Explaining Benford's Law*, Chapter 34 in The Scientist and Engineer's Guide to Digital Signal Processing, 1997. Republished in softcover by Newnes, 2002.

[Sor]   D. Sornette, *Critical Phenomena in Natural Sciences*, Springer, 2000.

[Sta]   W. H. Starbuck, *How Much Better Are the Most-Prestigious Journals? The Statistics of Academic Publication*, Organization Science **16**, no. 2 (2005), 180–200.

[StSh1]   E. Stein and R. Shakarchi, *Fourier Analysis: An Introduction*, Princeton University Press, Princeton, NJ, 2003.

[StSh2]   E. Stein and R. Shakarchi, *Complex Analysis*, Princeton University Press, Princeton, NJ, 2003.

[SS3]   E. Stein and R. Shakarchi, *Real Analysis: Measure Theory, Integration, and Hilbert Spaces*, Princeton University Press, Princeton, NJ, 2005.

[SpTh]   M. D. Springer and W. E. Thompson, *The Distribution of Products of Independent Random Variables*, SIAM J. Appl. Math. **14** (1966), 511–526.

[Sti]   G. Stigler, *The Distribution of Leading Digits in Statistical Tables*. Note, written 1945–1946, Stigler's 1975 address was Haskell Hall, University of Chicago, Chicago, IL 60637.

[Str]   K. Stromberg, *Probabilities on a Compact Group*, Trans. Amer. Math. Soc. **94** (1960), 295–309.

[Sug]   G. Sugihara, *Minimal Community Structure: An Explanation of Species Abundance Patterns*, American Naturalist **116** (1980), no. 6, 770–787.

[TeKaDu]   Y. Terawaki, T. Katsumi and V. Ducrocq, *Development of a Survival Model with Piecewise Weibull Baselines for the Analysis of Length of Productive Life of Holstein Cows in Japan*, Journal of Dairy Science **89** (2006), no. 10, 4058–4065.

[Ti]   H. Tijms, *Understanding Probability: Chance Rules in Everyday Lives*, Cambridge University Press, 2007.

[Töd]   K.-H. Tödter, *Benford's Law as an Indicator of Fraud in Economics*, German Economic Review **10**, no. 3 (2009), 339–351.

[TorFGS]   J. Torres, S. Fernández, A. Gamero and A. Sola, *How Do Numbers Begin? (The First Digit Law)*, European Journal of Physics **28** (2007), L17–L25.

[Tuc]   H. G. Tucker, *Absolute Continuity of Infinitely Divisible Distributions*, Pac. J. Math **12** (1962), 1125–1129.

[Tur]   P. R. Turner, *The Distribution of Leading Significant Digits*, IMA J. Numer. Anal. **2** (1982), no. 4, 407–412.

[Va]   H. Varian, *Benford's Law*, American Statistician **23** (1972), 65–66.

[Vos]   R. Voss, *Comment on "Linguistic Features of Noncoding DNA Sequences"*, Physical Review Letters **76** (1996), no. 11, 1978.

[VoxGo]   W. Voxman and R. Goetschel, Jr., *Advanced Calculus*, Mercer Dekker, New York, 1981.

[Wall]   G. K. Wallace, *The JPEG Still Picture Compression Standard*, Communications of the ACM **34** (1991).

[Walt]   W. Walter, *Ordinary Differential Equations*, Springer, 1998.

[Wash]   L. Washington, *Benford's Law for Fibonacci and Lucas Numbers*, Fibonacci Quart. **19** (1981), no. 2, 175–177.

[We]   W. Weibull, *A Statistical Distribution Function of Wide Applicability*, J. Appl. Mech. **18** (1951), 293–297.

[WCPBKO]   K. Wells, J. Chiverton, M. Partridge, M. Barry, H. Kadhem and B. Ott, *Quantifying the Partial Volume Effect in PET using Benford's Law*, IEEE Trans. Nucl. Sci. **54(5)**, Oct. 2007, 1616–1625.

[WerAa]   M. Wernick and J. Aarsvold, *Emission Tomography, The Fundamentals of PET and SPECT*, Elsevier Academic Press, 2004.

[Whi]   D. Whitford, *Proteins: Structure and Function*, John Wiley & Sons, Chichester, 2005.

[Wig1]   E. Wigner, *Results and Theory of Resonance Absorption*, Gatlinburg Conference on Neutron Physics by Time-of-Flight, Oak Ridge, National Lab. Report No. ORNL-2309, 1957; Elsevier Science B.V, 2001.

[Wig2]   E. Wigner, *On the Statistical Distribution of the Widths and Spacings of Nuclear Resonance Levels*, Mathematical Proceedings of the Cambridge Philosophical Society **47** (1951), 790–798.

[Wis]   J. Wishart, *The Generalized Product Moment Distribution in Samples from a Normal Multivariate Population*, Biometrika **20A** (1928), 32–52.

[Wój]   M. R. Wójcik, *A Characterization of Benford's Law through Generalized Scale-Invariance*, Mathematical Social Sciences **71** (2014), 1–5.

[WolAu]   H. Wolff and M. Auffhammer, *Endogenous Choice of Development Indicators: Are Development Countries Misclassified? Evidence from the HDI*, Agricultural and Resource Economics, UC Berkeley (2006).

[ZiBGS]   W. T. Ziemba, S. L. Brumelle, A. Gautier and S. L. Schwartz, *Dr. Z's 6/49 Lotto Guidebook*, Vancouver and Los Angeles: Dr. Z. Investments., 1986.

[Zip]   G. K. Zipf, *Human Behavior and the Principle of Least Effort*, Addison–Wesley Press, 1949.

[ZwJRR]   M. Zwietering, I. Jongenburger, F. Rombouts and K. van't Riet, *Modeling of the Bacterial Growth Curve*, Applied and Environmental Microbiology **56** (1990), 1875–1881.

[Zy]   A. Zygmund, *Trigonometrical Series*, vols. I and II, Cambridge University Press, Cambridge, 1968.

Index